Biodegradation Technology of Organic and Inorganic Pollutants

Edited by Kassio Ferreira Mendes,
Rodrigo Nogueira de Sousa
and Kamila Cabral Mielke

Published in London, United Kingdom

IntechOpen

Supporting open minds since 2005

Biodegradation Technology of Organic and Inorganic Pollutants
http://dx.doi.org/10.5772/intechopen.94650
Edited by Kassio Ferreira Mendes, Rodrigo Nogueira de Sousa and Kamila Cabral Mielke

Contributors
Shekh Md. Mamun Kabir, Joonseok Koh, Nibal Khaleel Mousa, Abdul-Jabbar Ali, Maha Hussien, Carlos M. Morales-Bautista, Selene Lagunas-Rivera, Sugey López-Martínez, Sebastian A. Ramos-Arcos, Edith G. González-Mondragón, Eduardo S. López-Hernández, Ana R. Rodríguez-Luna, Pedro Eulogio Cisterna Osorio, Barbara Faundez-Miño, Sandra Aparecida de Assis, Raquel Guimarães Benevides, Alison Borges Vitor, Geise Camila Ribeiro, Cleide Souza de Santana, Larissa Emanuelle Da Silva Almeida, Adelaide Perdigão, José Luís da Silva Pereira, Zwahruddeen Muhammad Salisu, Diya'uddeen Bashir Hasan, Yahaya Gambo Liman, Suleiman Mohammed Awal, Ukanah Suleiman Pendo, Mohammed Kabir Yakubu, Ishiaku Semo Umaru, Željko Savković, Miloš Stupar, Nikola Unković, Aleksandar Knežević, Jelena Vukojević, Milica Ljaljević Grbić, Ju-Chen Chia, Mahima Begum, Bhaswati Sarmah, Gayatri Goswami Kandali, Sontara Kalita, Ipsita Ojah, Raktim Bhagawati, Lipika Talukdar, Danushika C. Manatunga, Rohan S. S Dassanayake, Renuka N. Liyanage, Anita Verma, Madhuri Girdhar, Anand Mohan, Kopal Singh, Zeba Tabassum, Vaishali Arora, Babita Khosla, Aaltharackal Vikraman Meera, Manorama Thampatti KC, Jacob John, Bhadra Sudha, Abdulmajeed Sajeena, Selvanayaki Krishnasamy, Ramkumar Lakshmanan, Mythili Ravichandran, Muzafar Riyaz, Rauf Ahmad Shah, Kuppusamy Sivasankaran, Chandan Singh, Deepak Vyas, Eustace Fernando, Godfrey Kyazze, Pavithra Fernando, Ahmed Ahsan, Abubakar Lawal, Lukman Bola Bola Abdul'rauf, Melinda-Haydee Kovacs, Emoke Dalma Kovacs, Sushil Kumar Singh, Manish Singh Singh Rajput, Ravish Singh Rajput, Virendra Kumar, Ashutosh Singh Gautam, Sakshi Singh, Asmae Bouziani, Mohamed Yahya, Dibakar Roy, Dasari Sreekanth, Deepak Pawar, Himanshu Mahawar Himanshu Mahawar, Kamal K. Barman, Kassio Ferreira Mendes, Kamila Cabral Mielke, Rodrigo Nogueira de Sousa, Bruna Aparecida de Paula Medeiros

First published in London, United Kingdom, 2022 by IntechOpen
IntechOpen is the global imprint of INTECHOPEN LIMITED, registered in England and Wales, registration number: 11086078, 5 Princes Gate Court, London, SW7 2QJ, United Kingdom
Printed in Croatia

British Library Cataloguing-in-Publication Data
A catalogue record for this book is available from the British Library

Additional hard and PDF copies can be obtained from orders@intechopen.com

Biodegradation Technology of Organic and Inorganic Pollutants
Edited by Kassio Ferreira Mendes, Rodrigo Nogueira de Sousa and Kamila Cabral Mielke
p. cm.
Print ISBN 978-1-83968-895-9
Online ISBN 978-1-83968-896-6
eBook (PDF) ISBN 978-1-83968-897-3

For EU product safety concerns:
IN TECH d.o.o., Prolaz Marije Krucifikse Kozulić 3, 51000 Rijeka, Croatia, info@intechopen.com or visit our website at intechopen.com.

Meet the editors

Kassio Ferreira Mendes is a Professor of Biology and Integrated Management of Weeds, Department of Agronomy, Federal University of Viçosa, Brazil. He graduated from the State University of Mato Grosso in 2011 and received a Master's in Agronomy (Crop Science) from the Federal University of Viçosa in 2013. He obtained a doctorate and post-doctorate degrees in Sciences - Nuclear Energy in Agriculture (Chemistry in Agriculture and Environment) from the Center of Nuclear Energy in Agriculture, Luiz de Queiroz Campus, University of São Paulo, Brazil, in 2017 and 2019, respectively. He completed research fellowships at the Department of Soil, Water, and Climate, College of Food and Agricultural Sciences, University of Minnesota, USA, and the United States Department of Agriculture - Agricultural Research Service (USDA-ARS). Dr. Mendes is a member of the Brazilian Society of Weed Science.

Rodrigo Nogueira de Sousa obtained an undergraduate degree in Agronomic Engineering and a master's in Soil Science and Plant Nutrition from the Federal University of Viçosa, Brazil, in 2016 and 2018, respectively. He obtained a Ph.D. in Soil Science and Plant Nutrition from the University of São Paulo. From 2014 to 2015, he studied at North Carolina Agricultural and Technical State University, USA. He also completed an internship at the Department of Crop and Soil Sciences, North Carolina State University, USA, in 2015, for which he studied the management of nitrogen fertilization in corn crops. During his doctorate program, he studied the feasibility of alternative sources of phosphate fertilizers in tropical soils.

Kamila Cabral Mielke is an agronomic engineer who graduated from the Federal Institute of Education, Science and Technology of Rondônia, Brazil, in 2016. She obtained a master's degree in Tropical Agriculture from the Federal University of Espírito Santo, Brazil, and a Ph.D. in Crop Science from the Federal University of Viçosa, Brazil. She is currently a member of the Integrated Weed Management (IPM) research group where she studies the behavior and fate of herbicides in the environment, their impact on soil microbiological activity, and remediation of environments contaminated with herbicides using phytoremediation and carbonaceous materials such as biochar and bonechar.

Contents

Preface

Agricultural activities are fundamental for producing food around the world, but they are also major contributors to environmental pollution. They contaminate soil, water, and air matrices through the release of organic and inorganic pollutants. Pesticides (e.g., herbicides, fungicides, and insecticides), heavy metals, and antibiotics are among the most used pollutants in agriculture. Pesticides are potentially harmful substances due to their persistence, toxicity, and bioaccumulation in the environment. Many of these molecules can contaminate groundwater, mainly due to the lack of efficient treatment methods to reduce their concentration in water resources. However, it is important to point out that extensive industrial activity also contaminates the environment.

In contact with the environment, organic and inorganic pollutants are subject to physical and chemical processes that regulate their behavior and fate. Some molecules accumulate on the surface and subsurface of the soil, with direct consequences on microorganisms and mainly autotrophic organisms, including many plant species. Other molecules can reach waterways. Moreover, one must also consider the airborne displacement of herbicide molecules enhanced by physicochemical characteristics, unfavorable weather conditions, and inadequate application technology. Toxicity and persistence make pesticides a serious environmental concern, making it necessary to introduce additional technologies to adequately remove these compounds from the environment.

In view of this, and of the potential impact that these molecules can have when present in the environment, new techniques to remediate the presence of chemical molecules in the environmental matrix have been presented as efficient alternatives for the reduction of contamination and pollution caused by these molecules and their metabolites, as biotic and abiotic processes. The bioremediation of soils contaminated by organic and inorganic pollutants consists of reducing the levels of contaminants to levels that are safe and compatible with the protection of human health, either by preventing or hindering the spread of harmful substances to the environment.

This book presents a clear overview of the applications of different technologies for bioremediation of organic and inorganic pollutants that will help improve human life and ecosystems.

Kassio Ferreira Mendes and Kamila Cabral Mielke
Department of Agronomy,
Federal University of Viçosa,
Viçosa, Brazil

Rodrigo Nogueira de Sousa
Department of Soil Science,
University of São Paulo,
Piracicaba, Brazil

Section 1

Organic Pollutants

Chapter 1

Degradation Process of Herbicides in Biochar-Amended Soils: Impact on Persistence and Remediation

Kamila Cabral Mielke, Kassio Ferreira Mendes, Rodrigo Nogueira de Sousa and Bruna Aparecida de Paula Medeiros

Abstract

Biochar is a solid material derived from different feedstocks that is added to the soil for various agronomic and environmental purposes, such as nutrient sources and CO_2 emission mitigators. In modern agriculture, the application of herbicides directly in the soil is common for pre-emergent weed control; however, biochars may interfere in the degradation processes of these agrochemicals, increasing or decreasing their persistence. Long persistence is desirable for some herbicides in determined cultivation systems, especially in monoculture, but persistence is undesirable in crop rotation and/or succession systems because the subsequent cropping can be sensitive to the herbicide, causing carryover problems. Therefore, knowing the interactions of biochar-herbicide is essential, since these interactions depend on feedstock, pyrolysis conditions (production temperature), application rate, biochar aging, among other factors; and the physical-chemical characteristics of the herbicide. This chapter shows that the addition of biochar in the soil interferes in the persistence or remediation processes of the herbicide, and taking advantage of the agricultural and environmental benefits of biochars without compromising weed control requires a broad knowledge of the characteristics of biochar, soil, and herbicide and their interactions.

Keywords: bioavailability, sorption, weed control, pollution soil

1. Introduction

Herbicides are the pesticides most applied in modern agriculture for weed control worldwide, in pre-emergency, directly in the soil, or in post-emergence in leaves. Regardless of the application of herbicides, these reach the soil and may persist with residual effect (carryover) or contaminate the non-target organism and environment. The behavior of the herbicide in the soil is governed by the physico-chemical properties of the molecule and the soil and can have retention, transport, and transformation processes [1]. In transformation processes, the herbicide molecule is degraded into secondary compounds (metabolites) by physical (photodegradation), chemical, and biological processes (**Figure 1**) [2].

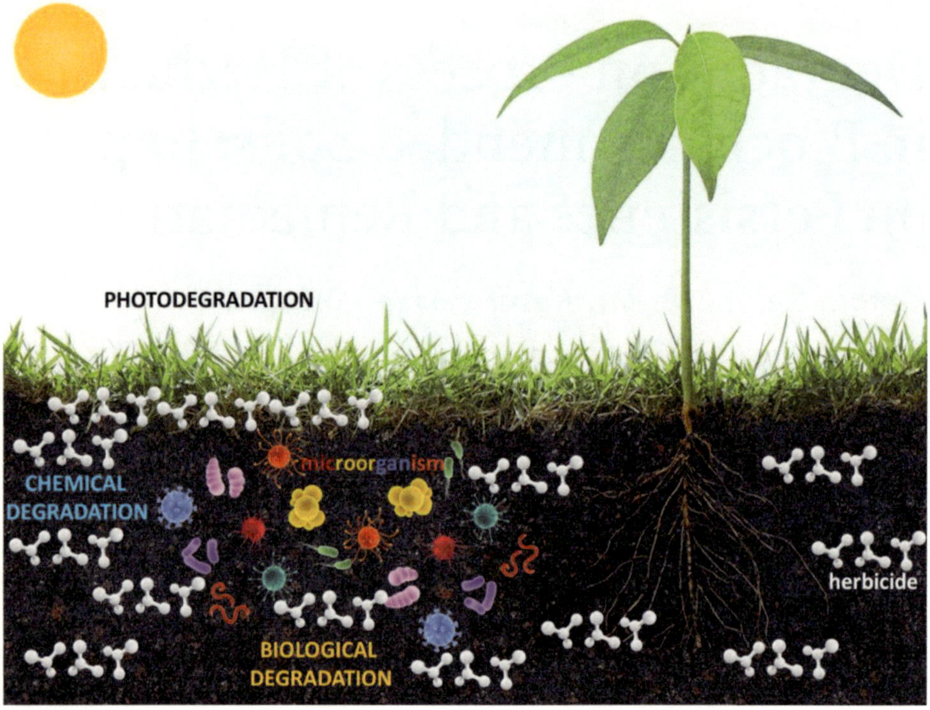

Figure 1.
Degradation process (chemical, biological, and photodegradation) of herbicides in the soil.

Biological degradation is the most common way to dissipate the herbicides in the environment, and it is carried out mainly by the soil microbiota which use the herbicide molecules as an energy source and transforms it into compounds without herbicidal action, the process is also known as detoxification [3, 4]. The chemical complexity of the herbicide determines the higher or lower facility of microorganisms to degrade the molecules, characterizing it in low or high persistence in the soil [5], being measured by degradation or dissipation half-life time (DT_{50}) in laboratory or field conditions, respectively [2].

The degradation of herbicides in the soil by microorganisms can be aerobic (with oxygen) or anaerobic (without oxygen). In the presence of oxygen, the herbicide is mineralized in CO_2 and water. Without oxygen, the herbicide is mineralized in CH_4, CO_2, and water [6]. The efficiency of aerobic degradation of herbicides is higher than the anaerobic. The aerobic bacteria oxygen act as an oxidizing agent, and they are present in the region of the soil where there is a higher content of organic matter (OM) and an excellent soil-water-air ratio for the microbiota [7]. In conditions of absence of oxygen, the herbicide can become more persistent in the soil and its degradation pathways are different from microorganisms with aerobic metabolism [8].

The addition of organic materials, like biochar, in the soil directly influences the microbial community, responsible for herbicide degradation [9]. Biochar is a carbonaceous material produced by different feedstocks in pyrolysis conditions with the limited presence of oxygen. Naturally, biochar is found in the anthropogenic soil, known as "Terra Preta de Índio", i.e., Amazonian Dark Earths in the Amazon, which gave rise to synthetic biochar produced worldwide [10]. Pyrolyzed feedstocks and pyrolysis conditions determine the physico-chemical properties of biochar, such as nutrient content, porosity, specific surface area, among others.

In agricultural soils, the biochar has been added to increase porosity, water-holding capacity, reduce acidity, sequester carbon, reduction of greenhouse gas emissions, plant growth promotion, improve soil fertility, and immobilize (remediation) herbicides by increasing sorption and microbial diversity [11]. This chapter showed that is possible to recommend the addition of biochar in the soil to interfere in the persistence or remediation processes of the herbicide.

2. Biochar characteristics

Biochar is the carbon-rich product resulting from the pyrolysis of organic residues such as wood, animal wastes, crop residues, and biosolids [12]. The feedstock usually determines the chemical composition, quantity of macropores, and nutrient content in biochar. Pyrolysis conditions (such as temperature, heating rate, and residence time) determine the morphology and surface structure changes in feedstock and C/H content [11]. The dominant properties affecting herbicide sorption and degradation by biochar include porosity, specific surface area, pH, functional groups, carbon content and aromatic structure, and mineralogical composition [13].

More porous structures and higher specific surface area will result in higher sorption capacities and lower degradation of herbicides [13]. Higher pH of biochar can accelerate the hydrolysis of organophosphorus and carbamate herbicides in the soil through the alkali catalysis mechanism [14]. Surface functional groups including carboxylic (–COOH), hydroxyl (–OH), lactonic, amide, and amine groups are essential for the sorption capacity of biochar [15, 16]. Carbon content and aromatic structure can increase herbicide sorption and reduce their bioavailability to be degraded [13]. The mineralogical composition can reduce the bioavailability of herbicides through surface chelation and/or surface acidity mechanisms [17].

Biochar amendment also affects the degradation of herbicides in the soil in several ways and the effects can be either stimulatory or suppressive [18]. Biochar may contain available nutrients that stimulate overall microbial activity and, thus, degradation of herbicides [19, 20]. However, the degradation of herbicides in biochar-amended soils is most commonly reduced because herbicide sorption increases [21]. Biochar also sorbs dissolved organic carbon (OC), which can contribute to co-metabolic biodegradation [22]. Some changes in the degradation rate can be a result of indirect effects of biochar amendment, e.g., changes in soil pH, albedo, and aeration [18].

3. Microbial diversity in biochar-amended soils

Soil correction with biochar can affect the soil microbiota in different ways: (1) It can provide an increase in the microbiota [23, 24]; (2) It can negatively affect the resident microbiota by the amount of organic substances (volatile compounds) formed in the production of biochar [25, 26]; or (3) It may not effect the soil microbiota [27, 28]. The possible interaction mechanisms of biochar and soil microbiota are exemplified in **Figure 2** [29, 30]. The physical–chemical structures of the biochar surface (macro and micropores, roughness, surface load, and hydrophobicity) are a refuge for the soil microbiota [31, 32], where microorganisms can find nutrients and ions adsorbed in biochar particles useful for their growth [29, 33]. In addition, biochars can contain significant amounts of organic substances (volatile organic compounds and free radicals) [34, 35], improve the soil's physical–chemical properties, which are important for microbial growth by modifying habitats (aeration, water content, and pH) [36], affect the enzymatic activity of the soil [37, 38],

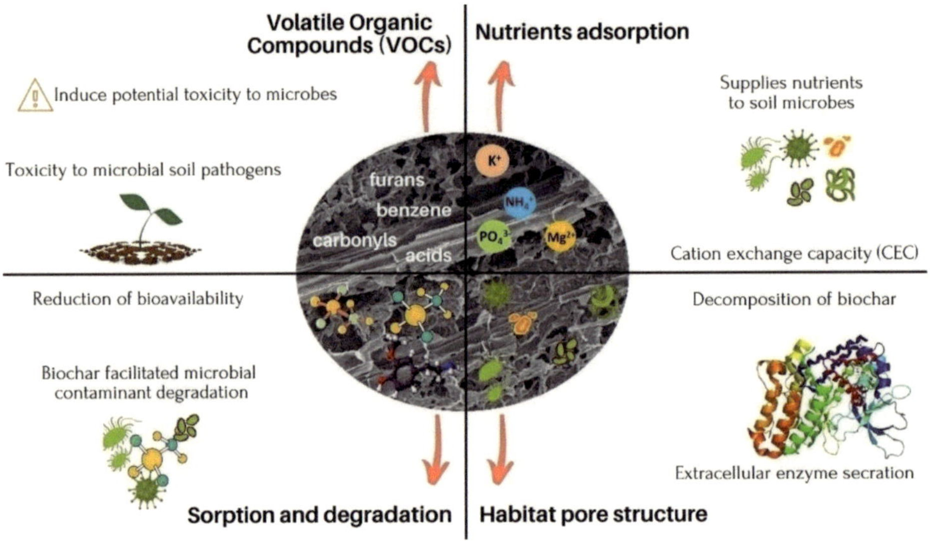

Figure 2.
Interactions between biochar and soil microbiota and environmental effects. Source: Adapted from Zhu et al. [29].

and increase the sorption of herbicides, reducing the bioavailability and toxicity of these agrochemicals for the soil microbiota [29, 39, 40].

Biochar-amended soil has a higher respiratory rate and microbial communities due to carbon mineralization by soil microorganisms [41]. Microbial biomass carbon and nitrogen increased by 18% and 63% with the application of 1% of sugarcane bagasse biochar [42]. The role of biochar nutrients in the biodegradation of coexisting dichlobenil and atrazine in soil by their respective bacterial degraders was evaluated. The degradation increased with increasing biochar content, due to nutritional stimulation on microbial activities [43]. The application of hardwood-derived biochar increased atrazine mineralization by stimulating atrazine-adapted microflora compared to unamended soil [19]. Soil amended with biochar derived from wheat straw increased the abundance and diversity rate of bacteria and fungi beneficial to plants in the rhizosphere of wheat seedlings [24]. In addition, these microorganisms use fomesafen as a source of nutrients, which favors their proliferation from the soil [24]. The change and proliferation of the soil microbiota with the addition of biochar is related to the chemical characteristics of biochar (mainly pH and nutrient content) and physical properties (pore size, pore-volume, and specific surface area), OM content, and water retention that provide favorable conditions for soil microbiota [28]. Although soil microbial biomass is generally benefited with the addition of biochar, the response depends on the type of raw material, pyrolysis temperature, and biochar application rate, since these factors directly interfere with the physical–chemical characteristics of biochar and consequently on the response of the microbiota in herbicide degradation. The proposed mechanisms involved in biochar and microbiota interactions require further studies to elucidate the impact of biochar on soil microbial activity.

4. Influence of biochar amendment in soil on the herbicide degradation

Herbicides are applied to the soil to control weeds during a certain time after application; however, long persistence may affect the subsequent crop, a process known as carryover. Therefore, the process of degradation of the herbicide is important for the dissipation of herbicides in the soil when the intention is the remediation of the product. However, under agronomic conditions in which a residual effect of the herbicide on the soil is desired for weed control, the addition of biochar can reduce the persistence of the product, consequently reducing its effectiveness in management [2, 9, 44].

The degradation of herbicide molecules into secondary compounds (metabolites) can occur by biotic (biological degradation) or abiotic (hydrolysis, reduction, oxidation, and photolysis) processes [45]. Biodegradation, carried out by the soil microbiota (bacteria, fungi, protozoa, and actinomycetes), is the main decomposition pathway for most herbicides [13, 46]. Microorganisms can use herbicide molecules as an energy source and transform them into compounds without herbicide action, a process known as catabolism, or through co-metabolism, in which herbicide degradation requires the presence of a growth substrate that is used as primary carbon and energy source [3, 47], i.e., microorganism does not obtain energy or benefit from the herbicide degradation. The transformation process is usually mediated by non-specific enzymes that are capable to transform various organic compounds [4]. Herbicides have varied susceptibility to microbial degradation depending on the complexity of the molecule that influences low or high persistence in the soil [5].

Microbial degradation generally reduces the DT_{50} of herbicides in the soil; however, the addition of biochar, according to studies performed, may increase or decrease the DT_{50} values, depending on the herbicide and pyrolyzed feedstock (**Table 1**). The high sorption capacity for herbicides in the biochar-amended soil decreases herbicide degradation, providing a higher DT_{50} than the unamended soil [46, 49]. For example, less atrazine degradation was observed in amended soils with sugarcane bagasse biochar (0.5% w/w) (**Table 1**), increasing in 15 days the DT_{50} of the herbicide in relation to unamended soil [49]. Flumioxazin DT_{50} increased by ~10 days when bamboo biochar (10% w/w) was added compared to unamended soil (**Table 1**) [51]. The DT_{50} of 2-methyl-4-chlorophenoxyaceticacid (MCPA) increased from 5.2 days (unamended soil) to 21.5 days in amended soil with 1% of wheat straw biochar [59].

The application of biochar can also increase soil microbial activity, improving herbicide degradation [29, 60]. The increase in microbial biomass may be due to the addition of available organic substrates, which are the main energy source readily available to soil microorganisms [55]. The high content of dissolved OC in the soil MO can reduce herbicide sorption by biochar particles, as dissolved OC competes with herbicide molecules to occupy available biochar sorption sites [61]. Biochar also sorbs dissolved OC, which can contribute to co-metabolic biodegradation [22]. Some changes in degradation rate may result from indirect effects of biochar amendment, e.g., changes in soil pH and aeration [18]. The highest degradation of oxyfluorfen was observed in amended soils with different rates of application of rice husk biochar, decreasing DT_{50} between 2 and 23 days compared to unamended soil (**Table 1**) [50]. Alachlor mineralization increased up to 50% using biochar derived from soybean stoves, sugarcane bagasse, and wood chips compared to unamended soil (**Table 1**) [58].

Photolysis and hydrolysis are the main abiotic processes involved in herbicide degradation [48, 62]. Photolysis or photodegradation occurs when herbicides are exposed to sunlight [63] and can be direct (a herbicide molecule absorbs light energy, is later excited and transformed) or indirect (species photochemically produced in the soil matrix react with the herbicide molecule triggering its degradation) [64]. Water degrades herbicides by dividing large molecules into smaller molecules, breaking them in the process called hydrolysis [65]. The hydrolysis of herbicides in the soil can be influenced by several factors such as dissolved ion

Location (Country)	Soil texture (%)				Feedstock	Application rate (%)[b]	Pyrolysis temperature (°C)	Herbicide	DT_{50} (biochar-amended soil)	DT_{50} (unamended soil)	References
	Sand	Silt	Clay	OM[a]							
Australia	n.a.	n.a.	36.1	2.4	Wheat straw	0.5	450	Atrazine	61.5[d]	76.5[d]	Nag et al. [44]
						1			57.9[d]		
	n.a.	n.a.	28	4.0		0.5			65.9[d]	68.6[d]	
						1			52.0[d]		
Australia	n.a.	n.a.	36.1	2.4	Wheat straw	0.5	450	Trifluralin	73.6[d]	75.4[d]	Nag et al. [44]
						1			71.1[d]		
	n.a.	n.a.	28	4.0		0.5			63.2[d]	66.4[d]	
						1			61.2[d]		
Brazil	n.a.	n.a.	n.a.	n.a.	Industrial-production of charcoal	3	350-550	Sulfometuron-methyl	52.1	36.6	Alvarez et al. [48]
						6			55.4		
China	n.a.	n.a.	n.a.	3.2	Sugarcane bagasse	0.2	500	Atrazine	38.5	28.1	Huang et al. [49]
						0.5			45.0		
				2.0		0.2			35.5	23.7	
						0.5			41.2		
				3.6		0.2			41.2	39.8	
						0.5			54.8		

Location (Country)	Soil texture (%)				Feedstock	Application rate (%)[b]	Pyrolysis temperature (°C)	Herbicide	DT_{50} (biochar-amended soil)	DT_{50} (unamended soil)	References
	Sand	Silt	Clay	OM[a]							
China	21.4	51.4	27.2	2.8	Coal	1.5	n.a.[c]	Isoproturon	53.3	54.6	Si et al. [46]
						5			60.8		
						8			71.4		
	27.9	33.6	38.5	1.5		1.5			67.9	16	
						5			102		
						8			136		
	44.9	39.5	15.6	1.2		1.5			58.2	15.2	
						5			88.9		
						8			107		
China	32.1	24.7	43.2	0.84	Rice husk	0.5	500	Oxyfluorfen	59	65	Wu et al. [50]
						1			57		
						2			53		
	73.2	12.3	14.5	0.98		0.5			104	108	
						1			85		
						2			77		
	55	23.1	21.9	2.2		0.5			43	45	
						1			42		
						2			35		

Location (Country)	Soil texture (%)				Feedstock	Application rate (%)[b]	Pyrolysis temperature (°C)	Herbicide	DT_{50} (biochar-amended soil)	DT_{50} (unamended soil)	References
	Sand	Silt	Clay	OM[a]							
China	77.4	4.4	18.2	1.5	Cornstalk	10	500	Flumioxazin	15.4	11.1	Chen et al. [51]
					Rice husk		500		16.7		
					Bamboo		700		23.2		
	21	17	62	1.7	Cornstalk		500		18.5	11.5	
					Rice husk		500		22.3		
					Bamboo		700		25.4		
	6.8	55.3	37.9	3.8	Cornstalk		500		20.5	15.4	
					Rice husk		500		22.6		
					Bamboo		700		29.2		
	3.7	64.7	31.6	4.3	Cornstalk		500		21.0	20.8	
					Rice husk		500		24.7		
					Bamboo		700		30.7		
Germany	30.1	62.5	7.8	1.2	Hardwood	0.1	500	Atrazine	74.0[d]	72.4[d]	Jablonowski et al. [19]
						1			72.4[d]		
						5			68.0[d]		
	24.4	25.2	50.3	3.1		0.1			53.0[d]	42.6[d]	
						1			49.8[d]		
						5			44.4[d]		

Location (Country)	Soil texture (%)				Feedstock	Application rate (%)[b]	Pyrolysis temperature (°C)	Herbicide	DT_{50} (biochar-amended soil)	DT_{50} (unamended soil)	References
	Sand	Silt	Clay	OM[a]							
India	56.6	29.6	13.8	n.a.	Rice straw	0.25	350	Bispyribac-sodium	10.7	27.1	Sharma et al. [39]
						0.5			11.5		
						1			12.1		
						0.25	550		8.8		
						0.5			9.9		
						1			11.2		
Latvia	89.2	8.9	1.9	n.a.	Wood chips	5.3	725	MCPA	1986[f]	94.5[f]	Muter et al. [52]
						4.1			3854[f]	11.1[f]	
					Wheat straw	5.3			1636[f]	94.5[f]	
						4.1			15.3[f]	11.1[f]	
Malaysia	40	21.5	37.9	0.99	Oil palm empty fruit bunches	1	300	Imazapic	46.2	34.6	Yavari et al. [53]
								Imazapyr	53.3	38.5	
					Rice husk			Imazapic	40.7	34.6	
								Imazapyr	46.3	38.5	

Location (Country)	Soil texture (%)				Feedstock	Application rate (%)[b]	Pyrolysis temperature (°C)	Herbicide	DT_{50} (biochar-amended soil)	DT_{50} (unamended soil)	References
	Sand	Silt	Clay	OM[a]							
Russia	3.1	30.4	66.5	n.a.	Woods (*Betula* sp. and *Piceaabies*)	1	400	Diuron	47	40	Zhelezova et al. [18]
						10			42		
						20			56		
						30			45		
						1		Glyphosate	187	17	
						10			151		
						20			131		
						30			51		
	83.7	8.8	7.5	n.a.		1		Diuron	58	112	
						10			33		
						20			35		
						30			40		
						1		Glyphosate	83	182	
						10			66		
						20			78		
						30			53		

Location (Country)	Soil texture (%)				Feedstock	Application rate (%)[b]	Pyrolysis temperature (°C)	Herbicide	DT_{50} (biochar-amended soil)	DT_{50} (unamended soil)	References
	Sand	Silt	Clay	OM[a]							
Spain	24	47	30	1.3	Hardwood (*Caryatomentosa*)	2	350	Clomazone	97	29	Gámiz et al. [54]
							400		77		
							700		99		
					Hardwood (*Caryaillinoinensis*)		350		107		
							400		65		
							700		67		
					Hardwood (*Caryatomentosa*)		350	Bispyribac-sodium	n.a.	21	
							400		n.a.		
							700		84		
					Hardwood (*Caryaillinoinensis*)		350		n.a.		
							400		n.a.		
							700		33		

Location (Country)	Soil texture (%)				Feedstock	Application rate (%)[b]	Pyrolysis temperature (°C)	Herbicide	DT_{50} (biochar-amended soil)	DT_{50} (unamended soil)	References
	Sand	Silt	Clay	OM[a]							
Spain	43	32	23	0.9	Olive mill waste	2.5	n.a.	Metribuzin	39	22	López-Piñeiro et al. [55]
						5			48		
					Olive mill waste plus leaves	2.5			13		
						5			17		
	53	32	14	0.6	Olive mill waste	2.5			49	35	
						5			52		
					Olive mill waste plus leaves	2.5			19		
						5			22		
	43	14	42	0.9	Olive mill waste	2.5			40	29	
						5			43		
					Olive mill waste plus leaves	2.5			18		
						5			16		
USA	n.a.	n.a.	n.a.	0.7	Sugarcane bagasse	0.2	350	Metribuzin	54	25	White Junior et al. [56]
					Sugarcane bagasse	0.1	700		25		
	n.a.	n.a.	n.a.	0.8	Sugarcane bagasse	0.2	350		74	57	
					Sugarcane bagasse	0.1	700		39		
	n.a.	n.a.	n.a.	1.2	Pine wood	0.4	400		39	28	
USA	22	55	23	>2	Mixed sawing	5	500	Acetochlor	34.5	9.7	Spokas et al. [57]
USA	n.a.	n.a.	n.a.	n.a.	Soybean waste	10	500	Alachlor	4.6[e]	10.4[e]	Mendes et al. [58]
					Sugarcane bagasse		350		3.4[e]		
					Wood bark (grape)		500		3.8[e]		

[a]*Organic Matter;* [b]*Application rate in relation to soil mass (ww^{-1});* [c]*Data not available;* [d]*Degradation (%);* [e]*Mineralization (%) to CO_2;* [f]*Herbicide concentration after incubation period (μg kg^{-1}).*

Table 1.
Effect of biochar amendment in soil on the degradation half-live time (DT_{50} - days) of different herbicides.

concentration, soil pH, and content of clays and metal oxides capable of catalyzing this herbicide degradation process [14, 66].

The application of biochar can influence the degradation of herbicides by hydrolysis and photolysis, since persistent free radicals existing or photogenerated in biochars can react with the herbicide by the activation of other free radicals such as hydroxyl, sulfate, anion, and superoxide [67, 68]. In addition, the increase in soil pH, the presence of active groups on the mineral surface of biochar, and the high sorption of herbicides have a direct effect on the chemical degradation processes of herbicides [64, 66]. Atrazine was hydrolyzed by 27.9% in the presence of biochar derived from pig manure (700°C) after 12 h due to the mineral surface and dissolved metal ions released from biochars that catalyze hydrolysis [66]. In contrast, imazapic and imazapyr were resistant to degradation by hydrolysis in amended soil with biochar derived from empty fruit bunch of oil palm and rice husk, and their DT_{50}'s increased by ~6 to 12 days because the photodegradation rate diminished [53] (**Table 1**). The addition of biochar to the soil at 1 or 5% inhibited the photodegradation of metribuzin and its metabolites deamino (DA), deaminodiketo (DADK), and diketometribuzin (DK), which increased their DT_{50}'s due to the immobilization of these compounds the surface layer of the biochar [64]. Therefore, the application of biochar has a direct impact on herbicide degradation processes and should be constantly examined for its application in the soil.

5. Factors affecting herbicide degradation in biochar-amended soils

The impact on the degradation of herbicides due to their high sorption in the biochar particles depends on the rate of biochar applied to the soil. The application of different rates of application of hardwood biochar in Rhodic Ferralsol soil increased atrazine degradation by 49% (0.1% of biochar), 51% (1.0% of biochar), and 62% (5.0% of biochar) after 88 days of incubation (**Table 1**) [19]. DT_{50} of isoproturon in unamended Alfisol was 16 days, however, when biochar was added at 1.5 and 5%, DT_{50} increased to 67 and 136 days, respectively (**Table 1**) [46], i.e., the persistence of isoproturon is prolonged as the rate of biochar added to the soil increases. DT_{50} of fomesafen increased from 34.6 days in unamended soil to 51, 83, and 160 days in amended soils with rice husk biochar at 0.5, 1, and 2%, respectively [61]. The increased persistence of fomesafen can be explained by the higher sorb capacity of biochar and, therefore, little bioavailability of the herbicide for microbial degradation.

Pyrolysis temperature defines the physicochemical characteristics of biochars [69]. Generally, biochar produced at relatively high pyrolysis temperatures (>500°C) presents an increase in specific surface area, microporosity, and hydrophobicity, improving herbicide sorption [70]. However, even with higher herbicide sorption capacity, degradation at high pyrolysis temperatures may be more intensified than low temperatures. The addition of sugarcane bagasse biochar produced at 700°C in clay soil decreased the DT_{50} of metribuzin from 57 (unamended soil) to 39 days, but when biochar was produced at 350°C, DT_{50} went from 57 to 74 days (**Table 1**) [56]. These conflicting results could be due to the impact of ash on the alkalinity of the soil amended with biochar produced at 700°C (20.3% of ash), which increased the soil pH and improved the conditions for the degradation of metribuzin, and to the greater amount of dissolved OC from biochar produced at 350°C (3.78 mg g^{-1}), which is more preferred by microorganisms as substrate, increasing the persistence of the herbicide. The variation in pyrolysis temperature of eucalyptus wood residue biochar affected the total hexazinone unavailable (mineralized + non-extractable residue) being higher for 850°C (46%) and 950°C

(49%) compared to biochar pyrolised at 650°C (33%) and 750°C (42%) [71]. The addition of biochar did not alter the mineralization of hexazinone, but it did reduce the bioavailability of this herbicide in the soil due to the greater amount of non-extracted residue, reducing the risk of environmental contamination [71].

Aging alters the properties of biochar, affecting the degradation of herbicides, however, these changes are not fully elucidated [72]. Glyphosate showed no variation in degradation in two tropical soils (Ultisol and Alfisol) amended with eucalyptus biochar aged [73]. The aging of soil-wood biochar mixtures (*Betula* sp. and *Piceaabies*) decreased glyphosate and diuron sorption compared to fresh biochar amended soil [18]. In addition, herbicide degradation was not affected by changes or biochar aging in the soils studied [18]. The degradation of S-metolachlor was not affected with the addition of three macadamia nutshell biochars aged [74]. The persistence of mesotrione in different soils amended with fresh and aged biochar was similar to unamended soils [75]. In contrast, the extractable amounts of picloram were 20 and 50% lower for soils amended with fresh and aged oak wood biochar, respectively, in relation to unamended soil [76]. The addition of 10% fresh biochar from the olive oil industry increased the DT_{50} of metribuzin from 20 (unamended soil) to 30.2 days, however, the DT_{50} decreased to 6.4 days with the addition of aged biochar, possibly because microorganisms in soil aged with biochar used metribuzin as a source of carbon and energy instead of the labile fraction of soil OM (**Table 1**) [55]. The effects of biochar on herbicide degradation in soils should not be generalized due to the different characteristics of biochars and the complexity of the soil system. The variation of temperature and application rate of biochar can bring different degradation responses for each herbicide studied. Furthermore, the aging of biochar in the soil can influence the bioavailability of herbicides in soil solution by altering the sorption capacity of the biochar; therefore, the conditions of pyrolysis, type of feedstock as well as aging must be taken into consideration when planning its use in agriculture and for soil remediation purposes [18].

6. Simultaneous use of herbicides and biochar

In an agricultural context, the property of biochar that offers potential for herbicide sorption (environmental remediation) can also decrease the efficacy of herbicides applied to the soil, influencing their bioavailability and susceptibility to leaching and consequently their degradation [77]. The bioavailability of diuron and microbial degradation was reduced in soils amended with rice straw biochar, which decreased the effectiveness of diuron to jungle rice (*Echinocloa colona*) control [78]. The addition of wheat straw biochar to the soil inactivated the herbicides atrazine and trifluralin, resulting in increased seed germination and biomass of annual ryegrass (*Lolium rigidum*). In this study, the efficacy of the herbicides for ryegrass control was achieved when the application doses were four times higher than recommended [44]. In a bioassay with *Echinochloa colona*, injuries 9 days after planting decreased with increasing application rates of rice straw biochar indicating that sorption of clomazone increased and directly influenced the bioavailability of herbicide in the soil [79]. The control efficiency of S-metolachlor was evaluated on green foxtail (*Setaria viridis)* in soil amended with wood biochar at different application rates (0, 0.5, 1, and 2%) [80]. *S. viridis* control at the highest application rate (2%) was lower than the other application rates evaluated, however, better than the control treatments (no herbicide) [80].

The biochar applied to soil also influences the soil physicochemical properties and the improved nutritional availability of these directly impacts crop growth and consequently weed growth [81]. Soil amended with walnut shell biochar

(5 Mg ha^{-1}) for 4 years was evaluated for weed control [82]. Weed density was dramatically higher in biochar-amended soils (60-78%) compared to unamended soil, being related to increased nutrient availability and improvements in soil physicochemical properties such as cation exchange capacity (CEC), density and porosity, increased soil aeration, and water retention. The application of 2 Mg ha^{-1} of cow bonechar prevented weed control by indaziflam which is related to the increase of soil fertility, especially the phosphorus and carbon content, and to the increase of pH because it is a basic material [83]. In addition, goosegrass (*Eleusine indica)* and crabgrass (*Digitaria horizontalis)* accounted for about 99.7% of the entire weed community infestation [83].

On the other hand, the decrease in efficacy depends on the characteristics of the herbicide evaluated. The dose of pretilachlor to inhibit 50% of E. colona emergence and biomass was higher in soil amended with rice-husk biochar, however, the effectiveness of pendimethalin in controlling E. colona was not influenced by the application rate of biochar [84]. The effectiveness on metribuzin in soils amended with biochar was evaluated by White Junior et al. [56]. The addition rates of biochar did not alter Palmer (Amaranthus palmeri) emergence, and it is possible that the residual activity was sufficient to reduce germination at any rate of biochar [56].

The addition of biochar to soil increases the sorption of different herbicides and reduces their effectiveness, which may result in the need for higher herbicide application rates, additional application times, or more weed control operations required [85]. Residual herbicides, applied in pre-emergence, can not provide good weed control regardless of soil type after biochar application. This does not necessarily mean that biochar should be avoided, however, when biochar is applied to the soil, management practices need to be adjusted to obtain appropriate weed control [86].

7. Conclusions

Modifying soil characteristics with biochar is a world-renowned emerging practice for either environmental and/or agronomic purposes, and the benefits these carbonaceous materials brig to the soil are clear. However, the pyrolysis conditions for biochar production directly interfere with the physical–chemical properties of the produced material, which govern the biochar-herbicide interactions. If the objective is to apply the herbicide in pre-emergence after the addition of biochar in the soil, care should be taken, as biochar can decrease or increase the persistence of the chemical product, interfering in the effectiveness of weed control over time. On the other hand, if the objective is herbicide remediation in contaminated soils, the interference of biochar in the bioavailability of the herbicide in the soil solution to increase soil microbiological diversity should be known.

Acknowledgements

The authors wish to thank the Coordination for the Improvement of Higher Education Personnel (CAPES - 88887.479265/2020-2100) and Foundation for Research Support of the State of Minas Gerais - Brazil (FAPEMIG - APQ-01378-21) for the financial support.

Conflict of interest

The authors declare no conflict of interest.

Author details

Kamila Cabral Mielke[1], Kassio Ferreira Mendes[1*], Rodrigo Nogueira de Sousa[2] and Bruna Aparecida de Paula Medeiros[1]

1 Department of Agronomy, Federal University of Viçosa, Viçosa, Minas Gerais, Brazil

2 Department of Soil Science, "Luiz de Queiroz" College of Agriculture, University of São Paulo, São Paulo, Brazil

*Address all correspondence to: kfmendes@ufv.br

References

[1] Arias-Estévez M, López-Periago E, Martínez-Carballo E, Simal-Gándara J, Mejuto JC, García-Río L. The mobility and degradation of pesticides in soils and the pollution of groundwater resources. Agriculture, Ecosystems and Environment. 2008;**123**:247-260

[2] Mendes KF, Mielke KC, Barcellos Júnior LH, de la Cruz RA, Sousa RN. Anaerobic and aerobic degradation studies of herbicides and radiorespirometry of microbial activity in soil. In: Radioisotopes in Weed Research. Boca Raton: CRC Press; 2021. pp. 95-125

[3] Maier RM. Microorganisms and organic pollutants. In: Environmental Microbiology. San Diego: Academic Press; 2000. pp. 363-402

[4] Reis FC, Tornisielo VL, Martins BAB, Souza AJ, Andrade PAM, Andreote FD, et al. Respiration induced by substrate and bacteria diversity after application of diuron, hexazinone, and sulfometuron-methyl alone and in mixture. Journal of Environmental Science and Health. 2019b;**54**:560-568

[5] Bending GD, Lincoln SD, Edmondson RN. Spatial variation in the degradation rate of the pesticides isoproturon, azoxystrobin and diflufenican in soil and its relationship with chemical and microbial properties. Environmental Pollution. 2006;**139**:279-287

[6] Organization for Economic Co-Operation and Development (OECD). Guidelines for Testing of Chemicals - Aerobic and anaerobic transformation in soil. Vol. Test 307. Paris, France: OECD; 2002. p. 17

[7] Gebler L, Spadoto CA. Comportamento ambiental dos herbicidas. In: Manual de manejo e controle de plantas daninhas. Passo Fundo, RS, Brazil: Embrapa Trigo; 2008. pp. 39-69

[8] Wang W, Wang Y, Li Z, Wang H, Yu Z, Lu L, et al. Studies on the anoxic dissipation and metabolism of pyribambenz propyl (ZJ0273) in soils using position-specific radiolabeling. Sci Total Environ. 2014;**472**:582-589

[9] Takeshita V, Mendes KF, Alonso FG, Tornisielo VL. Effect of organic matter on the behavior and control effectiveness of herbicides in soil. Planta Daninha. 2019;**37**:e019214401

[10] Hilbert K, Soentgen J. From the "Terra Preta de Indio" to the "Terra Preta do Gringo": A History of Knowledge of the Amazonian Dark Earths. In: Ecosystem and Biodiversity of Amazonia. London: IntechOpen; 2021

[11] Ahmad M, Rajapaksha AU, Lim JE, Zhang M, Bolan N, Mohan D, et al. Biochar as a sorbent for contaminant management in soil and water: A review. Chemosphere. 2014;**99**:19-33

[12] Lehmann J, Joseph S. Biochar for environmental management: Science, technology and implementation. Vol. 1. London: Earthscan Publications; 2015

[13] Liu Y, Lonappan L, Brar SK, Yang S. Impact of biochar amendment in agricultural soils on the sorption, desorption, and degradation of pesticides: A review. Sci Total Environ. 2018;**645**:60-70

[14] Zhang P, Wu JY, Li LI, Liu Y, Sun HW, Sun TH. Sorption and catalytic hydrolysis of carbaryl on pig-manure-derived biochars. J Agro-Environ Sci. 2012;**31**:416-421

[15] Li H, Dong X, da Silva EB, de Oliveira LM, Chen Y, Ma LQ. Mechanisms of metal sorption by biochars: biochar characteristics and

modifications. Chemosphere. 2017;**178**: 466-478

[16] Antón-Herrero R, García-Delgado C, Alonso-Izquierdo M, García-Rodríguez G, Cuevas J, Eymar E. Comparative adsorption of tetracyclines on biochars and stevensite: looking for the most effective adsorbent. Applied Clay Science. 2018;**160**:162-172

[17] Wei J, Furrer G, Kaufmann S, Schulin R. Influence of clay minerals on the hydrolysis of carbamate pesticides. Environmental Science & Technology. 2001;**35**:2226-2232

[18] Zhelezova A, Cederlund H, Stenström J. Effect of biochar amendment and ageing on adsorption and degradation of two herbicides. Water, Air, and Soil Pollution. 2017;**228**:216

[19] Jablonowski ND, Borchard N, Zajkoska P, Fernández-Bayo JD, Martinazzo R, Berns AE, et al. Biochar-mediated [^{14}C] atrazine mineralization in atrazine-adapted soils from Belgium and Brazil. Journal of Agricultural and Food Chemistry. 2013;**61**:512-516

[20] Safaei-Khorram M, Zhang Q, Lin D, Zheng Y, Fang H, Yu Y. Biochar: a review of its impact on pesticide behavior in soil environments and its potential applications. Journal of Environmental Sciences. 2016;**44**:269-279

[21] Beesley L, Moreno-Jiménez E, Gomez-Eyles JL, Harris E, Robinson B, Sizmur T. A review of biochars' potential role in the remediation, revegetation and restoration of contaminated soils. Environmental Pollution. 2011;**159**:3269-3282

[22] Lin Y, Munroe P, Joseph S, Kimber S, Van Zwieten L. Nanoscale organo-mineral reactions of biochars in ferrosol: An investigation using microscopy. Plant and Soil. 2012;**357**:369-380

[23] Sun D, Meng J, Liang H, Yang E, Huang Y, Chen W, et al. Effect of volatile organic compounds absorbed to fresh biochar on survival of *Bacillus mucilaginosus* and structure of soil microbial communities. Journal of Soils and Sediments. 2015;**15**:271-281

[24] Meng L, Sun T, Li M, Saleem M, Zhang Q, Wang C. Soil-applied biochar increases microbial diversity and wheat plant performance under herbicide fomesafen stress. Ecotoxicology and Environmental Safety. 2019;**171**:75-83

[25] Spokas KA, Novak JM, Stewart CE, Cantrell KB, Uchimiya M, DuSaire MG, et al. Qualitative analysis of volatile organic compounds on biochar. Chemosphere. 2011;**85**:869-882

[26] Sun D, Lan Y, Xu EG, Meng J, Chen W. Biochar as a novel niche for culturing microbial communities in composting. Waste Management. 2016;**54**:93-100

[27] Noyce GL, Basiliko N, Fulthorpe R, Sackett TE, Thomas SC. Soil microbial responses over 2 years following biochar addition to a north temperate forest. Biology and Fertility of Soils. 2015;**51**: 649-659

[28] Ma H, Egamberdieva D, Wirth S, Bellingrath-Kimura SD. Effect of biochar and irrigation on soybean-rhizobium symbiotic performance and soil enzymatic activity in field rhizosphere. Agronomy. 2019;**9**:626

[29] Zhu X, Chen B, Zhu L, Xing B. Effects and mechanisms of biochar-microbe interactions in soil improvement and pollution remediation: A review. Environmental Pollution. 2017;**227**:98-115

[30] Palansooriya KN, Wong JTF, Hashimoto Y, Huang L, Rinklebe J, Chang SX, et al. Response of microbial communities to biochar-amended soils: a critical review. Biochar. 2019;**1**:3-22

[31] Noyce GL, Winsborough C, Fulthorpe R, Basiliko N. The

microbiomes and metagenomes of forest biochars. Scientific Reports. 2016;**6**:1-12

[32] Rummel CD, Jahnke A, Gorokhova E, Kühnel D, Schmitt-Jansen M. Impacts of biofilm formation on the fate and potential effects of microplastic in the aquatic environment. Environmental Science & Technology Letters. 2017;**4**:258-267

[33] DeLuca TH, Gundale MJ, MacKenzie MD, Jones DL. Biochar effects on soil nutrient transformations. In: Biochar for Environmental Management. London, UK: Routledge. 2015. pp. 453-486

[34] Ghidotti M, Fabbri D, Hornung A. Profiles of volatile organic compounds in biochar: insights into process conditions and quality assessment. ACS Sustainable Chemistry & Engineering. 2017;**5**:510-517

[35] Li J, Li Q, Qian C, Wang X, Lan Y, Wang B, et al. Volatile organic compounds analysis and characterization on activated biochar prepared from rice husk. International journal of Environmental Science and Technology. 2019;**16**:7653-7662

[36] Quilliam RS, Glanville HC, Wade SC, Jones DL. Life in the 'charosphere'–Does biochar in agricultural soil provide a significant habitat for microorganisms? Soil Biology and Biochemistry. 2013;**65**:287-293

[37] Lehmann J, Rillig MC, Thies J, Masiello CA, Hockaday WC, Crowley D. Biochar effects on soil biota–a review. Soil Biology and Biochemistry. 2011;**43**:1812-1836

[38] Pukalchik M, Mercl F, Terekhova V, Tlustoš P. Biochar, wood ash and humic substances mitigating trace elements stress in contaminated sandy loam soil: Evidence from an integrative approach. Chemosphere. 2018;**203**:228-238

[39] Sharma N, Kaur P, Jain D, Bhullar MS. In-vitro evaluation of rice straw biochars' effect on bispyribac-sodium dissipation and microbial activity in soil. Ecotoxicology and Environmental Safety. 2020;**191**:110204

[40] Egamberdieva D, Jabbarov Z, Arora NK, Wirth S, Bellingrath-Kimura SD. Biochar mitigates effects of pesticides on soil biological activities. Environ Sustain. 2021;**4**:335-342

[41] Pei J, Zhuang S, Cui J, Li J, Li B, Wu J, et al. Biochar decreased the temperature sensitivity of soil carbon decomposition in a paddy field. Agriculture, Ecosystems and Environment. 2017;**249**:156-164

[42] Irfan M, Hussain Q, Khan KS, Akmal M, Ijaz SS, Hayat R, et al. Response of soil microbial biomass and enzymatic activity to biochar amendment in the organic carbon deficient arid soil: A 2-year field study. Arabian Journal of Geosciences. 2019;**12**:95

[43] Qiu Y, Pang H, Zhou Z, Zhang P, Feng Y, Sheng GD. Competitive biodegradation of dichlobenil and atrazine coexisting in soil amended with a char and citrate. Environmental Pollution. 2009;**157**:2964-2969

[44] Nag SK, Kookana R, Smith L, Krull E, Macdonald LM, Gill G. Poor efficacy of herbicides in biochar-amended soils as affected by their chemistry and mode of action. Chemosphere. 2011;**84**:1572-1577

[45] Mendes KF, Sousa RN, Soares MB, Viana DG, Souza AJ. Sorption and desorption studies of herbicides in the soil by batch equilibrium and stirred flow methods. In: Radioisotopes in Weed Research. Boca Raton: CRC Press; 2021. pp. 17-61

[46] Si Y, Wang M, Tian C, Zhou J, Zhou D. Effect of charcoal amendment on adsorption, leaching and degradation

of isoproturon in soils. Journal of Contaminant Hydrology. 2011;**123**:75-81

[47] Fritsche W, Hofrichter M. Aerobic degradation by microorganisms. In: Biotechnology. Germany: Wiley – VCH; 2008. pp. 1-24

[48] Alvarez DO, Mendes KF, Tosi M, Souza LF, Cedano JCC, Souza FNP, et al. Sorption-desorption and biodegradation of sulfometuron-methyl and its effects on the bacterial communities in Amazonian soils amended with aged biochar. Ecotoxicology and Environmental Safety. 2021;**207**:111222

[49] Huang H, Zhang C, Zhang P, Cao M, Xu G, Wu H, et al. Effects of biochar amendment on the sorption and degradation of atrazine in different soils. Soil Sediment Contam: An Int J. 2018;**27**:643-657

[50] Wu C, Liu X, Wu X, Dong F, Xu J, Zheng Y. Sorption, degradation and bioavailability of oxyfluorfen in biochar-amended soils. Sci Total Environ. 2019;**658**:87-94

[51] Chen Y, Lan T, Li J, Yang G, Zhang K, Hu D. Effects of biochar produced from cornstalk, rice husk and bamboo on degradation of flumioxazin in soil. Soil Sediment Contam: An Int J. 2021:1-15

[52] Muter O, Berzins A, Strikauska S, Pugajeva I, Bartkevics V, Dobele G, et al. The effects of woodchip-and straw-derived biochars on the persistence of the herbicide 4-chloro-2-methylphenoxyacetic acid (MCPA) in soils. Ecotoxicology and Environmental Safety. 2014;**109**:93-100

[53] Yavari S, Sapari NB, Malakahmad A, Yavari S. Degradation of imazapic and imazapyr herbicides in the presence of optimized oil palm empty fruit bunch and rice husk biochars in soil. Journal of Hazardous Materials. 2019;**366**:636-642

[54] Gámiz B, Velarde P, Spokas KA, Hermosín MC, Cox L. Biochar soil additions affect herbicide fate: importance of application timing and feedstock species. Journal of Agricultural and Food Chemistry. 2017;**65**:3109-3117

[55] López-Piñeiro A, Peña D, Albarrán A, Becerra D, Sánchez-Llerena J. Sorption, leaching and persistence of metribuzin in Mediterranean soils amended with olive mill waste of different degrees of organic matter maturity. Journal of Environmental Management. 2013;**122**:76-84

[56] White PM Jr, Potter TL, Lima IM. Sugarcane and pinewood biochar effects on activity and aerobic soil dissipation of metribuzin and pendimethalin. Industrial Crops and Products. 2015;**74**:737-744

[57] Spokas KA, Koskinen WC, Baker JM, Reicosky DC. Impacts of woodchip biochar additions on greenhouse gas production and sorption/degradation of two herbicides in a Minnesota soil. Chemosphere. 2009;**77**:574-581

[58] Mendes KF, Hall KE, Spokas KA, Koskinen WC, Tornisielo VL. Evaluating agricultural management effects on alachlor availability: Tillage, green manure, and biochar. Agronomy. 2017;**7**:64

[59] Tatarková V, Hiller E, Vaculík M. Impact of wheat straw biochar addition to soil on the sorption, leaching, dissipation of the herbicide (4-chloro-2-methylphenoxy) acetic acid and the growth of sunflower (*Helianthus annuus* L.). Ecotoxicology and Environmental Safety. 2013;**92**:215-221

[60] Ge X, Cao Y, Zhou B, Wang X, Yang Z, Li MH. Biochar addition increases subsurface soil microbial biomass buft has limited effects on soil CO2 emissions in subtropical moso

bamboo plantations. Appl Soil Eco. 2019;**142**:155-165

[61] Khorram MS, Lin D, Zhang Q, Zheng Y, Fang H, Yu Y. Effects of aging process on adsorption–desorption and bioavailability of fomesafen in an agricultural soil amended with rice hull biochar. Journal of Environmental Sciences. 2016;**56**:180-191

[62] Khalid S, Shahid M, Murtaza B, Bibi I, Naeem MA, Niazi NK. A critical review of different factors governing the fate of pesticides in soil under biochar application. Sci Total Environ. 2020;**711**:134645

[63] Sandín-España P, Sevilla-Moran B, Lopez-Goti C, Mateo-Miranda MM, Alonso-Prados JL. Rapid photodegradation of clethodim and sethoxydim herbicides in soil and plant surface model systems. Arabian Journal of Chemistry. 2016;**9**:694-703

[64] Haskis P, Mantzos N, Hela D, Patakioutas G, Konstantinou I. Effect of biochar on the mobility and photodegradation of metribuzin and metabolites in soil–biochar thin-layer chromatography plates. International Journal of Environmental Analytical Chemistry. 2019;**99**:310-327

[65] Varjani S, Kumar G, Rene ER. Developments in biochar application for pesticide remediation: current knowledge and future research directions. Journal of Environmental Management. 2019;**232**:505-513

[66] Zhang P, Sun H, Yu L, Sun T. Adsorption and catalytic hydrolysis of carbaryl and atrazine on pig manure-derived biochars: impact of structural properties of biochars. Journal of Hazardous Materials. 2013; **244**:217-224

[67] Yang J, Pan B, Li H, Liao S, Zhang D, Wu M, et al. Degradation of p-nitrophenol on biochars: role of persistent free radicals. Environmental Science & Technology. 2016;**50**:694-700

[68] Zhang P, Sun H, Min L, Ren C. Biochars change the sorption and degradation of thiacloprid in soil: insights into chemical and biological mechanisms. Environmental Pollution. 2018;**236**:158-167

[69] Sun K, Keiluweit M, Kleber M, Pan Z, Xing B. Sorption of fluorinated herbicides to plant biomass-derived biochars as a function of molecular structure. Bioresource Technology. 2011;**102**:9897-9903

[70] Shinogi Y, Kanri Y. Pyrolysis of plant, animal and human waste: physical and chemical characterization of the pyrolytic products. Bioresource Technology. 2003;**90**:241-247

[71] Fernandes BCC, Mendes KF, Tornisielo VL, Teófilo TMS, Takeshita V, PSF d C, et al. Effect of pyrolysis temperature on eucalyptus wood residues biochar on availability and transport of hexazinone in soil. International journal of Environmental Science and Technology. 2021;**19**:499-514

[72] Martin SM, Kookana RS, Van Zwieten L, Krull E. Marked changes in herbicide sorption–desorption upon ageing of biochars in soil. Journal of Hazardous Materials. 2012;**231**:70-78

[73] Junqueira LV, Mendes KF, Sousa RND, Almeida CDS, Alonso FG, Tornisielo VL. Sorption-desorption isotherms and biodegradation of glyphosate in two tropical soils aged with eucalyptus biochar. Archives of Agronomy and Soil Science. 2020;**66**:1651-1667

[74] Trigo C, Spokas KA, Hall KE, Cox L, Koskinen WC. Metolachlor sorption and degradation in soil amended with fresh and aged biochars. Journal of Agricultural and Food Chemistry. 2016;**64**:3141-3149

[75] Gámiz B, Velarde P, Spokas KA, Cox L. Dynamic effect of fresh and aged biochar on the behavior of the herbicide mesotrione in soils. Journal of Agricultural and Food Chemistry. 2019b;**67**:9450-9459

[76] Gámiz B, Velarde P, Spokas KA, Celis R, Cox L. Changes in sorption and bioavailability of herbicides in soil amended with fresh and aged biochar. Geoderma. 2019a;**337**:341-349

[77] Cabrera A, Cox L, Spokas KURT, Hermosín MC, Cornejo J, Koskinen WC. Influence of biochar amendments on the sorption–desorption of aminocyclopyrachlor, bentazone and pyraclostrobin pesticides to an agricultural soil. Sci Total Environ. 2014;**470**:438-443

[78] Yang Y, Sheng G, Huang M. Bioavailability of diuron in soil containing wheat-straw-derived char. Sci Total Environ. 2006;**354**:170-178

[79] Xu C, Liu W, Sheng GD. Burned rice straw reduces the availability of clomazone to barnyardgrass. Sci Total Environ. 2008;**392**:284-289

[80] Graber ER, Tsechansky L, Gerstl Z, Lew B. High surface area biochar negatively impacts herbicide efficacy. Plant and Soil. 2012;**353**:95-106

[81] Genesio L, Miglietta F, Baronti S, Vaccari FP. Biochar increases vineyard productivity without affecting grape quality: Results from a four years field experiment in Tuscany. Agriculture, Ecosystems and Environment. 2015;**201**:20-25

[82] Khorram MS, Zhang G, Fatemi A, Kiefer R, Mahmood A, Jafarnia S, et al. Effect of walnut shell biochars on soil quality, crop yields, and weed dynamics in a 4-year field experiment. Environmental Science and Pollution Research. 2020;**27**:18510-18520

[83] Mendes KF, Furtado IF, Sousa RND, Lima ADC, Mielke KC, Brochado MGDS. Cow bonechar decreases indaziflam pre-emergence herbicidal activity in tropical soil. Journal of Environmental Science and Health, Part B. 2021;**56**:532-539

[84] Chauhan BS. Rice husk biochar influences seedling emergence of jungle rice (*Echinochloa colona*) and herbicide efficacy. American Journal of Plant Sciences. 2013;**04**:1345-1350

[85] Clay SA, Krack KK, Bruggeman SA, Papiernik S, Schumacher TE. Maize, switchgrass, and ponderosa pine biochar added to soil increased herbicide sorption and decreased herbicide efficacy. J Environ Sci Health Part B. 2016;**51**:497-507

[86] Soni N, Ferrell JA, Devkota P, Mulvaney MJ. Biochar Effects on Weed Management. Vol. 3. Florida, EUA: UF/IFAS Extension University of Florida; 2021

Chapter 2

Effects of Biochar in Soil and Water Remediation: A Review

Adelaide Perdigão and José Luís da Silva Pereira

Abstract

In the last decades increased global environmental concerns to water and soils pollution. The main concerns are related to the contamination of the ecosystem, food security, and human health since many of the contaminants present in soil and water (residues of pesticides and antibiotics, genes of resistance to antibiotics, and heavy metals) are absorbed by plants and enter the food chain. Remediation of the contaminated water and soil to ensure sustainable water supply and food production is urgently needed. The use of biochar can have a positive effect on this remediation process. There are several studies that demonstrate the biochar's ability to block/reduce the contaminating effect of pesticides, antibiotic residues, antibiotic resistance genes, and heavy metals. The objective of this chapter is to carry out a comprehensive review of the effect of using biochar on the availability/transmission of these contaminants to the soil and food supply chain.

Keywords: antibiotics, biochar, environment, food, human health, heavy metals

1. Introduction

Currently, water and soil pollution is a global concern due to its negative effect on ecological safety and health risks [1, 2]. Soil contamination through inorganic and organic contaminants is a well-known problem [1, 3]. Contamination of agricultural soils is due to the long-term application of pesticides, fertilizers, plastic film, wastewater irrigation, sewage application, and other activities [4]. Organic contaminants, such as organochlorine pesticides (OCPs), polychlorinated biphenyls (PCBs), phthalate esters (PAEs), and polycyclic aromatic hydrocarbons (PAHs), are characterized by high toxicity, persistence, and bioaccumulation in the environment [5]. Soil is a major reservoir for a variety of pollutants and is a secondary emission source of contaminants to groundwater and surface water [4].

In the last years has been increasing interest amongst the scientific community in the development of technologies to remediate contaminated sites [6]. Soil remediation techniques are categorized as physical, chemical, or biological, based on the specific nature of the remediation mechanism employed [1]. Physical and chemical methods are costly, inefficient, and result in further pollution, especially in the case of chemical methods [1]. Adsorption is one of the most efficient biological methods for removing contaminants from water and wastewater [7]. The mechanism and capacity of adsorption are influenced by many factors: physicochemical properties of the adsorbent; type and nature of the adsorbate; the related affinity of the adsorbate for the adsorbent; and process conditions.

Therefore, the use of biological materials derived from organic material is considered an eco-friendly and sustainable approach [8]. The application of proper amendments such as biochar is a known efficacious and environmentally friendly method for reducing the availability and mobility of potentially toxic elements in contaminated soil via *in situ* immobilization [3, 9]. Biochar is a sensible and robust material for the enhancement of soil fertility and management of contaminated soils for sustainable agriculture and mitigation of climate change [1, 10]. Is a carbon-rich material that can be prepared from various organic waste feedstocks [1], like from various biomass, both woody (primarily residues from forestry and trees) and non-woody (agricultural crops and residues, animal waste, urban and industrial solid waste). The term biochar is used to designate carbonaceous materials, produced from biological sources, due to the incomplete combustion of fossil fuels and vegetation and constitute an important carbon sink due to their stability to microbial and chemical degradation [11]. Thus, biochar is a porous carbonaceous material largely containing carbon jointly with the inorganic components of the biomass utilized, such as alkali and alkaline earth metals.

For these organic wastes to be transformed into biochar, they are subjected to pyrolysis, gasification, or hydrothermal carbonization. These are the most common methods for biochar preparation [12, 13]. Biochar properties are highly dependent on the temperature (300–1000°C), time of pyrolysis, final acidity, and feedstock from which the biochar is made [14]. According to the residues types, biochar can present diverse physiochemical properties biochar's content of volatile matter, dissolved organic matter, ash, and carbon [1, 15]. Additionally, the pyrolysis condition may influence the physicochemical and chemical properties of biochar [16, 17]. However, there are characteristics that are always present, such as rich carbon content, high cation exchange capacity, large surface area, and stability structure [1, 13]. These characteristics, namely specific surface area, porosity, and cation exchange capacity, are responsible for the high adsorption capacity that gives it the ability to remove organic pollutants and heavy metals [13]. Also, the solution pH has a great influence on the adsorption capacity of biochars [16]. Moreover, biochar will improve the soil's biological activities, nutrient retention, water-retention capacity, an increase of pH value, and amount of soil organic matter.

In last year's, the utilization of biochar has been widely used in environmental applications such as soil remediation and water remediation. According to Krasucka et al. [7] the use of "green", low cost, or sustainable biochar for contaminant sorption yields economic and environmental benefits, furthermore, agrees with global trends in generating a circular economy and sustainable development. Biochar has been a common material for environmental contamination due to the widespread availability of its raw materials, its simple preparation process, low cost, and strong adsorption performance [18].

The application of biochar can improve soils that pose abiotic stresses because of the presence of heavy metals, salt, or organic contaminants [1]. Biochars produced from pyrolysis of biomass materials have received worldwide attention due to their broad usage in contaminant adsorption, soil remediation, and wastewater treatment [19]. Biochar application can reduce the bioavailability and mobility of soil pollutants including pesticides, antibiotics, heavy metal, and antibiotic resistance genes in soil microorganisms [20].

2. Soil and water contaminants

Soil is one of the main resources that human beings rely on to survive, and also the material repository of biogeochemical cycles. The remediation of contaminated soils, in order to protect human health and achieve sustainable development, has become the goal of the scientific community. Remediating contaminated soils to protect human health and to achieve sustainable development has become a desirable goal [21].

The main soil contaminants that most concern society are the following: pesticides, antibiotic residues, antibiotic resistance genes (ARGs), and heavy metals.

2.1 Pesticides

Pesticide is any substance or mixture of substances intended for preventing, destroying, repelling, or mitigating any pest (insects, mice, and other animals, weeds, or microorganisms) [11]. Pesticides are widely used to prevent or control the growth of crop pests in agricultural practices. Their leaching and accumulation in the soil and water is a threat to the environment and human health.

Pesticides are one of the persistent organic pollutants due to their recalcitrant, persistent, and bioaccumulative nature and due to their ability to be carried for long distances to affect remote uninhabited parts of the globe [11].

The *in situ* application of an adsorbent amendment in contaminated soils is a new and cost-effective alternative for remediation of pesticide-contaminated soils [11]. Biochar has been reported efficient for the reduction of pesticides bioavailability in a soil environment with additional benefits of increased soil fertility [22], giving you adsorption potential.

The strong adsorption potential of biochar has resulted in decreased availability of various agrochemicals (insecticides, fungicides, acaricides, and others) used for pest control in various crops [1]. The effect of biochars on the fate of pesticide contaminants in soil depends on the biochars type and properties, which in turn is affected by the feedstock and pyrolysis conditions [22]. Pesticide adsorption on biochars was affected by their aromaticity, polarity, pore-volume, pore diameter, pH, and surface acidity [23]. More porous structures and higher surface area will result in higher sorption capacities [24]. Studies carried out by Liu et al. [25] showed that an increase in pyrolysis temperature results in a larger pore volume and surface area. The pore volume of biochar increased from 0.056 to 0.099 $cm^3\ g^{-1}$ with a temperature from 500 to 900°C, while surface area increased from 25.4 to 67.6 $m^2\ g^{-1}$ [26]. The composition of biochar also affects its properties. For example, the surface area of plant biochar (112–642 $m^2\ g^{-1}$) such as oak wood, maize stover, and pine needle is generally much higher than that of pig manure (3.32–20.5 $m^2\ g^{-1}$), and biosolid biochar (50.9–94.2 $m^2\ g^{-1}$) such as poultry and turkey litter [25, 27, 28], similarly, pine needle biochar shows higher porosity (0.076–1.90 $cm^3\ g^{-1}$) than biosolid biochar (0.053–0.068 $cm^3\ g^{-1}$) derived at 500–700°C [28]. In general, it can be stated that feedstock with high lignin content produces biochar with macroporous structures, while feedstock with high cellulose content mainly produces biochar with microporous structures [24].

The pyrolysis temperature of biochars produced from various biomasses (biosolids, agricultural residues, and animal manure) also interferes with its pH, generally, the increase in temperature results in a higher pH of the biochar, and higher pH of biochar can accelerate the hydrolysis of organophosphorus and carbamate pesticides in the soil through alkali catalysis mechanism [24].

Biochar increases microbial activity by providing available carbon and other nutrients which ultimately accelerate the biodegradation of pesticides in soil [22].

Several studies have been performed to investigate the effect of biochar addition for pesticide remediation. Biochar proved to be efficient in the sorption of pesticides such as atrazine, terbuthylazine, pyrimethanil, or bentazone [29–32]. Using biochar irreversible adsorption of pesticides to soils has been as shown due to entrapment into biochar micropores, surface-specific adsorption, and partitioning into condensed structures [11]. Adsorption of triazine herbicides by biochar was primarily affected by the mesoporosity and microporosity of biochars [33]. Liu et al. [34] compared plant (soybean, corn stalk, and rice stalk) and manure (poultry, cattle, and pig) derived biochars for atrazine adsorption and concluded that, in general, plant-based biochars were more effective in removing atrazine and the removal efficiency increased with elevation in high temperatures treatments and biochar pH.

When studying the application of biochar for sulfamethazine absorption in soils treated with and without biochar derived from an invasive cucumber plant (*Sicyos angulatus* L.), it was found that, after the application of biochar, 86% of the sulfamethazine was reduced in the soil enriched with 5 mg kg^{-1}, whereas only 63% of sulfamethazine was reduced in soil enriched with 50 mg kg^{-1} of sulfamethazine pesticide [35]. Vithanage et al. [36] also used cucumber biochar (*Sicyos angulatus* L.) for sulfamethazine removal in clayey and sandy sand soils under different conditions of pH and sulfamethazine loads. High temperature pyrolyzed biochar (700°C) showed a high degree of sulfamethazine retention. Maximum sulfamethazine retention was observed at pH 3.0, possibly due to π-π electron donor-acceptor interactions and electrostatic cation exchange. As the pH rises to 5.0–7.0, cation exchange was the main sorption mechanism. Biochar was able to maintain up to 89% and 82% increase in sulfamethazine retention in sandy loam and sandy clay soil, respectively.

The use of biochar in water pesticides remediation also proved to be efficient. Mandal et al. [23] affirm that rice straw biochars are efficient to adsorb atrazine and imidacloprid from water and have great potential as the next-generation low-cost adsorbent to prevent contamination of groundwater and minimize the environmental impact caused by the pesticides.

However, biochar application studies for contaminated soil remediation have mainly been conducted in laboratories, greenhouses, or small plot experiments, so large-scale field experiments are needed before implementing remediation projects on an operational scale [11]. Resuming, it can be concluded that pyrolysis temperature affects biochar properties and consequently pesticide sorption capacity. Biochar from high pyrolysis temperatures increases pesticide sorption in soils, reducing their soil mobility and bioavailability.

2.2 Antibiotic residues

The widely abuse of antibiotics has led to unpredictable residue releasing in the ecosystem and caused a series of environmental problems [37]. The ecological toxicity of antibiotics is a subject of growing concern because residues of antibiotics in the environment may pose adverse effects on the ecosystem [19]. Thus, antibiotics have a very low degradation rate in the natural environment and can persist for long periods [38]. Residues of antibiotics are detected in wastewater, liquid manure, surface waters, groundwater, soil and plants, drinking water, and food [7]. The application of animal manure in agricultural land is a major way through which antibiotics enter soils that are used for food production; subsequently, they can be transported to other environmental compartments and enter the food chain [39]. The behavior of antibiotics and their transport in manure is related to their physicochemical properties as well as the abiotic properties of the environment.

The content of antibiotic residues in environmental samples (water, sediment, and soil) is seasonal, in the autumn the amounts of tested drugs were higher than in the summer because higher temperatures causing degradation of the drug and frequent rainfall increases the dilution [7]. Irrespective of the different dissipation pathways for antibiotics in manure, adsorption is one of the key processes controlling the fate, mobility, and reactivity of antibiotics [12]. Adsorption is predominantly, especially in the removal of antibiotics from the environment [7].

Really, the residues of the antibiotics cannot remove effectively in traditional biological wastewater treatment facilities [37]. The methods for antibiotics removal from water include bio-electrochemical systems, heterogeneous photocatalytic method, advanced oxidation process, microbiological method, and adsorption method [40]. Biochars are highly adsorptive with the potential for sequestration of environmental organic contaminants and proved suitable for mitigating effects of antibiotics residues in manure, which effects propagate when manure is added to the soil [39].

Biochar strongly sorbs several antibiotics, including oxytetracycline, ciprofloxacin, tetracycline hydrochloride, doxycycline hydrochloride, and fluoroquinolone [16] in an aqueous solution. The efficiency of removal of the three antibiotics increased with increasing biochar dosage up to 1.2 g L^{-1}. In experimental studies developed by Peng et al. [41], on the adsorption of seven antibiotics in the environmental concentration of aqueous solutions by carbon-based materials have been observed and showed that the carbon-based materials have good adsorption properties for antibiotics in the actual concentration of environments, by which the highest removal efficiency of antibiotics can be up to 100%.

Biochar produced at high pyrolytic temperatures had greater adsorption capacity for antibiotic residues [17]. The efficiency of removal of tetracycline hydrochloride, doxycycline hydrochloride, and ciprofloxacin was enhanced with increasing pyrolysis temperature and biochar dosage [7, 14]. In studies developed by Zeng et al. [16], the best adsorption was obtained with biomass pyrolyzed at 700°C. Regarding antibiotic adsorption, the characteristics of the biochar produced play an essential role, namely its sorption parameters, aromaticity, hydrophilicity, and group density surface oxygen functionals or ash content [7].

2.3 Antibiotic resistance genes

Antibiotic resistance genes (ARGs) are the means through which bacteria become super antibiotic-resistant bacteria [42]. In last year's dissemination and persistence of antibiotic resistance genes (ARGs), received increasing concerns like one of the biggest threats to global public health and food security [43, 44]. The spread of antibiotic-resistant pathogens is a growing problem in the world is considered a type of emerging contaminant.

According to World Health Organization [45] antibiotic resistance is one of the most critical human health challenges of the next century and heralded the need for "a global strategy to contain resistance". In agreement with the last UN Global Environmental Outlook [46] ARGs as classified as the main pollutants for water sources, and call for the application of effective policies for its control in water bodies worldwide. Soil amendments with animal manure are potential sources of ARGs in soil, water, and plants [47]. Animal manures, commonly applied to the soil, support the spread of ARGs from soil to humans via the food chains [44, 48]. Remediation of agricultural soils polluted with ARGs is important for protecting food safety and human health [20, 44].

Biochar amendments have proven able to decrease the relative abundance of multiple subtypes of antibiotic resistance genes (ARGs) [19] because could affect their dissemination and fate in the environment [49]. Recently, a few studies also attempted to use biochar to alleviate ARG pollution in soil [49, 50]. Several studies demonstrated that biochar efficiently mitigated ARG pollution to some extent in soils [51] but not all the biochars consistently showed a positive effect.

In studies developed by Chen et al. [20] reported that biochar of rice straw application apparently reduced the abundance of 131 ARGs in non-planted soil, but little effect was found in a *Brassica chinensis* L. planted soil. In this study, soils were amended with or without 0.5% biochar (w/w). The addition of dissolved biochar effectively inhibited the increase of the conjugative transfer frequency of ARGs between bacteria in studies developed by Liu et al. [18] and Lian et al. [49] affirm that the replication of ARGs was also greatly inhibited after interacting with biochar.

Cui et al. [52] found that wheat straw biochar increased the abundance of tet and sul genes in a soil-plant system. These studies showed the uncertainties about the effectiveness of biochar in remediating soil ARG pollution. This uncertainty is related to the properties of the biochar, as the properties of biochar may affect the conjugative transfer of ARGs between bacteria namely feedstock and pyrolytic conditions [53]. Also, the number of heavy metals and antibiotics present in the soil determines the evolution of ARGs in the soil [44, 52], being interesting the development of biochars with excellent adsorption capacities of heavy metals and antibiotics to prevent soil pollution by ARGs [44].

Relative to heavy metals and other organic pollutants the interaction between biochar and ARGs is much less explored [49]. Therefore, further studies are needed to confirm the effect of different biochar on ARGs and even develop alternative strategies to improve the efficiency of biochar in the remediation of ARG soil pollution.

2.4 Heavy metals

Soil contamination with heavy metals has become a global environmental-health concern [1, 14, 54]. Their high level in the soil cause hazardous effects on soil quality, fertility, food safety, and human health [14]. The entry of soil-borne heavy metals into the food chain depends on the amount and source of heavy metals input, the properties of the soil, the rate and magnitude of uptake by plants [6].

The main aim for researchers and environmentalists is to stop the entry of metals and metalloids into the food chain for better human health [8]. Indiscriminate waste disposal practices have led to significant build-up in soils of a wide range of heavy metals [6]. Sources of heavy metals in soil could be anthropogenic activities, such as mining and smelting, wastewater irrigation, exhaust emissions, and sludge applications [6, 55].

Heavy metals could impact soil fertility, microbial activities, biodiversity, crop yields and pose risks to human health due to dietary exposure [6]. Biochar serves as a faster, more efficient, and environmentally friendly alternative to heavy metals in the agricultural soil [14].

The ability of biochar to remediate soil contaminated by heavy metals is linked to their facility to immobilize pollutants [55]. Immobilization is the conversion of soluble and potentially soluble forms of heavy metals to geochemically stable solid phases. Adsorption, ion exchange, complexation, and precipitation are the major mechanisms reducing heavy metal bioavailability in soils [15, 56, 57]. Biochar could reduce the bioavailability of heavy metals in the soil through cation exchange, complexation, and other related effects [15]. the addition of biochar to soil binds and/or precipitates heavy metals in soil and reduces their accumulation in plants [58].

The remediation effect depends on the characteristics of both biochar and soil and their interactions [6, 55]. Biochar applications could decrease the mobility/bioavailability of heavy metals in soils and their accumulation in plants [55].

The evolution of heavy metals in soil solutions altered by different biochars under vibrant redox conditions is a challenge and more investigation is necessary to comprehend the influence of the biochar on the dynamic forces of the heavy metals contaminated soils [1]. There are several studies that demonstrate the effectiveness of biochar in the treatment of soils contaminated by heavy metals. Thus, studies developed by Zeng et al. [16] reported that maize straw biochar reduced the availability of Cd in soil by transforming this heavy metal into a state of lower availability. Using apricot shell and the apple tree derived biochar [9] found that the labile fractions of Cd in smelter-contaminated soil decreased. Also, in studies developed by [59] using biochar from sewage sludge, this one showed up effective in immobilizing non-essential heavy metals for plants. Li et al. [60] applied 3% of soybean-straw-derived biochar (hydrothermal carbon at 350°C) to the As and Cd co-contaminated farmland soil, which reduced the bioaccumulation of As in rice plants by 88% and the treatment effects on Cd were similar. A field experiment made by Zheng et al. [61] for Cd, showed that, when the application rates of soybean-straw-derived biochar and rice-straw-derived biochar were 20 ton ha^{-1}, the content of Cd in rice roots, rice shoots, rice husks, and rice grains decreased by 25.0–44.1% and 19.9–44.2%, and 46.2–70.6% and 25.8–70.9%, respectively.

The ability of biochar to immobilize heavy metals is related to its ability to change soil pH, biochar typically contains alkaline components, which can considerably improve soil pH [58]. For example, Van Poucke et al. [62] showed a significant negative correlation between the pH value increase and the soil exchangeable Cd content after biochar application. In studies developed by Fang et al. [63] was found that the addition of sludge-derived biochar to contaminated soil could increase the soil pH and reduce the effective concentration of Pb, Cd, Ni, and Cr in soil.

Studies developed by Li et al. [64] evidenced that the heavy metal concentrations in biochar were diluted by adding antibiotic mycelial residue, which led to lower toxic inputs to the environment, moreover, heavy metals were transformed to more stable fractions after co-pyrolysis. Liu et al. [18] confirm that the addition of dissolved biochar is an effective measure to control copper pollution in water, as the combination of humic acid-like components in dissolved biochar with Cu(II) significantly reduces the concentration of Cu(II) in water.

The hydrological conditions in the natural field can influence the effect of biochar on the stabilization of contaminated soil [65]. Studies developed by dos Santos et al. [66], showed the biochar efficiency in removal methylene blue from water. About 83% of methylene blue removal was achieved within 30 minutes of equilibration time.

Braghiroli et al. [67] showed that biochar high aromaticity and porosity are essential for the sorption of organic contaminants, while the presence of oxygen-containing functional groups and optimum pH are crucial for the sorption of inorganic contaminants, especially metals. Thus, biochar can be proficiently adopted for metals adsorption from polluted water [68], showing to be a promising option for the treatment of contaminants in water, but further research is required to evaluate its performance with real effluents containing contaminants of emerging concern [67].

3. Conclusions

Biochar is widely used in wastewater treatment and soil remediation and shows great potential in ameliorating the toxicological effect of antibiotics, pesticides, antibiotic resistance genes, and heavy metals. Biochar efficacy for reducing the availability and mobility of potentially toxic elements in soil and water is dependent on the properties and chemical structure of the contaminants, as well as the adsorption process conditions; however, it is primarily dependent on biochar physicochemical properties. Biochar properties are predominantly determined by the feedstock type used and pyrolysis temperature. Hence, to obtain an efficient and selective biochar adsorbent, it is essential to select appropriate feedstocks and production conditions. However, research gaps still exist in the development of practical methods for preparing and applying different biochars that target specific heavy metals, and more research is needed to expand knowledge in this area.

Acknowledgements

This research was funded by National Funds by FCT—Portuguese Foundation for Science and Technology, under the project UIDB/04033/2020 and UIDB/00681/2020, CERNAS-IPV.

Conflict of interest

The authors declare no conflict of interest.

Author details

Adelaide Perdigão[1,2]* and José Luís da Silva Pereira[2,3]

1 CERNAS-IPV Research Centre, Polytechnic Institute of Viseu, Viseu, Portugal

2 Agrarian School of Viseu, Polytechnic Institute of Viseu, Viseu, Portugal

3 CITAB—Centre for the Research and Technology of Agro-Environmental and Biological Sciences, University of Trás-os-Montes and Alto Douro, Vila Real, Portugal

*Address all correspondence to: aperdigao25@gmail.com

References

[1] Murtaza G, Ditta A, Ullah N, Usman M, Ahmed Z. Biochar for the management of nutrient impoverished and metal contaminated soils: Preparation, applications, and prospects. Journal of Soil Science and Plant Nutrition. 2021;**21**(3):2191-2213

[2] Rostami S, Azhdarpoor A. The application of plant growth regulators to improve phytoremediation of contaminated soils: A review. Chemosphere. 2019;**220**:818-827

[3] Moradi N, Karimi A. Fe-modified common reed biochar reduced cadmium (Cd) mobility and enhanced microbial activity in a contaminated calcareous soil. Journal of Soil Science and Plant Nutrition. 2021;**21**(1):329-340

[4] Sun J, Pan L, Tsang DCW, Zhan Y, Zhu L, Li X. Organic contamination and remediation in the agricultural soils of China: A critical review. Science of the Total Environment. 2018;**615**:724-740

[5] Sun J, Pan L, Zhan Y, Lu H, Tsang DCW, Liu W, et al. Contamination of phthalate esters, organochlorine pesticides and polybrominated diphenyl ethers in agricultural soils from the Yangtze River Delta of China. Science of the Total Environment. 2016;**544**:670-676

[6] Bolan N, Kunhikrishnan A, Thangarajan R, Kumpiene J, Park J, Makino T, et al. Remediation of heavy metal(loid)s contaminated soils—To mobilize or to immobilize? Journal of Hazardous Materials. 2014;**266**:141-166

[7] Krasucka P, Pan B, Sik Ok Y, Mohan D, Sarkar B, Oleszczuk P. Engineered biochar—A sustainable solution for the removal of antibiotics from water. Chemical Engineering Journal. 2021;**405**:126926

[8] Mehmood S, Wang X, Ahmed W, Imtiaz M, Ditta A, Rizwan M, et al. Removal mechanisms of slag against potentially toxic elements in soil and plants for sustainable agriculture development: A critical review. Sustainability. 2021;**13**(9):5255

[9] Ali A, Shaheen SM, Guo D, Li Y, Xiao R, Wahid F, et al. Apricot shell- and apple tree-derived biochar affect the fractionation and bioavailability of Zn and Cd as well as the microbial activity in smelter contaminated soil. Environmental Pollution. 2020;**264**:114773

[10] Pereira JLS, Figueiredo V, Pinto AFMA, Silva MEF, Brás I, Perdigão A, et al. Effects of biochar and clinoptilolite on composition and gaseous emissions during the storage of separated liquid fraction of pig slurry. Applied Sciences. 2020;**10**(16):5652

[11] Morillo E, Villaverde J. Advanced technologies for the remediation of pesticide-contaminated soils. Science of the Total Environment. 2017;**586**: 576-597

[12] Guo M, Song W, Tian J. Biochar-facilitated soil remediation: Mechanisms and efficacy variations. Frontiers in Environmental Science. 2020;**8**:183

[13] Wang J, Wang S. Preparation, modification and environmental application of biochar: A review. Journal of Cleaner Production. 2019;**227**: 1002-1022

[14] Awad M, Liu Z, Skalicky M, Dessoky ES, Brestic M, Mbarki S, et al. Fractionation of heavy metals in multi-contaminated soil treated with biochar using the sequential extraction procedure. Biomolecules. 2021;**11**(3):448

[15] Xu C, Zhao J, Yang W, He L, Wei W, Tan X, et al. Evaluation of biochar pyrolyzed from kitchen waste, corn straw, and peanut hulls on immobilization of Pb and Cd in

contaminated soil. Environmental Pollution. 2020;**261**:114133

[16] Zeng Z, Tian S, Liu Y, Tan X, Zeng G, Jiang L, et al. Comparative study of rice husk biochars for aqueous antibiotics removal. Journal of Chemical Technology and Biotechnology. 2018;**93**(4):1075-1084

[17] Ahmad M, Lee SS, Rajapaksha AU, Vithanage M, Zhang M, Cho JS, et al. Trichloroethylene adsorption by pine needle biochars produced at various pyrolysis temperatures. Bioresource Technology. 2013;**143**:615-622

[18] Liu X, Wang D, Wang L, Tang J. Dissolved biochar eliminates the effect of Cu(II) on the transfer of antibiotic resistance genes between bacteria. Journal of Hazardous Materials. 2022;**424**:127251

[19] He G, Jiang X, Yao L, Liu G, Yang Y, Jiang Y, et al. Effects of tetracycline on nitrogen and carbon cycling rates and microbial abundance in sediments with and without biochar amendment. Chemosphere. 2021;**270**:129509

[20] Chen Q-L, Fan X-T, Zhu D, An X-L, Su J-Q, Cui L. Effect of biochar amendment on the alleviation of antibiotic resistance in soil and phyllosphere of *Brassica chinensis* L. Soil Biology and Biochemistry. 2018;**119**: 74-82

[21] Cheng M, Zeng G, Huang D, Lai C, Xu P, Zhang C, et al. Hydroxyl radicals based advanced oxidation processes (AOPs) for remediation of soils contaminated with organic compounds: A review. Chemical Engineering Journal. 2016;**284**:582-598

[22] Varjani S, Kumar G, Rene ER. Developments in biochar application for pesticide remediation: Current knowledge and future research directions. Journal of Environmental Management. 2019;**232**:505-513

[23] Mandal A, Singh N, Purakayastha TJ. Characterization of pesticide sorption behaviour of slow pyrolysis biochars as low cost adsorbent for atrazine and imidacloprid removal. Science of the Total Environment. 2017;**577**:376-385

[24] Liu Y, Lonappan L, Brar SK, Yang S. Impact of biochar amendment in agricultural soils on the sorption, desorption, and degradation of pesticides: A review. Science of the Total Environment. 2018;**645**:60-70

[25] Liu Y, Yao S, Wang Y, Lu H, Brar SK, Yang S. Bio- and hydrochars from rice straw and pig manure: Inter-comparison. Bioresource Technology. 2017;**235**:332-337

[26] Chen T, Zhang Y, Wang H, Lu W, Zhou Z, Zhang Y, et al. Influence of pyrolysis temperature on characteristics and heavy metal adsorptive performance of biochar derived from municipal sewage sludge. Bioresource Technology. 2014;**164**:47-54

[27] Cantrell KB, Hunt PG, Uchimiya M, Novak JM, Ro KS. Impact of pyrolysis temperature and manure source on physicochemical characteristics of biochar. Bioresource Technology. 2012;**107**:419-428

[28] Li H, Dong X, da Silva EB, de Oliveira LM, Chen Y, Ma LQ. Mechanisms of metal sorption by biochars: Biochar characteristics and modifications. Chemosphere. 2017; **178**:466-478

[29] Wang H, Lin K, Hou Z, Richardson B, Gan J. Sorption of the herbicide terbuthylazine in two New Zealand forest soils amended with biosolids and biochars. Journal of Soils and Sediments. 2010;**7**:283-289

[30] Dairy-Manure Derived Biochar Effectively Sorbs Lead and Atrazine|Environmental Science &

Technology [Internet]. Disponível em: https://pubs.acs.org/doi/abs/10.1021/es803092k [citado 29 de Setembro de 2021]

[31] Cabrera A, Cox L, Spokas K, Hermosín MC, Cornejo J, Koskinen WC. Influence of biochar amendments on the sorption–desorption of aminocyclopyrachlor, bentazone and pyraclostrobin pesticides to an agricultural soil. Science of the Total Environment. 2014;**470-471**:438-443

[32] Reduced plant uptake of pesticides with biochar additions to soil—ScienceDirect [Internet]. Disponível em: https://www.sciencedirect.com/science/article/pii/S0045653509004226 [citado 29 de Setembro de 2021]

[33] π+−π Interactions between (Hetero) aromatic Amine Cations and the Graphitic Surfaces of Pyrogenic Carbonaceous Materials|Environmental Science & Technology [Internet]. Disponível em: https://pubs.acs.org/doi/abs/10.1021/es5043029 [citado 29 de Setembro de 2021]

[34] Liu N, Charrua AB, Weng C-H, Yuan X, Ding F. Characterization of biochars derived from agriculture wastes and their adsorptive removal of atrazine from aqueous solution: A comparative study. Bioresource Technology. 2015;**198**:55-62

[35] Rajapaksha AU, Vithanage M, Lim JE, Ahmed MBM, Zhang M, Lee SS, et al. Invasive plant-derived biochar inhibits sulfamethazine uptake by lettuce in soil. Chemosphere. 2014;**111**:500-504

[36] Vithanage M, Rajapaksha AU, Tang X, Thiele-Bruhn S, Kim KH, Lee S-E, et al. Sorption and transport of sulfamethazine in agricultural soils amended with invasive-plant-derived biochar. Journal of Environmental Management. 2014;**141**:95-103

[37] Li H, Hu J, Yao L, Shen Q, An L, Wang X. Ultrahigh adsorbability towards different antibiotic residues on fore-modified self-functionalized biochar: Competitive adsorption and mechanism studies. Journal of Hazardous Materials. 2020;**390**:122127

[38] Huang A, Yan M, Lin J, Xu L, Gong H, Gong H. A review of processes for removing antibiotics from breeding wastewater. International Journal of Environmental Research and Public Health. 2021;**18**(9):4909

[39] Ngigi AN, Ok YS, Thiele-Bruhn S. Biochar affects the dissipation of antibiotics and abundance of antibiotic resistance genes in pig manure. Bioresource Technology. 2020;**315**:123782

[40] Wang H, Lou X, Hu Q, Sun T. Adsorption of antibiotics from water by using Chinese herbal medicine residues derived biochar: Preparation and properties studies. Journal of Molecular Liquids. 2021;**325**:114967

[41] Peng B, Chen L, Que C, Yang K, Deng F, Deng X, et al. Adsorption of antibiotics on graphene and biochar in aqueous solutions induced by π-π interactions. Scientific Reports. 2016;**6**(1):31920

[42] Identification and quantification of bacterial genomes carrying antibiotic resistance genes and virulence factor genes for aquatic microbiological risk assessment—ScienceDirect [Internet]. Disponível em: https://www.sciencedirect.com/science/article/pii/S0043135419309340 [citado 29 de Setembro de 2021]

[43] Hernando-Amado S, Coque TM, Baquero F, Martínez JL. Defining and combating antibiotic resistance from one health and global health perspectives. Nature Microbiology. 2019;**4**(9):1432-1442

[44] Zheng H, Feng N, Yang T, Shi M, Wang X, Zhang Q, et al. Individual and

combined applications of biochar and pyroligneous acid mitigate dissemination of antibiotic resistance genes in agricultural soil. Science of the Total Environment. 2021;**796**:148962

[45] Antibiotic resistance [Internet]. Disponível em: https://www.who.int/news-room/fact-sheets/detail/antibiotic-resistance [citado 29 de Setembro de 2021]

[46] UN Environment, editor. Global Environment Outlook—GEO-6: Healthy Planet, Healthy People. 1st ed. Cambridge: Cambridge University Press; 2019. Disponível em: https://www.cambridge.org/core/product/identifier/9781108627146/type/book [citado 9 de Fevereiro de 2021]

[47] Xu Y, Li H, Shi R, Lv J, Li B, Yang F, et al. Antibiotic resistance genes in different animal manures and their derived organic fertilizer. Environmental Sciences Europe. 2020;**32**(1):102

[48] Lu Y, Li J, Meng J, Zhang J, Zhuang H, Zheng G, et al. Long-term biogas slurry application increased antibiotics accumulation and antibiotic resistance genes (ARGs) spread in agricultural soils with different properties. Science of the Total Environment. 2021;**759**:143473

[49] Lian F, Yu W, Zhou Q, Gu S, Wang Z, Xing B. Size matters: Nano-biochar triggers decomposition and transformation inhibition of antibiotic resistance genes in aqueous environments. Environmental Science & Technology. 2020;**54**(14):8821-8829

[50] Ding J, Yin Y, Sun A-Q, Lassen SB, Li G, Zhu D, et al. Effects of biochar amendments on antibiotic resistome of the soil and collembolan gut. Journal of Hazardous Materials. 2019;**377**:186-194

[51] Ye M, Sun M, Zhao Y, Jiao W, Xia B, Liu M, et al. Targeted inactivation of antibiotic-resistant *Escherichia coli* and *Pseudomonas aeruginosa* in a soil-lettuce system by combined polyvalent bacteriophage and biochar treatment. Environmental Pollution. 2018;**241**: 978-987

[52] Cui E-P, Gao F, Liu Y, Fan X-Y, Li Z-Y, Du Z-J, et al. Amendment soil with biochar to control antibiotic resistance genes under unconventional water resources irrigation: Proceed with caution. Environmental Pollution. 2018;**240**:475-484

[53] Liu X, Wang D, Tang J, Liu F, Wang L. Effect of dissolved biochar on the transfer of antibiotic resistance genes between bacteria. Environmental Pollution. 2021;**288**:117718

[54] Rizwan MS, Imtiaz M, Zhu J, Yousaf B, Hussain M, Ali L, et al. Immobilization of Pb and Cu by organic and inorganic amendments in contaminated soil. Geoderma. 2021;**385**:114803

[55] He L, Zhong H, Liu G, Dai Z, Brookes PC, Xu J. Remediation of heavy metal contaminated soils by biochar: Mechanisms, potential risks and applications in China. Environmental Pollution. 2019;**252**:846-855

[56] Cheng S, Chen T, Xu W, Huang J, Jiang S, Yan B. Application research of biochar for the remediation of soil heavy metals contamination: A review. Molecules. 2020;**25**(14):3167

[57] Boni MR, Marzeddu S, Tatti F, Raboni M, Mancini G, Luciano A, et al. Experimental and numerical study of biochar fixed bed column for the adsorption of arsenic from aqueous solutions. Water. 2021;**13**(7):915

[58] Wei L, Huang Y, Huang L, Huang Q, Li Y, Li X, et al. Combined biochar and soda residues increases maize yields and decreases grain Cd/Pb in a highly Cd/Pb-polluted acid Udults soil.

Agriculture, Ecosystems and Environment. 2021;**306**:107198

[59] Chagas JKM, de Figueiredo CC, da Silva J, Paz-Ferreiro J. The residual effect of sewage sludge biochar on soil availability and bioaccumulation of heavy metals: Evidence from a three-year field experiment. Journal of Environmental Management. 2021;**279**:111824

[60] Li G, Khan S, Ibrahim M, Sun T-R, Tang J-F, Cotner JB, et al. Biochars induced modification of dissolved organic matter (DOM) in soil and its impact on mobility and bioaccumulation of arsenic and cadmium. Journal of Hazardous Materials. 2018;**348**:100-108

[61] Zheng R, Chen Z, Cai C, Tie B, Liu X, Reid BJ, et al. Mitigating heavy metal accumulation into rice (*Oryza sativa* L.) using biochar amendment—A field experiment in Hunan, China. Environmental Science and Pollution Research. 2015;**22**(14):11097-11108

[62] Van Poucke R, Ainsworth J, Maeseele M, Ok YS, Meers E, Tack FMG. Chemical stabilization of Cd-contaminated soil using biochar. Applied Geochemistry. 2018;**88**:122-130

[63] Fang S, Tsang DCW, Zhou F, Zhang W, Qiu R. Stabilization of cationic and anionic metal species in contaminated soils using sludge-derived biochar. Chemosphere. 2016;**149**: 263-271

[64] Li C, Xie S, You F, Zhu X, Li J, Xu X, et al. Heavy metal stabilization and improved biochar generation via pyrolysis of hydrothermally treated sewage sludge with antibiotic mycelial residue. Waste Management. 2021;**119**: 152-161

[65] Shentu J, Li X, Han R, Chen Q, Shen D, Qi S. Effect of site hydrological conditions and soil aggregate sizes on the stabilization of heavy metals (Cu, Ni, Pb, Zn) by biochar. Science of the Total Environment. 2022;**802**: 149949

[66] dos Santos KJL, dos Santos GEDS, de Sá ÍMGL, Ide AH, Duarte JLDS, de Carvalho SHV, et al. Wodyetia bifurcata biochar for methylene blue removal from aqueous matrix. Bioresource Technology. 2019;**293**:122093

[67] Braghiroli FL, Bouafif H, Neculita CM, Koubaa A. Activated biochar as an effective sorbent for organic and inorganic contaminants in water. Water, Air, and Soil Pollution. 2018;**229**(7):230

[68] dos Santos GEDS, Lins PVDS, Oliveira LMTDM, da Silva EO, Anastopoulos I, Erto A, et al. Layered double hydroxides/biochar composites as adsorbents for water remediation applications: Recent trends and perspectives. Journal of Cleaner Production. 2021;**284**:124755

Chapter 3

Pesticide Residues: Impacts on Fauna and the Environment

Muzafar Riyaz, Rauf Ahmad Shah and Kuppusamy Sivasankaran

Abstract

Pesticide residues are the traces of pesticide compounds that remain on or in the crop, water, soil and air after the application. Pesticide residues get into the environment as a result of application or by accident and can be found in the air, water and soil. Pesticide residues, if present in air, soil and water can pose a serious threat to biological diversity and human health. After depositing in the environment, the pesticides start to break down and forms metabolites that are more or less toxic. Pesticide residues decline as the pesticide breaks down over time, therefore the levels of residues are highest immediately after the application and diminish as the crops continue to grow. When exposed to sunlight or microorganisms in the soil, most pesticides degrade easily however, the utmost number of pesticides after application scatter into non-target areas or leach into groundwater or move in surface runoff by misuse and misapplication while handling or spraying. The impact of widespread usage of chemical pesticides has made an uncountable number of effects on human health, environment and other life forms and has turned into a serious issue across the globe. The present study aims to present an introduction to the environmental pesticide residues and various aspects highlighting their impact on nature and biodiversity.

Keywords: Pesticide, Residues, Environment, Contamination, Human Health

1. Introduction

To ensure food safety around the world, it is important to build up all necessary measures to boost crop production. A crop loss due to pests is the biggest challenge our agriculture sector faces today. A decrease in crop yield from pest damage is one of the significant errands to guarantee crop productivity. Pesticides assume an imperative job in boosting rural profitability. The advantages of pesticides involve increased crop yield, expanded benefits for agriculturists and the counteractive action to crop diseases. Pesticides help farmers to overcome work costs by diminishing the measure of time required to control weeds and pests from fields. Pesticides are the chemical compounds that are used to control various pests and disease-spreading vectors like mosquitoes, ticks and household pests such as rats and cockroaches. The majority of the pesticides are used in agriculture to control various types of insect pests as well as non-insect pests like ticks and mites, weeds and fungal infestations and other crop diseases. Pesticides assume a comparative job in control the pests and enhance the crop from notorious pests thereby boosting the economy of a country. However, with the rise in global population, the crops are being cultivated on a large scale resulted in unrestricted utilization of pesticides. Pesticides have been linked to various environmental contaminations like soil, water and air [1]. In addition to control insect pests, weeds, vectors and other household pests, there has been a great impact of pesticide use on beneficial insects like pollinators, birds, fishes, non-target plants and on human health as well [2]. While utilizing the pesticides, the residues can remain in the environment for a long period and can be dispersed over a long distance. While spraying these chemical pesticides, a series of reactions can undergo; plants can take up pesticides through leaves and roots, the atmosphere can take up pesticides as vapors carried off as drift, pesticides can get ingested by insects, worms and microorganisms [3]. Soil is one of the final destinations of pesticides after application. Depending on the physical–chemical characteristics of the pesticide and soil, the pesticides may be sorbed the particles or be leached and/or carried on the surface by the rains reaching subterranean waters and rivers [4]. Pesticide residues in soil and water can pose a threat to biological diversity and human health. After getting deposited in the environment, the pesticides start to break down and forms metabolites that are more or less toxic [5]. Abiotic and biotic transformations play an important role in removing the pesticide residues from the environment. Environmental degradation of pesticides involves biotic transformation processes facilitated by microorganisms or plants and by abiotic processes such as chemical and photochemical reactions [6]. When a pesticide is applied on the crops the pesticide residues remain in the environment even when a farmer follows all label instructions. Pesticides can cause both acute and chronic effects on human health and the farmers are the most susceptible to intoxication. The pesticides especially the insecticides which are designed to control the insect pests have caused an unaccountable damage among the non-target insect pests which include insect pollinators such as honey bees, bumblebees, syrphid flies and insect predators which check the insect pest populations in an ecosystem, therefore breaching the protocols of insect food chains and food webs [7]. The avian fauna including some of the top predatory birds has also threatened by the large-scale utilization of chemical pesticides including the DDT, which is banned in more than 40 countries, however, very persistent and its residues are found to this day [8]. Fish diversity and other aquatic creatures both animals and plants are also affected by pesticides. The entry of pesticides into water bodies is because of man-made or by natural activities, therefore, can pose a serious threat to aquatic life. Agriculture is the main source of food across the planet and to ensure crop productivity, pesticides are indispensable, however, contamination of the environment raises concerns. The impact of chemical pesticides including health ailments among farmers and environmental contaminations has been reported from all parts of the world from both developed nations to developing nations [9].

2. Pesticides residues

The term Pesticide includes all of the following; herbicide, insecticide, nematicides, acaracide, rodenticide, bactericide, fungicide, insect repellent, disinfectant and so on. The most commonly used pesticides are fungicides which account for 80% of all pesticides used. Most pesticides are intended to serve as plant protection products which in general protect plants from weeds, fungi or insects. Target pests can include insects, plant pathogens, weeds, mollusks, birds that destroy crops, cause nuisance or spread diseases. Although pesticides have benefits, most of the pesticides utilized in farm fields or in residential areas to control disease vectors have several drawbacks such as potential toxicity to humans and other organisms.

With the large-scale utilization, they still hold the potential to contaminate our ecosystems, pollute soil, water, air, impact wildlife, beneficial pollinators and human health. Pesticides have physical–chemical properties that will inflate their behavior in the environment. These are the properties of pesticides which after application can cause short-term or long-term effects on the environment and other organisms as well by either persisting at a long period or by drifting to places other than target sites [10].

a. Persistence: - How long the pesticide remains active in the environment.

b. Mobility: - How easily the pesticide can move from where it is applied.

c. Non-target toxicity: - How toxic is the pesticide to other organisms other than a pest.

d. Volume of use: - How much of that pesticide is used in the environment.

A number of properties of pesticides can affect their behavior in their environment and can cause multiple numbers of environmental contaminations which include Persistence, Degradation Bio-accumulation, Volatility, Adsorption and Absorption (**Figure 1**) [11]. Sooner or later, pesticides are broken down in the environment by a process called Degradation. Depending upon the nature of pesticide and environmental conditions of a particular area, the process of degradation can be rapid or deliberate. However, microorganisms present in the soil, chemical reactions and sunlight play a key role in the degradation of pesticides. On the other hand, Pesticide molecules can be a food source of microbes while taking the advantage of moist and warm soils; microbes can turn the pesticide molecules into carbon

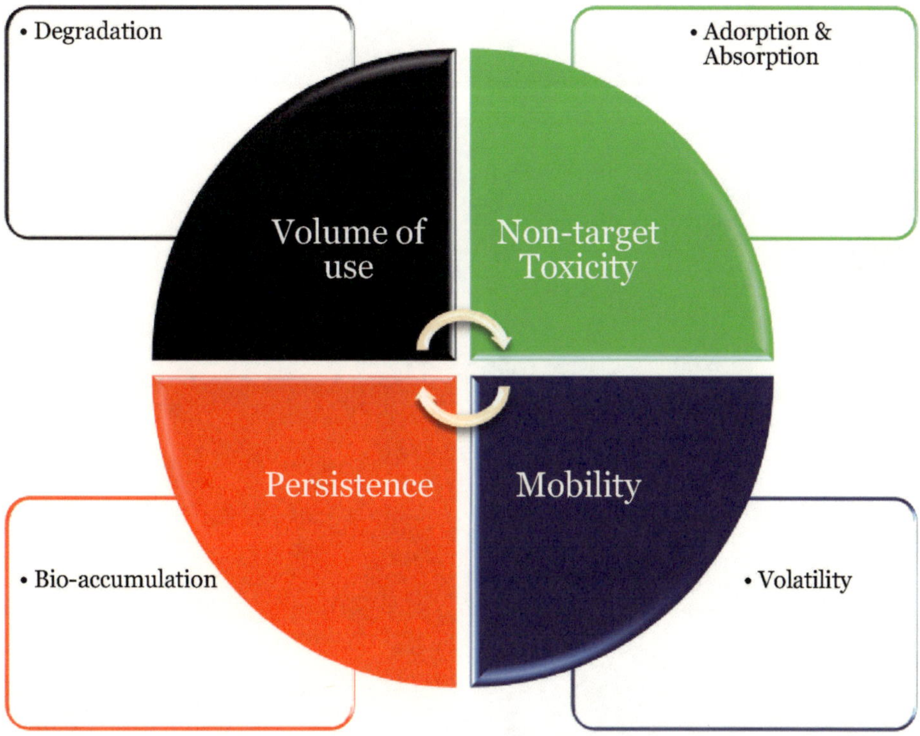

Figure 1.
Factors affecting pesticide and its degree of risk to the environment. (Designed in MS OFFICE POWERPOINT by Muzafar Riyaz).

dioxide and water. Some pesticides such as Chlordane and DDT do not break down quickly, this class of pesticides are called persistent pesticides [12]. Persistence can be greater in heavy clay or organic soil than in sandy soil. Some pesticides after intake through food, water or air may accumulate or build up in body tissues or body fat of humans and animals by a process called Bio-accumulation. These fat-soluble pesticides such as DDT are stores in the body's fat, and when the fatty tissues are used for energy, the compounds are released and cause acute poisoning [13]. If the organism cannot eliminate the pesticides from its body, there is a chance that more pesticide compounds will store in the fat cells, if the organism is exposed to the pesticides routinely. In the 1940s scientists found residues of the man-made chlorinated hydrocarbon pesticide, DDT in human fat which was an alarming issue back then, as many chlorinated hydrocarbon pesticides do not degrade readily and

Figure 2.
Drift can cause pesticides to travel away from the target site with vapors or dust particles while application. (Photo Muzafar Riyaz 2021).

because they accumulate in fat, they move from one organism to another upward in the food chain all the way to humans [14]. Small levels of these types of pesticides in water and soil can magnify into a significant hazard to predators at the top of the food chain. When exposed to air or evaporate, pesticides may change into a vapor by a process called volatility. Once a pesticide evaporates, it is carried for miles simultaneously with the dust particles in the air (**Figure 2**). Pesticides can bind onto soil particles and organic matter by a process called adsorption. In adsorption, A pesticide can bind to the surface of soil particle similarly to that of magnetic attraction. The most adsorptive soils are clay and soils which are having a high concentration of organic matter. Since these pesticides are bound tightly with the soils, there is a very low chance for pesticides to leach with water and therefore they cannot move downward through soil and will less likely to reach groundwater. Water or wind can be the cause of the erosion of pesticides tightly adsorbed to the soil and not be so readily degraded by soil microorganisms. Pesticides can be taken up by the flora and faunal species including insects by a process called absorption. The fate of the pesticides can be determined by a combination of properties and not by a single property. The crusade of pesticides in an environment is very complex as after the application the pesticides can move by some natural processes such as drift, surface runoff, leaching and soil erosion (**Figure 3**) [15].

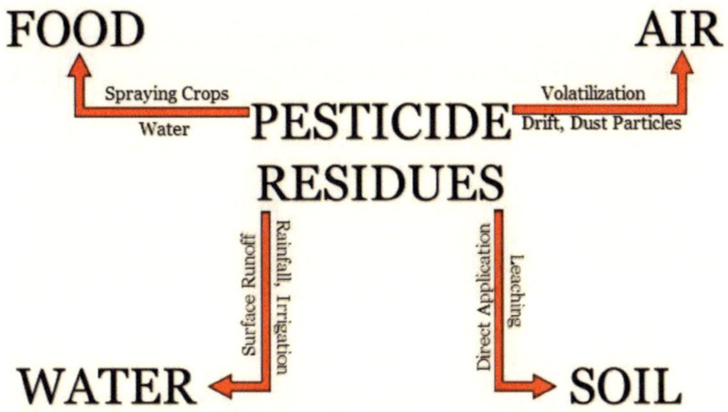

Figure 3.
Movement of pesticide residues in the environment. (Designed in MS OFFICE POWERPOINT by Muzafar Riyaz).

2.1 Air

For over 12 days, pesticides may thoroughly orbit the globe and particles will abide in the air for around seven days; at a height of 6 km, for 30 days; at a height of 30 km, for two years [16]. The purpose of pesticide application is to control a pest population. Ideally, pesticide application should impact only the target organism and have little or no impact on other organisms in the environment. However, many pesticide applications have the potential to affect non-target organisms and move beyond the application site. The potential for a pesticide to contaminate the environment depends in large part on the nature of the pesticide, its ability to break down in a given substrate, type of formulation, application rate, frequency of application and environmental conditions [17, 18]. A pesticide can change its nature from liquid form to a vapor by a process called Volatilization. Pesticides become airborne in many ways, including volatilization, drift or through movement as dust borne particles. Volatility increases with increase in temperature, wind speed and humidity. Applying pesticides in cooler temperatures (below 29°C), or above wind speeds of (10 m/hour) or when the humidity is less high, it's likely for pesticides to volatilize and move off the target sites. After application, volatilized pesticides such as Methyl-bromide drift off the application site and locomote into the atmosphere and taking advantage of air currents can relocate to longer distances as highly volatile compounds vaporize or evaporate at low temperatures [19]. On the other hand, volatility is a useful property in the application of pesticides which aids a pesticide to disperse across the farm field or application site or target area and therefore increase the exposure of pests to the pesticide however, it also can lead to exposure for non-target organisms. Environmental conditions such as air movement, relative humidity and temperature also influence volatilization. A number of pesticides, such as emulsifiable concentrate formulations and all of the fumigants, are classified as volatile organic compounds (VOCs) because they readily volatilize into the atmosphere. With the help of Sunlight, VOCs react with nitrogen oxides to produce ozone can which can contribute to smog and cause respiratory and plant injuries. Drift, on the other hand, refers to the airborne movement of pesticides away from the treatment site during application. Drift can damage plants away from the application site, reduce the effectiveness of a pesticide and cause environmental contamination such as water pollution. Drift is most serious when applications are made in windy conditions. Low relative humidity and high temperatures increase the potential for drift by causing spray droplets to evaporate faster [20]. Air temperature also contributes to pesticide drift by creating inversion layers near the soil surface; as a result of which warmer air layers trap cool air layers. At the time of pesticide application, fine spray droplets and pesticide vapors can be trapped by the inversion layers which can form a concentrated cloud with the ability to move from the treatment site. During pesticide application, droplet size also plays an important role in the movement of spray particles away from the application site. Small droplets fall through the air slowly and have a great potential to drift therefore while large droplets fall faster and are more likely to fall to the ground. Applications that release the pesticide as close to the target site as possible reduce drift. Spray pressure also affects drift by influencing the size of spray droplets; higher pressure decreases droplet size and increases drift. After application, fine particles of pesticides may drift off while splashing dust formulations and liquid droplets may stick to the soil particles and later be transported by the wind into the atmosphere [21].

2.2 Soil

Pesticide characteristics like water solubility, tendency to adsorb to the soil, pesticide persistence and soil characteristics like clay, sand and organic matter are important in determining the fate of the chemicals in the environment. Pesticides may be directly applied to the soil surface, incorporated into the top few inches of soil, or applied through chemigation (**Figure 4**). Once pesticides are present, the soil acts as a reservoir from which persistent pesticides can move into the bodies of invertebrates, be taken up by plants, pass into air or water or break down. After contact with the soil, pesticides are influenced by many factors, including adsorption rate, soil texture, organic matter content in the soil, microorganisms and the presence of water. The soil type influences pesticide persistence and leaching as the tendency for pesticides to be adsorbed vary with the proportion of clay and organic matter in the soil: the higher the percentage of clay and organic matter in the soil, the greater the number of adsorption sites as clay and inorganic matter increase the binding because they have more positive and negative charge sites. It also decreases the potential of a pesticide to move down through the soil, therefore the residues stay in the soil for longer periods of time without moving [22]. Pesticides tend to stay

Figure 4.
Wet soil due to fresh pesticide application in a farm field. (Photo Muzafar Riyaz 2021).

longer in soils with high clay content and organic matter. The amount of water in the soil affects the persistence of pesticides, when more water is added there is a high chance of pesticide release from the soil particle as the water can and force it onto a solution. Usually, the half-life of the pesticide is a used parameter by which the persistence of a pesticide can be measured. A half-life of a pesticide is the period that takes 50% of the pesticide to break down in the environment; the longer the half-life, the greater the possibility for movement of a pesticide before it degrades [23]. On contrary, adsorption refers to the tendency of pesticides to become attached to soil particles. After their release into the environment, pesticides undergo a series of reactions that transform the original compound into various degradation products. Comparing the parent compound, the breakdown products of pesticides may be more toxic, less toxic or equally. Chemically induced transformations of pesticides occur through hydrolysis, photodegradation, microbial degradation and oxidation–reduction. The beneficial soil microorganisms and their associated biotransformation in the soils have been adversely affected by the pesticide residues. The pesticides have also resulted in inactivation of nitrogen-fixing and phosphorus-solubilizing microorganisms soils. A number of studies have shown that some pesticides disturb molecular interactions between plants and N-fixing rhizobacteria and consequently inhibit the vital process of biological nitrogen fixation. Pesticide residue can also reduce activities of soil enzymes that are key indicators of soil health [24–26].

2.3 Water

Water is the basis of life and only a tiny share of all the water on earth is fresh and renewed by the water cycle. Less than 1% of the water is left for drinking, agriculture, industry and nature. Another potential fate of the pesticide residues in the environment is moving into the water. The potential for movement is greater for pesticides that have a long persistence rate while other factors may include the tendency to adsorb to soil and high-water solubility. Lower adsorption can be a potential cause for pesticides to leach or move in the water. However, some pesticides that adsorb to soil particles, such as pyrethroid insecticides can be washed into surface water when soil and sediment erode. The water solubility of a pesticide affects the ease with which it leaches into soil or moves with surface runoff water [27]. Surface water and groundwater contamination can be closely connected and water-soluble pesticides by a problem in both. Surface water contamination occurs through a direct application (usually by accident) or through drift or runoff. Runoff is one of the most common ways that surface water can become contaminated. During pesticide application from a particular area, the movement of water and dissolved or suspended matter move into surface water or onto neighboring land. However, it's likely to occur when heavy rainfall or irrigation takes place after an application. Groundwater contamination can happen in several ways. Pesticides contaminate groundwater through direct entry and by leaching through the soil. Any opening in the soil will be the cause of direct entry of pesticides into groundwater, as it allows water (or contaminants) to detour the soil's natural filtration agents such as plant roots, burrows, abandoned wells etc. Spilling pesticides while mixing them near a well, pumping water into pesticide application equipment without using air gaps or backflow prevention devices and injecting pesticides into an irrigation system without a backflow prevention device can cause groundwater contamination [28]. Ground water has more possible chances to get contaminated than surface water by the pesticide residues as most surface waters (except deep lakes) have a rapid turnover rate, which means that fresh water dilutes the concentration of the contaminant quickly. On contrary, most surface waters contain free oxygen, which enhances the rate at which pesticides are broken down by microorganisms [29]. Another cause of the movement of pesticides is leaching, which makes a passage for a pesticide to move in water descending through the soil as a result of rainwater or irrigation water which percolates between the soil particles, carrying water-soluble pesticides with it. Nonpoint source pollution, as a result of normal applications on a farm field, orchard, or other wide areas over time, occurs when a small amount of pesticide enters groundwater from any location. Point source pollution, due to pesticide mishandling or from improperly constructed disposal sites or holding facilities, would include large quantities of contaminants entering groundwater at small defined locations. Pesticides that are more mobile in the soil and are resistant to degradation can easily settle down in the groundwater. Shallow water tables beneath treated areas are more susceptible to contamination because pesticides pass through less soil and therefore do not degrade much.

2.4 Food

Pesticides are considered important for protecting harvests and ensuring our food supply. All pesticides contain active substances which are essential ingredients that enable them to function. This can be a chemical or a microorganism such as a bacterium or a virus. In some cases, the chemical works by making the crop less palatable for pests. However, the pesticides work by simply killing or damaging the insect pests, weeds, fungi and so on. In some cases, small amounts called residues can find their way into food that humans eat [30]. These residues could be harmful if they exceed certain levels. There are many ways in which pesticide residues can get into our food [31]. Residues in treated crops can be carried from the field into the food by direct application of pesticides on crops till the time of harvest. Pesticide residues can get into the water supply or they can contaminate soil and animal feed, therefore, find their way into our food indirectly. The human food chain is also affected by the pesticide residues left in crops soil and water. Intake of pesticide residues in the body has been connected to birth imperfection, danger to the embryo, disease, hereditary deformities, neurotoxicity and endocrine disruption [32].

Pesticide residues can pose a risk to the heath of end consumers, if residue levels are too high. This is maintained by through Maximum residue levels (MRLs) which are the highest amounts of an individual pesticide that is permitted to be present. Pesticide residues are identified and quantified by comparing the sample extract to a calibration standard solution and analyzing them by liquid or gas chromatography coupled with mass spectroscopy. Once pesticides are demonstrated to be safe for the consumers, they have MRLs set for them which are determined based on rigorous evaluations. A maximum residue level is the maximum amount of residue that is legally permitted in food measured in milligrams of substance per kilogram of food based on good agricultural practices. MRLs are set far below levels that could possibly pose a risk to human health. Since MRLs are not safety limits but trading standards, these are not determined by the industry. However, MRLs are determined by independent government agencies which fully review each active substance present in pesticides. A number of reasons by which MRLs can surpass their limit of 3–5% which include; the incorrect way of pesticide application or exceptional climatic or crop conditions have occurred [33–35].

3. Impact of pesticide residues

The intemperate utilization of pesticides has made catastrophic concerns about the fatalistic consequences on human prosperity and a large number of pesticides are not degradable; they hold on in the soil, drain to ground and surface water and defile the more extensive environment [36]. Pesticide use around the world

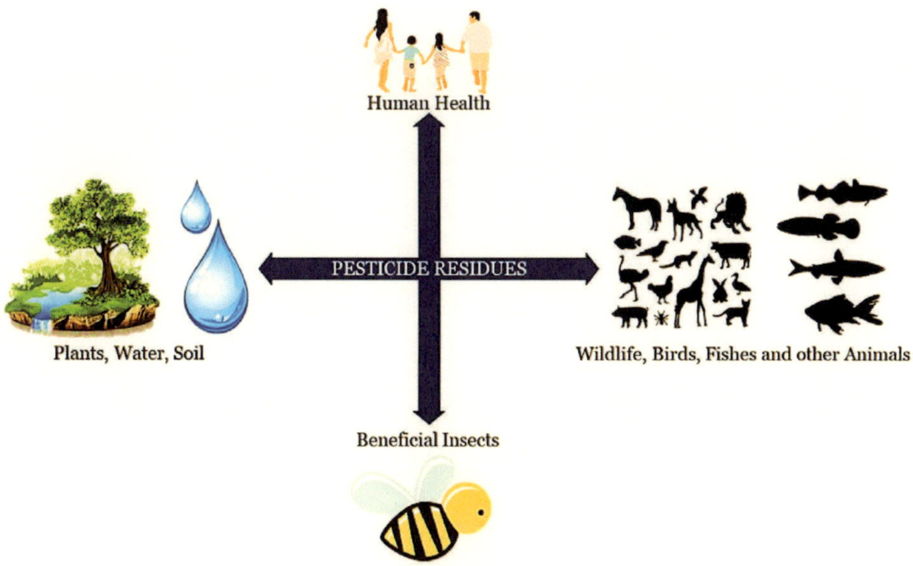

Figure 5.
Impact of pesticide residues on environment and different life forms. (Designed in MS OFFICE POWERPOINT by Muzafar Riyaz).

S. No	Name (Trade name)	Chemical Formula	Impacts on		References
			Human Health	Other Animals	
I	**Chlorinated Hydrocarbons**				
1	DDT	$C_{14}H_9Cl_5$	Cancer, Nervous system disorders, Respiratory damage, Reproductive organs, Immune system and endocrine disruptions, Birth defects	Central nervous system of Insects and other Animals, eggshell production in birds, Wildlife, Aquatic life including Fishes, sealions etc.	[52–55]
2	Methoxychlor	$C_6H_{15}Cl_3O_2$	Cancer, Central nervous depression, diarrhea, damage to liver, kidney, and heart	Physiological disruptions in animals, fishes and birds especially aquatic birds.	[56]
3	Chlorobenzilate	$C_{16}H_{14}Cl_2O_3$	Carcinogenic, Genotoxic, Eye damage	Toxic to Insects including honey bees, birds and fishes.	[57, 58]
4	BHC	$C_6H_6Cl_6$	Highly carcinogenic, dermatitis, psoriasis, burning, rashes,	Effects on domestic animals, wildlife and aquatic organisms.	[59, 60]
5	Toxaphene	$C_{10}H_{10}Cl_8$	Carcinogenic, Immune system failure, Reproductive organ damage, DNA damage,	Physiological disruptions in animals, adult reduction & egg shortening in birds, wildlife and fishes.	[61, 62]
6	Aldrin	$C_{12}H_8Cl_6$	Systemic, neurological, reproductive/developmental, immunological, genotoxic and tumorigenic.	Physiological disruption in Birds, toxic to aquatic animals, wildlife, domestic animals.	[63–65]
7	Dieldrin	$C_{12}H_8Cl_6O$	Carcinogenic, neurological, reproductive/developmental, immunological and genotoxic.	Carcinogenic, Highly toxic to birds, wildlife and other animals. Physiological disruptions in aquatic animals.	[66, 67]
8	Endosulfan	$C_9H_6Cl_6O_2$ S	Cancer, Acute and chronic toxicity, respiratory failure, endocrine disruption, reproductive failure, DNA damage	Physiological, developmental, neurotoxic disruptions in birds, fishes, wildlife and other aquatic organisms.	[68–72]
II	**Organophosphates**				
9	Chlorfenvinphos	$C_{12}H_{14}O_4Cl_3P$	Developmental, reproductive, and immunologic effects	Physiological disruptions in animals, effects on the aquatic animals and birds.	[73, 74]
10	Methyl parathion	$C_8N_{10}NO_5PS$	Headaches, nausea, night-waking, diarrhea, difficulty breathing, mental confusion, nervous system, cardiovascular and reproductive system	Toxic to earthworms, fishes, insects, and other aquatic organisms. Physiological and metabolic disruptions in Fishes.	[75–77]
11	Diazinon	$C_{12}H_{21}N_2O_3PS$	Cancer, Reproductive system, Acute and chronic toxicity, respiratory failure, endocrine disruption	Highly toxic to birds and bees and other insects, birds and fishes.	[78–81]

S. No	Name (Trade name)	Chemical Formula	Impacts on		References
			Human Health	**Other Animals**	
12	Ethion	$C_9H_{22}O_4P_2S_4$	Clinical toxicity, abdominal pain, diarrhea, vomiting, respiratory problems and undue secretions	Toxic to fishes, birds, aquatic organism, wildlife and domestic animals.	[82–84]
13	Malathion	$C_{10}H_{19}O_6PS_2$	Liver, kidney, testis, ovaries, lung, pancreas, blood, genotoxic and carcinogenic	Toxic to fishes, birds, aquatic creatures, domestic pets, and other animals.	[85–87]
III	**Carbamates**				
15	Carbaryl	$C_{12}H_{11}NO_2$	Neurological, Reproductive, Immunological disorders, possible carcinogen.	Moderate to high effect on birds, fishes and other animals.	[88, 89]
16	Aminocarb	$C_{11}H_{16}O_2N_2$	Cholinesterase inhibition, effects on the nervous system, sometimes death.	Toxic to birds, mammals and other animals including fishes and wildlife	[90, 91]
17	Carbofuran	$C_7H_{15}NO_3$	Body weakness, abdominal pain, blurred vision, nausea, sweating, muscle shuddering, coordination dysfunctions, respiratory and nervous system disorders.	Highly toxic to birds, Toxic to aquatic animals including fishes, other animals and non-target terrestrial creatures.	[92–94]
18	Aldicarb	$C_7H_{14}N_2O_2$ S	Headache, nausea, sweating, diarrhea, coordination system disruptions and sometimes death.	Toxic to aquatic organisms including fishes, toxic to birds and other organisms.	[95–97]
IV	**Pyrethroids**				
19	Cypermethrin	$C_{22}H_{19}Cl_2NO_3$	Neurotoxic, Hepatoxic, effects on behavior, molecular level and reproductive system.	Toxic to birds, aquatic organism including fishes and other creatures.	[98–100]
20	Deltamethrin	$C_{22}H_{19}Br_2NO_3$	Paranesthesia, Unwanted sensations, burning and partial numbness, "pins and needles", skin problems.	Toxic to domestic animals, aquatic organisms and terrestrial animals and plants.	[101–103]

Table 1.
Impact of some commonly used synthetic pesticides on human health and other animals.

has brought about various instances of acute and chronic poisoning, with impacts of proliferating peril on human wellbeing, from delicate effects to death [37]. Exposure to pesticides normally occurs while preparing the spray solutions and while showering the pesticides on crops. Proceeded with an introduction to sub-deadly amounts of pesticides for a protracted timeframe, may result in unending health-related issues among people [38]. Comparative health impacts are reliant upon the nature of the substance, the quantity received, course of the entrance, for example, intake by breath, ingestion or skin assimilation and individual perceptivity. Due to pesticides, there are possible incidences of several chronic diseases and disorders, such as cancer, diabetes, respiratory failures and fertility issues examined by several studies [39]. Different investigations have uncovered a connection between pesticide use and sarcomas, numerous myelomas, malignant growth of the prostate, pancreas, lungs, ovaries, the breast, gonads, liver, kidneys, alimentary tracts and brain [40–42]. As indicated by a 2017 European Food Safety Authority report, 44% of food samples conventionally produced contained one or more significant residues [43]. Pesticides have been linked to a wide range of human health hazards ranging from short-term impacts such as headache and nausea to chronic impacts like cancer, reproductive harm and endocrine disruption [44]. Chronic health effects may occur years after even minimal exposure to pesticides in the environment or result from the pesticide residues which gets transported to humans through the food and water. Pesticides have been linked to many types of cancers among humans. Some of the most prevalent forms include leukemia, non-Hodgkin's lymphoma, brain, bone and breast, ovarian, prostate, testicular and liver cancers [45]. Mounting evidence suggests that exposure to pesticides disrupts the endocrine system [46]. As the highest number of pesticides are synthetic chemicals, they can elicit a physiological reaction after getting an entry into a plant or animal body, which means if the pesticide can kill a creature; humans, domestic animals, pets, beneficial insect diversity such as pollinators and predators [47], birds, aquatic animals and plants, wildlife [48], non-target plants and our surrounding environment will also get affected by these chemical pesticides (**Figure 5**). Apart from all these consequences, pesticides can contaminate air, water and soil which in turn can be a cause of ailing human health across the globe [49–51]. The impact of various classes of pesticides on human beings and other animals have been listed in (**Table 1**).

4. Discussion and conclusion: strategies and alternatives

Pesticide movement can be reduced in these natural processes by developing strategic farming practices. To reduce pesticide drift, farmers can be provided with such spray nozzles that produce larger spray droplets or lowering the boom of a sprayer. Surface runoff of pesticides can be reduced by no-till or minimum tillage practices which can also reduce pesticide movement via soil erosion. Leaching is the movement of water down through the soil, potentially to tile lines and surface waters or groundwater. Adsorptive pesticides are less likely to leach because they stick to organic matter. Increasing the amount of organic matter in the soils, manures and crop residues can be a better alternative to farmers. Pesticides can also be prevented to enter the environment by handling pesticides with care. Farmers should be encouraged to store the pesticides properly so they do not contaminate organisms or the environment and when the pesticide application is done, one can dispose of the empty containers to pesticide container collection sites. An Integrated Pest Management (IPM) can help in control pests without pesticides. However, if the pesticide is being used, reading of pesticide label and checking environmental precautions should be made mandatory for farmers. The benefits of pesticide invention and application have saved our world from hunger and have caused direct adverse threats to nature as well. Considering the rich biodiversity of our planet it is impossible to assess the effect of pesticides on every organism and to conclude that any pesticide is completely safe. From the results of the recent studies, it is evident that we are yet to learn the unknown effects of pesticides on life forms and the physical world. The result of pesticide use for many decades has taught humans to search for a solution to the drastic impact it has created on nature. The remedy to the unintentional persistence of toxicity of pesticides in the environment came from nature itself with pesticide degrading microbes and also by the observation of abiotic degradation in the environment. The unexplored aspects of pesticide toxicity and their biotic and abiotic degradation in qualitative and quantitative aspects in air, water, soil and living beings need to be addressed. The identification of pesticide degrading microbes and intentional application of these organisms through bioremediation and comprehensive research using innovative technologies will create a revolution for a safer tomorrow. The present study aims to encourage people to the importance of alternative methods to solve the problem of the present and future generations. As proved by many, the remedy to the problem is the usage of alternative methods like biological control and biopesticides. The agents of biological control, as well as compounds for biopesticides, are to be comprehensively explored and utilized. The awareness of the people especially the farmers is the first step towards this movement. The easy and cheap access to biological pesticides and biological control measures are to be studied and made available to the public through government and non-governmental organizations. The present study thus finds an important place in the process of conservation and protection of nature and natural resources and of human health towards a bright future.

Acknowledgements

The authors wish to thank Muzafar Riyaz (Corresponding author) for carrying out extensive field works to study the adverse effects of pesticides on farmers health and other organisms.

Conflict of interest

The authors declare no conflict of interest.

Author details

Muzafar Riyaz*, Rauf Ahmad Shah and Kuppusamy Sivasankaran
Entomology Research Institute, Loyola College, Chennai, Tamil Nadu, India

*Address all correspondence to: bhatmuzaffar471@gmail.com

References

[1] Carvalho FP. Pesticides, environment, and food safety. Food and Energy Security. 2017 May;6(2):48-60. https://doi.org/10.1002/fes3.108

[2] EEA. 2013. Late lessons from early warnings: science, precaution, innovation. European Environment Agency, Report No 1/2013. EEA, Copenhagen.

[3] Ware GW. Effects of pesticides on nontarget organisms. Residue Reviews. 1980:173-201.

[4] Riyaz M, Iqbal WA, Sivasankaran K, Ignacimuthu S. Impact on Farmers' Health Due to the Pesticide Exposure in the Agrarian Zones of Kashmir Valley: A Review. Acta Scientific Agriculture. 2020, 4 (2), 16-22.

[5] Tiryaki O, Temur C. The fate of pesticide in the environment. Journal of Biological and Environmental Sciences. 2010;4(10):29-38.

[6] Ragnarsdottir KV. Environmental fate and toxicology of organophosphate pesticides. Journal of the Geological Society. 2000 Jul 1;157(4):859-876. https://doi.org/10.1144/jgs.157.4.859

[7] Ndakidemi B, Mtei K, Ndakidemi PA. Impacts of synthetic and botanical pesticides on beneficial insects. Agricultural Sciences. 2016;7(06):364. http://dx.doi.org/10.4236/as.2016.76038

[8] Mitra A, Chatterjee C, Mandal FB. Synthetic chemical pesticides and their effects on birds. Research Journal of Environmental Toxicology. 2011;5(2):81-96. http://dx.doi.org/10.3923/rjet.2011.81.96

[9] Popp J, Pető K, Nagy J. Pesticide productivity and food security. A review. Agronomy for sustainable development. 2013 Jan 1;33(1):243-255. http://dx.doi.org/10.1007/s13593-012-0105-x

[10] Gavrilescu M. Fate of pesticides in the environment and its bioremediation. Engineering in life sciences. 2005 Dec;5(6):497-526. https://doi.org/10.1002/elsc.200520098

[11] Pereira VJ, da Cunha JP, de Morais TP, de Oliveira JP, de Morais JB. Physical-chemical properties of pesticides: concepts, applications, and interactions with the environment. Bioscience Journal. 2016 Jun 1;32(3). https://doi.org/10.14393/BJ-v32n3a2016-31533

[12] Edwards CA, Adams RS. Persistent pesticides in the environment. Critical Reviews in Environmental Science and Technology. 1970 Jan 1;1(1-4):7-67.

[13] Vighi M, Matthies M, Solomon KR. Critical assessment of pendimethalin in terms of persistence, bioaccumulation, toxicity, and potential for long-range transport. Journal of Toxicology and Environmental Health, Part B. 2017 Jan 2;20(1):1-21. https://doi.org/10.1080/10937404.2016.1222320

[14] Morgan DP, Roan CC. The metabolism of DDT in man. In Essays in toxicology 1974 Jan 1 (Vol. 5, pp. 39-97). Elsevier.

[15] Arias-Estévez M, López-Periago E, Martínez-Carballo E, Simal-Gándara J, Mejuto JC, García-Río L. The mobility and degradation of pesticides in soils and the pollution of groundwater resources. Agriculture, Ecosystems & Environment. 2008 Feb 1;123(4):247-260. https://doi.org/10.1016/j.agee.2007.07.011

[16] Miller GT. Living in the environment: an introduction to environmental science. Wadsworth publishing company; 1990.

[17] Kosikowska M, Biziuk M. Review of the determination of pesticide residues

in ambient air. TrAC Trends in Analytical Chemistry. 2010 Oct 1;29(9):1064-1072. https://doi.org/10.1016/j.trac.2010.06.008

[18] Breidenbach AW. Pesticide residues in air and water. Archives of Environmental Health: An International Journal. 1965 Jun 1;10(6):827-830. https://doi.org/10.1080/00039896.1965.10664105

[19] Spencer WF, Farmer WJ, Cliath MM. Pesticide volatilization. In Residue Reviews 1973 (pp. 1-47). Springer, New York, NY. https://doi.org/10.1007/978-1-4613-9377-1_1

[20] Woodrow JE, Gibson KA, Seiber JN. Pesticides and related toxicants in the atmosphere. Reviews of Environmental Contamination and Toxicology Volume 247. 2018:147-96. https://doi.org/10.1007/398_2018_19

[21] Van den Berg F, Kubiak R, Benjey WG, Majewski MS, Yates SR, Reeves GL, Smelt JH, Van der Linden AM. Emission of pesticides into the air. InFate of Pesticides in the Atmosphere: Implications for Environmental Risk Assessment. 1999 (pp. 195-218). Springer, Dordrecht. https://doi.org/10.1007/978-94-017-1536-2_9

[22] Gevao B, Semple KT, Jones KC. Bound pesticide residues in soils: a review. Environmental pollution. 2000 Apr 1;108(1):3-14. https://doi.org/10.1016/S0269-7491(99)00197-9

[23] Kah M, Beulke S, Brown CD. Factors influencing degradation of pesticides in soil. Journal of agricultural and food chemistry. 2007 May 30;55(11):4487-4492. https://doi.org/10.1021/jf0635356

[24] Hussain S, Siddique T, Saleem M, Arshad M, Khalid A. Impact of pesticides on soil microbial diversity, enzymes, and biochemical reactions. Advances in agronomy. 2009 Jan 1;102:159-200. https://doi.org/10.1016/S0065-2113(09)01005-0

[25] Ataikiru TL, Okpokwasili GS, Okerentugba PO. Impact of pesticides on microbial diversity and enzymes in soil. South Asian Journal of Research in Microbiology. 2019 Aug 13:1-6. https://doi.org/10.9734/sajrm/2019/v4i230104

[26] Chowdhury A, Pradhan S, Saha M, Sanyal N. Impact of pesticides on soil microbiological parameters and possible bioremediation strategies. Indian Journal of microbiology. 2008 Mar 1;48(1):114-127. https://doi.org/10.1007/s12088-008-0011-8

[27] Pérez-Lucas G, Vela N, El Aatik A, Navarro S. Environmental risk of groundwater pollution by pesticide leaching through the soil profile. In Pesticides-use and misuse and their impact in the environment. 2018 Nov 29. IntechOpen. https://doi.org/10.5772/intechopen.82418

[28] Sankhla MS, Kumari M, Sharma K, Kushwah RS, Kumar R. Water contamination through pesticide & their toxic effect on human health. International Journal for Research in Applied Science and Engineering Technology. 2018 Feb 1;6(1):967-970.

[29] Aydinalp C, Porca MM. The effects of pesticides in water resources. Journal of Central European Agriculture. 2004 Apr 16;5(1):5-12.

[30] Fan AM, Jackson RJ. Pesticides and food safety. Regulatory toxicology and pharmacology. 1989 Apr 1;9(2):158-174. https://doi.org/10.1016/0273-2300(89)90033-0

[31] Abelson PH. Pesticides and food. Science. 1993 Feb 26;259(5099):1235-1236.

[32] Kim KH, Kabir E, Jahan SA. Exposure to pesticides and the

associated human health effects. Science of the Total Environment. 2017 Jan 1;575: 525-535. https://doi.org/10.1016/j.scitotenv.2016.09.009

[33] MacLachlan DJ, Hamilton D. Estimation methods for maximum residue limits for pesticides. Regulatory Toxicology and Pharmacology. 2010 Nov 1;58(2):208-218. https://doi.org/10.1016/j.yrtph.2010.05.012

[34] Ambrus A, Yang YZ. Global harmonization of maximum residue limits for pesticides. Journal of agricultural and food chemistry. 2016 Jan 13;64(1):30-35. https://doi.org/10.1021/jf505347z

[35] Yeung MT, Kerr WA, Coomber B, Lantz M, McConnell A. Why maximum residue limits for pesticides are an important international issue. In Declining International Cooperation on Pesticide Regulation. 2017 (pp. 1-9). Palgrave Macmillan, Cham.

[36] Butler PA. Monitoring pesticide pollution. Bioscience. 1969 Oct 1;19(10):889-891. https://doi.org/10.2307/1294712

[37] Nicolopoulou-Stamati P, Maipas S, Kotampasi C, Stamatis P, Hens L. Chemical pesticides and human health: the urgent need for a new concept in agriculture. Frontiers in public health. 2016 Jul 18;4:148. https://doi.org/10.3389/fpubh.2016.00148

[38] Tomer V, Sangha JK, Ramya HG. Pesticide: An appraisal on human health implications. Proceedings of the National Academy of Sciences, India Section B: Biological Sciences. 2015 Jun;85(2):451-463. https://doi.org/10.1007/s40011-014-0388-6

[39] Mostafalou S, Abdollahi M. Pesticides: an update of human exposure and toxicity. Archives of toxicology. 2017 Feb;91(2):549-599. https://doi.org/10.1007/s00204-016-1849-x

[40] Alavanja MC, Hoppin JA, Kamel F. Health effects of chronic pesticide exposure: cancer and neurotoxicity. Annual Reviews of Public Health. 2004 Apr 21;25:155-197. https://doi.org/10.1146/annurev.publhealth.25.101802.123020

[41] Valcke M, Levasseur ME, da Silva AS, Wesseling C. Pesticide exposures and chronic kidney disease of unknown etiology: an epidemiologic review. Environmental Health. 2017 Dec;16(1):1-20. https://doi.org/10.1186/s12940-017-0254-0

[42] Lee WJ. Pesticide exposure and health. Journal of Environmental Health Sciences. 2011;37(2):81-93. https://doi.org/10.5668/JEHS.2011.37.2.081

[43] European Food Safety Authority and European Centre for Disease Prevention and Control (EFSA and ECDC). The European Union summary report on trends and sources of zoonoses, zoonotic agents and food-borne outbreaks in 2017. EFSa Journal. 2018 Dec;16(12):e05500. https://doi.org/10.2903/j.efsa.2018.5500

[44] Sabarwal A, Kumar K, Singh RP. Hazardous effects of chemical pesticides on human health–Cancer and other associated disorders. Environmental toxicology and pharmacology. 2018 Oct 1;63: 103-114. https://doi.org/10.1016/j.etap.2018.08.018

[45] Bassil KL, Vakil C, Sanborn M, Cole DC, Kaur JS, Kerr KJ. Cancer health effects of pesticides: systematic review. Canadian Family Physician. 2007 Oct 1;53(10):1704-1711.

[46] Mnif W, Hassine AI, Bouaziz A, Bartegi A, Thomas O, Roig B. Effect of endocrine disruptor pesticides: a review. International journal of environmental research and public health. 2011 Jun;8(6):2265-2303. https://doi.org/10.3390/ijerph8062265

[47] Sanchez-Bayo F, Goka K. Impacts of pesticides on honey bees. Beekeeping and Bee Conservation-Advances in Research. 2016 May 20;4: 77-97. IntechOpen. https://doi.org/10.5772/62487

[48] Moore NW. Effects of pesticides on wildlife. Proceedings of the Royal Society of London. Series B. Biological Sciences. 1967 Feb 21;167(1007):128-133. https://doi.org/10.1098/rspb.1967.0017

[49] Costa LG. Toxic effects of pesticides. Casarett and Doull's toxicology: the basic science of poisons. 2008;8: 883-930.

[50] Ndakidemi B, Mtei K, Ndakidemi PA. Impacts of synthetic and botanical pesticides on beneficial insects. Agricultural Sciences. 2016;7(06):364. http://dx.doi.org/10.4236/as.2016.76038

[51] Bertolote JM, Fleischmann A, Eddleston M, Gunnell D. Deaths from pesticide poisoning: a global response. The British Journal of Psychiatry. 2006 Sep;189(3):201-203. https://doi.org/10.1192/bjp.bp.105.020834

[52] Thuy TT. Effects of DDT on environment and human health. Journal of Education and Social Sciences. 2015; 2:108-114.

[53] Cohn BA, La Merrill M, Krigbaum NY, Yeh G, Park JS, Zimmermann L, Cirillo PM. DDT exposure in utero and breast cancer. The Journal of Clinical Endocrinology & Metabolism. 2015 Aug 1;100(8):2865-2872. https://doi.org/10.1210/jc.2015-1841

[54] Byard JL, Paulsen SC, Tjeerdema RS, Chiavelli D. DDT, chlordane, toxaphene and PCB residues in Newport Bay and Watershed: assessment of hazard to wildlife and human health. Reviews of Environmental Contamination and Toxicology Volume 235. 2015:49-168. https://doi.org/10.1007/978-3-319-10861-2_3

[55] Ware GW. Effects of DDT on reproduction in higher animals. Residue reviews. 1975:119-140. https://doi.org/10.1007/978-1-4612-9863-2_5

[56] Chen G. Methoxychlor. In: Encyclopedia of Toxicology, Wexler, P (Ed.); 3rd Edition. 2014, (pp. 254-255). Academic Press. https://doi.org/10.1016/B978-0-12-386454-3.00162-7

[57] Lewis KA, Tzilivakis J, Warner DJ, Green A. An international database for pesticide risk assessments and management. Human and Ecological Risk Assessment: An International Journal. 2016 May 18;22(4):1050-1064. https://doi.org/10.1080/10807039.2015.1133242

[58] Janz DM. Chlorobenzilate. In: Encyclopedia of Toxicology, Wexler, P (Ed.); 3rd Edition. 2014. (pp. 874-875). Academic Press. https://doi.org/10.1016/B978-0-12-386454-3.00113-5

[59] Loomis D, Guyton K, Grosse Y, El Ghissasi F, Bouvard V, Benbrahim-Tallaa L, Guha N, Mattock H, Straif K, IARC L, International Agency for Research on Cancer Monograph Working Group. Carcinogenicity of lindane, DDT, and 2, 4-dichlorophenoxyacetic acid. The Lancet. Oncology. 2015 Aug;16(8):891-892. https://doi.org/10.1016/S1470-2045(15)00081-9

[60] Shirakawa M. Experimental Studies on the Toxicity of Benzene Hexachloride (Bhc) Iv. Effects of Lindane Vapor to Insects and Mammals. The Kurume Medical Journal. 1959 May 31;5(4):230-244. https://doi.org/10.2739/kurumemedj.5.230

[61] Wallace, D.R. Toxaphene. In: Encyclopedia of Toxicology, Wexler, P

(Ed.); 3rd Edition. 2014. (pp. 606-609). Academic Press. https://doi.org/10.1016/B978-0-12-386454-3.00202-5

[62] Jongbloed RH, Visschedijk AJ, Van Dokkum HP, Laane RW. Toxaphene: An analysis of possible problems in the aquatic environment. RIKZ/2000.010. 2000.

[63] US-EPA (United States Environmental Protection Agency). Health Effects Support Document for Aldrin/Dieldrin. Office of Water (4304T), Health and Ecological Criteria Division, Washington DC, USA, 2003.

[64] Brown VK, Richardson A, Robinson J, Stevenson DE. The effects of aldrin and dieldrin on birds. Food and cosmetics toxicology. 1965 Oct;3(4):675-679. https://doi.org/10.1016/s0015-6264(65)80263-2

[65] Honeycutt M, Shirley S. Aldrin. In: Encyclopedia of Toxicology, Wexler, P (Ed.); 3rd Edition. 2014. (pp. 126-129). Academic Press. https://doi.org/10.1016/B978-0-12-386454-3.00094-4

[66] Honeycutt M, Shirley S. Dieldrin. In: Encyclopedia of Toxicology, Wexler, P (Ed.); 3rd Edition. 2014. (pp107-110). Academic Press. https://doi.org/10.1016/B978-0-12-386454-3.00132-9

[67] Scott TG, Willis YL, Ellis JA. Some effects of a field application of dieldrin on wildlife. The Journal of Wildlife Management. 1959 Oct 1;23(4):409-427. https://doi.org/10.2307/3796489

[68] Singh P, Volger B, Gordon, E. Endosulfan. In: Encyclopedia of Toxicology, Wexler, P (Ed.); 3rd Edition. 2014. (pp. 341-343). Academic Press. https://doi.org/10.1016/B978-0-12-386454-3.00141-X

[69] Sebastian R, Raghavan SC. Induction of DNA damage and erroneous repair can explain genomic instability caused by endosulfan. Carcinogenesis. 2016 Oct 1;37(10):929-940. https://doi.org/10.1093/carcin/bgw081

[70] Naqvi SM, Vaishnavi C. Bioaccumulative potential and toxicity of endosulfan insecticide to non-target animals. Comparative Biochemistry and Physiology Part C: Comparative Pharmacology. 1993 Jul 1;105(3):347-361. https://doi.org/10.1016/0742-8413(93)90071-R

[71] Prakash PJ, Geetha R, Krishnappa H, Sulaiman SM, Rao KV. Effect of Endosulfan 35% EC on the egg laying and egg shell thickness in Japanese quails. Research Journal of Environmental Toxicology. 2009;3(3):140-146.

[72] Sutherland TD, Horne I, Weir KM, Russell RJ, Oakeshott JG. Toxicity and residues of endosulfan isomers. Reviews of environmental contamination and toxicology. 2004:99-113. https://doi.org/10.1007/978-1-4419-9100-3_4

[73] Reed NR, Koshlukova S. Chlorfenvinphos. In: Encyclopedia of Toxicology, Wexler, P (Ed.); 3rd Edition. 2014. (pp. 851-854). Academic Press. https://doi.org/10.1016/B978-0-12-386454-3.01110-6

[74] Dorsey AS, Kueberuwa SS. Toxicological Profile for Chlorfenvinphos. Public Health Service, Agency for Toxic Substances and Disease Registry, U.S. Department of Health and Human Services. 1997:1-220.

[75] Edwards FL, Tchounwou PB. Environmental toxicology and health effects associated with methyl parathion exposure–a scientific review. International Journal of Environmental Research and Public Health. 2005 Dec;2(3):430-441. https://doi.org/10.3390/ijerph2005030007

[76] Suthar S. Toxicity of methyl parathion on growth and reproduction

of three ecologically different tropical earthworms. International Journal of Environmental Science and Technology. 2014 Feb 1;11(1):191-198. https://doi.org/10.1007/s13762-012-0154-3

[77] Marina Y, Tabche LM, García CM. Bioaccumulation of methyl parathion and its toxicology in several species of the freshwater community in Ignacio Ramirez dam in Mexico. Ecotoxicology and environmental safety. 1997 Oct 1;38(1):53-62. https://doi.org/10.1006/eesa.1997.1551

[78] Beane Freeman LE, Bonner MR, Blair A, Hoppin JA, Sandler DP, Lubin JH, Dosemeci M, Lynch CF, Knott C, Alavanja MC. Cancer incidence among male pesticide applicators in the Agricultural Health Study cohort exposed to diazinon. American journal of epidemiology. 2005 Dec 1;162(11):1070-1079. https://doi.org/10.1093/aje/kwi321

[79] Harchegani AB, Rahmani A, Tahmasbpour E, Kabootaraki HB, Rostami H, Shahriary A. Mechanisms of diazinon effects on impaired spermatogenesis and male infertility. Toxicology and industrial health. 2018 Sep;34(9):653-664. https://doi.org/10.1177/0748233718778665

[80] Sheffield SR, Lochmiller RL. Effects of field exposure to diazinon on small mammals inhabiting a semienclosed prairie grassland ecosystem. I. Ecological and reproductive effects. Environmental Toxicology and Chemistry: An International Journal. 2001 Feb;20(2):284-296. https://doi.org/10.1002/etc.5620200209

[81] Clarke Z. Diazinon. In: xPharm: The Comprehensive Pharmacology Reference. 2007. (pp. 1-4). Elsevier. https://doi.org/10.1016/B978-008055232-3.61621-6

[82] Dewan A, Patel AB, Pal RR, Jani UJ, Singel VC, Panchal MD. Mass ethion poisoning with high mortality. Clinical Toxicology. 2008 Jan 1;46(1):85-88. https://doi.org/10.1080/15563650701517251

[83] Prasanna C, Anithasmruthi C, Venkatarathnamma V. A Study on Oxygen Consumption in Freshwater Fish Labeo Rohita Exposed to Lethal and Sub Lethal Concentrations of Ethion 50% Ec. Indian Journal of Forensic Medicine & Toxicology. 2020 Oct 1;14(4).

[84] Fry DM. Reproductive effects in birds exposed to pesticides and industrial chemicals. Environmental health perspectives. 1995 Oct;103(suppl 7):165-171. https://doi.org/10.1289/ehp.95103s7165

[85] Badr AM. Organophosphate toxicity: updates of malathion potential toxic effects in mammals and potential treatments. Environmental Science and Pollution Research. 2020 Jul; 27:26036-26057. https://doi.org/10.1007/s11356-020-08937-4

[86] Jira D, Janousek S, Pikula J, Vitula F, Kejlova K. Toxicity hazard of organophosphate insecticide malathion identified by in vitro methods. Neuroendocrinology Letters. 2012 Jan 1; 33:3-9.

[87] Deka S, Mahanta R. Malathion toxicity on fish-a review. International Journal of Current Research. 2016;8(12):44120-44128.

[88] Reed NR, Koshlukova S. Carbaryl. In: Encyclopedia of Toxicology, Wexler, P (Ed.); 3rd Edition. 2014. (pp. 668-672). Academic Press. https://doi.org/10.1016/B978-0-12-386454-3.00107-X

[89] Morais S, Dias E, Pereira ML. Carbamates: human exposure and health effects. The impact of pesticides. 2012:21-38.

[90] Rodgers KE. Immunotoxicity of Pesticides. In: Handbook of Pesticide Toxicology, Krieger, R. I. and Krieger, W. C. (Eds.); 2ndEdition. 2001. (pp. 769-782). Academic Press.

[91] Weinberger P, Greenhalgh R. Some adjuvant effects on the fate of fenitrothion and aminocarb. Environmental Toxicology and Chemistry: An International Journal. 1984 Apr;3(2):325-334. https://doi.org/10.1002/etc.5620030214

[92] Gupta RC. Carbofuran toxicity. Journal of Toxicology and Environmental Health, Part A Current Issues. 1994 Dec 1;43(4):383-418. https://doi.org/10.1080/15287399409531931

[93] Freedman B. Pesticides. In: Environmental Ecology, Freedman, B. (Ed.); 2nd Edition. 1995. (pp. 213-277). Academic Press.

[94] Dobšíková R. Acute toxicity of carbofuran to selected species of aquatic and terrestrial organisms. Plant Protection Science. 2003;39(3):103. https://doi.org/10.17221/3865-PPS

[95] Baron RL, Merriam TL. Toxicology of aldicarb. Reviews of environmental contamination and toxicology. 1988: 1-70. https://doi.org/10.1007/978-1-4612-3876-8_1

[96] Pant J, Tewari H, Gill TS. Effects of aldicarb on the blood and tissues of a freshwater fish. Bulletin of environmental contamination and toxicology. 1987 Jan 1;38(1):36-41. https://doi.org/10.1007/BF01606554

[97] Moore DR, Teed RS, Rodney SI, Thompson RP, Fischer DL. Refined avian risk assessment for aldicarb in the United States. Integrated Environmental Assessment and Management: An International Journal. 2010 Jan;6(1): 83-101. https://doi.org/10.1897/IEAM_2009-022.1

[98] Aman S, Bhuvnesh Y, Shipra R, Baljeet Y. Cypermethrin toxicity: a review. Journal of Forensic Sciences & Criminal Investigation. 2018;9(4): 555767. https://doi.org/10.19080/JFSCI.2018.09.555767

[99] Suzan AA. The pathological effect of cypermethrin on domestic pigeons (Culumba livia gaddi) at Basrah City/Southern Iraq. International Journal of Poultry Science. 2012 Apr 1;11(4): 302-310.

[100] Tiwari S, Tiwari R, Singh A. Impact of cypermethrin on fingerlings of common edible carp (Labeo rohita). The Scientific World Journal. 2012 Jan 1;2012. https://doi.org/10.1100/2012/291395

[101] Doi H, Kikuchi H, Murai H, Kawano Y, Shigeto H, Ohyagi Y, Kira J. Motor neuron disorder simulating ALS induced by chronic inhalation of pyrethroid insecticides. Neurology. 2006 Nov 28;67(10):1894-1895. https://doi.org/10.1212/01.wnl.0000244489.65670.9f

[102] Elias P. The use of deltamethrin on farm animals. In Insecticides-development of safer and more effective technologies. Chapter 18; In: Insecticides-Development of Safer and More Effective Technologies; Trdan, S. (Ed.). 2013. (pp. 495-503). IntechOpen. https://doi.org/10.5772/54839

[103] Chrustek A, Hołyńska-Iwan I, Dziembowska I, Bogusiewicz J, Wróblewski M, Cwynar A, Olszewska-Słonina D. Current research on the safety of pyrethroids used as insecticides. Medicina. 2018 Sep;54(4):61.

Chapter 4

Persistant Organic Pollutants in Soil and Its Phytoremediation

Mahima Begum, Bhaswati Sarmah,
Gayatri Goswami Kandali, Sontara Kalita, Ipsita Ojha,
Raktim Bhagawati and Lipika Talukdar

Abstract

Persistent organic pollutants (POPs) of soil mainly exhibit toxic characteristics that posses hazard to whole mankind. These toxic pollutants includes several group of compound *viz.*, polychlorinated biphenyls, polybrominated biphenyls, polychlorinated dibenzofurans, polycyclic aromatic hydrocarbons, organophosphorus and carbamate insecticides, herbicides and organic fuels, especially gasoline and diesel. They can also be complex mixture of organic chemicals, heavy metals and microbes from septic systems, animal wastes and other sources of organic inputs. Phytoremediation is an emerging technology which can be used for remediation of soil from organic pollutants. In this chapter an attempt has been made to discuss about the sources of organic pollutants, factors that influenced the uptake of organic pollutants by plants, the different mechanism responsible for organic pollutants, phytoremediation of organic pollutants and their advantages and limitation.

Keywords: Persistent organic pollutants (POPs), Phytoremediation, Soil beneficial microbes

1. Introduction

Land and water are the two crucial pillars of natural resources on which the sustainability of agriculture and the continued existence of civilization rely. Unfortunately, both have been drastically degraded due to various natural (leaching, mineralization, volcanic eruption, etc.) as well as anthropogenic (industrial waste, chemical agriculture, smelting, mining) activities.

Out of different component of soil degradation, the organic pollutant (OP) in soil is considered as an important cause that poses serious environmental damage as well as several health hazards to mankind. Generally, organic pollutants persist in the soil in very low concentrations and keep accruing over long period of time. Though steadily increasing, these low concentrations of organic pollutant in the affected soil, makes a times constrained toxicological study difficult. These organic pollutants are both lipophilic and hydrophobic in nature [1] and these organic pollutant may be deposited in the soil in every geographical area of earth [2] through spontaneous processes of nature like forest fires, volcanic eruptions *etc*, or by some anthropogenic practices. These organic pollutants entered into plant system through different plants mechanism. However, some of the organic component of the wastes is biodegradable, but heavy metals and metalloids are an emerging threat due to their long-term persistence in the environment. By adopting some phytoremediation process the effect of organic pollutants to the environments could be alleviated to some extent [3].

2. Organic pollutants in soil: sources and its effect on environment

The natural sources of organic pollutants are those that occur spontaneously without human involvement. Apart from the erosion of materials from the soil, organic pollutants in the soil may be sourced from spontaneous atmospheric sedimentation after forest fires. The forest fires which occur in high vegetation areas are a major source of organic pollutants in soil. Polycyclic aromatic hydrocarbons, anubiquitous organic pollutant, are considered to be carcinogenic in nature and hazardous to humans [4]. They are released by the burning of vegetation/biomass [5–7] and remain either absorbed in the surface soil or are mobilized due to rain water percolating through the soil [8, 9]. Several other organo-halogen compounds may be formed in the soil due to the burning of flora and fauna due to similar spontaneous sources like volcanic eruption and other geogenic causes [10, 11].

The anthropogenic sources of organic pollutants can be developed through several ways. Agricultural practices may be an anthropogenic source of organic pollutants due to contamination by several point source pollutants or diffused source pollutants. Fertilizers or pesticides which are the direct inputs in an agricultural field are the source of point source organic pollutants. Atmospheric deposition and flooding form an indirect means of pollution to the soil and are referred to as diffused organic pollutants. Ever since the advent of conscious agriculture, fertilizers and pesticides have existed to reduce and prevent any loss to the crop as well as to increase the productivity [12]. With the growth in global population, demand for food is increasing but due to the limited availability of new agricultural land, intensification of agricultural production will be required [13].

Organic fertilizers have revamped the agricultural production system, especially as people are becoming more health and nutrition conscious. These organic fertilizers are a great means for producing organic products while improving the overall health of the soil by enriching it with organic carbon and slow release of nutrients. Organic fertilizers can be prepared from compost, animal waste, municipal wastes, sewage and waste water [14]. These materials appear to have a more environment friendly disposal and recycling option [15, 16]. However, in the long run, we may find that there are certain loopholes associated with the management of organic fertilizers as well.

Organic manures prepared from animal waste may contain increased levels of copper and zinc which are added as a part of animal feed and are in turn reflected in their fecal material [17, 18]. These excess of these elements in the soil acts as pollutants and associated with risks to the agricultural production [19, 20]. Concerns over organic pollutants from organic manures rise when the manures are the sources of antimicrobials in the soil after incomplete metabolism in the animal/human body [21, 22]. Due to the increased concentration of the antimicrobials in the soil after treatment with organic manures, several resistant strains may develop and accumulate in the soils which are again recycled to the human/animal body posing a great health risk worldwide [23–25].

Biological wastes, such as waste water, municipal solid waste compost, green waste and food waste from households can be manufactured into organic fertilizers by fermentation and composting. However, recent studies have found that such fertilizer can be a source of bio-solids and micro-plastic particles that are very challenging to remove [26]. Bio-solids contain high concentrations of organic matter and biogenic compounds, especially nitrogen and phosphorus, necessary for plant growth and have been tested to be appropriate for use as fertilizer [27]. However, bio-solids contaminated with lipophilic trace elements when applied to land are one of the most important soil contributors of trace elements in soils [28–30]. Bio-solids are also a source of nano- and micro-plastics. It is estimated that of all the micro-plastics that go through the wastewater treatment plant, 95 percent is contained in the bio-solids [31]. Besides trace elements, wastewater sludge and bio-solids can be contaminated with POPs including polychlorinated dibenzo-p-dioxins and dibenzo furans (PCDD/F), poly chlorinated biphenyls (PCBs), chlorinated paraffin (CPs) and perfluorinated alkylated substances (PFASs) likeperfluoro octane sulfonate (PFOS) or perfluorooctanoic acid (PFOA), which has resulted in the pollution of agricultural soils [32–34].

The first pesticides were based on inorganic chemicals such as nitrogen, sulfur, copper, mercury and arsenic compounds [35–37]. However, midway toward the 20th century, when the world evidenced a major shift in agriculture with the beginning of green revolution, the inorganic pesticides were replaced by the organic compounds. These organic pesticides, since then, have been used continuously in agriculture with commercialized in the global market [12].

Organic pesticides are washed off the sprayed plants or seeds by rainfall or irrigation and deposited in the soil [38]. Agricultural soils are also frequently affected by accidental releases of pesticides from leaking [39]. The inappropriate disposal of unwanted or out of date pesticides, pesticide packaging and the cleaning of application equipment can also cause pollution. Most of the pesticides, though organic in nature, are not degraded and persist in the soil owing to their long half-lives. These organic pesticides and their residues may accrue in soils [40–43] and may cause detrimental effects on the animals and humans over a long period of time [44, 45]. Some volatile compounds may be transported over distances and deposit in a non-native soil as well [46].

3. Factors affecting uptake of organic pollutants by plants

Contamination of soil environment with heavy metal accumulation has become a rife across the globe. Phytoremediation has emerged out to be quite effective in this aspect. It involves growing of plants to purge contaminants from soil without hampering its regular growth and development. Literature reported by several scholars states that the mechanism of phytostabilization, rhizodegradation, rhizofiltration, phytodegradation and phytovolatilization [47] are effectual for eliminating organic contaminants from the lithosphere. The uptake of organic pollutants by plants is determined by various components. An understanding of these factors is beneficial to upgrade the uptake capabilities of the crop physiology.

3.1 Plant species

The absorption of organic contaminant by plants includes a series of complex reactions. The absorption of a compound is influenced by different attributes of the plant as well as properties of the element. The plant species should have vigorous growth rate, high biomass, substantial root system and resistance to excessive concentration of polluting metals [48]. The identification of plant species acceptable for heavy metal accretion into their system along with effective growth and development with the conventional management practices is an important pre requisite for the uptake of organic compound from a highly degraded environment. The burning of the crop after harvest gains energy and recycles the metal from the ash, as a result of which it gets removed from the soil system. In a green house experiment conducted by Ampiah-Bonney *et al.* [49], it was observed that *Leersiaoryzoides*, a type of terrestrial plant could maintain the high arsenic uptake up to 6 weeks of study in its system in addition to producing good yield. Cho-Ruk and his co-workers [50] studied the test crop (*Alternantheraphyloxeroides*) for uptake of lead into its physiology and found that the characteristic stolons and huge fibrous root system provided larger surface area for better assimilation of the metal. The efficacy of the process was noted to be around 30–80%. The Brassica species also have excellent mechanism for uptake of cadmium and lead from the soil solution by releasing root exudates that forms complexes with these metals, thereby reducing their mobility in the environment [51].

3.2 Properties of medium

The absorption of pollutants by crops also depends on the medium. The contaminants exist in adynamic state between soil particles, in between air and water [52] of the media. It has been reported that pH and redox potential of the medium as well as presence of electrolytes hold utmost importance in the bioavailability of organic compounds into the soil solution which facilitate its uptake by the plant root system. The content of organic matter in soil is again a vital environmental factor affecting the absorption of non-ionic organic compounds by the roots from soil. The package and practices of the crops are developed accordingly to escalate the phyto-extraction and phyto-stabilization processes. In an investigation carried out by Marques et al. [53], it was found that heavy metal availability in the medium reduced by 80% after treatment of polluted soils with compost. The amount of lead taken up by the plant was highly reduced after application of lime which increased the soil pH to 6.5–7.0 as observed by Traunfeld and Clement [54].

3.3 Rhizosphere chemistry

The rhizosphere chemistry regulates the concentration of soluble cations within the region of the soil influenced by root secretions and microorganisms. It is also affected by the concentration of ions present for possible absorption by plants [55]. The root ecology can assimilate pollutants and reserve or mobilize them inside the plant tissue. The organic molecules enter the root cell either through apoplastic pathway or symplastic pathway. This process of rhizo filtration prevents leaching of heavy metals to freshwater bodies and groundwater table. The diverse microbial community present in the rhizosphere further enhances the breakdown of complex organic compounds into simpler substances by releasing certain enzymes. These along with the root exudates liberated by the plant system helps in rhizo-degradation of the contaminants. Sunflower and Indian mustard have been found to have massive fibrous root habitat which makes them favorable terrestrial candidates for metal removal through rhizo filtration [56].

3.4 Incorporation of amendments

There is also a great possibility of improving the rapid absorption of heavy metals by plants through the use of chelating agents, natural zeolites, lime and other amendments. They make the contaminants available in the solution which in turn increases their absorption by the crop. The compounds often remain sorbed onto the clay mineral lattice which makes it unavailable for absorption. Consequently, sudden change in soil environmental quality leads to groundwater contamination. The ligand group of the chelating compounds undergo ion exchange and form complexes at the exchange sites of the soil minerals liberating the organic pollutants into the system for uptake by plants. A laboratory study performed by Roy *et al.* [57] reported that exposing plants to EDTA for an extended period of time strengthen the metal translocation in plant anatomy altogether improving the phyto-extraction process.

3.5 Properties of the contaminants

The pathway through which the organic compounds penetrate the plant body is related to the physicochemical property of each element such as lack of affinity for water, dissolution in water and vapor pressure [52]. The solubility of the pollutants in the water is highly dependent on the time to which the metals can be retained in the medium as well as the interactivity with other elements and substancesin the medium [47]. Most of the contaminants are hydrophobic in nature which allow them to accumulate in aerial plants of the plant. The phytovolatilization occurs at relatively low concentration keeping the air pollution free *B. juncea* and *Brassica napus*have provided excellent results for phytovolatilization of soils tainted with selenium [58].

3.6 Environmental conditions

Abiotic factors like temperature, humidity, stress condition, rainfall also affect the uptake mechanism of organic pollutants by plants. For instance, Merkl *et al.* [59] observed an increase in the diameter of the rootand reduction in root length due to its impermeability in dry soil under drought stress condition. This limits the absorption of heavy metals by plants making them prone to run-off and soil erosion.

4. Mechanisms of organic pollutants uptake by plants

Plant absorbed the organic pollutants such as hormones, polychlorinated dibenzo-*p*-dioxins and polychlorinated dibenzofurans (PCDD/Fs), polychlorinated biphenyls (PCBs) and antibiotics, herbicides and bisphenol A (BPA) *etc.* Most of the organic pollutants are manmade xenobiotic released to environment as spills, wood treatment, explosives, pesticides, herbicides, industrial chemical, petrochemical *etc.* [60]. These organic pollutants are absorbed by higher plants through root and (or) leaves. Plant roots can absorb PCBs, PCDD/Fs, herbicides, antibiotics and BPA; whereas above ground plant parts especially leaves can absorb PCBs, PCDD/Fs and herbicides if these organic pollutants come in direct contact as liquid or in vapor form from the atmosphere [61].

Leaves absorbed different kind of organic contaminants from atmosphere *via* stomata and cuticle. The stomata of leaf are abundant on abaxial side of a leaf and are mainly involved in absorption the organic substances than the thicker cuticular layer of the adaxial side. Stomata of leaf allow easy passage of gases and liquid form of organic pollutant. The degree of opening of aperture of stomata and its number on leaf determine the permeability of gaseous from of organic pollutants. However, moisture on the leaf surface, surface tension of the liquid contaminants (eg. pesticides, herbicides, liquid aerosol *etc.*) and morphology of stomata determine the permeability of liquid organic pollutant through stomata [62]. After entry of gas molecules through the aperture, these are transported to other plant parts*via* the phloem [63].

Although uptake of organic pollutants is absorbed by plants either from air or soil, but roots play the major role in the absorption of organic pollutants from soil. Generally, organic pollutants are of low volatility, so root tissues is the first site of contact between plant and the organic pollutants in contaminated soil or water [61]. Some lipophillic organic pollutants are passively adsorbed to the lignin of cell wall of plant surface or root that come in contact with the contaminant [64] and thus phytostabilize the pollutants and prevent their entry to groundwater through leaching or to air by volatilization or into the food chain. Again these contaminates are easily passed through the cuticle free non-suberized cell walls of root hairs from the surrounding environment unlike cuticular layer of leaf. Plant roots absorb organic pollutant inside in two phases *viz.* uptake of the substances from surrounding soil and water into root (first phase) and are subsequently distributed and accumulated in different parts of the plant (second phase) [62]. In the first phase organic pollutants are taken up by plant root across the cell membrane by passive process of diffusion [63, 65]. The root concentration factor *i.e.* the ratio of a pollutant concentration in plant root as compared to external solution determines the movement of pollutant to the root [52, 60]. The hydrophobicity of the organic pollutant is one of the factors affecting its uptake. After uptake organic pollutants are translocated to different plant parts [66]. There are two kinds of transport pathways in higher plants *i.e.* short distance transport (intracellular and intercellular transport) and long distance transport (conducting tissue transport). In studies of mechanism of organic pollutant uptake, different researchers have revealed that these pollutants after uptake by root penetrate through free intercellular space (apoplast) or cell to cell movement *via* plasma desmata (symplastic way) along with water and enter the root xylem transport tissues [52, 67]. In case of compounds that move in the apoplast of the root cortex need active transportation through plasma membranes of endodermal cells *i.e.* symplast to move to the xylem where subsequent translocation of compounds occurs [52, 68]. For long distance transport to other parts of the plant *i.e.* to leaves translocation of organic pollutants is necessary. The organic pollutants that move in symplastic way (cell to cell) in the root enter into the root xylem from root symplast by simple diffusion similar to the uptake process [60]. The ratio of a compound concentration in the xylem sap to the external solution known as transpiration stream concentration factor determines the translocation of organic compound [52, 60, 68]. Flow of compound along with water from root to shoot is influenced by transpiration pull which is more at high atmospheric temperature, low relative humidity with moderate wind flow and good amount of light.

Generally, organic pollutants are less toxic to the plant as get conjugated and stored or degraded enzymatically after entry to the cell and are less reactive. Depending on the properties of organic pollutants, these may be degraded in the plant root zone or uptake by plant followed by different processes like degradation, sequestration and volatilization. According to "green liver concept" organic pollutants or xenobioticsare metabolized by plants similarly as mammalian live function. Organic pollutants are gone through three phases; chemical modification, conjugation and compartmentation [60, 69, 70]. The detoxification process involves enzyme catalyzing reactions (oxidation, reduction, hydrolysis, conjugation *etc.*). Chemical modification includes functionalization (initial transformation) *i.e.* by enzymatic oxidation, reduction, hydrolysis *etc*. ahydrophobic organic pollutant receives a hydrophilic functional group like carboxyl, amino, hydroxyl *etc.* to attain polarity which boosts toxicant molecules' reactivity and affinity to enzyme for further transformation and conjugation.

Generally, a huge part of organic toxicant undergo conjugation, aprocess of coupling of the toxicants with intracellular endogenous compounds such as amino acids, proteins, organic acids, different carbohydrate molecules, lignin *etc.* [62]. Intermediates of initial transformation or original pollutants containing function group are liable to conjugation with different cellular compounds [60, 62]. For example, oxidative transformation of organic herbicide such as atrazine by creating a hydroxyl side group and this transformation is catalyzed by cytochrome P450 monooxygenasesenzyme [71]. Creation of such side group support the process of conjugation and these conjugates are very less toxic as compared to parent compounds. Immediate temporary detoxification of organic pollutants encompasses conjugation followed by compartmentation of conjugates in the vacuole (soluble conjugates that couple with sugar, amino acids, peptides *etc.*) and by sequestration on cell wall (insoluble conjugates that couple with lignin, cellulose, pectin, starch *etc.*) where these can cause least harm to vital cellular activity [60, 62, 69, 70, 72].

Plants do not possess any special excretion mechanism to keep away contaminants conjugate from vital cell constituents and activities, therefore depend on active transport of these conjugate complexes to vacuole and cell wall using ATP dependent glutathione pump [73]. Glutathione plays an important role conjugation and sequestration of organic toxicants [74]. One example of functionalization followed by conjugation and compartmentalization is vacuole deposition of 2,4-D after hydroxylation and conjugation with glucose and malonyl residues [62]. Organic compounds move by simple diffusion from xylem to symplast of shoot and then to leaf. In the leaf cell compartmentation of pollutant occurs similarly as in root cell [69, 72]. Epidermis and trichomes are the part if these compounds or conjugates of pollutants are stored or accumulated at tissue levels in leaves [75, 76].

Degradation or decomposition of organic pollutants both in root and (or) shoot tissue is one of the important step of organic pollutant transformation and phytoremediation. Degradation process is enzyme catalyzed process. It results either into complete mineralization of organic pollutant to CO_2, water and other simple molecules or partial degradation to more stable intermediate (for conjugation and sequestration) that can be further stored in the plant [64]. Enzymes directly involve in the degradation of organic pollutants are dehalogenases, peroxidases, phenoloxidases, ascorbatoxidase, catalase, carboxylesterases, peroxygenases, nitrilases, Esterases, phosphatases, mono- and dioxygenases, nitroreductases*etc.* Enzymes catalyze conjugation arecytochrome P450-containing monooxygenases, Glutathione-S-transferases, malonyl-O-transferase, glucosyl-O-transferaseetc.

5. Phytoremediation technology: advantages and limitations

Phytoremediation is an *in-situ* approach in which standing green plants extract, stabilize and degrade contaminants from polluted sites. It is an emerging technology that exploits the plant's natural absorption capacity and subsequent detoxification of heavy metals and other pollutants. Some of the plants used in phytoremediation of contamination like heavy metals and other organic pollutants are listed in **Table 1**.

The phytoremediation processes includes phytoextraction, **phytostabilisation, phytovolatilization,** phytodegradation, phytoaccumulation, rhizofiltration and rhizodegradation. Among these, phytostabilisation also provides the additional benefits like waste stabilization, minimal soil erosion, and hydraulic control [83].

Phytoremdiation works best in shallow contaminated soils. Vegetation with rhizosphere depth of less than 10 feets are more efficient. Good results are obtained in places with low levels of existing pollution. A wide range of contaminants like hydro-tolerant heavy metals (nickel, zinc, arsenic, selenium, copper, cadmium etc.),

Plant	Contaminants	Process of removal	Sustainable bioenergy approach	References
Jatropha curcas	Cd	Phytoremediation	Bioenergy production	Marques and Nascimento [77]
Canola, oat, wheat	Cd	Phytoremediation	Biogas production	Zhang et al. [78]
King grass (*Pennisetum americanum*, *Pennisetum purpureum*)	Cd	Phytoremediation	Bioenergy (biomass) production	Zhang et al. [79]
Water hyacinth	Inorganic nutrients	Phytoremediation	Biogas production	Wang and Calderon [80]
Poplars (Populus spp.) and willows (Salix spp.)	fertilizers, inorganic metals and metalloids, petrochemical compounds, soluble radionuclides	Phytoremediation	Bioenergy (biomass) production	Licht and Isebrands [81]
Sunflowers (*Helianthus annuus*)	Pb, Zn and Cd	Phytoremediation	Oil yielding	Angelova et al. [82]

Table 1.
List of plants suitable for phytoremediation along with their bioenergy approach.

radioactive nuclides, petroleum products, pesticide residues and radioactive nuclides are targeted [84]. The efficiency is also determined by the pollutant's hydrophobicity nature. If the pollutant strongly prefers organic material, then it becomes very difficult to separate the pollutants from the compounds. Extreme hydrophilic contaminants remain in the solution and pass through plant tissues without significant accumulation.

6. Steps in phytoremediation

6.1 Selection of plants and plant density

Plant is selected on the basis of the nature of contaminant, soil characteristics and local climatic parameters. Generally, plants with heavy biomass (> 3 tons/ acre) are chosen. When targeted area of remediation is groundwater, deep rooted trees like willow, cotton woods and poplar are planted in rows perpendicular to the flow of water. Some monitoring wells are placed in the surrounding areas.

6.2 Irrigation and soil amendment practices

Flooding encourages the dissolution of contaminants and increases net evapotranspiration. Simultaneously, pH of the soil may alter which require additional adjustments. Efficiency of phytoextraction can be increased by using chelating agents. Ethylenediaminetetraacetic acid (EDTA) forms chelate complexes with heavy metals and radionuclides that keep them in the solution. This helps is easy absorption by vegetation.

6.3 Agronomic practices

It includes the following processes.

6.3.1 Inorganic amendments

In a trial conducted by Vamerali and his co-workers expected that cement acted by capping pollutants, lime by raising pH, and iron sulphate by immobilizing As [85]. Lime and cement at small rates (1%) did reduce the mobility of Pb, Cu and Zn, but not of Cd [86]. Due to their relatively small active rate, they concluded that cement and lime can be applied cheaply on a large scale, with some attention to lime, which raises pH and As mobilization [87]. The response of fertilizers toward phytoextraction of heavy metals was found to be plant specific [88].

6.3.2 Organic amendments

The removals of some metals were enhanced by manure, a fact suggesting that organic matter plays an active role in soil pore-water metal mobility. This response was probably caused by increases in metal influx [89] and the chelating ability of humic acids.

6.3.3 Plowing

Plowing has shown to reduce the impact of metal pollution in plants. Plowing has shown to reduce the impact of metal pollution in plants [88].

Intercropping.

Intercropping with *C. crepidioides*, *Galinsogaparviflora*, *Solanum nigrum* and *Solanum orientalis* significantly decreased Cd contents in shoots of grape seedlings by 78.7%, 12.7%, 29.8% and 26.5%, respectively [90].

6.3.4 Monitoring

Sampling of soil/ water is practised at definite intervals. A differential contaminant concentration ensures the efficacy of the process. Subsequent modifications are made if the process is too slow. This is a 'feedback loop' that may necessitate the alteration or modification of the previous steps.

6.3.5 Harvesting

After harvesting, the hazardous biomass may be composted or incinerated which provide heat and electricity.

6.4 Advantages of phytoremediation

i. The pollutants are phytostabilized in the rhizosphere which prevent runoff into nearby water bodies and agricultural lands.

ii. Phytoremediation uses green plants and natural resources which makes it less expensive than other industrial methods. It is a passive technology that saves a lot input and maintenance costs and suitable for remediation of large areas. Zadrow [91] performed a comparative study between the costs of remediating 500 ppm lead polluted soil through conventional means

Contaminant and matrix	Conventional application	Projected costs	Treatment	Costs	Savings
Lead in soil, (1 acre)	Extraction, harvest, disposal	$150 K-$250 K	Excavate and land drill	$500 K	50–60%
Solvents in ground water (2.5 acres)	Degradation and hydraulic control	$200 K for installation and initial maintenance	Pump and treatment	$700 K annual	50% cost saving by 3rd year

Table 2.
Estimated savings using phytoremediation over other conventional methods [92].

(excavation, disposal) and phytoremediation. He stated that costs of excavation and disposal were $300,000 per acre, while phytormediation costs $110,000 per acre (approx.). Thus phyto-remediation is estimated to cost effective (**Table 2**).

iii. Phytoremediation sites are esthetically more pleasing that other system.

iv. Most of the hyper accumulators' plants have shallow root zones and remediate the soils within the depth of agricultural importance. Hence, it is ideal for restoring agricultural soils contaminated by dispersed contaminants from industrial waste outlets [93].

v. Ash (incinerated biomass) containing higher metal content can be processed to separate the metal from it. For example, 'A. murale' can be processed to separate nickel if its content is above 20% [94].

It has certain limitation such as phytoremediation technology requires more on-field results to be embraced as a mainstream technology for remediation of polluted soils by government agencies so that the benefits of this emerging technology are utilized and also it is a slow process and takes a long time (3–4 years) to meet the clean-up goals. The waste biomass is a biohazard and must be handled carefully. Sometimes improper handling and elevated post-harvest handling costs are notable setbacks of this technology [84].

Phytoremediation can be enhanced by the assistance of chelating agents like EDTA and EDGA. However, significant results had been seen only when larger quantities of chelating substances were applied and a potential threat of chelate enhanced metal leaching and groundwater contamination is a serious concern. The addition of EDTA has been shown to increase metal shoot: root ratio with the cost of lower net root and shoot biomass production [95]. Alternatively, a biodegradable chelating agent like EDDS in hot solution (90°C) can be used in substitution to chemical enhanced phytoremediation to reduce chemical leaching [96].

Author details

Mahima Begum*, Bhaswati Sarmah, Gayatri Goswami Kandali, Sontara Kalita, Ipsita Ojha, Raktim Bhagawati and Lipika Talukdar
Assam Agricultural University, Jorhat, Assam, India

*Address all correspondence to: mahimabegum25@rediffmail.com

References

[1] Bajaj, S., K.D. Singh (2015): Biodegradation of persistent organic pollutants in soil, water and pristine sites by cold-adapted microorganisms: Mini review. *Int. Biodet & Biodegra.,100*: 98-105

[2] Ritter, L., K. Solomon, J. Forget, M. Stemeroff, C. O'Leary, (1995). A review of Persistent Organic Pollutants. Internat. Program on Chemical Safety, Geneva: 1-149

[3] Fuentes, M., C. Benimeli, S. Cuozzo, M. Amoroso (2010): Isolation of pesticide degrading actinomycetes from a contaminated site: bacterial growth, removal and dechlorination of organochlorine pesticides. Int. Biodeterior. Biodegrad., 64, 434-441

[4] Menichini, E., F. Monfredini, Monitoring of carcinogenic PAHs in air under mild-warm ambient temperatures (2003): relative importance of vapour- and particulate-phase analyses in assessing exposure and risk. *Intern. J. Environ. Anal.* Chem.; 83(11) : 897-908

[5] García-Falcón, M. S., B. Soto-González, J. Simal-Gándara(2006): Evolution of the concentrations of polycyclic aromatic hydrocarbons in burnt wood land soils. *Environ. Sci. and techno.*, 40(3): 759-763

[6] Kim, E. J., J. E. Oh, Y. S. Chang(2003): Effects of forest fire on the level and distribution of PCDD/Fs and PAHs in soil. Sci. of the Tot. Environ., 311(1-3): 177-189.

[7] Vergnoux, A., L. Malleret, L. Asia, P. Doumenq, and F. Theraulaz (2011) : Impact of forest fires on PAH level and distribution in soils. Environ. res., 111(2) :193-198

[8] Estrellan, C. R., F. Iino (2010) : Toxic emissions from open burning. Chemos., 80(3): 193-207

[9] Gabos, S., M. G. Ikonomou, D. Schopflocher, B. R. Fowler, J. White. E. Prepas, W. Chen (2001): Characteristics of PAHs, PCDD/Fs and PCBs in sediment following forest fires in northern Alberta. Chemosph., 43 (4-7): 709-719

[10] Gribble, G. W (2003): The diversity of naturally produced organohalogens. Chemosph., 52(2): 289-297

[11] Gunasekara, A. S., B. Xing (2003): Sorption and desorption of naphthalene by soil organic matter: importance of aromatic and aliphatic components. J. of Environ. Qual., 32(1):240-246

[12] Sánchez-Bayo, F (2011): Impacts of agricultural pesticides on terrestrial ecosystems. Eco. impacts of toxic chem., 63-87

[13] Pingali, P. L (2012): Green revolution: impacts, limits, and the path ahead. Proceedings of the National Academy of Sciences., 109(31): 12302-12308

[14] Khan, M. N., M. Mobin, Z. K. Abbas, and S. A. Alamri (2018): Fertilizers and their contaminants in soils, surface and groundwater. Encyclop. of the Anthropo., 5: 225-240

[15] Gottschall, N., E. Topp, M. Edwards, M. Payne, S. Kleywegt, D. R. Lapen (2017): Brominated flame retardants and perfluoroalkyl acids in groundwater, tile drainage, soil, and crop grain following a high application of municipal bio-solids to a field. Sci. of the tot. Environ., 574, 1345-1359

[16] Petersen, S. O., K. Henriksen, G. K. Mortensen, P. H. Krogh, K. K. Brandt, J. Sorensen C., Gron (2003): Recycling of sewage sludge and household compost to arable land: fate and effects of organic contaminants, and impact on soil

fertility. *Soil and Till.* Rese., 72(2): 139-152

[17] Wang, W., W. Zhang, X. Wang, C. Lei, R. Tang, F. Zhang, and F. Zhu (2017): Tracing heavy metals in 'swine manure-maggot-chicken'production chain. Scient. rep., 7(1): 1-9

[18] Zhang, F., Y. Li, M. Yang, and W. Li (2012): Content of heavy metals in animal feeds and manures from farms of different scales in northeast China. Internat. J. of environ. resear. and public health., 9(8): 2658-2668

[19] Mantovi, P., G. Bonazzi, E. Maestri, N. Marmiroli (2003): Accumulation of copper and zinc from liquid manure in agricultural soils and crop plants. Plant and soil., 250(2): 249-257

[20] Xiong, X., L. Yanxia, L. Wei, L. Chunye, H. Wei, and Y. Ming (2010): Copper content in animal manures and potential risk of soil copper pollution with animal manure use in agriculture. Resources, Conservation and Recycling, 54(11), 985-990

[21] Daghrir, R., P. Drogui (2013): Tetracycline antibiotics in the environment: a review. Environmental chemistry letter., 11(3): 209-227

[22] Wu, L., X. Pan, L. Chen, Y. Huang, Y. Teng, Y. Luo, and P. Christie (2013): Occurrence and distribution of heavy metals and tetracyclines in agricultural soils after typical land use change in east China. Environ. Sci. and Pollut. Res., 20(12): 8342-8354

[23] Berendonk, T.U., C.M. Manaia, C. Merlin, D. Fatta-Kassinos, E. Cytryn, F. Walsh, H. Burgmann, H. Sorum, M. Norstrom, M.N. Pons, N. Kreuzinger, P. Huovinen, S. Stefani, T. Schwartz, V. Kisand, F. Baquero, J.L. Martinez (2015): Tackling antibiotic resistance: the environmentalframework. Nat. Revi. Microbiology., 13(5): 310-317.

[24] Cyco, M., A. Mrozik, and Z. Piotrowska-Seget (2019) : Antibiotics in the Soil Environment-Degradationand Their Impact on Microbial Activity and Diversity. Frontiers in Microbiology, 10.

[25] Grenni, P., V. Ancona, A. Barra Caracciolo (2018): Ecological effects of antibiotics on naturalecosystems: A review. Microchem. J., 136: 25-39

[26] Weithmann, N., J.N. Moller, M.G.J. Loder, S. Piehl, C. Laforsch, and R. Freitag (2018): Organicfertilizer as a vehicle for the entry of microplastic into the environment. Science Advances, 4(4):eaap8060

[27] Rai, P.K., S.S. Lee, M. Zhang, Y.F. Tsang, K.H. Kim: Heavy metals in food crops (2019): Health risks, fate, mechanisms, and management. *Environ. Interna.*, 125: 365-385

[28] Bhattacharyya, P., A. Chakraborty, K. Chakrabarti, S. Tripathy, M.A. Powell (2005): Chromiumuptake by rice and accumulation in soil amended with municipal solid waste compost. Chemosp., 60(10): 1481-1486

[29] Madrid, F., R. López, F. Cabrera (2007) :Metal accumulation in soil after application of municipalsolid waste compost under intensive farming conditions. *Agricu., Ecosys. and Enviro.*, 119(3):249-256

[30] Srivastava, V., S.A. Ismail, P. Singh, R.P. Singh (2015) : Urban solid waste management in thedeveloping world with emphasis on India: challenges and opportunities. Reviews in Environ. Sci. and Biotechn., 14(2): 317-337

[31] Ziajahromi, S., P.A. Neale, and F.D.L. Leusch (2016): Wastewater treatment plant effluent as a source ofmicroplastics: review of the fate, chemical interactions and potential risks to aquatic organisms. Water Science and Technology: A J. Inter. Assoc. on Water Pollut. Resear., 74(10): 2253-2269

[32] Rideout, K., Teschke, K (2004): Potential for increased human food borne exposure to PCDD/F when recycling sewage sludge on agricultural land. Environ. health persp., 112(9): 959-969

[33] Weber, R., C. Herold, H. Hollert, J. Kamphues, M. Blepp, and K. Ballschmiter (2018): Reviewing therelevance of dioxin and PCB sources for food from animal origin and the need for their inventory, control and management. Environ. Sci. Euro., 30(1): 42

[34] Zeng, L., H. Li, T. Wang, Y. Gao, K. Xiao, Y. Du, Y. Wang, and G. Jiang (2013): Behavior, fate, and mass loading of short chain chlorinated paraffins in an advanced municipal sewage treatment plant. Environ. sci. and techno., 47(2): 732-740

[35] Brambilla, G., W. D'Hollander, F.Oliaei, T. Stahl, R. Weber (2015): Pathways and factors for foodsafety and food security at PFOS contaminated sites within a problem based learning approach. Chemosph., 129: 192-202

[36] Umlauf, G., E.H. Christoph, R. Savolainen, H. Skejo, J. Clemens, H. Goldbach, H. Scherer, and L. Lanzini (2004): PCDD/Fs and dioxin-like PCBs in soil after 42 years of bio waste application. Organohalogen Compd, 66: 1340-1345

[37] Venkatesan, A.K., R.U. Halden (2013): National inventory of per fluoroalkyl substances in archived US biosolids from the 2001 EPA National Sewage Sludge Survey. J. Hazard mater., 252: 413-418

[38] Sitaramaraju, S., N.V.V.S.D. Prasad, V. Chengareddy, E. Narayana (2014): Impact of pesticides used for crop production on the environment. J. of Chem. and Pharma. Sci.

[39] Kumar, P., K.H. Kim, A. Deep (2015): Recent advancements in sensing techniques based on functional materials for organophosphate pesticides. Biosen. and Bioelects., 70: 469-481

[40] Baez, M.E., J. Espinoza, R. Silva, E. Fuentes (2015): Sorption-desorption behavior of pesticides and their degradation products in volcanic and non-volcanic soils: interpretation of interactions through two-way principal component analysis. Enviro. Sci and Pollut. Resea., 22 (11): 8576-8585

[41] Jablonowski, N.D., A. Linden, S. Koppchen, B. Thiele, D. Hofmann, W. Mittelstaedt, T. Pütz, P. Burauel (2012): Long-term persistence of various 14C-labeled pesticides in soils. Environ. Pollut., 168: 29-36

[42] Navarro, S., N. Vela, and G. Navarro (2007): Review. An overview on the environmental behavior of pesticide residues in soils. Spanish Journal of Agricultural Research, 5(3): 357-375

[43] Turgut, C., O. Erdogan, D. Ates, C. Gokbulut, and T.J. Cutright (2010): Persistence and behavior ofpesticides in cotton production in Turkish soils. Environ. Moni. and Assess., 162(1): 201-208

[44] Damalas, C.A., I.G. Eleftherohorinos (2011): Pesticide Exposure, Safety Issues, and Risk Assessment Indicators. Inter. J. Environ. Res and Public Health., 8(5): 1402-1419

[45] Kim, K.H., E. Kabir, S.A. Jahan (2017) : Exposure to pesticides and the associated human healtheffects. Sci. of the Tot. Environ., 575: 525-535

[46] Tripathi, V., S.A. Edrisi, R. Chaurasia, K.K. Pandey, D. Dinesh, R. Srivastava, P. Srivastava, and P.C. Abhilash (2019): Restoring HCHs polluted land as one of the priority activities during the UN-International Decade on Ecosystem Restoration

(2021-2030): A call for global action. Sci. of the tot. Environ., 689: 1304-1315

[47] Tanglahu, B.V., S.R.S. Abdullah, H. Basri, M. Idrish, N. Anuar (2011). A Review on Heavy Metals (As, Pb, and Hg) Uptake by Plants through Phytoremediation. Int. J. Chem. Eng. 1-31

[48] Chibuike, G.U., S.C. Obiora (2014): Heavy metal polluted soils: Effect on plants and bioremediation methods. *Appl. and Enviro Soil Sci.*, 1-12

[49] Ampiah-Bonney, R.J., J. F. Tyson, G.R. Lanza (2007): Phytoextractionof arsenic from soil by *Leersiaoryzoides*. Inter. J. Phyto., 9(1): 31-40

[50] Cho-Ruk, K., J.Kurukote, P. Supprung, and S. Vetayasuporn (2006): Perennial plants in the phytoremediation of leadconta minatedsoils. Biotech., **5**(1): 1-4

[51] Van Ginneken, L., E. Meers, R. Guisson Ruttens, K. Elst, F.M.G. Tack, J. Vangronsveld, L. Diels, W. Dejonghe (2007) : Phytoremediation for heavy metal-contaminated soils combined with bioenergy production. J. Environ. Engine and Landscape Manage., 15(4):227-236

[52] Hellstrom, A (2004): Uptake of organic pollutants in plants.*Uppsala, Swedish University of Agricultural Sciences.*, 1-40

[53] Marques, A. P. G. C., R. S. Oliveira A. O. S. S., Rangel, and P. M. L. Castro (2008) : Application of manure and compost to contaminated soil and its effect on zinc accumulation by Solanumnigrum inoculated with arbuscular mycorhizal fungi. Environmental Pollution, 151(3): 608-620

[54] Traunfeld, J.H. and Clement, D.L. (2001). Lead in Garden Soils. Home and Garden. Maryland Cooperative Extention, University of Maryland.http://www.hgic.umd.edu/media/documents/hg18.pdf

[55] Girdhar, M., N.R. Sharma, H. Rehman, A. Kumar, A. Mohan (2014): Comparative assessment for hyperaccumulatory and phytoremediation capability of three wild weeds. Biotechno., 4: 579-589

[56] Prasad, M. N. V., and H.M. De Oliveira Freitas (2003): Metal hyper accumulation in plants biodiversity prospecting for phytoremediation technology. Electro. J. Biotec., 6(3): 110-146

[57] Roy, S., S. Labelle, P. Mehta, A. Mihoc, N. Fortin, C. Masson, R. Leblanc, G. Chateauneuf, C. Sura, C. Gallipeau, C. Olsen (2005): Phytoremediation of heavy metal and PAH- contaminated brownfield sites. Plant and Soil, 272(1-2): 277-290

[58] Ba-nuelos, G.S., H.A .Ajwa, B. Mackey, L.Wu, C. Cook, S. Akohoue, S. Zambruzuski (1997) : Evaluation of different plant species used for phytoremediation of high soil selenium. J. Environ. Qua., **26**(3): 639-646

[59] Merkl, N., R. Schultze-Kraft, C. Infante (2005) : Phytoremediation in the tropics—influence of heavy crude oil on root morphological characteristics of graminoids. Environ. Polluti., 138(1): 86-91

[60] Pilon-Smits, E(2005) : Phytoremediation. *Ann. Review of Plant Bio.,* **56**:15-39

[61] Cheng, Z., F. Yao, L. Yuan-wang, C. Hui-qing, L. Zhao-jun, and X. Jian-ming (2017): Uptake and translocation of organic pollutants in plants: A review. J. Integ. Agri., 16(8): 1659-1668

[62] Kvesitadze, E., T. Sadunishvili, G. Kvesitadze (2009): Mechanisms of organic contaminants uptake and

degradation in plants. World Academy of Science, *Enginee and Techn.*, **55**: 458-468

[63] Calderon-Preciado, D., Q. Renault, V. Matamoros, N. R. Cañameras, J. M. Bayona(2012): Uptake of organic emergent contaminants in spath and lettuce: an in vitro experiment. J. Agric and food chem., **60**(8): 2000-2007

[64] McCutcheon, S.C., J.L. Schnoor (2003): Overview of phytotransformation and control of wastes. In *Phytoremediation: Transformation and Control of Contaminants*, ed. SC McCutcheon and JL Schnoor, New York:Wiley, pp. 3-58.

[65] Trapp, S., C. N. Legind (2011): Uptake of organic contaminants from soil into vegetables and fruits. Dealing with contaminated sites, Springer, pp. 369-408

[66] Lin, H., S. Tao, Q. Zuo, and R. M. Coveney (2007): Uptake of polycyclic aromatic hydrocarbons by maize plants. *Environ.l Pollut.*, **148**: 614-619

[67] Kvesitadze, G., G. Khatisashvili, T. Sadunishvili, and E. Kvesitadze (2015): Plants for remediation: Uptake, translocation andtransformation of organic pollutants. In: Öztürk M, AshrafM, Aksoy A, Ahmad M S A, Hakeem K R, eds., *Plants,Pollutants and Remediation*. Springer Netherlands, USA,pp. 241-305

[68] Trapp, S., C.McFarlane: Plant Contamination (1995): Modeling and simulation of organicprocesses, Lewis Publisher, Boca Raton, pp. 254

[69] Burken, J.G: Uptake and metabolism of organic compounds (2003): green-liver model. In *Phytoremediation:Transformation and Control of Contaminants*, ed. SC McCutcheon, JL Schnoor, New York:Wiley., 59-84

[70] Sandermann, H (1994).: Higher plant metabolism of xenobiotics: the"green liver" concept. Pharmacoge., 4: 225-241

[71] Coleman, J.O.D., M.M.A. Blake-Kalff, T.G.E. Davies (1997): Detoxification of xenobiotics byplants: chemical modification and vacuolarcompartmentation. Trends Plant Sci., **2**:144-151

[72] Cobbett, C.S., P.B. Goldsbrough (2000) : Mechanisms of metal resistance: phytochelatins and metallothioneins. In *Phytoremediation of toxic metals using plants to clean up the environment*, ed. I Raskin, BD Ensley, New York:Wiley, pp. 247-71

[73] Martinova, E (1993): An ATP-dependent glutathione-S-conjugate "export" pump in the vacuolar membrane of plants. Nature., **364**: 247-249

[74] Marrs, K.A (1996): The functions and regulation of glutathione s-transferases in plants. Annual Review of Plant Physio. and Plant Molec. Bio., **47**:127-158

[75] Hale, K.L., S. McGrath, E. Lombi, S. Stack, N. Terry(2001): Molybdenum sequestration in *Brassica*: a role for anthocyanins. Plant Physio., 126:1391-1402

[76] Kupper, H., F. Zhao, S.P. McGrath (1999): Cellular compartmentation of zinc in leaves of the hyperaccumulat or *Thlaspicaerulescens*. Plant Physio., 119:305-311

[77] Marques, M.C., A. Nascimento(2013): Analysis of chlorophyll fluorescence spectra for the monitoring of Cd toxicity in a bio-energy crop (Jatropha curcas). J Photochem Photobiol 127:88-93

[78] Zhang, H., Y. Tian, L .Wang, L. Zhang, L. Dai(2013) : Ecophysiological

characteristics and biogas production of cadmium-contaminated crops. Bioresour. Technol., 146:628-636

[79] Zhang, X., X. Zhang, B. Gao, H. Xia, H. Li, J. Li (2014): Effect of cadmium on growth, photosynthesis, mineral nutrition and metal accumulation of an energy crop, king grass (*Pennisetum americanum*, *P. purpureum*). Biomass Bioe., 67:179-187

[80] Wang, Z., M.M. Calderon (2012): Environmental and economic analysis of application of water hyacinth for eutrophic water treatment coupled with biogas production. J Environ Manag 110:246-253

[81] Licht, L., J.G. Isebrands (2005): Linking phytoremediated pollutant removal to biomass economic opportunities. Biotech. Bioeng., 28:203-218

[82] Angelova V., R. Ivanova, K. Ivanov, M.N. Perifanova-Nemska, G.I. Uzunova (2012): Potential of sunflower (*Helianthus annuus* L.) for phytoremediation of soils contaminated with heavy metals. In: BALWOIS-Ohrid, Republic of Macedonia, pp 1-11

[83] Cunningham, S.D., W.R. Berti, J.W. Huang (1995): Phytoremediation of contaminated soils. Trends Biotech., 13: 393-397

[84] Sharma, H.D., K.R. Reddy (2004). : "Geo environmental Engineering." Jon Wiley & Sons, Hoboken, New Jersey, 478-485

[85] Hartley, W., R. Edwards, N.W. Lepp (2004): Arsenic and heavy metal mobility in iron oxide-amended contaminated soils as evaluated by short and long-term leaching tests. *Environ. Pollut.*131:495-504

[86] Bandiera, M., N.M. Dickinson, W. Hartley, T. Vamerali (2010): Remediation of canal sediment exposed to continued wetting and drying: effect on metal mobility by inorganic amendment addition. Page 135 in Proc. 7th Int. Phytotechnol. Conf., Parma, Italy.

[87] Moreno-Jiménez E., E. Esteban, J.M. Peñalosa (2012): The fate of arsenic in soil-plant systems. Rev. *Environ. Contam. Toxicol*. 215:1-37

[88] Vamerali, T. M., Luca. B, F. Marianna, D. Guido, L. Nicholas, M. Paola, Giuliano, Z. Giuseppe (2012): Advances in agronomic management of phytoremediation: Methods and results from a 10-year study of metal-polluted soils. Ita. J. Agron., 7. 10.4081/ija.2012.e42.

[89] Halim, M., P. Conte, A. Piccolo(2003) : Potential availability of heavy metals to phytoextraction from contaminated soils induced by exogenous humic substances. Chemosh. 52:265-275

[90] Hua Lin., X. Zhang, Jun Chen, Liang Liang, Li-Heng Liu (2019): Phytoremediation potential of *Leersia hexandra* Swartz of copper contaminated soil and its enhancement by using agronomic management practices, Ecolog. Enginee., 0925-8574,

[91] Zadrow, J.J (1999): "Recent Applications of Phytoremediation Technologies." Remediation, spring; 29-36

[92] Rock, S.A., P.G. Sayre (1998): "Phytoremediation of Hazardous Wastes: Potential Regulatory Acceptability." Remediation, autumn; 5-17

[93] Mudhoo, A. (2011). "Phytoremediation of Cadmium: A Green Approach." Eds.Rashmi Sangi,Vandan Singh "Green chemistry for environmental remediation" 661-698

[94] Chaney, R .L., L. Broadhurst, T. Centofanti (2010): "Phytoremediation

of Soil Trace Elements" Trace elements in soil. https://doi.org/10.1002/9781444319477.ch14

[95] Romkens, Paul., L. J. Bouwman, J.D. Cathrina. (2002): Potentials and drawbacks of chelate-enhanced phytoremediation of soils. Environ. Pollut. (Barking, Essex : 1987). 116. 109-21. 10.1016/S0269-7491(01)00150-60

[96] Luo, C., Z.G. Baker, Li. Alan, Xiang-Dong (2006) : A novel strategy using biodegradable EDDS for the chemically enhanced phytoextraction of soils contaminated with heavy metals. *Chin. J. of Geo.* 25. 10.1007/s11104-006-0059-3. DOI: 10.1007/s11104-006-0059-3

Chapter 5

Bacillus megaterium Biodegradation Glyphosate

Nibal Khaleel Mousa, Abdul-Jabbar Ali and Maha Hussein

Abstract

The *Bacillus megaterium* ability was evaluated in this paper to degrade the Glyphosate. organophosphorus pesticides, The bacteria re-cultured that isolated from other researches of Baghdad soils and morphological identification and biochemical tests besides by selectivity media. The (5 and 25) ppm showed the highest growth results were within two days to two months on mineral salt media. The highest glyphosate degradation ratio % were (70) % per 25 ppm/two months. Incubation period Increasing led to highest glyphosate degradation ratio% at (25) ppm led to conclusion that bacteria digestive the pesticides as carbon and nitrogen sources and will be well harvest it form contaminated areas.

Keywords: bacterium, bioremediation, glyphosate, HPLC

1. Introduction

Glyphosate is an organophosphate which is a heterogeneous compound, utilized on the grasses leaves and broadleaf plants due its non-selectivity pesticides. In the seventies was the first registration in the United State by Monsanto (Roundup) [1]. Glyphosate stops a shikimic acid enzyme pathway which is essential for some microorganisms and plants. In trace analysis, Glyphosate is a difficult herbicide, has a good water solubility that causes difficulty in determination its physical and chemical properties [2].

The microorganism ability to eliminate pollutants is one of the bioremediation methods [3]. Bioremediation is applied as an auxiliary strategy due to its inoffensive to ecology, economic worth, reduce environmental poison, and validation [4–8]. Metabolic is one of the microorganism degradation methods across pesticides in soils, also catabolism strategies and the enzymes of co-metabolism [9]. Organic-pesticides fate in the ecology system can be marked by utilizing biodegradation as a major agent. The study purpose is to achieve and inspect the domestic bacterial separated on broken down different glyphosate concentrations that can be remaindered in soil and detect residues concentration from bacteria digestive via HPLC after extraction.

2. Material and methods

2.1 Materials

Chemosate. The trade name of "Glyphosate", bought from the local market. Materials were available and supply in Remediation Pollutants Center. The Mineral

Weight (g)	Compounds	Note
0.2	KH_2PO_4	Sterilized separately at 125 °C/25 min (part A)
0.5	K_2HPO_4	
1	$(NH_4)_2SO_4$	All mixed and added to part A of 1-liter flask, and adjust (pH 7.0 ± 0.3)
0.2	$MgSO_4{\cdot}7H_2O$	
0.2	NaCl	
0.05	$CaCl_2{\cdot}2H_2O$	
0.025	$FeSO_4{\cdot}7H_2O$	
0.005	Na_2MoO_4	
0.005	$MnSO_4$	

Table 1.
The mineral salt media.

Weight (g)	Compounds	Note
10	Glucose	• 1liter distillate water
0.5	Yeast extract	• adjust (pH 7.0 ± 0.3)
0.25	$MgSO_4{\cdot}7H_2O$	• incubate28-30 °C/48 h.
15	Agar	
0.1	$CaCl_2{\cdot}2H_2O$	

Table 2.
Sperber media structure.

Salt Media (MSM), was used in growing *B. megaterium* to investigate glyphosate degradation, **Table 1** [10]. As the only carbon source, Flasks (125 ml) were supplemented with Glyphosate. The Final Concentration of Glyphosate with 0.5 were (5 to 25) ppm from bacteria re- culture in comparison with control.

2.2 Re-growth and identification of *Bacillus megaterium*

Bacillus megaterium was growth and kept in incubation of our laboratory from other studies [3] and to re-identification by selective media, was by Sperber's Medium [10, 11], **Table 2**, By the spectrophotometer OD $_{600,}$ then the hydrolysis capacity measured for different period (2,5,7,14,21,30,60) days [12].

2.3 Glyphosate degradation ratio via *Bacillus megaterium*

The degradation ratio % investigated for (1-2) months, extracted through added equal volume each ethyl acetate and MSM which utilized as a reagent, for twice time, centrifuged at 3000 rpm/10 minutes, filtered then anhydrous sodium sulfate utilized as dry factor followed glass-fiber paper (Whatman GF/B) [10]. In Eq. (1), measured the degradation ratio:

$$P = \left(1 - \frac{C1}{C0}\right) x100\% \tag{1}$$

P = the rate of degradation of Glyphosate,

HPLC condition analysis	
UV–Vis detector	254 nm
Manual Injector Equipped	20-μL loop
Column/Stationary phase	C-18*
Mobile Phase	Acetic acid (1%) + methanol (6:4 (v/v).
Flow Rate	1.0 ml /min
Temperature	23–25 °C
ZORBAX (5 μm, 150 mm × 4.6 mmi.d.)	

Table 3.
HPLC conditions.

C1 = Glyphosate dose in sample.
C0 = control [13].

2.4 Metabolite analysis

Each ethyl acetate extraction was analyzed by HPLC condition, **Table 3** [14], calculates the final concentration [15] of glyphosate used Eq. (2):

$$Pest.con = Asa \times Cs \,/\, As \times Csa \tag{2}$$

Pest.con = pesticide concentration in sample (mg/L)
Asa = sample peak area
Csa = sample concentration, mg/L
As = standard. peak.area
Cs = standard concentration, mg/L

3. Results and discussion

3.1 Re-growth and identification of *Bacillus megaterium*

Besides using selective media, Sperber Medium, **Table 4** shows the Test of Morphological, and **Table 5** represents the biochemical tests.

3.2 *Bacillus megaterium* hydrolyzes and bacteria growth

3.2.1 Growth of B. megaterium

In **Figure 1**, the results show that the highest *B. megaterium* growth was in (two months) for both (5, 25) ppm (0.164, 0.167) sequentially, while the 15 ppm show in 60 days, the highest growth is (0.215) in comparison with others when used glyphosate as carbon sources.

3.2.2 Degradation rate%

The highest degradation rate% for Glyphosate by *B. megaterium* in comparison among concentration was for both the 5-25 ppm in 60 days reached (70.01-70.9) %, **Figure 2**.

Morphological tests	
Spore shape	Rod-like/ flagella spores
Colonies	Round to irregular /yellow to brown or black after prolonged incubation
Motility	+
Gram stain	+
Aerobic	+
Temperature	3-20 °C/ 35-45 °C,optimun30°C
pH	5.7- 7

Table 4.
The tests of morphological.

Biochemical tests			
Catalase	+	Nitrate reduction / Degradation of tyrosine	+/—
Starch Hydrolysis	+	Casein hydrolysis	+
Citrate utilization	+	Indol/ Methyl Red	—
Esculin hydrolysis	+	Arginine dihydrolase	—
Gelatin hydrolysis	+	Tryptophan deaminase	—
Oxidase	+	Hydrolysis Urea	—

Table 5.
The tests of biochemical.

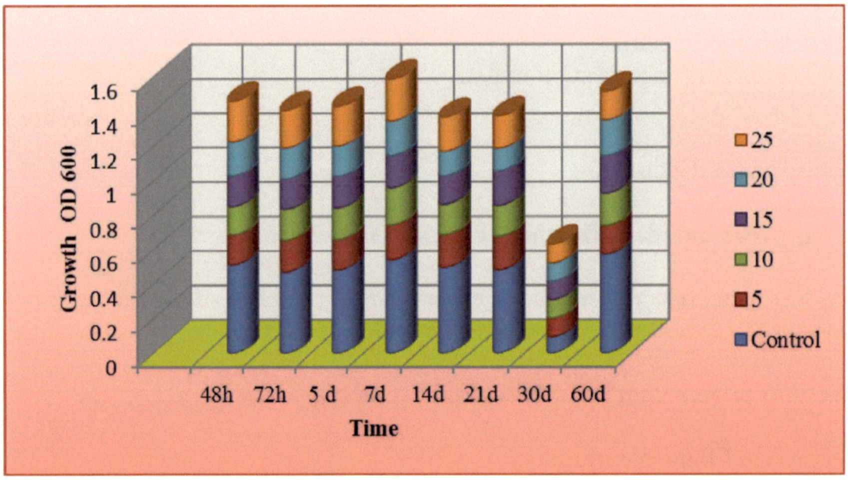

Figure 1.
Culture of B. megaterium *on Mineral Salt Media with Glyphosate.*

3.2.3 Glyphosate residues by HPLC test

The *B. megaterium* growth on different concentration in MSM at 30 °C, in comparison with control in **Figures 3** and **4**. Decreasing in glyphosate in 30 days for concentration (5,10) ppm (7, 8)% is the best peak area, while the 25 ppm showed 28%, while the results glyphosate peak area two months incubation, led to (20, 15) ppm in comparative with control. When comparing among HPLC, **Figures 5** and **6**

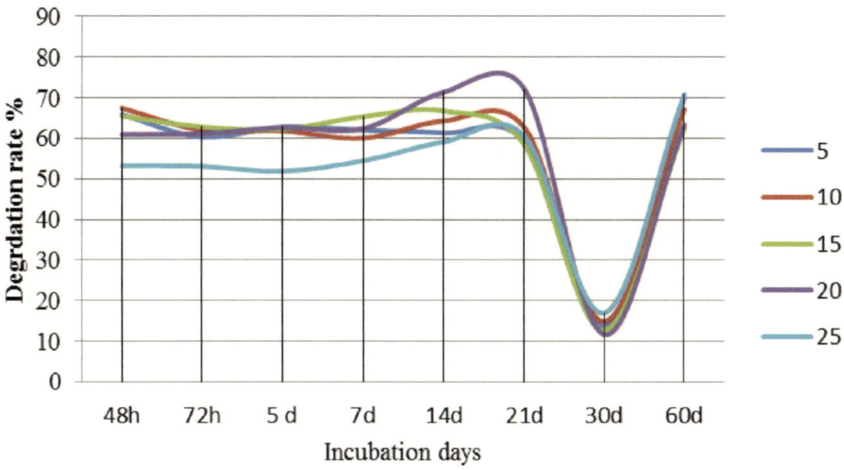

Figure 2.
The degradation rate of Glyphosate in MSM in Comparative with control.

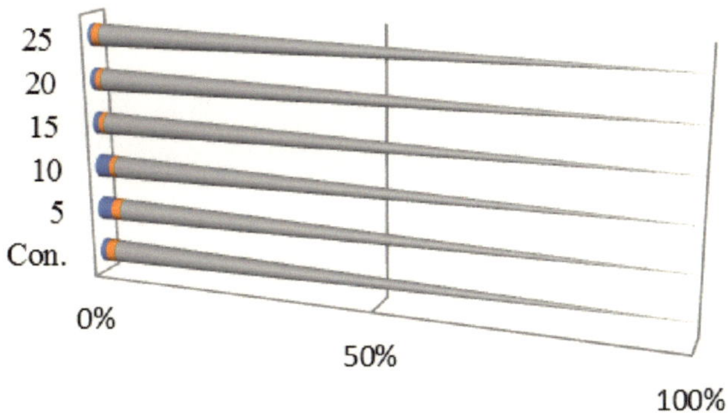

	Con.	5	10	15	20	25
Retention Time	4.9	5.7	6.2	6.2	5.9	5.3
peak height	7.792	3.792	2.437	5.319	6.141	10.988
peak area	443.557	212.476	230.015	535.95	612.048	778.009

Figure 3.
Peak height, peak area and retention time of Glyphosate after incubation B.M. 30 days in MSM.

improved that the *Bacteria* degradation ratio% positive results with all glyphosate concentrations via incubating two months incubation. However, the best was for (5, 25) ppm for each degradation ration% and the HPLC.

Organophosphorus pesticides microbial degradation and the bioremediation progress development for contaminated soils depend on the introduction of microbe's biodegrading [16]. *B. megaterium* shows the degradation rate% and highest growth in two months cultivation time for both five and twenty five concentration. The degradation pesticide ability like Chlorpyrifos (600 mg-L1) via *B. megaterium* in ten days incubation was 81% [17] and 73% for 20 ppm during three weeks. On the other hand, *Bacillus megaterium* 99% in seven days, degradation ratio towards atrazine (50 ppm) and Chlorpyrifos in (1-2) weeks [18, 19]. Monocrotophos (MCP), 83% degradation ratio reached led sub-products to CO_2, NH_4, and (HPO_3) [20]. Each bacterium produces enzymes to analyze

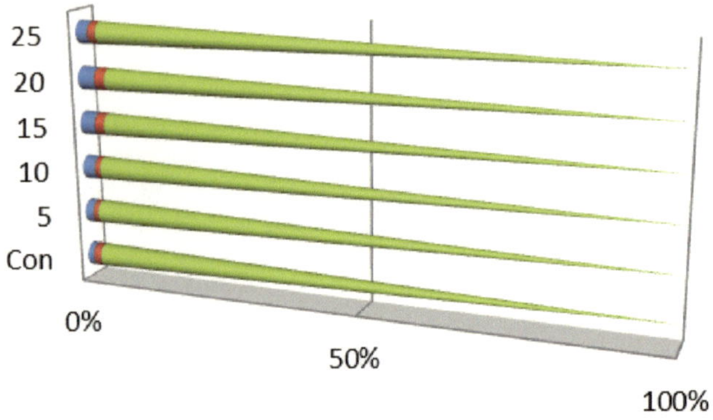

	Con	5	10	15	20	25
Retention Time	5.5	5.8	5.6	5.4	5.5	6.1
Peak Heght	6.898	3.794	2.414	3.486	3.441	3.903
Peak Area	463.7	372.577	236.264	188.52	170.858	260.242

Figure 4.
Peak height, peak area and retention of Glyphosate after incubation B.M. 60 days in MSM.

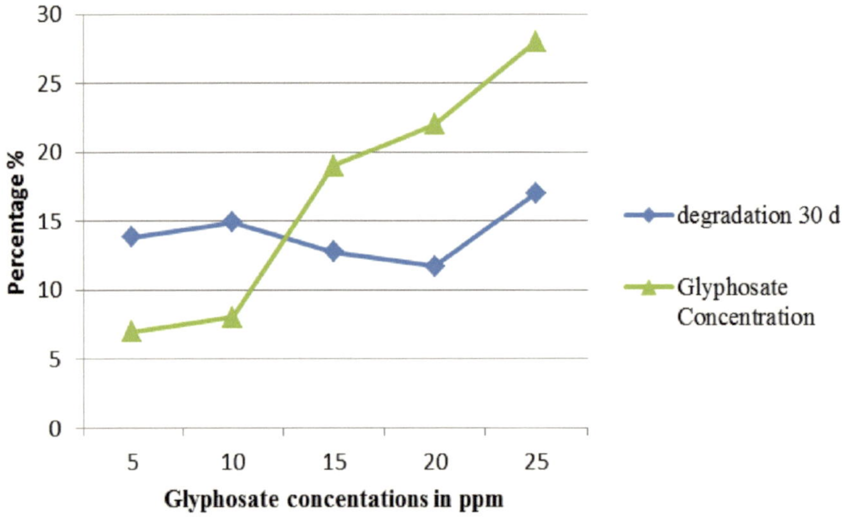

Figure 5.
Comparative among the degradation ratio% and Glyphosate concentrations via HPLC in 30 days incubation on MSM.

compounds and supplying its self with elements in need for re-build enzyme and cell body. One of *Bacillus megaterium* enzyme is β-Amylase [21] that consist from carbon and nitrogen mainly, so the bacteria most often organic-phosphors compounds, Glyphosate used to supply a single element (carbon, phosphorus or sulfur) and [22]. Due to that the concentration of glyphosate generally reduces along with bacteria growth. Phosphate ester and phosphonate are the analysis hydrolysis results of phosphorus. The esters groups have they have many vulnerable to hydrolysis sites. Beside hydrolysis, the oxidation also one of the major reactions, and the alkylation,dealkylation [23]. Detoxification is one of microbial degradation principles via hydrolysis phosphor bonds with oxygen (P-O-alkyl, P-O-aryl [24].

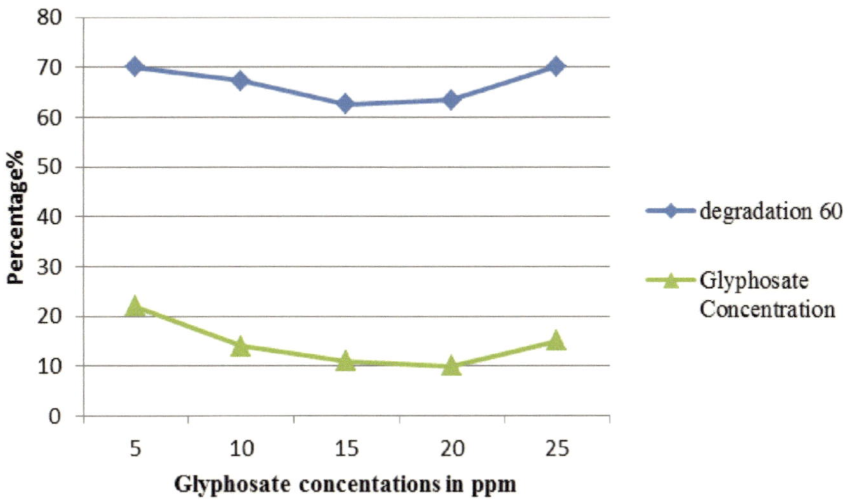

Figure 6.
Comparative among the degradation ratio% and Glyphosate concentration via HPLC in 60 days incubation on MSM.

4. Conclusion

In this study, *Bacillus megaterium* improved the best results for growth was in 48 h while in two months, had the same growth each five and twenty five concentration. The degradation rate % ability was the best in (5,25) ppm/two months reached (70-71)%. The Glyphosate degradation ratio% increasing equally with the increasing the incubation to two months,the best was for five and twenty five ppm, each the HPLC and Degradation ratio%. The conclusion is the *B. megaterium* utilized Glyphosate as supplier for elements sulfur, carbon, nitrogen and phosphorus and cultivated highly from culture could be well exploited for biodegradation from its pollutants sites.

Author details

Nibal Khaleel Mousa*, Abdul-Jabbar Ali and Maha Hussein
Remediation Pollutants Center, Environment and Water Directorate, Ministry of Science and Technology, Baghdad, Iraq

*Address all correspondence to: nibal.mousa@mail.co.uk

References

[1] Valavanidis A. Glyphosate. The Most Widely Used Herbicide. *Scientific Reviews.* 2018.Chem-toxecotox. org. pages:41.www.chem-tox-ecotox.org/ Scientific Reviews

[2] Kaczyński, P, Łozowicka, B,:Liquid chromatographic determination of glyphosate and aminomethylphosphonic acid residues in rapeseed with MS/MS detection or derivatization/fluorescence detection. Open Chem., 2015; 13: 1011-1019. DOI: 10.1515/chem-2015-0107

[3] Ibrahim GAG, Amin MK, Hassan AA, El-Sheikh. EA: Identification Of Pesticides Degrading Bacteria Isolated from Egyptian Soil. Zagazig J. Agric. Res., 2015; 42 (5),1129-1143.

[4] Finley SD, Broadbelt LJ, Hatzimanikatis V: In Silico Feasibility of Novel Biodegradation Pathways for 1,2,4-1140 Ibrahim, *et al.* Trichlorobenzene. *BMC Systems Biology*, 2010; 4 (7): 4-14. https://doi.org/10.1186/1752-0509-4-7

[5] Comeau Y, GreerCW, Samson R: Role of inoculum preparation and density on the bioremediation of 2,4-D contaminated soil by bioaugmentation. *Appl. Microbiol. Technol.*, 1993; 38: 681-687. DOI: https://doi.org/10.1007/BF00182810

[6] Ghassempour A, Mohammad A, Najafi F, Rajabzadeh M.:Monitoring of the pesticide diazinon in soil, stem and surface water of rice fields. *Anal. Sci.*,2002; 18: 779-783.DOI https://doi.org/10.2116/analsci.18.779

[7] Yasouri FN : Plasmid-mediated degradation of diazinon by three bacterial strains Pseudomonas sp., Flavobacterium sp. and Agrobacterium sp. Asian J. Chem.,2006: 18: 2437-2444.

[8] Sørensen SR, Albers CN, Aamand J,: Rapid mineralization of the phenylurea herbicide diuron by Variovorax sp. strain SRS16 in pure culture and within a two-member consortium. Appl. Environ. Microbiol.,2008:74: 2332-2340. DOI:10.1128/AEM.02687-07

[9] El-Sheikh EA, Ashour, MB. : Biodegradation Technology for Pesticide Toxicity Elimination. In: Bioremediation Technology-Recent Advances, M.H. Fulker (Editor). Capital Publishing Company, New Delhi,2010: 162-205. DOI: https://doi.org/10.1007/978-90-481-3678-0_6

[10] Sperber JI, Solubilization of mineral phosphate by soil bacteria. Nature, 1957:180: 994-995

[11] Nieminen T, Rintaluoma N, Andersson M, Taimisto AM, Ali-Vehmas T, Seppälä A, Priha O, Salkinoja-Salonen M.:Toxinogenic *Bacillus pumilus* and Bacillus licheniformis from mastitic milk. Vet Microbiol,2007:124(3-4):329-339. DOI:https:// doi.org/10.1016/j.vetmic.2007.05.015

[12] Ortiz-Hernández ML, Sánchez-Salinas E.: Biodegradation Of The Organophosphate Pesticide Tetrachlorvinphos By Bacteria Isolated From Agricultural Soils In México. Rev. *Int. Contam. Ambient*.2010: 26 (1) 27-38.

[13] Tang M,You M.: Isolation, identification and characterization of a novel triazophos-degrading Bacillus sp. (TAP-1). Microbiological Research .2012:167:299-305.https://doi.org/10.1016/j.micres.2011.10.004

[14] Islas G, Rodriguez JA, Mendoza -Huizar, LH, Pérez-Moreno F, Gabriela Carrillo, E.: Determination Of Glyphosate and aminomethylphosphonic acid in soils by HPLC with pre-column derivatization

using 1,2-Naphthoquinone-4-Sulfonate. Journal of Liquid Chromatography & Related Technologies.2014:37:1298-1309. https://doi.org/10.1080/10826076.2013.789801

[15] Alehagen M: Development of a method for determination of pesticide residues in honey using liquid chromatography-tandem mass spectrometry. Master Thesis, Swedish University of Agricultural Sciences, Department of Food Science, Uppsala.2011: pages, 42.

[16] Semple KT, Reid BJ, Fermor TR: Impact of composting strategies on the treatment of soils contaminated with organic pollutants. Environ Pollut. 2001:112:269-283. https://doi.org/10.1016/S0269-7491(00)00099-3

[17] Shweta N, Jadhav SK, Keshavkant S:Bacillus megaterium: A potential swimmer and an efficient bio-degrader of an organophosphorus pesticide. International Conference on Environmental Microbiology and Microbial Ecology and International Conference on Ecology and Ecosystems.2017:7, Issue 2; page 84. DOI: 10.4172/2157-7625-C1-029

[18] Zhu J, Fu L, Jin C, Meng Z, Yang N: Study on the Isolation of Two Atrazine-Degrading Bacteria and the Development of a Microbial Agent. Microorganisms . 2019:7,80:1-11.https://doi.org/10.3390/microorganisms7030080

[19] Chandrashekar MA, Supreeth M, Soumya Pai K, Ramesh SKC, Geetha N, Puttaraju HR, Raju NS : Biodegradation Of Organophosphorous Pesticide, Chlorpyrifos By Soil Bacterium - Bacillus Megateriumrc 88. Asian Jr. of Microbiol. Biotech. Env. Sc.2017: 19(1), 127-133.

[20] Bhadbhade BJ, Sarnaik SS, Kanekar PP: Biomineralization of an organophosphorus pesticide, Monocrotophos, by soil bacteria. J Appl Microbiol.2002:93(2):224-34.https://doi.org/10.1046/j.13652672.2002.0168 0.x

[21] Taniguchi H, Honnda Y: Amylases, Encyclopedia of Microbiology (3rd Edition), 2009: Pages 159-173.

[22] Singh BK, Kuhad RC, Singh A, Lal R, Triapthi KK: Biochemical and molecular basis of pesticide degradation by microorganisms. Crit Rev Biotechnol.1999: 19: 197-225. https://doi.org/10.1080/0738-859991229242

[23] Singh P. Garg A, Raman M, Agrawal D.: Effect of replacing barley grain with wheat bran on intake and utilization of nutrients in adult sheep. Small Rumin. Res.1999: 31 (3): 215-219.https://doi.org/10.1016/S0921-4488(98) 00145-X

[24] Mousa, N.K., Gatae, I. H., Hasan.A K 2019. Biodegradation of (N-phosphonomethyl)glycine Utilizing *Bacillus subtilis* using different incubation periods. International Conference on Agricultural Sciences, IOP Conf. Series: Earth and Environmental Science 388,012080.IOP Publishing. doi:10.1088/1755-1315/388/1/012080

Chapter 6

Biodegradation by Fungi for Humans and Plants Nutrition

Chandan Singh and Deepak Vyas

Abstract

Fungi being achlorophyllous depends on other living organisms for their food either being parasite or saprophyte. Saprophytic fungi are good biodegraders. Through their enzymatic batteries, they can degrade any organic substances. Most of the time during the processes of degradation, macrofungi (mushrooms) are occurred as per the climatic conditions prevailing in the particular locations. Micro and macrofungi are considered a good source of human nutrition and medicine since time immemorial. Some of the fungi which are commonly known as mycorrhizae facilitate nutrients to more than 90% of green plants. Fungi play a basic role in plant physiology and help in the biosynthesis of different plant hormones that provides the flexibility of plant to withstand adverse environmental stress, the whole fungi are more friend than foe.

Keywords: fungal biodegradation, residues, tree internet, mushrooms, biocontrol

1. Introduction

Biodegradation is defined as the biologically catalyzed reduction in the complexity of chemical compounds. Indeed, biodegradation is the process by which organic substances are broken down into smaller compounds by living microbial organisms. When biodegradation is complete, the process is called "mineralization". However, in most cases, the term biodegradation is generally used to describe almost any biologically mediated change in a substrate [1]. Fungal diversity is globally estimated to 1.5 million species and consists of an incredibly diverse group of organisms. Organisms studied by mycologists include members of the fungal Kingdom but also others like Protozoa e.g. slime molds [2]. Biodegradation by fungi is also known as mycodegradation. Likewise, bioremediation in which fungi are employed is sometimes called mycoremediation [3]. Fungi are parasitic, saprophytic, mutualistic, and decomposers and grow faster on their substrate and synthesis metabolites to adjust with all the adverse condition and competitor, therefore it has several secondary metabolites, these metabolites serve as a treasure for the new source of potential drugs for human health and the plant health [4]. Fungi have ancient application in human health and nutrition, it produces several enzymes like cellulase, lipase, ligninolytic enzymes, catalase, laccase, etc., alkaloids, pigments, aroma, and flavors, and used in biological control of nematodes, in plants pest control, health benefits by edible fungi [5, 6]. The diversity of fungi play important role in the environment as it acts as decomposers and recycles the organic matter in nature, provides nutrition to plants through mycorrhization [7, 8], and the enzymes secreted by fungi are investigated for the production of the different by-products out of waste and sludges. Many filamentous fungi are now investigated for the production of biofertilizers [9, 10]. Fungi produce numerous secondary metabolites that are used for human benefit. Despite the benefit of fungi for human health and plant health, it has a negative effect too. The different aspect of fungi effects on human and plant is all due to the potential of the fungi to utilizes the recalcitrants wastes through a process called as biodegradation. In this chapter, the different aspects of fungal biodegradation and its relation to human and plant nutrition have been highlighted.

2. Mechanism of fungal degradation

Fungi have numerous enzyme system and occur under the various climatic condition on a variety of substrates, the mode of nutritious in fungi is always heterotrophic, therefore being a heterotrophic organism they obtain their nutrition either through parasitic or saprotrophic mode and to do so, they employ a series of enzyme reaction on the substrate they grow since the biomass (dead bodies of plants and animals) are complex in chemical composition they are made available to fungi nutrition by converting it to the simpler form [3]. Generally, plant residues are lignocellulosic in nature which is a very complex molecule to digest by any organism, but fungi with the potential to produce enzymes that digest these residues to simpler, the mechanism of fungal degradations depend on the type of substrates and the enzymatic system, for example, lignin degradation by the white-rot fungi is an oxidative process and phenol oxidases are the key enzymes, manganese peroxidases, lignin peroxidases, laccase from the white-rot fungi have been found to play a significant role in lignin degradations [11] (discussed in Section 3.1.), white-rot fungi degrades lignin to use it as a sole carbon and energy source, and it is generally believed that lignin breaks down is necessary to gain access to cellulose and hemicelluloses of the substrates [11].

3. Fungi-mediated biodegradation

Fungal has the potential to degrade organic materials naturally, considering this capacity of fungi to convert organic residue to different simple products are harnessed to produce valuable products for mankind's, which is used under control condition for the production of desired products by humans like production of bread, wine, medicine, and other industrial application, among this cultivation of edible and medicinal mushroom on organic residue is an example of fungal mediated biodegradation. The conversion of lignocellulosic residue to value products involves multi-steps which includes [11]:

a. Pretreatment (mechanical, chemicals, or biological).

b. Hydrolysis of polymers to produce readily metabolizable molecules (eg. Hexose or pentose sugars).

c. Use of these molecules to support microbial growth or to produce chemical products.

d. Separation and purification.

Edible mushroom cultivation is a mediated fungal degradation for the production of non-consumable residue into the consumable source of nutrition riched food, mushrooms are fleshy and saprophytic fungus that utilizes wood trunks of trees, decaying organic matter, and damp soil rich in organic substances for their growth. Cultivation of mushrooms can be viewed as an effective way to utilize bioresources left in agricultural residues and environmental protection strategy [12, 13]. Cultivation of any type of mushroom implies principles of microbiology, environmental engineering, and solid-state fermentation in the conversion of domestic agricultural, industrial, forestry wastes into food for humans. *Pleurotus* mushrooms are simplest and are easily cultivable on the agric residue available on the agric farm, different types of substrate have been used for the cultivations to increase the yield [14–16]. *Pleurotus* is a genus of edible mushrooms widely cultivated throughout the world in a variety of substrates and conditions. This genus consists of more than 200 saprophytic species distributed worldwide in temperate and tropical environments and the most common species of *Pleurotus* genera (Oyster mushroom), are *P. ostreatus*, *P. djamor*, *P. citrinopileatus,* and *P. eryngii.* Among the common substrate used are wheat straw, sawdust, paddy straw, corn cob, sugarcane bagasse, ground nutshell, etc. [15, 17, 18]. *Pleurotus* produces the enzyme system to degraded the lignocellulosic components of the substrates and made them available for the mushroom for their metabolism which makes the mushroom a rich source of protein, dietary fiber, vitamins, and minerals [14, 19, 20]. Apart from oyster mushroom cultivation different variety have been adopted worldwide for the cultivation on the large scale both for medicine and the nutritions like species of *Agaricus, Lentinus*, *Calocybe*, *Volvariella*, *Auricularia*, *Ganoderma*, *Trametes,* etc., are some of the examples of fungal mediated biodegradation for production and utilization of wastes.

3.1 Biodegradation of agricultural waste by fungi

Tons of agricultural residue generated each year from the cropland and some of these residues are used as animal feed and others for industrial use but the majority of the residue is burned in the crop field causing environmental pollution, but using fungal species these are converted into compost or either used for the production of the edible mushroom. Since fungi possess a proficient hydrolytic system that is capable to convert lignocellulosic material to essential metabolites in the form of mushrooms. Usually, fungi (micro and macrofungi) secrete enzymes, including cellulases (cellobiohydrolases, endoglucanases), hemicellulases (xylanases), and β-glycosidases [21, 22]. The recent developments in our understanding of the genetics, physiology, and biochemistry of fungi, has led to the exploitation of fungi for the preparation of different agriculture and industrial products of economic importance [4], therefore the agric residue which is rich in lignocellulose consists of lignin, hemicellulose, and cellulose [11, 23] can potentially be converted into different value-added products as depicted in (**Figure 1**) including biofuels, chemicals, animal feed, textile and laundry, pulp and paper. Production of ethanol and other alternative fuels from lignocellulosic biomass can reduce urban air pollution, decrease the release of carbon dioxide into the atmosphere, and provide new markets for agricultural wastes [21].

As the lignocellulosic biomass is made up of complex carbohydrates, which is a source of the sugars that can be processed to obtain ethanol, but due to the recalcitrant nature of the lignocellulose biomass it is very difficult to produce ethanol out of this biomass, production of ethanol from these biomass involves series of step to convert complex cellulose to simple sugars one of the major steps in the production step of ethanol is pretreatment of the recalcitrants biomass, which raise the cost of production of ethanol. The three methods physical,

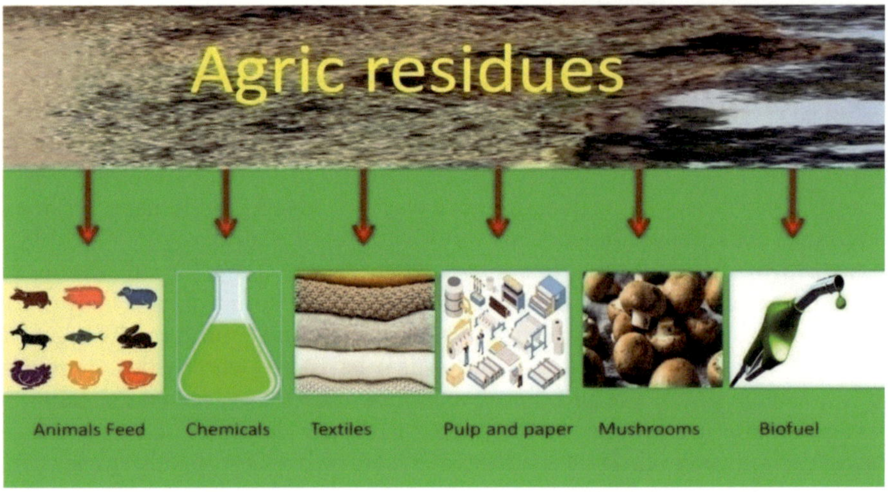

Figure 1.
Different value-added products from agric residues.

chemical, and biological methods are used in the pretreatment of the biomass. Pretreatment is done to digest the lignocellulose to produce simpler sugars that are further converted into bioethanol [24]. The biological methods of treatment using microbes are eco-friendly and produce clean fuel, among all microbes fungi have great potential to convert recalcitrants into simpler sugars through enzymatic and hydrolytic methods [25]. The existence of the enzymatic system in the fungi serves as the treasure of the Novo enzyme source for the finding of candidates of the enzyme to digest lignocelluloses, enzymes like catalases, laccase, hemicellulases, ligninases, pectinases play a crucial in the digestion of the recalcitrants biomass [26]. The effectiveness of a biological pretreatment is determined by several factors like composition of biomass, inoculum concentration, aeration rate, moisture content, incubation time, incubation temperature, pH, and the fungi species involved [27]. The most common mechanism of pretreatment is illustrated in **Figure 2**.

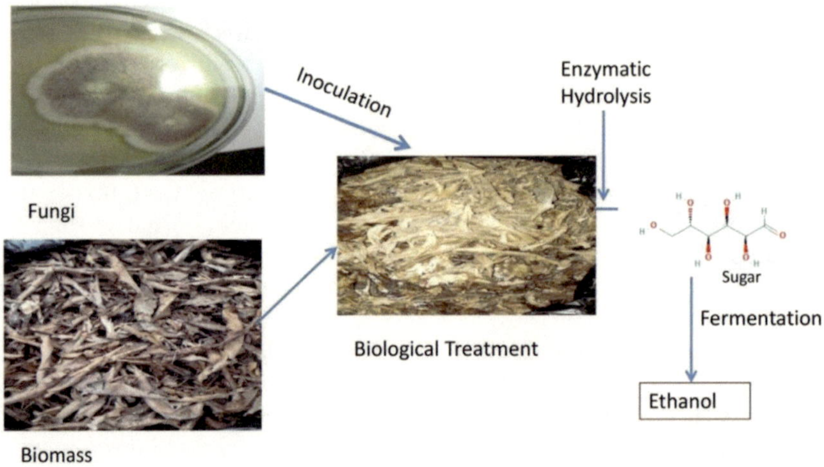

Figure 2.
Schematic diagram for biological treatment by fungi of lignocellulosic biomass.

3.2 Wood decaying fungi

Wood decaying or xylophagous fungi are those fungi that decompose wood dead or alive. These fungi in nature break down complex molecules of deadwood and branches and return their nutrients to the soil [2, 28]. Wood decaying fungi cause brown rot, soft rot, and white rot, with a different set of enzymes, and

Sl. No	White-rot fungi	Enzyme/s	Function	Reference
1.	*Phaberochaete chrysosporium*	Lignin peroxidases	Degradation of Lignin & lignin-like compound, biomineralization of lignin	[11, 23]
2.	*Trametes versicolor*	Lignin peroxidase, cellulase Manganese peroxidases	Degradation of Lignin & lignin-like compound act as a wood decomposer, biomineralization of lignin	[11, 23, 29]
3	*Pleurotus sp*	Lignin peroxidase, Versatile peroxidase, Laccase, Manganese peroxidases	Lignin degradation, production of biofuel, environmental remediation	[23, 34, 35]
4.	*Bjerkandera adjusta*	Lignin peroxidase, Versatile peroxidase, pro-duce some accessory enzymes for H_2O_2 production.	licentious enzyme having Lignin peroxidase acts as a wood decomposer, biomineralization of lignin.	[23, 35]
6.	*Pycoporus cinnabarius*	Laccases, lipase, protease	Biomineralization of lignin and Cellulose, degradation of dyes	[29, 35, 36]
7.	*Schizophyllum commune*	Cellobiose dehydrogenase,	Biomineralization of lignin and cellulose	[29, 35, 36]
8.	*Inonotus hispidus*	Laccases, manganese peroxidases, and lignin peroxidases	Lignin-modifying, Dye decolorization.	[5, 23]
9.	*Lentinus tigrinus*	Laccases, manganese peroxidases, and lignin peroxidases	Degradation of dyes, Biomineralization of lignin and cellulose.	[5, 23]
10	*Ganoderma sp*	Laccases, manganese peroxidases, and lignin peroxidases	Lignin-modifying Dye decolourization	[5, 23]
11	*Coriolus sp*	Laccases, manganese peroxidases, and lignin peroxidases	Lignin-modifying, Dye decolorization.	[5, 23]
12	*Irpex sp*	Laccases, manganese peroxidases, and lignin peroxidases	Lignin-modifying, Dye decolourization	[5, 23]
13	*Laccaria fraterna*	Laccases, manganese peroxidases, and lignin peroxidases	Lignin-modifying, Dye decolorization	[36]
14	*Lentinus polychrous*	Laccases, manganese peroxidases, and lignin peroxidases	Lignin-modifying, Dye decolorization,	[11, 36]

Table 1.
White-rot fungi and lignin degradation enzyme produced by them.

based on the type of decay they cause they are classified as brown-rot fungi, soft-rot fungi, and white-rot fungi [29]. Brown-rot fungi produce hydrogen peroxide that breaks down cellulose and due to this the fungus removes all the cellulose compound from the wood and left the wood brown in color. Soft-rot fungi secrete the cellulolytic enzyme and break down the cellulose of the wood, these fungi are less aggressive than the white-rot fungi, whereas the white-rot fungi break both lignin and cellulose of the wood [5]. White-rot fungi produce laccase enzyme in higher concentrations and therefore they have been investigated for various use in mycoremediation, biofuel production, medical industries [5]. Fungi belonging to ascomycetes and basidiomycetes are generally white rot-fungi and have the potential to produce numerous enzymes of economic importance and hence white-rot fungi are extensively investigated to support the human lifestyle [29], however, these fungi also have adverse effects on our household wooden gadgets.

The wood and the leaf litter biomass have mostly lignocelluloses as the major components produced by the plant photosynthesis and represent the most abundant renewable resource in the soil [11]. Lignocellulose is consists of three types of polymers, cellulose, hemicellulose, and lignin which are strongly linked together by chemical bonds like non-covalent forces and by covalent cross-linkage, very small amount of lignocellulose produced by-product in agriculture or used in industries, and the remaining are considered as residue [11, 30]. Different microorganism degrades residue for the carbon as a source of energy, however, filamentous fungi evolved with the ability to degrade lignin to CO_2 and other compounds as discussed in the above paragraph [31]. Some other lignocelluloses degrading fungi like brown-rot and soft-rot fungi rapidly modify the lignin content and these fungi collectively play an important role in the carbon cycle [1, 11]. Apart from degrading lignin the white-rot fungi also degrade a variety of persistent environmental pollutants [30, 32, 33] like aromatic hydrocarbon, aliphatics hydrocarbon, cyanide compound, pesticides, fungicides, etc. This ability of white-rot fungi is only due to the strong oxidative activity and low substrate specificity of their ligninolytic enzyme (**Table 1**). As discussed above the white-rot fungi most commonly the macrofungi belongings to ascomycetes and basidiomycetes served as the integrated part in an ecosystem service as they produce various enzymatic mechanisms that degrade the wood and helps in maintaining the forest soil healths. Many fungi are the source of food for invertebrates in the forest, the organisms that feed on the fungi are termed as the fungivore. The ecological function of the fungi is strongly linked with the wood decay dynamics in the forest ecosystem [37], in **Table 1** the fungus listed with the enzymes are some most common enzymes produced by these fungi when come in contact with the woods and forest debris to degrade the recalcitrants forest wood and debris, the listed fungi in **Table 1** is mostly common in the forest ecosystem that acts as the decomposers.

4. Role of fungi in human nutrition

Fungi degrade different biomass and obtain nutrition for their growth and development. Fungi occur in diverse climates under different challenging conditions, and therefore they synthesized secondary metabolites to pace with the challenging threats of their life and the secondary metabolites have a broad range from antibiotic to mycotoxin. Fungi mainly use three pathways to synthesized metabolites: i. the mevalonic acid pathway (synthesize terpenoids, steroids, etc), ii. the shikimic acid pathway (synthesize aromatic amino acids, alkaloids, etc), and iii. The acetate pathway (synthesize polyketides, fatty acids, etc). Metabolites have beneficial effects on human health like antioxidants, antibiotics, immunity booster, potent anticancer agents, and reduce stress, etc.

Mushrooms (filamentous fungi) with fruiting bodies show a huge number of pharmacological aspects in human health. Macrofungus like *Ganoderma sp.* and *Cordyceps sp.* used in the traditional medicines, *Ganoderma* (Reishi) mushroom has also been commonly referred to as the "mushroom of immortality", "ten-thousand-year mushroom", "mushroom of spiritual potency", and "spirit plant by the Chinese monk [38, 39]. Generally, Mushrooms seemed to be used for food, medicine, poison, and in spiritual mushroom practices in religious rituals across the world since at least 5000 BC [40]. Gordon Wasson (father of modern Ethno mycology) supposed that the Soma plant used in religious ceremonies, over 4000 years ago, before the beginning of the Christian era, by the people who called themselves "Aryans" was a mushroom [41, 42]. The Vedic juice called "soma rasa" is said to bestow divine qualities on the soul of the consumer,

Macrofungi	Nutritional value-Protein(g/100 g)	Carbohydrate (g/100 g)	Medicinal value
Pleurotus	17–42	37–48	Anticancer, antioxidant,antitumor, antiviral, antibacterial, antidiabetic,antihypercholesterolemic, eye health, anti-arthritic, immunomodulatory, hepatoprotective,anti-obesity
Agaricus	56.3	37.5	Anticancer, antidiabetic, antihypercholesterolemic, immunomodulatory, hepatoprotective, antiviral, antimutagenic
Tricholoma	18.1–30.5	31.1–52.3	Antihypercholesterolemic, anti-aging
Lentinus	26.3	65.1	Anticancer, immunomodulatory
Hericium	22.3	57.0	Antihypercholesterolemic

Table 2.
Nutraceutical value of some edible mushrooms.

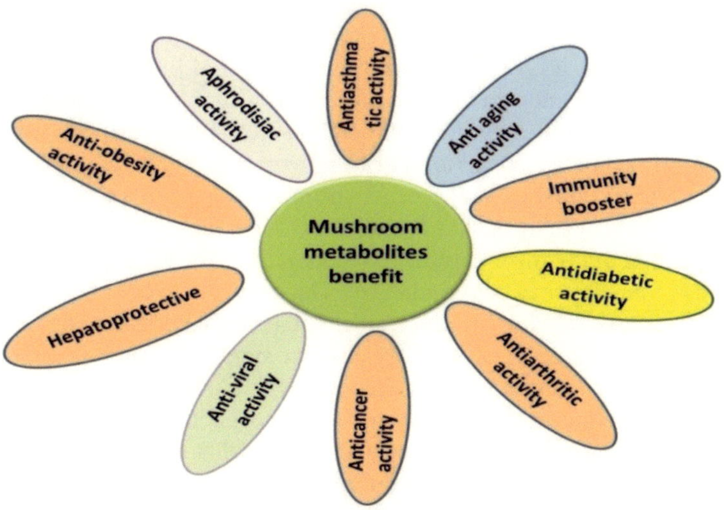

Figure 3.
Schematic illustration of the therapeutic application of secondary metabolites of mushrooms.

even immortality [40, 42]. Mushrooms are considered one of the delicious foods and are commonly produced worldwide. They have been an essential part of the human diet and are used as both food and medicine for centuries. As shown in **Table 2**, they are a rich source of nutrients and bioactive compounds such as carbohydrates, fibers, proteins, vitamins, minerals, and have enormous medicinal attributes such as antibacterial, antiviral, antioxidant, anticancerous, and hypocholesterolemic which are valuable for human health [43]. The mushrooms are rich in protein and carbohydrate content, whereas low in lipid content. They contain essential amino acids, which help in meeting the needs of amino acids in the human body (**Figure 3**). They are also rich in many essential unsaturated fatty acids, such as linoleic and oleic acids, which are necessary for the proper functioning of the body. Apart from this, they contain many essential minerals, which are responsible for the proper metabolism of many pathways. Mushrooms being achlorophyllous therefore, they grow on decayed organic matters, rich in lignin, cellulose, and other important carbohydrates. It is economical, rich in pharmacological properties, easy to cultivate, requires low resources and area, and can be grown all over the world. Nutritional, medicinal, bioremediation, and biodegradation aspects of mushrooms are increasing day by day and have got strong focused in investigation for their hidden metabolic compounds [44, 45].

5. Production of organic fertilizer and biofertilizer using fungi

The substances that fertilize the soil can be called fertilizers. Fertilizer is widely used to supply essential nutrients for plants to increase yield. Many types of fertilizers existed such as inorganic, organic, and biofertilizer, fertilizers that provide nutrients in inorganic forms are called mineral fertilizers, and those derived from plants or animal residue are considered organic fertilizers whereas biofertilizers are products containing living cells of different types of microorganisms that have the ability to mobilize nutritionally important elements from non-usable form to usable form through the biological process [46]. Biofertilizer is well known for its application in sustainable agricultural practices it is a cost-effective, eco-friendly, and renewable source for plant nutrition [10, 47], the ability of the fungi to restrict the growth of other microbes and have potential to control the disease of the plant is termed broadly as biocontrol agents, many fungi like *Trichoderma harzianum*, *Ampelomyces quisqualis*, *Chaetomium globosum*, *C. cupreum*, *Gliocladium virens*, *Coniothyrium minitans*, etc. are some common example of biocontrol agents and many of the micro and microfungi made the availability of the complex nutritional compounds into a simpler form which is used by plant easily such fungi acts as symbiotic agent. The symbiosis of fungi with the plants helps plants to fix nitrogen, solubilizes phosphate, and other complex compounds of the micro and macronutrient present in the soil as recalcitrants form. Based on the several abilities of beneficial fungi many formulations have been made for the application of the fungal-based biofertilizer in arable soil [48]. Details account of fungi in plant nutrition has been discussed below, production of fungal biofertilizer using mycorrhizal fungi is very selective since AM fungi are obligate symbiotic, they can not be grown without plant host on synthetic media, hence it is produced in association with the host plant. Mass production by pot culture is the most common method used in the production of AM based biofertilizer, no matter what method or formulation is used but for the success of the formulation depends on (a) economic viability of production (b) retention of the inoculum viability after formulation (c) handling and dispersal capacity during application [48, 49]. The formulations are availably available in the form of powder, pellets, gell beads [50]. There are several AM

Figure 4.
Gradual conversion of leaf litter into organic leaf compost, arrow indicates the gradual conversion of leaf into matured compost.

fungi formulations but the efficiency of the applications depends on the products, conditions of the environment, bulking agents, and other variables [3, 10, 47, 48]. However, to produce organic fertilizer the enzymatic system of fungi is used to convert biodegradable substances into compost, in nature, the fungi decompose the recalcitrants substrates into simple form and helps in nutrient recycling. These facts of fungi are used in the production of organic fertilizers, for example, leaf compost where the dead leaf convert into dark brown organic fertilizer see **Figure 4**.

Trichoderma viride is a filamentous fungus widely used as a biofertilizer as a biocontrol agent, this fungus nowadays has gained global market attention as a biofertilizer. *T.viride* acts as an antagonistic fungus, it is effective in controlling seed-borne pathogens as well as soil-borne pathogens. The working mechanism of *Trichoderma* as a biocontrol agent is either direct or indirect, *Trichoderma* restricted phytopathogens growth indirectly by competing for nutrients and space, by modifying the environmental conditions, by promoting plant growth and by plant defensive mechanism and antibiosis, and directly *Trichoderma* controls phytopathogens by a mechanism such as mycoparasitism [51].

5.1 Fungi in plant nutrition

The saprotrophic and AM fungi provide nutrition to the plants.AM fungi increase the growth and productivity of the plant by increasing nutrient uptake, mycorrhizae form mutualistic symbiotic relationships with plant roots of more than 80% of land plants including many important crops and forest tree species [52–54]. The two dominant types of mycorrhizae are ectomycorrhizae (ECM) and arbuscular mycorrhizae (AM) which can improve water and nutrient uptake and provide protection from pathogens but only a few families of plants can form functional associations with both AM and ECM fungi. However, AM fungi are most commonly found in the rhizosphere roots of a wide range of herbaceous and woody plants [49, 55]. ECM fungi help the growth and development of trees because the roots colonized with ectomycorrhiza can absorb and accumulate nitrogen, phosphorus, potassium, and calcium more rapidly and over a longer period than nonmycorrhizal roots [56]. ECM fungi help to break down the complex minerals and organic substances in the soil and transfer nutrients to the tree [54, 57]. AM fungi are a widespread group and are found from the arctic to the tropics and are present in most agricultural and natural ecosystems with different forms and structures see **Figure 5**.

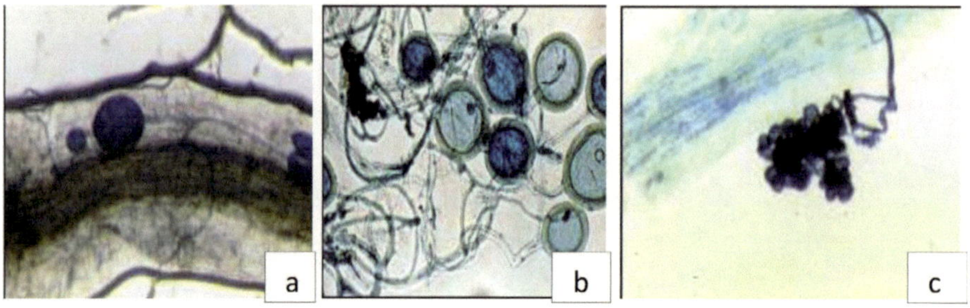

Figure 5.
Endomycorrhizae characterized by structures formed in the root cortex region (a) hyphae, vesicles, and spores in soybean roots (b) external spores (c) cluster of auxiliary cells outside the root.

Cultivars	Spent Substrate		literature
	SMC	SMS	
Amaranthus hybridus	+	—	[60]
Bush Beans, collards, squash, tomato	+	—	[61]
Salvia officinalis	+	—	[62]
Capsicum, Tomato, Cauliflower, Pea, Potato, Ginger, Garlic, Wheat, Paddy, Maize, and Apple.	+	+	[63]
Tomato, Garlic, Onion, Brinjal, Cauliflower, Wheat, Capsicum, maize, Pea, Spinach, Broccoli, Lettuce	+	—	[64]
Beetroot, Cucumber, vineyards, barley, marigold, green gram, sweet potato.	+	+	[65]

Table 3.
Showing application of SMS and SMC in the different horticultural cultivars (source Singh et al. [41]); the positive indicates 'use' and a negative sign indicates 'not known'.

Apart from the fungal biofertilizer and AM fungi, the spent left out after mushroom harvest are a good source of plant nutrition, the spent are commonly known as the Spent mushroom compost (SMC) or spent mushroom substrates (SMS) depending upon process and mushroom species of cultivation, mushroom compost has a major role in the integrated farm system [41, 58, 59]. The application of SMC and SMS have been evaluated on different cultivars which impart significant result in production and yield (**Table 3**).

SMC has organic matters that make them to be used in a large variety of crops including legumes crops such as *Vigna unguiculata* [66], chickpea, pea, soyabeans [63].

5.2 Wood wide web

The evolution of the plant from the water to land some 500 million years ago could not have possible without striking up a relationship with fungi, today almost all plants depend on symbiotic mycorrhizal fungi and some completely depend on the fungal partners and lost the ability of photosynthesis (eg- *Voyria tenella*) and therefore these type of plant obtain their food from neighboring photosynthetic plant through the shared fungal network, shared network of mycorrhizal develops due to the ability of both plant and mycorrhizal fungi to form a relationship with multiple partners [67]. The plants which depend on fungi for its whole life are called as mycoheterotrophs, most of the *orchid* lives as mycoheterotrophs when they are young and later starts photosynthesis as grew older which is known as 'take now, pay later'. A wide range of minerals is transported between plants through a shared network of mycorrhizal fungi. The notion that plants can 'talk' to one another is recently accepted in the mainstream of science, though the plant can not move but they pass their massages through chemical signal(volatile organic compounds) to another plant via a network of connection which is termed as the common mycorrhizal networks (CMN) formed by mycorrhizal fungi and different plants, and this is described as the below-ground internet network which is colloquially called as The Wood Wide Web [68]. The CMN integrate multiple plants and fungal species, that interact, provides feedbacks and adapt, and form complex adaptive social networks, the formation of these networks influenced by various factors, however, CMN provides communication facilities between plants principally by two forms of CMN- arbuscular and ectomycorrhizal, not only CMN provides connectivity among the plants but also provides facilities for the uptake of nutrition and distribution of minerals among the plants. CMN plays a crucial role in soil ecosystem management. It has been established that if a plant is attacked by the plant pathogens it sends the signal to the neighbor plants so that they can prepare themselves to fight against the pathogen by activating various defense mechanisms [68, 69]. Fungal hyphae as a network cable have far-reaching potential in the betterment of plant health systems, but not everything that is transmitted between plants is beneficial to individual plants because toxin (allelopathic chemical) may also be transported via a mycelial network. The whole system is very complex and holistic and research has emerged to understand this at the molecular level.

6. Conclusion

Biodegradation is a natural process executed by microbes. Among the microbes, fungi played vital role in the degradations of complex molecular substances into simpler ones. Here we have attempted to explore how fungi can solve the problem related to human and plant nutrition. Nutrition is a basic requirement of all living organisms. Fungi being heterotrophs act as scavengers utilize waste and bring nutritious food for humans as well as nutrients for the plants. Although many fungi are beneficial for humans and plants but there are notorious fungi which cause disease in human and plants that we have not touched upon.

Conflict of interest

The authors declare no conflict of interest.

Author details

Chandan Singh and Deepak Vyas*
Lab of Microbial Technology and Plant Pathology, Department of Botany, Dr. Harisingh Gour Vishwavidyalaya, Sagar, Madhya Pradesh, India

*Address all correspondence to: deepaknvyas64@gmail.com

References

[1] Joutey NT, Bahafid W, Sayel H, Ghachtouli N El. Biodegradation: Involved Microorganisms and Genetically Engineered Microorganisms. In: Chamy R, Rosenkranz F, editors. Biodegrad. - Life Sci., IntechOpen; 2013. http://dx.doi.org/10.5772/56194 291.

[2] Frac M, Hannula SE, Belka M, Jędryczka M. Fungal biodiversity and their role in soil health. Front Microbiol 2018;9:1-9. https://doi.org/10.3389/fmicb.2018.00707.

[3] Ellegaard-Jensen L. Fungal Degradation of Pesticides - Construction of Microbial Consortia for Bioremediation. University of Copenhagen, Denmark, 2012.

[4] Yuvaraj M, Ramasamy M. Role of Fungi in Agriculture. In: Mirmajlessi SM, Radhakrishnan R, editors. Biostimulants Plant Sci., IntechOpen; 2020. http://dx.doi.org/10.5772/intechopen.89718.

[5] Bentil JA, Thygesen A, Mensah M, Lange L, Meyer AS. Cellulase production by white-rot basidiomycetous fungi: Solid-state versus submerged cultivation. Appl Microbiol Biotechnol 2018;102:5827-5839. https://doi.org/10.1007/s00253-018-9072-8.

[6] Baldrian P. Fungal laccases-occurrence and properties. FEMS Microbiol Rev 2006;30:215-242. https://doi.org/10.1111/j.1574-4976.2005.00010.x.

[7] Vyas D, Mishra M, Shukla A. Mycorrhizospheric interactions in medicinal plants and phytoprotection against *Fusarium*. J Bot Soc Univ Saugor 2015;45.

[8] Singh M, Vyas D, Singh PK. Interaction of soil microbes with mycorrhizal fungi in tomato. Arch Phytopathol Plant Prot 2014;47:737-743. https://doi.org/10.1080/03235408.2013.820389.

[9] Singh PK, Vyas D. *Trichoderma* species: The history and evaluation of current concepts of biological control. J Bot Soc Univ Sagar 2016;46.

[10] Odoh CK, Eze CN, Obi CJ, Anyah F, Egbe K, Unah UV, et al. Fungal biofertilizers for sustainable agricultural productivity. In: Yadav AN, Mishra S, Kour D, Yadav N, Kumar a, editors. Agric. Important fungi sustain. Agric. Fungal biol., vol. 1, Springer Nature Switzerland AG 2020; 2020, p. 199-237. https://doi.org/10.1007/978-3-030-45971-0.

[11] Sánchez C. Lignocellulosic residues: Biodegradation and bioconversion by fungi. Biotechnol Adv 2009;27:185-194. https://doi.org/10.1016/j.biotechadv.2008.11.001.

[12] Philippoussis A, Zervakis G, Diamantopoulou P. Bioconversion of agricultural lignocellulosic wastes through the cultivation of the edible mushrooms *Agrocybe aegerita, Volvariella volvacea* and *Pleurotus spp.* World J Microbiol Biotechnol 2001;17:191-200. https://doi.org/10.1023/A:1016685530312.

[13] Chang S-T, Miles PG. Edible Mushrooms and Their Cultivation. First. Delhi, India: CBS Publisher and Distributors, 485, Jain Bhawan, Bhola Nath Nagar, Shahdara(Copyright CRC Press,Inc., of Boca Raton, Florida, USA.); 1993.

[14] Hoa HT, Wang CL, Wang CH. The effects of different substrates on the growth, yield, and nutritional composition of two oyster mushrooms (*Pleurotus ostreatus* and *Pleurotus cystidiosus*). Mycobiology

2015;43:423-434. https://doi.org/10.5941/MYCO.2015.43.4.423.

[15] Dehariya P, Vyas D. Evaluation of supplementation of *Daucus carota* on growth parameter and yield of *Pleurotus sajor-caju*. Int J Agric Food Sci Technol 2020;14:23-30.

[16] Jain AK, Vyas D. Cultivation of three *Pleurotus* species on different substrates. J Basic Appl Mycol 2003;2:88-89.

[17] Dehariya P, Vyas D. Effect of different agro waste substrates and their combinations in the yield and biological efficiency of *Pleurotus sajor caju*. ISOR J Pharm Biol Sci 2013;8:60-64.

[18] Vyas D, Chaubey A, Dehariya P. Biodiversity of mushrooms in Patharia forest of Sagar (M.P.)-III. Int J Biodivers Conserv 2014;6:600-607. https://doi.org/10.5897/IJBC2014.0681.

[19] Chaubey A, Dehariya P, Vyas D. Substrate availability and suitability for the growth and better yield of *Pleurotus djamor.* Int J Mushroom Res 2010;19: 36-39.

[20] Draganova D, Valcheva I, Kuzmanova Y, Naydenov M. Effect of wheat straw and cellulose degrading fungi of genus *Trichoderma* on soil respiration and cellulase, betaglucosidase and soil carbon content. Agric Sci Technol 2018;10:349-353. https://doi.org/10.15547/10.15547/ast.2018.04.064.

[21] Soliman SA, El-Zawahry YA, El-Mougith AA. Fungal Biodegradation of Agro-Industrial Waste. In: Ven T van de, Kadla J, editors. Cellul. – Biomass Convers., IntechOpen; 2013. http://dx.doi.org/10.5772/56464.

[22] Vyas A, Vyas D, Vyas K. Screening of extracellular cellulase producing fungi from different lignocellulosic wastes. J Basic Appl Mycol 2003;2:14-16.

[23] Akhtar N, Goyal D, Goyal A. Biodegradation of Cellulose and Agricultural Waste Material. Adv. Biodegrad. Bioremediation Ind. Waste, Taylor & Francis Group,LLC; 2015. https://doi.org/10.1201/b18218-9.

[24] Saritha M, Arora A, Lata. Biological pretreatment of lignocellulosic substrates for enhanced delignification and enzymatic digestibility. Indian J Microbiol 2012;52:122-130. https://doi.org/10.1007/s12088-011-0199-x.

[25] Ummalyma SB, Supriya RD, Sindhu R, Binod P, Nair RB, Pandey A, et al. Biological pretreatment of lignocellulosic biomass-current trends and future perspectives. In: Basile A, Dalena F, editors. Second Third Gener. Feed. Evol. Biofuels, Elsevier; 2019, p. 197-212. https://doi.org/10.1016/B978-0-12-815162-4.00007-0.

[26] Dashtban M, Schraft H, Syed TA, Qin W. Fungal biodegradation and enzymatic modification of lignin. Int J Biochem Mol Biol 2010;1:36-50.

[27] Chaurasia B. Biological pretreatment of lignocellulosic biomass (water hyacinth) with different fungus for enzymatic hydrolysis and bio-ethanol production resource: Advantages, future work and prospects. Acta Sci Agric 2019;3:89-96.

[28] Ortega GM, Martinez EO, Betancourt D, Gonzalez AE, Otero MA. Bioconversion of sugar cane crop residues with white-rot fungi *Pleurotus* sp. World J Microbiol Biotechnol 1992;8:402-405.

[29] Kim YS, Singh AP. Micromorphological characteristics of wood biodegradation in wet environments : A review. Int Assoc Wood Anat 2000;21:135-155. https://doi.org/10.1163/22941932-90000241.

[30] Chukwuma OB, Rafatullah M, Tajarudin HA, Ismail N.

Lignocellulolytic enzymes in biotechnological and industrial processes: A review. Sustainability 2020;12:1-31. https://doi.org/10.3390/su12187282.

[31] Rodríguez J. Lignin biodegradation by the ascomycete *Chrysonilia sitophila*. Appl Biochem Biotechnol - Part A Enzym Eng Biotechnol 1997;62:233-242. https://doi.org/10.1007/BF02787999.

[32] Jambon I, Thijs S, Weyens N, Vangronsveld J. Harnessing plant-bacteria-fungi interactions to improve plant growth and degradation of organic pollutants. J Plant Interact 2018;13:119-130. https://doi.org/10.1080/17429145.2018.1441450.

[33] Ceci A, Pinzari F, Russo F, Persiani AM, Gadd GM. Roles of saprotrophic fungi in biodegradation or transformation of organic and inorganic pollutants in co-contaminated sites. Appl Microbiol Biotechnol 2019;103:53-68. https://doi.org/10.1007/s00253-018-9451-1.

[34] Gulzar ABM, Vandana UK, Paul P, Mazumder PB. The Role of Mushrooms in Biodegradation and Decolorization of Dyes. In: Passari AK, Sánchez S, editors. An Introd. to Mushroom, IntechOpen; 2020. http://dx.doi.org/10.5772/intechopen.90737.

[35] Deshmukh R, Khardenavis AA, Purohit HJ. Diverse metabolic capacities of fungi for bioremediation. Indian J Microbiol 2016;56:247-264. https://doi.org/10.1007/s12088-016-0584-6.

[36] Jebapriya GR, Gnanadoss JJ. Bioremediation of textile dye using white rot fungi: A review. Int J Curr Res Rev 2013;05:5.

[37] Marcot BG. A review of the role of fungi in wood decay of Forest ecosystems. United States Dep Agric 2017:1-32.

[38] Singh C, Pathak P, Chaudhary N, Rathi A, Vyas D. *Ganoderma lucidum*: Cultivation and Production of Ganoderic and Lucidenic Acid. In: Dehariya P, editor. Recent Trends Mushroom Biol., Global books Organisation; 2021.

[39] Nahata A. *Ganoderma lucidum*: A potent medicinal mushroom with numerous health benefits. Pharm Anal Acta 2013;04. https://doi.org/10.4172/2153-2435.1000e159.

[40] Panda AK, Swain KC. Traditional uses and medicinal potential of *Cordyceps sinensis* of Sikkim. J Ayurveda Integr Med 2011;2:9-13. https://doi.org/10.4103/0975-9476.78183.

[41] Singh C, Pathak P, Chaudhary N, Rathi A, Dehariya P, Vyas D. Mushrooms and mushroom composts in integrated farm management. Res J Agric Sci 2020;11:1436-1443.

[42] Gordon WR. Divine mushroom of immortality. In: T.Frust P, editor. Flesh Gods, vol. 3, Praeger Publisher, New York; 1968, p. 3-4. https://doi.org/10.2307/1385058.

[43] Rathi A, Singh C, Pathak P, Chaudhary N, Vyas D. A Thematic Approach on *Cordyceps*. In: Dehariya P, editor. Recent Trends Mushroom Biol., Global books Organisation; 2021.

[44] Vaseem H, Singh VK, Singh MP. Heavy metal pollution due to coal washery effluent and its decontamination using a macrofungus, *Pleurotus ostreatus*. Ecotoxicol Environ Saf 2017;145:42-49. https://doi.org/10.1016/j.ecoenv.2017.07.001.

[45] Singh VK, Singh MP. Bioremediation of vegetable and agrowastes by *Pleurotus ostreatus*: A novel strategy to produce edible mushroom with enhanced yield and nutrition. Cell Mol Biol 2014;60:2-6. https://doi.org/10.14715/cmb/2014.60.5.2.

[46] Lakshman HC, Channabasava A. Biofertilizers and Biopesticides. First. Jaipur,India: Pointers Publishers; 2014.

[47] Pandey VC, Singh V. Exploring the Potential and Opportunities of Current Tools for Removal of Hazardous Materials From Environments. Elsevier Inc.; 2018. https://doi.org/10.1016/B978-0-12-813912-7.00020-X.

[48] Kaewchai S, Soytong K, Hyde KD. Mycofungicides and fungal biofertilizers. Fungal Divers 2009;38:25-50.

[49] Vyas D. Arbuscular Mycorrhizal Fungi: A Natural Symbiont. J Bot Soc Univ Sagar 2016;46.

[50] Vyas D, Singh M, Singh pradeep K. Arbuscular mycorrhizal fungi: The symbiotic bioengineers. J Bot Soc Univ Sagar 2015;45:2229-7170.

[51] Dehariya K, Sheikh IA, Dubey MK, Ahirwar S, Shukla A, Singh V, et al. Interactive effect of *Trichoderma* species with glomus intraradices in growth promotion and wilt disease suppression of *Cajanus cajan*. Int J Adv Res JournalwwwJournalijarCom Int J Adv Res 2013;1:867-873.

[52] Vyas D, Gupta RK. Effect of edaphic factors on the diversity of Vam fungi. Int J Res Biosci Agric Technol 2014;1:14-25. https://doi.org/10.29369/ijrbat.2014.02.ii.0091.

[53] Jha A, Vyas D, Kumar A, Shukla A, Salunkhe O. Soil moisture levels affect mycorrhization during early stages of development of agroforestry plants. Biol Fertil Soils 2012;49:545-554. https://doi.org/10.1007/s00374-012-0744-8.

[54] Nicolás C, Martin-Bertelsen T, Floudas D, Bentzer J, Smits M, Johansson T, et al. The soil organic matter decomposition mechanisms in ectomycorrhizal fungi are tuned for liberating soil organic nitrogen. ISME J 2019;13:977-988. https://doi.org/10.1038/s41396-018-0331-6.

[55] Shukla A, Kumar A, Jha A, Dhyani SK, Vyas D. Cumulative effects of tree-based intercropping on arbuscular mycorrhizal fungi. Biol Fertil Soils 2012;48:899-909. https://doi.org/10.1007/s00374-012-0682-5.

[56] Shukla A, Vyas D, Jha A. Soil depth: An overriding factor for distribution of arbuscular mycorrhizal fungi. J Soil Sci Plant Nutr 2013;13:0-0. https://doi.org/10.4067/s0718-95162013005000003.

[57] Giovannetti M, Avio L, Fortuna P, Pellegrino E, Sbrana C, Strani P. At the root of the wood wide web: Self recognition and non-self incompatibility in mycorrhizal networks. Plant Signal Behav 2006;1:1-5. https://doi.org/10.4161/psb.1.1.2277.

[58] Pathak P, Singh C, Chaudhary N, Rathi A, Vyas D. Fertilizing with spent mushroom compost. In: Dehariya P, editor. Recent Trends Mushroom Biol. 1st ed., Delhi: Global books Organisation; 2021, p. 175-186.

[59] Pathak P, Singh C, Chaudhary N, Vyas D. Application of biochar, leaf compost, and spent mushroom compost for tomato growth in alternative to chemical fertilizer. Res J Agric Sci 2020;11:1362-1366,.

[60] Jonathan SG, Oyetunji OJ, Asemoloye MA. Supplementation of spent mushroom compost (SMC) of *Pleurotus ostreatus* (Jackuin Ex. Fr.) Kummer as a soil amendment for the growth of *Amaranthus hybridus Lin*. A Nigerian green vegetable. Biotechnol An Indian J 2012. https://doi.org/BTAIJ, 6(12), 2012 [396-403].

[61] Stephens JM, Bennett DL. Mushroom compost AS a soil amendment for vegetable gardens. Proc Fla State Hort Soc 1989;102:108-111.

[62] Castro RL. Spent oyster mushroom substrate in a mix with organic soil for plant pot cultivation. Mcologia Apl Int 2008;20:17-26.

[63] Sagar MP, Ahlawat OP, Raj D, Vijay B, Indurani C. Indigenous technical knowledge about the use of spent mushroom substrate. Indian J Tradit Knowl 2009;8:242-248.

[64] Ahlawat OP, Sagar MP. Management of Spent Mushroom Substrate.National Research Centre for Mushroom (ICAR) Chambaghat, Solan. vol. 213. 2007.

[65] Rinker DL. Spent Mushroom Substrate Uses. In: Pardo-Gimenez DC and A, editor. Edible Med. Mushrooms. First edit, John Wiley & Sons Ltd; 2017, p. 427-54. https://doi.org/10.1002/9781119149446.ch20.

[66] Cowpea LW, Prabu M, Jeyanthi C, Kumuthakalavalli R. Spent mushroom substrate : An enriched organic manure for improving the spent mushroom substrate : An enriched organic manure for improving the yield of *Vigna unguiculata* [L] Walp (Cowpea) leguminous crop. Scrut Int Res J Agric Plant Biotechnol Bio Prod 2014;1.

[67] Sheldrake M. Hackers of the wood wide web: A visual guide. In: Aloi G, editor. Antennae, Antennae: The Journal of Nature in Visual Culture; 2020.

[68] Rhodes CJ. The whispering world of plants: 'The wood wide web.' Sci Prog 2017;100:331-337. https://doi.org/10.3184/003685017X14968299580423.

[69] Beiler KJ, Durall DM, Simard SW, Maxwell SA, Kretzer AM. Architecture of the wood-wide web: *Rhizopogon spp.* genets link multiple Douglas-fir cohorts. New Phytol 2010;185:543-553. https://doi.org/10.1111/j.1469-8137.2009.03069.x.

Chapter 7

Aquatic Plants as Bioremediators in Pollution Abatement of Heavy Metals

Aaltharackal Vikraman Meera, Manorama Thampatti KC, Jacob John, Bhadra Sudha and Abdulmajeed Sajeena

Abstract

Over use of chemical inputs and exploitation of natural resources have degraded our ecosystem to a large extent. Our water bodies are drastically affected, especially due to the impact of heavy metal loading. The biomagnification that results from these difficult to degrade metals is naturally affecting the human health. The physical and chemical methods commonly employed for water purification are not only highly expensive but also further aggravate the pollution problem. Hence, all efforts must be taken to exploit the emerging green technology approach in pollution remediation. Several aquatic plants have specific affinity towards heavy metals and they flourish well in this contaminated environment. The common mechanisms of phytoremediation and varied type of aquatic plants with high remediation potential are reviewed in this chapter.

Keywords: pollution, aquatic macrophytes, phytoremediation, hyperaccumulators

1. Introduction

Industrialization, urbanization and over exploitation of precious natural resources have resulted in much degradation of our environment. The dire need for promotion of intensive cultivation to satisfy primary human needs led to over dependence on chemical resources. This in turn, caused much degradation to our ecosystem mainly through environmental pollution. Among the natural resources, the worst affected are water resources. 97% of hydrosphere is covered by saltwater, leaving only mere 3% fresh water, of which hardly 1.5% is available for ready use [1]. The entire world is relying on this meager resource for daily consumption, irrigation, industrial purposes, power and other diverse uses. Injudicious human activities including disposal of sewage and wastes have caused great impact on water bodies all over the world. Wetlands act as sink for contaminants and thereby reduce the impact of point and non-point sources of pollution [2]. But drastic reduction in water inflow has been resulted due to fragmentation of water bodies and irreversible conversion to satisfy human needs.

Heavy metal pollution in water bodies is a serious environmental problem, threatening not only the aquatic ecosystems, but also human health. Over the years, the main sources of metal pollution have shifted from mining and manufacturing to rock weathering and waste discharge [3]. There are several reports on the deleterious effects of biomagnification of heavy metals within aquatic organisms and its impact on human nervous, reproductive and cardio vascular systems [4]. Disposal of plastic wastes, batteries, fertilizer materials, untreated industrial effluents etc. releases heavy metals including Cd into the aquatic environment which causes several causalities like osteoporosis, kidney failure, infertility and improper brain development. Globally, majority of surface water bodies are highly polluted with heavy metals like As, Co, Cr and Ni, with levels exceeding WHO and USEPA guideines and have evoked much concern among the government agencies and public [5].

As heavy metals are non-biodegradable, removal of these metals from the aquatic system is the only remedy available for decontamination [6]. The conventional methods usually employed to remove the metals from a polluted system like coagulation, flocculation, osmosis, stabilization etc. are highly expensive. In addition, they further aggravate deterioration with the release of chemicals being used and hence these methods are not at all environmentally safe [7, 8]. But, a new method of decontamination employing green plants is fast emerging, referred as phytoremediation, which is specifically suited for wetland restoration. The plants growing in the contaminated areas will absorb the elements from the sediment/soil/water by roots. The absorbed elements travel from root through cell sap and finally get precipitated in vacuole or cell membrane, thereby reduces the level of contaminants in sediment/soil/water [9]. Such aquatic plant species and adsorbents can be included in land management plans to reduce human risks. This method is relatively cheap and very successful over other methods [10].

2. Phytoremediation: a bio-decontamination approach

The concept of extraction of metals by macrophytes was actually given by Chaney [11]. Efficiency of macrophytes to extract metals from contaminated site depends on the metal hyperaccumulation capacity and biomass production. The selection of particular plant species for phytoremediation depends on the following characteristics:

i. native to the particular ecosystem.

ii. well flourishing nature and high biomass yield.

iii. ability to uptake large amount of metals.

iv. transportation of metals to aboveground plant portion.

v. mechanism to tolerate metal toxicity.

In addition, factors like pH, light intensity and nutrient availability influences the plant growth and thus, phytoremediation potential [12–16]. Agronomic practices for soil and crop management and improved genetic engineering technologies to enhance metal tolerance and translocation can affect the remediation mechanism. Exsituas well asinsitu methods of phytoremediation are there:*exsitu method* involves excavation of contaminated soil followed by its treatment and also shifting the soil for land filling;*insitu method* is less laborious and more cost effective and commonly employs mechanisms like phytoextraction and phytostabilization [17].

3. Mechanisms of phytoremediation

Depending upon the process by which plants/microbes are removing or reducing the toxic effect of contaminants from the soil and water, phytoremediation technology can be broadly classified as follows:

a. Phytoextraction or phytoaccumulation –This refers to the uptake and translocation of metal contaminants in the soil by plant roots with subsequent transport to the aerial plant organs. Certain plants called hyperaccumulators absorb unusually large amounts of metals in comparison to other plants and concentrate them in the aerial portions [11, 18–20].

b. Phytosequestration–The phytochemicals that are released into the rhizosphere may form complex association with the contaminants, sequestering them in the root zone and thus reducing their mobility This prevents further transport to soil, water and air. The complexation can also occur with the aid of transport proteins on root surface or through sequestration in the vacuoles of root cells [21].

c. Rhizofiltration - It is the adsorption or precipitation of contaminants onto plant roots or absorption into the roots that are in solution surrounding the root zone. The acclimatized plants against contamination are planted in the contaminated area and the roots extract the contaminants along with water. As the roots become saturated with contaminants, they are harvested and incinerated [22–25].

d. Phytodegradation or phytotransformation– Here, organic pollutants are converted by internal or secreted enzymes into compounds with reduced toxicity. The metabolic processes, with the aid of enzymes within the plant or secreted externally, result in the degradation of pollutants and may be incorporated into the plant tissues or used as nutrients [20, 26, 27].

e. Rhizodegradation–Microbial activity in the rhizosphere results in the breakdown of contaminants, leading to their phytoremediation. Compared to phytodegradation it is a much slower process. Microflora (yeast, fungi, or bacteria) utilize the organic substrates for nutrition and energy [28, 29].

f. Phytostabilization–The particular plant species involved helps in the immobilization of contaminants through absorption and accumulation by roots, adsorption onto roots, or precipitation within the root zone. This results in reduction in mobility of contaminants and migration to ground water or air is blocked, which in turn hinders their bioavailability [30, 31].

g. Phytovolatalization –It is the uptake and transpiration of contaminant by a plant, with the release of that contaminant or its modified form to the atmosphere. In this process, the soluble contaminants are taken up along with water by the roots, transported to the leaves, and volatized into the atmosphere through the stomata. For *eg*., volatilization of mercury (Hg) by conversion to the elemental form in transgenic *Arabidopsis* and yellow poplars containing modified bacterial mercuric reductase (*merA*) [32–34].

Among the different methods of phytoremediation, phytoextraction by hyperaccumulators is the most efficient one as it helps in removal of the phytoextracted biomass from contaminated sites. But phytoremediation cannot be used as a primary treatment method for highly contaminated areas with heavy metals like Cd, Zn, Cr and Pb, because of the prolonged time taken for the complete clean up. The dominant families that include hyperaccumulators are Asteraceae, Brassicaceae, Caryophyllaceae, Cyperaceae, Cunouniaceae, Fabaceae, Flacourtiaceae, Lamiaceae, Poaceae, Violaceae, and Euphobiaceae. Brassicaceae has the largest number of taxa *viz*. 11 genera and 87 species. *Thlaspi* species are known to hyperaccumulate more than one metal *viz.,T. caerulescence* - Cd, Ni, Pb, and Zn; *T. goesingense* - Ni and Zn and *T. ochroleucum* - Ni and Zn and *T. rotundifolium* - Ni, Pb and Zn. Aquatic plants in freshwater, marine and estuarine systems act as receptacle for several metals. Several aquatic macrophyteslike*Eichhorneacrassipes, Hydrillaverticillata, Typhaangustata,* etc. can remove Zn, Cu, Pb, Ni and Cd from lakes and maintain water quality.

4. Phytoremediation by aquatic macrophytes

Aquatic macrophytes constitute a group of taxonomically diverse macroscopic plants whose life cycle takes place completely or periodically in the aquatic environment. They play a dominant role in maintaining the ecosystem biodiversity, represented by 33 orders and 88 families, numbering about 2614 species in 412 plant genera. The wide adaptation in their growing habits help them to classify as emergent, floating-leaved, free-floating, submerged and marginal plants [35, 36].

i. *Emergent*macrophytes: They grow in shallow littoral waters and form aerial leaves, suited for life in environments where the soil is saturated with water (wetlands, marshes, swamps, flooded areas), and their root and rhizome systems are often adapted for constantly anaerobic sediments, rooted in the lake bottom, but their leaves and stems extend out of water.

ii. eg. *Phragmitesaustralis, Typhaaugustifolia, Limnocharisflava.*

iii. *Floating-leaved*macrophytes: Their roots are attached to the ground and possess floating or aerial reproductive organs eg. *Nymphaea* sp., *Nupharlutea*, *Potamogetonnatans.*

iv. *Free floating*macrophytes:They float on the surface of pond with roots hanging in water and possess well developed root system or very short roots. The reproductive organs of these plants are floating and aerial. Eg. *L. minor*, *Eich horneacrassipes,Salviniamolesta.*

v. *Submerged*macrophytes: Such plants complete their life cycle fully under the water surface. Some are rooted plants with most of their vegetative portion below the water surface. eg. *Vallisneria* sp., *Myriophyllum* sp.

vi. *Marginal*macrophytes:They grow around the margins where the water is shallow. Eg. *Rhizophora*sp., *Cyperus* sp.

In the given **Table 1**, some common aquatic macrophytesand their specificity for particular elements are detailed.

These macrophytes have the ability to concentrate metals both in the root and aerial parts, without causing any toxic symptoms on plant growth. In general,

Macrophyte group	Plant species	Heavy/toxic metal	References
Emergent	*L. flava*	Pb	[37]
		Pb, Cd	[38–40]
		Fe	[41]
	Typha sp.	Al	[42, 43]
	R. fluitans	Pb, Mn and Zn	[44]
	Scirpus sp	Pb	[45, 46]
	C. esculenta	Pb, Cd	[40, 47]
Floating-leaved	*N. nucifera*	Cd	[48, 49]
	Nymphaea sp.	Pb, Cd	[50, 51]
Free floating	*Eichhornea crassipes*	Al,Pb, Cd, Fe, S	[49, 52–61]
	P. stratiotes	Al, Fe	[52, 62]
	Salvinia polyrrhiza	Fe	[53, 62]
	Azollapinnata	Cd	[63]
Submerged	*C. demersum*	Pb	[64]
	Potamogeton scrispus	Pb, Cd	[65]
	V. spiralis	Al, Fe, Si, Mn, Pb	[66, 67]
	H. verticillata	Al, Fe, Si, Mn	[66]
	A. pinnata	Al, Fe, Si, Mn	[66]
	R. rotundifolia	Pb	[46]
	Myriophyllum intermedium	Pb	[46]
Marginal	*Cynadon*sp.	Al, Pb, Cd, Fe	[68–72]
	Commolina bengalensis	Fe, Al	[68]
	A. philoxeroides	Pb	[56, 73]
	S. trilobata	Pb	[40, 74]

Table 1.
Common aquatic macrophytes and their phytoremediation potential.

the submerged and floating macrophytes have the potential to accumulate more metals than emergent ones. Rhizofiltration offers much scope in the purification of heavily contaminated precious water resources, a big boon for eco restoration of aquatic systems.

5. Indices to estimate hyperaccumulation potential

The hyperaccumulation potential of macrophytesare determined primarily based on two indices *viz.*,bio concentration factor (BCF) and translocation factor (TF). BCF is defined as the ability of a plant to accumulate a particular metal in its plant part with respect to its concentration in the soil substrate while TF is the ratio of metal concentration in shoot to that in the root. BCF more than oneindicates that the plant is an accumulator while less than one, means the plant is an excluder. Hyperaccumulators are plants that contain more than 10,000 mg kg^{-1} of Zn and Mn; 1000 $mgkg^{-1}$ of Cu, Cr, Pb, Ni, Co and 100 mg kg^{-1} of Cd and other rare metals, in the dry matter [75].

A high value for TF indicate the efficiency of the plant to translocate metals from the root to shoot and such plants (TF > 1) are referred as hyperaccumulators. They possess the phytoextraction ability to remove contaminants from the growth medium to the above ground portions and the biomass can be uprooted and removed. Aquaticmacrophytes, especially floating macrophytes, have the potential to concentrate metals more in the roots. Based on BCF and TF, the hyperaccumulation potential of *E. crassipes*and *A. philorexoides*for Cd has been proved beyond doubt, whereas higher BCF and lower TF is an indication of phytostabilisation effect eg. *L. flava* and *C. dactylon.*

6. Mechanisms of heavy metal tolerance by macrophytes

Accumulation of heavy metals inside the plant body results in certain physiological changes and synthesis of certain enzymes to tolerate the metal stress. Major changes that occur inside the plant cell to activate metal absorption include enhancement in the bioavailability of metal in the rhizosphere region leading to an increased uptake of metal towards the plasma membrane. Inside the cell wall, chelation of metal may occur by binding with various proteins like phytochelatin or, metallothionein or form a bond with the cell wall or get sequestered into the cell vacuole [76, 77].

Acidification of rhizosphere by the action of plasma membrane proton pumps and secretion of ligands capable of chelating the metal helps in desorption of metals from the soil matrix. Soluble metals can enter into the root symplast by crossing the plasma membrane of the root endodermal cells or they can enter the root apoplast through the space between cells. Excluder plants survive by enhancing specificity for the essential element or pumping the toxic metal back out of the plant. On reaching the xylem, the metal will get transported alongwith xylem sap towards the leaves and get deposited there. The cell tissue where the metal get deposited, vary with the hyperaccumulator species as shown by *T. caerulescens* and *Arabidopses halleri - T. caerulescens* has preferential adsorption for Zn in the epidermis over mesophyll cells while the reverse for *Arabidopses halleri* [78].

At any point along the pathway, the metal could be converted to a less toxic form by chemical conversion or complexation. Various oxidation states of toxic elements have very different uptake, transport, and sequestration or toxicity characteristics in plants. Two major chelating peptides present in plants include metallothioneins and phytochelatins. Sequestration of metals in sites away from where the cellular processes are likely to be get disrupted will result in their deposition. The most prominent site is cell vacuole, for that metal or metal- ligand complex must cross the vacuolar membrane. Metal ions may also get bonded with negative charges on cell wall leading to their sequestration in the cell wall.

7. Conclusions

It is high time that the water bodies be conserved for ecological sustenance and well-being of the future generation. Aquatic plants can play a vital role in the purification of contaminated lakes, rivers and ponds, which make them fit for human consumption and irrigation purposes. The nature and extent of amelioration varies with particular plant species. They are specifically adapted to tolerate heavy/ toxic metal concentration in their ecosystems.

Author details

Aaltharackal Vikraman Meera[1*], Manorama Thampatti KC[2], Jacob John[1], Bhadra Sudha[1] and Abdulmajeed Sajeena[1]

1 Integrated Farming System Research Station, Kerala Agricultural University, Karamana, India

2 Department of Soil Science and Agricultural Chemistry, College of Agriculture, Kerala Agricultural University, Vellayani, India

*Address all correspondence to: meera.av@kau.in

References

[1] https://earthdata.nasa.gov/learn/toolkits/freshwater-availability

[2] Bystorm O, Andersson H, and Gren I. Economic criteria for using wetlands as nitrogen sinks under uncertainty. Ecol. Econ. 2000;35 (1): 35-45

[3] Zhou Q, Yang N, Li, Y, Ren B, Ding X, Bian H, Yao X. Total concentrations and sources of heavy metal pollution in global river and lake water bodies from 1972 to 2017. Global Ecol. Conservation. 2020;22: 1-11.

[4] Mishra S, Srivastava S, Tripathi RD, Kumar R, Seth CS, Gupta DK. Lead detoxification by coontail (Ceratophylumdemersum L.) involves indication of phytochelatins and antioxidant systemin response to its accumulation. Chemosphere 2006;65 (6): 1027-1039.

[5] Kumar V, Parihar RD, Sharma A, Bakshi P, Sidhu GPS, Bali AS, Karaouzas I, Bhardwaj R, and Thukral AK. Global evaluation of heavy metal content in surface water bodies: A meta-analysis using heavy metal pollution indices and multivariate statistical analyses. 2019.124364,ISSN 0045-6535. https://doi.org/10.1016/j.chemosphere.2019.124364.

[6] Cheng S, Grosse W, Karrenbrock F, Thoennessen M. Efficiency of constructed wetlands in decontamination of water polluted by heavy metals. Ecol. Eng. 2002;8:317-325

[7] Rai PK. Heavy metal phytoremediation from aquatic ecosystems with special reference to macrophyte. Crit. Rev. Environ. Sci. Technol. 2009; 39: 118-125.

[8] Tangahu, BV, Abdullah SRS, Basri H, Idris M, Anuar N, Mukhlisin M. A Review on Heavy Metals (As, Pb, and Hg) Uptake by Plants through Phytoremediation. *Int. J. Chem. Eng.* 2011;20(11): 31-41.

[9] Cunningham,S.D. and Ow, D.W. Promises and prospects of phytoremediation. *Plant Physiol.* 1996;110: 715-719

[10] Yan A, Wang Y, Tan SN, MohdYusof ML, Ghosh S and Chen Z . Phytoremediation: A Promising Approach for Revegetation of Heavy Metal-Polluted Land. Front. Plant Sci. 2020;11:359. doi: 10.3389/fpls.2020.00359.

[11] Chaney, R.L. Plant uptake of inorganic waste constituents. In: Parr, J. F., Marsh, P. B. and Kla, J. M. (eds), Land Treatment of Hazardous Wastes. Noyes Data Corp., Park Ridge, New Jersey, 1983. pp 50-76.

[12] Ali H, Khan E, Sajad MA. Phytoremediation of heavy metals—Concepts and applications. Chemosphere. 2013;91: 869-881.

[13] Liu J, Wen Z, Penghang QU, Mingxin and Wang. Cadmium tolerance and accumulation in fifteen wetland plant species from cadmium-polluted water in constructed wetlands. *Front. Environ. Sci. Eng.* 2014: 1 -7.

[14] Zhang X, Zhang X, Gao B, Li Z, Xia H, Li H, Li J. Effect of cadmium on growth, photosynthesis, mineral nutrition and metal accumulation of an energy crop, king grass (*Pennisetumamericanum* and *P. purpureum*). *Biomass Bioenergy* 2014;67: 179-187.

[15] Reeves RD, Baker AJ, Ja_ré, T, Erskine PD, Echevarria G, van der Ent A. A global database for plantsthathyperaccumulate metal and metalloid trace elements. New Phytol. 2018.218: 407-411.

[16] Neina D. The role of soil pH in plant nutrition and soil remediation. *Applied Environ. Soil Sci.* 2019 Article ID 5794869 https://doi.org/10.1155/2019/5794869

[17] Sheoran V, Sheoran, A. Poonia, P. Factors affecting Phytoextraction: A Review. Pedosphere. 2016; 26: 148-166.

[18] Baker AJM, McGrath SP, Sidoli CMD, Reeves RD. The possibility of in-situ heavy-metal decontaminationof polluted soils using crops of metal-accumulating plants. *Res. Conserve. Recycl.* 1994;11: 41-49.

[19] Brooks RR, Chambers MF, Nicks LJ, Robinson BH. Phytomining. *Trends Plant Sci.* 1998;1: 359-362.

[20] Salt DE, Smith RD, Ruskin I. Phytoremediation. *Annu. Rev. Plant Physiol. Plant Mol. Biol.* 1998;49: 643-668.

[21] Cunningham SD, Shann JR, Crowley DE, Anderson TA. Phytoremediation of contaminated water and soil. In: Kruger EL, Anderson TA, Coats JR. (eds), *Phytoremediation of Soil and Water Contaminants.* American Chemical Society, Washington, D.C. pp. 1997;133-151.

[22] Dushenkov V, Kumar PBAN, Motto H, Raskin I. 1995. Rhizofiltration: the use of plants to remove heavy metals from aqueous streams. *Environ. Sci. Technol.* 29:1239-1245

[23] Zhu YL, Zayed AM, Quian JH, DeSouza M, Terry N. Phytoaccumulation of trace elements by wetland plants: II. water hyacinth. J. Environ. Qual. 1999;28: 339-344.

[24] Raskin I, Ensley BD. *Phytoremediation of Toxic Metals: Using Plants to Clean Up the Environment* John Wiley and Sons,Inc., New York, 2000;pp 53-70.

[25] Gardea-Torresdey JL, de la Rosa G, Peralta-Videa JR. X. Use of phytofiltration technologies in the removal of heavy metals: a review. *Pure Appl. Chem.* 2004;76(4): 801-803.

[26] Black H. Absorbing possibilities: Phytoremediation. *Environ. Health Prespect* 1995; 103(12): 1106-1108.

[27] Suresh B, Ravishankar G. Phytoremediation - A novel and promising approach for environmental clean-up. *Crit. Rev. Biotech.* 2004;24: 97-124.

[28] Kuiper I, Lagendijk EL, Bloemberg GV, Lugtenberg BJJ. Rhizoremediation: a beneficial plant-microbe interaction. *Mol. Plant-Microbe Interact.* 2004;17: 6-15.

[29] Yadav SK. Heavy metals toxicity in plants: an overview on the role of glutathione and phytochelatins in heavy metal stress tolerance of plants. S. *Afr. J. Bot.* 2010;76: 167-179.

[30] Berti WR, Cunningham SD Phytostabilization of metals. In: Raskin, I. and Ensley, B. (eds), *Phytoremediation of Toxic Metals: Using Plants to Clean Up the Environment*. Wiley Interscience, New York, 2000;pp. 71-88.

[31] Stoltz E, Greger M. Accumulation properties of As, Cd, Cu, Pb and Zn by four wetland plant species growing on submerged mine tailings. *Environ. Exp. Bot.* 2002;47(3), 271-280.

[32] Heaton ACP, Rugh CL, Wang N, Meagher RB. Phytoremediation of mercury- and methyl mercury polluted soils using genetically engineered plants. *J. Soil Contam.* 1998;74: 497-510.

[33] Rugh CL, Senecoff JF, Meagher RB, Merkle SA. Development of transgenic yellow poplar for mercury phytoremediation. *Nat. Biotechnol.* 1998;16: 925-928.

[34] Dushenkov D. Trends in phytoremediation of radionuclides. *Plant and Soil* 2003.;249: 167-175.

[35] Chambers PA, Lacoul P, Murphy KJ, Thomaz SM. Global diversity of aquatic macrophytes in freshwater. Hydrobiologia, 2008; 198: 9-26.

[36] Lesiv MS, Polishchuk AI, Antonyak HL. Aquatic macrophytes: Ecological features and functions. Biol. Stud. 2020;14(2): 79-94.

[37] RachmadiartiF, Soehono LA, Utomo WH, Yanuwiyadi B, Fallowfield H. Resistance of yellow velvetleaf (*Limnocharisflava* (L.) Buch.) exposed to lead. *J. Appl. Environ. Biol. Sci.* 2012;2(6): 210-215.

[38] Bindhu T, Sumi MM, Ramasamy, EV. Decontamination of water polluted by heavy metals with taro (*Colocasiaesculenta*) cultured in a hydroponic NFT system. *Environ.* 2010;30: 35-44.

[39] Anning AK, Percy E, Korsah, Addo-Fordjour P. Phytoremediation of waste water with *Limnocharisflava, Thaliageniculata* and *Typhalatifolia* in constructed wetlands. *Int. J. Phytoremediation* 2013;15(5): 452-464.

[40] Meera AV. Phytoremediation of inorganic contaminants in Vellayani wetland ecosystem. Ph.d Thesis, Kerala Agricultural University; 2017

[41] Kamrudzaman AN, Zakariaa MH, Aziza RA, Faizal MA, Jalil. Removal of iron (Fe) from landfill leachate using horizontal and vertical subsurface flow constructed wetland system planted with *Limnocharisflava*. *Int. J . Chem. Environ. Eng.* 2012;3(2): 15 – 20.

[42] Hegazy AK, Abdel-Ghani NT, El-Chaghaby A. Phytoremediation of industrial wastewater potentiality by *Typhadomingensis. Int. J. Environ. Sci. Tech*. 2011;8 (3): 639-648.

[43] Kumari A, Lala B, Rai UN. Assessment of native plant species for phytoremediation of heavy metals growing in the vicinity of NTPC sites, Kahalgaon, India. *Int. J. Phytoremediation* 2016;18 (6): 592-597.

[44] Othman R, Hanifah NA, Ramya R, BtMohdHatta FA, Sulaiman WS, Yaman M, Baharuddin ZMB. Sequestration rate of heavy metal contaminants using *Ricciafluitans* as potential phytoremediation agent in polluted aquatic ecosystem. Int. J. Sustain. Energy Environ. Res. 2014;3(4): 185-192.

[45] Tangahu BV, Rozaimah SA, Basri SRS, Idris HM, Anuar N, Mukhlisin M. Phytotoxicity of wastewater containing lead (Pb) effects *Scirpusgrossus. Int. J. Phytoremediation*. 2013; 15(8): 814-826.

[46] Marbaniang D, Chaturvedi SS. A study on lead uptake and phytoremediation potential of three aquatic macrophytes of Meghalaya, India. Int. J. Sci. Res. Publ. 2014;4(6): 224-241.

[47] Madera-Parra CA, Peña-Salamanca EJ, Peña MR, Rousseau DPL, Lens PNL. Phytoremediation of landfill leachate with *Colocasiaesculenta*, *Gynerumsagittatum* and *Heliconiapsittacorum* in constructed wetlands. *Int. J. Phytoremediation*. 2015;17 (1-6): 16-24.

[48] Mishra V, Pathak V, Tripathi B. Accumulation of cadmium and copper from aqueous solutions using Indian lotus (*Nelumbonucifera*). *AmbioJ. Human Environ*. 2009;38(2): 110-115.

[49] Kamal K. Evaluation of aquatic pollution and identification of phytoremediators in Vellayani lake. M.Sc.(Ag) thesis, Kerala Agricultural University, Thrissur; 2011.

[50] Schor-Fumbarov T, Keilin Z, Tel-Or E. Characterization of cadmium uptake by the water lily *Nymphaea aurora*. *Int. J. Phytoremediation* 2003;5(2): 169-179.

[51] Shuaibu UOA, Nasiru A. Phytoremediation of trace metals in Shadawanka stream of Bauchi Metropolis, Nigeria. *Universal J. Environ. Res. Technol.* 2011;1(2): 176 – 181.

[52] Klumpp A, Bauer K, Franz-Gerstein C, Max de Menezes. Variation of nutrient and metal concentrations in aquatic macrophytes along the Rio Cachoeira in Bahia (Brazil). *Environ. Int.* 2002;28: 165– 171.

[53] Mishra VK, Tripathi BD. Concurrent removal and accumulation of heavy metals by the three aquatic macrophytes. *Biores. Technol*. 2008;99:7091-7097.

[54] KAU [Kerala Agricultural University]. Bioremediation of inorganic contaminants of rice based wetland ecosystems of Kuttanad, Kerala. *Final Report of ICAR Adhoc Project.* Kerala Agricultural University, Thrissur, Kerala; 2009.

[55] Narain S, Ojha CSP, Mishra SK, Chaube UC, Sharma PK. Cadmium and chromium removal by aquatic plant. *Int. J. Environ. Sci*. 2011;1(6): 1297-1304.

[56] Nan H, Zheng J, Ding D, Li G, Yin J, Chen X, Yu J. Screening of native hyperaccumulators at the Huayuan river contaminated by heavy metals. *Bioremediation. J*. 2013;17(1): 21-29.

[57] Sukumaran D. Phytoremediation of heavy metals from industrial effluent using constructed wetland technology. *Appl. Ecol. Environ. Sci*. 2013;1 (5): 92-97.

[58] Khankhane PJ, Sushilkumar, H.S. Bisen. Heavy metal extracting potential of common aquatic weeds. *Ind. J. Weed Sci*. 2014;46(4): 361-363.

[59] Das S, Goswami S, Talukda, AD. Physiological responses of water hyacinth, *Eichhorniacrassipes* (Mart.) Solms, to cadmium and its phytoremediation potential. *Turkish. J. Biol.* 2016;40: 84-94.

[60] Meera AV, Thampatti MKC. *Nelumbonucifera* as an ideal macrophyte for phytoremediation of toxic metals in contaminated wetland. *Adv. Life Sci*. 2016;5(9): 3562 -3565.

[61] Thampatti KCM, Beena VI, Usha PB. Aquatic macrophytes for phytomining of iron from rice based acid sulphate wetland ecosystems of Kuttanad. *J. Indian Soc. Coastal Agric. Res*. 2016;34(2):1-6.

[62] Preetha SS, Kaladevi V. Phytoremediation of heavy metals using aquatic macrophytes. *World J. Environ. Biosci*. 2014;3 (1): 34-41.

[63] Rai PK. Phytoremediation of Hg and Cd from industrial effluents using an aquatic free floating macrophyte. *Azollapinnata. Int. J. Phytoremediation*. 2008;10: 430-439.

[64] Keskinkan O, Goksu MZL, Basibuyuk M, Forster CF. Heavy metal adsorption properties of a submerged aquatic plant (*Ceratophyllumdemersum*). *Biores. Technol*. 2004;92:197-200.

[65] Norouznia H, Hamidian AM. Phytoremediation efficiency of pondweed (*Potamogetoncrispus*) in removing heavy metals (Cu, Cr, Pb, As and Cd) from water of Anzali wetland. *Int. J. Aquat.Biol*. 2014;2(4): 206-214.

[66] Kumar NJI, Sreenivas S, Rana BC. EDAX- analysis of mud of four ponds from central Gujarat. *Indian Bot. Cont*. 1989;1: 75-76.

[67] Kumar NJI, Soni H, Kumar RN, Bhatt I. Macrophytes in phytoremediation of heavy metal contaminated water and sediments in Pariyej community reserve, Gujarat, India. *Turkish J. fish. Aquat. Sci.* 2008;8: 193-200.

[68] KAU [Kerala Agricultural University]. Bioremediation of inorganic contaminants of rice based wetland ecosystems of Kuttanad, Kerala. *Annual Report of ICAR Adhoc Project.* Kerala Agricultural University, Thrissur, Kerala;2008.

[69] Soleimani M, Hajabbasi MA, Afyuni M, Charkhabi AH, Shariatmadari, H. Bioaccumulation of nickel and lead by bermuda grass (*Cynodondactylon*) and tall fescue (*Festucaarundinacea*) from two contaminated soils. Caspian J. Environ. Sci. 2009;7 (2): 59-70.

[70] Yetneberk A, KassayeSalbu B, Skipperud L, Einset J. High tolerance of aluminum in the grass species *Cynodonaethiopicus*. *Acta Physiol. Plant*2013;35:1749-1761.

[71] Kumar A, Maiti AJ, Das, R. An assessment of metals in fly ash and their translocation and bioaccumulation in perennial grass growing at the reclaimed open cast mine. *Int. J. Environ. Res.* 2015;9 (3): 1089-1096.

[72] Mahmoud E, Ghoneim AM. Effect of polluted water on soil, sediments and plant contamination by heavy metals in El-Mahla El-Kobra, Egypt. *Solid Earth* 2016;34: 1-23.

[73] Bingzhong, Shi G, Yexu, Jinzhao H.U., Quinsong, XU. Physiological response of *Alternantherap hiloxeroides* (Mart) Griseb leaves to cadmium stress. *Environ. Pollut.* 2007;147(3): 800-803.

[74] Patel M, Nerkar B, Baghel PS, Pandey, B. Phytoremediation of chemicals by *Wedeliatrilobata*, *Tecomastans* and *Tageteserecta*. *Indian J. Sci. Res*. 2014;4(1): 165-169.

[75] Baker AJM. Brooks R R. Terrestrial higher plants which hyperaccumulate metallic elements- a review of their distribution, ecology and phytochemistry. *Biorecovery* 1989.1: 81-126.

[76] Hall JL. Cellular mechanism for heavy metal detoxification and tolerance. *J. Exp. Bot.* 2002;53(366): 1 – 11.

[77] Yadav R, Arora P, Kumar S, Chaudhury A. Perspectives for genetic engineering of poplars for enhanced phytoremediation abilities. *Ecotoxicol.* 2010. 19: 1574-1588.

[78] Kupper H, Zhao F, McGrath SP. Cellular compartmentation of zinc in leaves of the hyperaccumulator *Thlaspicaerulescens*. *Plant Physiol.* 1999; 119:305-311.

Chapter 8

Expedited Biodegradation of Organic Pollutants and Refractory Compounds Using Bio-Electrochemical Systems

Eustace Fernando, Godfrey Kyazze, Ahmed Ahsan and Pavithra Fernando

Abstract

Biodegradation of xenobiotics is often considered to be a slow process. This is especially true if the xenobiotic in question is polymeric in nature, contains many chemical substituent groups or generally exhibits high level of toxicity to environmental microbiota. Due to this observed slow kinetics of degradation, removal of many xenobiotics from contaminated environments using conventional bioremediation technologies is a difficult problem. To alleviate this, alternative technologies showing improved kinetics of biodegradation are sought by the scientific community. One such promising approach is the usage of the novel technology of bio-electrochemical systems for improved degradation of xenobiotics. Due to the newness of this technology and affiliated methods, not much information about its usage for biodegradation of xenobiotics is available in literature. Therefore, this chapter aims to address that gap and bring about a comprehensive analysis on the usage of bio-electrochemical systems for rapid removal of xenobiotic contaminants from the environment.

Keywords: bio-electrochemical systems, bioremediation, biodegradation, pollutants, xenobiotics

1. Introduction

Bioelectrochemical systems (BES) are devices that drive electrochemical reactions using biological agents as catalysts. Primarily, there are two types of BES. They are microbial fuel cells (MFCs) and Microbial electrolysis cells (MECs). Both depend on two electrodes where anodic and cathodic electrochemical reactions occur and both types are driven by microorganisms. In MFC systems, organic substrates, most often organic pollutants are microbially broken down at the anodic compartment. The electrons released are harnessed by the anode electrode and are driven towards an external circuit where it can produce a usable current. These electrons are then accepted by a cathode electrode where it is coupled with protons and an electron acceptor such as molecular oxygen to produce water (**Figure 1a**). MFCs contain a proton exchange membrane to exchange protons between anode

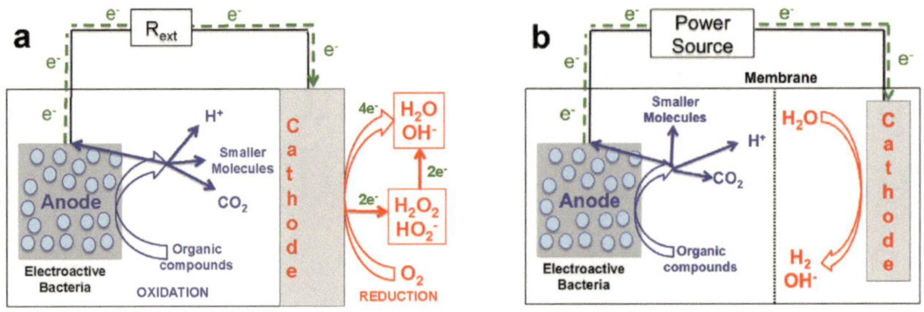

Figure 1.
Schematic diagrams showing the operating principles of (a) MFC and (b) MEC systems (modified figure adapted from [1], with permission).

and cathode chambers. On the contrary, MEC systems rely on a small exogenous supply of electrons to the cathode or anode electrode to produce combustible gases hydrogen or methane. It can therefore be termed as an MFC system run in reverse to produce gases with fuel value or other value added products. Both types of BES come in many different designs, shapes and sizes [2–4]. Out of the two BES types, MFCs are the most ubiquitously used type for laboratory and pilot scale studies and full-scale operation.

The primary utility of MFCs is to be used as a wastewater treatment technology and simultaneous recovery of electrical energy as a fringe advantage. However, the amount of recoverable electricity still remains low for large-scale practical applications of MFCs as electricity generation units. Until a research breakthrough in MFC electrical power generation is made, the focus therefore, firmly lies within wastewater and waste treatment capabilities of MFCs. One of the main benefits of MFCs in this regard is that they are demonstrated to possess better degradation kinetics for biodegradation of many recalcitrant organic pollutant types over their conventional counterpart systems such as activated sludge systems, anaerobic sludge blankets and constructed wetland systems [5–7]. Some of the refractory organic pollutants such as azo dyes and nitrophenolic compounds are known to be problematic due to their poor or non-degradable nature in conventional wastewater treatment systems such as activated sludge systems. These compounds are therefore known as highly recalcitrant organic pollutants. Earlier studies that use of MFCs however, have demonstrated that some of these compounds can be effectively degraded, transformed or mineralized into simpler intermediates with the use of BES. This chapter examines the utility of such BES systems for enhanced degradation and removal of such recalcitrant environmental pollutants.

2. Removal of azo dyes in BES

Azo dyes by far, are the most widely studied class of environmental pollutant remediated by electro active microorganisms. Azo dyes are characterized by one or more of the azo moieties, which are of oxidative nature. Therefore, the most obvious conversion mechanism of azo dyes by electro active microorganisms is by the azo pollutant acting as an electron acceptor and undergoing reduction into their constituent amines. Azo moieties are flanked by R groups which may contain various electron-rich or electron-poor substituent groups, which would influence the redox potential of the dye [8]. In recent years, Microbial Fuel Cell (MFC) technology has been explored extensively for their innovative features and environmental benefits [9]. At the anode, organic co-substrate is oxidized by electrochemically active microorganisms. Subsequently, the microorganisms transfer the electrons

BES type	Azo dye	Inoculum type	Power output/ consumption	Pollutant removal efficiency	References
Activated carbon packed bioelectrochemical reactor	Reactive red 272	Cow ruminal content microbiota	0.045 mA cm^2	95% (±3.5%) in a half-residence time of 1 hour	[14]
Microbial Fuel Cell (MFC)	Orange II	Anaerobic sludge	206.2 ± 3.1 mW/m^2	> 80%	[15]
Constructed Wetland Microbial fuel cell (CW-MFC)	Acid Red 18 (AR18)	Mixed culture sludge	1.58 mW/m^2	96%	[16]
Constructed Wetland Microbial fuel cell (CW-MFC	Acid Orange 7 (AO7)	Mixed culture sludge	1.13 mW/m^2	67%	[16]
Constructed Wetland Microbial fuel cell (CW-MFC	Congo Red (CR)	Mixed culture sludge	1.02 mW/m^2	60%	[16]
Microbial fuel cell coupled constructed wetland (CW-MFC)	Methyl Orange (MO)	Anaerobic sludge	81 mW/m^3	87.6%	[17]
Electrolysis cell coupled with microbial fuel cell (EC-MFC)	Methyl Red (MR)	Anaerobic microbiota	0.56 V	89.3%	[18]
Microbial fuel cell (MFC)	Orange G (OG)	Activated sludge	91.1 mW/m^2	97.4%	[19]
Microbial fuel cell (MFC)	Acid Orange 7 (AO7)	*Shewanella oneidensis* strain 14063 and Vibrio fischeri strain 13938	19.3 mW m^2	>98% within 30 h	[13]
Microbial fuel cell (MFC)	mixed dyes (reactive red 21 (RR21) and reactive orange 16 (RO16))	*Pseudomonas aeruginosa* 23 N1	940.61 ± 5 mW/m^2	~87%	[20]
Microbial fuel cell (MFC)	Reactive Orange 16	*Pseudomonas aeruginosa*	2887 ± 13 $\mu W/m^2$	98 ± 1.2%	[21]
Microbial fuel cell (MFC)	Reactive Black 5	*Pseudomonas aeruginosa*	1906 ± 7 $\mu W/m^2$	95 ± 2%	[21]
photoelectrocatalytic microbial fuel cell (photo-MFC)	methyl orange (MO)	anaerobic sludge	0.119 W/m^2	84.5%	[22]

Table 1.
BES studies involving azo dyes and the bioelectrochemical characteristics of the BES systems during azo dye remediation.

resulting from this oxidation to the anode via extracellular electron transfer which then passes through an external circuit to the cathode, thus producing current. Protons migrate through an ion exchange membrane to the cathode where they combine with azo dye and electrons and lead to the degradation of azo bond. Reduction of azo bond results in the formation of colorless and biodegradable aromatic amines [10]. It has been demonstrated in earlier studies that BES can be

BES type	Matrix	Pollutant	Microorganisms	Increase in maximal Power density normalized by anodic or cathodic surface area (measured in mW/m^2)	Increase in Maximal Current density normalized by anodic or cathodic surface area (measured in mA/m^2)	Maximal Voltage normalized by anodic or cathodic surface area (measured in mV)	References
MFC	Synthetic Medium	Benzene, phenanthrene	*P. aeruginosa, S. oneidensis* and Undefined cultures	1.25	18	65	[28]
MFC	Domestic waste water	Real Field Petroleum Sludge	Bacteria in Anaerobic sludge	20.6			[26]
SMFC	Saline Soil	TPHs, n-alkanes & PAHs	Polluted soil containing a microbial consortium		299		[29]
SMFC	Saline Soil	C8-C 40 n-alkanes, 16 PAHs	Petroleum contaminated soil containing *Geobacteraceae* sp. and *Escherichia* sp.	37	102	366	[30]
MFC	Soil	TPHs, 16 PAHs, C8-C40 n-alkanes	Microorganisms in polluted soil	—	—	—	[31]
MFC/MFC + anoxic cathode	wastewater	Petrochemical wastewater (BOD 5/COD: 0.36).	aerobic consortium & anaerobic consortium from effluent treatment plants		17.1 mW/m^2 (anoxic), 14.3 mW/m^2 (aerobic)	OCV: 248 and 280 mV	[32]
SMFC	Sediment	PAHs	*Caldisericum* sp. & other sediment bacteria			0.4–0.6 V	[33]
SMFC	Sediment	Benzo(a)pyrene	Aerobic genera (*Vogesella*, *Pseudomonas*, *Flavobacterium* and *Rhizobium*) and anaerobic genera (*Longilinea*, *Bellilinea*, *Desulfobacca* and *Anaeromyxobacter*)		17	61–65 mV,	[34]

Table 2.
BES studies involving PAHs and the bioelectrochemical characteristics of the BES systems during PAH remediation.

successfully coupled to other wastewater treatment technologies such as activated sludge systems and up-flow anaerobic sludge blanket reactors (UASBs) in order to fully mineralize azo dye pollutants [6, 11]. The use of BES has shown higher kinetic rates of azo dye biotransformation rates compared to other conventional treatment methods [12, 13]. Many other studies have hitherto demonstrated this ability of BES to effectively biodegrade many different azo dyes (**Table 1**).

3. Removal of polycyclic aromatic hydrocarbons (PAHs) in BES

Incomplete combustion of organic matter and fuels leads to creation of PAHs. Additionally, various industrial processes use large amounts of PAH compounds such and naphthalene, phenanthrene and chrysene. These PAHs are degraded into compounds which are less toxic in nature. This degradation takes place while these PAHs are being used as a source of carbon and energy in solid or liquid media. However, there is no evidence for the existence of a single microorganism species which can work on PAHs of several types. This is the reason why using microbial consortia can be very useful. Usage of microbial consortium is advantageous because, they are cheap, show higher applicability and produce lower negative environmental impacts [23].

Degradation activities of PAHs are extremely long, taking up to years or at least months [24]. All PAHs are "fat-loving" or "water-hating chemicals [25]. They become less available for biodegradation in water, as they have a tendency to adsorb to organic molecules. Even though various bioremediation techniques are used to treat PAHs in polluted environments, they have limitations due to the negative effect caused by activity and diversity of indigenous hydrocarbon degrading bacteria, low abundance, slow growth rates, less availability in aqueous solutions among others. Further problems can also be created due to addition of co-substrates and nutrients to improve indigenous microbial activity due to high cost & and the fact that the added chemicals having the ability to act upon anything other than the target compounds. Hence, using Microbial fuel cells remains as a viable solution circumvent some of these problems [24].

Less complex compounds can be formed from recalcitrant PAHs, if bio electrochemical activity can increase the metabolism of PAH degraders closer to the electrodes [26]. When a microbial fuel cell with a single chambered air-cathode working upon a Winogradsky solution was used, detoxification activity by microflora of a specialized variety growing on PAHs in that Winogradsky solution was found. Microbial Fuel Cells not only shows improved degradation activities, but it also show higher amount of microbial detoxification activity [23]. Inocula extracted from complex microbial communities such as soil or sediments to drive electrochemical reactions in BES and to degrade xenobiotic pollutants may confer other benefits such as simultaneous nitrogen and phosphorus removal from wastewater [27]. Many previous studies have demonstrated that various BES types employing many different electrochemically active microorganisms are effective in biodegradation of a range of PAH pollutants (**Table 2**).

4. Removal of pesticides and their residues using BES

Pesticides are defined as substances or mixtures designed for destroying and mitigating any group of pests such as insects and plants. During the past decades, different types of pesticides had widely been used in agriculture for high yield productions. Applications of pesticides have secured almost one-third of crop production in the world. Pesticides have led to the improvement of food production to secure the demands of an ever increasing human population [35]. Majority of pesticides and their residues can cause unintended splash damage to other organisms than the target pest. The widespread use of pesticides not only brought adverse influence on agro ecosystems but also caused alteration in physiological processes of non-target organisms [36].

The over and misuse of pesticide has precedence to immense health problems, economic loss and various environmental problems. The resultant health problems of pesticides include cancer, birth defects, reproductive problems, liver, kidney, and neural problems among others. In many developing countries majority of pesticides are associated with adverse effect on human health and environment due to the overuse of pesticide. On the other hand the overuse of pesticide also results in the environmental pollution such as water and soil pollution and cause imbalance of ecosystem [37].

Presently, the soil bioelectrochemical remediation system (BERS) can effectively remove pesticides, e.g., hexachlorobenzene [38], isoproturon [39] and Metalachlor [40] from soils by means of native functional microorganisms, with the advantage of the production of bioelectricity. This new technology has been paid more attention by the researchers and engineers in the field of environmental management. In the soil BERS, microorganisms is in charge of system performance since degraders and electroactive bacteria were responsible for pollutant degradation and electricity generation. Thus, the insights to the structure and evolution of microbial community in the remediation process of soil BERS is the most important for revealing the mechanism of microbial electrochemical degradation [41]. **Table 3** summarizes a range of different studies that have successfully utilized BES to bio-transform or biodegrade a variety of pesticides.

5. Removal of polychlorinated biphenyls (PCBs) using BES

Polychlorinated biphenyls (PCBs) are among the persistent organic pollutants which mainly accumulate in soils and sediments [46]. These lipophilic compounds are of major concern due to their low biodegradation, high bio-concentration and persistence in environment [47]. Polychlorinated biphenyls (PCBs) is a kind of light yellow or deep yellow oily liquid with the properties of insulating ability, thermal stability, resistance to acids, oxidation, hydrolysis, and flame resistance. Due to these unique physical and chemical properties, PCBs were widely used in many products especially in transformers and power capacitors [48]. Polychlorinated biphenyls are a subset of the synthetic organic chemicals known as polychlorinated hydrocarbons. They are composed of two attached benzene rings and multiple bonded chlorine atoms. PCBs consist of 209 isomers and congeners, with a 10,000 fold range in n-octanol/water partition coefficient (Kow, indicator of lipophilicity and then, of potential bioaccumulation) values between the mono-substituted chlorobiphenyl and the fully-substituted decachlorobiphenyl [49].

It is known that PCBs induce a wide variety of toxic responses in human, wildlife, plants and laboratory animals. They can penetrate the human body through skin contact, by inhalation of PCBs contaminated vapors and by consuming food contaminated with PCB residues [50]. Toxicological studies of PCBs in human were found to increase the rate of melanoma, gall bladder cancer, brain cancer, liver cancer, biliary tract cancer, gastrointestinal tract cancer and possibly connected to breast cancer. The people exposed to high levels of PCBs through skin contact or by consumption experience skin irritations like severe acne and rashes, nose and lungs infections and eye issues [51]. The uptake of PCBs and their negative

BES type	Pesticide	Inoculum type	Power output/ consumption	Pollutant removal efficiency	References
Single chambered microbial fuel cell (MFC)	Pentachlorophenol	Anaerobic culture	872.7 mWm^{-2}	66%	[42]
Dual chambered microbial fuel cell (MFC)	Pentachlorophenol	Anaerobic culture	1468.85 $mW\ m^{-2}$	89%	[42]
Single-compartment electrochemical flow cell	Oxyfluorfen	*Achromobacter denitrificans, Achromobacter xylosoxidans, Pseudomonas putida*, *Pseudomonas oryzihabitans* and *Brevibacterium casei*.	122 $kWh\ m^{-3}$	90%	[43]
Two-chamber MFC	Oxyfluorfen	Activated sludge	0.7 V OCV	77%	[44]
Microbial fuel cells (MFCs)	Hexachlorobenzene (HCB)	Anaerobic sludge	70.8 mW/m^2	71.14%	[45]

Table 3.
BES studies involving pesticides and the bioelectrochemical characteristics of the BES systems during pesticide remediation.

effects on plants has also been examined. For instance, PCBs firmly bind to the soil organic matter and provides low bioavailability for plants and microorganisms. But accumulation of PCBs in soil is perilous to all kinds of organisms including plants. The higher PCB concentrations affect the biosynthesis, ultrastructure of plant cell, membrane stability and plant DNA. Also, the plants that are highly exposed to PCBs result in reduction of photosynthesis, water/nutrients uptake and show visible symptoms of growth inhibition, browning of root tips and even death [52].

Anaerobic biodegradation is a major process for PCBs removal in sediments. However, a lack of or deficiency in terminal electron acceptors such as sulfate and nitrate can result in a decreased removal rate of bioremediation or no biodegradation at all [53]. Therefore, bio-stimulation by introducing electron acceptors could improve the microbiological activity in sediments [54].

Sediment microbial fuel cells (SMFCs), as a bio-electrochemical system, have been demonstrated to enhance the biodegradation of organic compounds, i.e. PCB. Xu et al. found that the combined application of a sediment microbial fuel cell (SMFC) and surfactant led to the highest removal efficiencies (43.26% of PCBs) after 60 days of operation, and produced the maximum power output (0.821 V of voltage and 18.30 W m^{-3} of power density) [55]. The degradation of polychlorinated biphenyls (PCBs) by sediment microbial fuel cell (SMFC) with/without nano-scale zero-valent iron (NZVI) addition was investigated by [56]. It was found that the combined application led to the highest removal efficiencies of PCBs (37.55 ± 1.11%) and TOC (49.72 ± 1.54%) in all circumstances and produced a higher power density (108.89 mW/m^2) [56]

6. Removal of cellulose using BES

Cellulose is a β-1,4-linkaged polymer of Glucose. Cellulosic biomass, usually obtained from agricultural and industrial activities, can be considered as one of the highly available sources of renewable energy on earth. Cellulosic materials can be used in a process of sustainable energy production by transforming them to H_2, ethanol, and methane. But due to economic and technical problems, it remains difficult to produce such products [57].

One of the main major downsides of Cellulose is that it's a moderately recalcitrant compound. It can be converted to different types of fuels such as bioethanol, methane and Hydrogen, but the energy recovery and treatment in waste water is difficult because of its low density in terms of energy and highly recalcitrant nature [58]. Another problem is that, cellulose is adhered with lignin or/and Hemicellulose. Lignin may cause problems by not allowing cellulases to digest cellulose material due to this tight adherence. However microbial fuel cells (MFCs) can be used to convert cellulosic materials to Electrical energy and it has been found to be an effective technology which can be used as an alternative. The microorganism or microorganisms must be able to breakdown cellulose anaerobically to sugar monomers and they should also be electrochemically active in order to directly convert liberated sugars to electricity in a Microbial Fuel Cell. Here, they should use an anode as an alternative electron acceptor. Metabolites of cellulose hydrolysis are oxidized in this manner to produce direct electrical energy [57].

Anaerobic microorganisms of facultative varieties can be found in the rumen. Through fermentation or anaerobic respiration, they hydrolyze cellulose while conserving energy significantly. Hence, Rumen microbial consortium can be used in Microbial Fuel Cells to generate electricity from cellulose [57]. Cellulose is different from other organic materials because, it needs this consortium to metabolize it [58]. When the Rumen consortia was used in Microbial Fuel cells, it showed the

BES type	Material used	Inoculum type	Maximal power density observed	Cell E.M.F	References
Two chambered compartment MFC	Microcrystalline Cellulose	Rumen microorganism from fistulated Holstein cow	55 mW/m^2	313 mV – at maximal level 475 mV – at stable level	[57]
Two-chamber microbial fuel cells	Particulate MN301 cellulose	Geobacter sulfurreducens (ATCC 51573) and Clostridium cellulolyticum (ATCC 35319) cultures	83 mW/m^2	350 mV – stable level	[58]
Two-chamber microbial fuel cells	Carboxymethyl cellulose	Geobacter sulfurreducens (ATCC 51573) and Clostridium cellulolyticum (ATCC 35319) cultures	153 mW/m^2	470 mV – stable level	[58]

Table 4.
BES studies involving cellulose and the bioelectrochemical characteristics of the BES systems during cellulose remediation.

presence of respiratory anaerobes & hydrolytic enzymes which involved in cellulose hydrolysis. Phylogenetic analysis was also used to understand this relationship [57].

Clostridium cellulolyticum and *Geobacter sulfurreducens* were used in one study to demonstrate the production of electricity in Microbial Fuel cells which used particulate MN301 Cellulose. Fluorescent *in situ* hybridization and quantitative PCR showed that most *Clostridium* cells were attached to cellulose which was there as a suspension. In the meantime, *Geobacter* cells were attached to the electrode. However, in contrast, use of soluble carboxymethyl cellulose caused bacteria to remain as suspension and biofilm. This explains that it is possible to improve electricity conversion from solution by decanting supernatant and settling Cellulose, as Cellulose hydrolysis can be improved by optimizing the reactor operations. Cellulose transformation performance can be increased with further research. The performance of the reactor with increasing biocatalyst density should be further analyzed. Hence, the effectiveness of decant and settling at cellulolytic density of higher concentration is one part which needs further focus to identify cellulose conversion extent [58]. Several previous studies have demonstrated the effectiveness of various BES types to biodegrade and convert the energy content of cellulose and its derivatives into electrical energy (**Table 4**).

7. Removal of lignin its derivatives in BES

Lignin is a phenolic polymer which is heterogenous in nature. It is made of renewable source of aromatic chemicals. Main components of Lignin are three alcohols namely phydroxyphenyl guaiacyl and syringyl. Different types of lignin in different plant components like grass, softwood and hardwood contain different amounts of methoxy groups depending on the degree of these three types of alcohols present in them. Microbial Fuel Cell (MFC) is a type of biological decomposition method which can be used to degrade lignin efficiently [59]. Microbial fuel cells are devices which produce electricity from catalysis of microorganisms by using chemical energy as their energy source [60]. They contain a cathode chamber which is aerobic and an anode chamber which is anaerobic. Both of these chambers are connected either by a salt bridge or a membrane where the protons can be exchanged. In the anode chamber, microorganisms like fermenting bacteria produce smaller fermentation products from larger organic molecules. CO_2, protons, and electrons are produced by oxidation using anaerobic bacteria afterwards. The circuit is completed by electrons passing to the anode interface which are then transferred to the cathode through an external wire. The cathode reduces the oxygen molecules to either form water or H_2O_2. Here, H_2O_2 is produced by a two-step reaction consisting of a two-electron reaction [59]. The lignin undergoes hydrogen peroxide mediated oxidative depolymerization when introduced to the cathode. In aerobic, low pH conditions this oxidative reaction will proceed better as oxygen is required for H_2O_2 production. An effective method for H_2O_2 production has already been observed recently via a microbial electrochemical cell [61].

8. Removal of micropollutants using BES

Micropollutants are natural or anthropogenic residue organic refractory compounds such as pharmaceutical compounds and their metabolic intermediates, antibiotics, personal care products, pesticides and industrial chemicals. These micropollutant compounds are present in polluted water in sub-microgram per liter levels [62]. In addition to their refractory nature, their presence such low concentrations in affected environments pose additional challenges for conventional biodegradation processes. Furthermore, some micropollutants such as endocrine disrupting chemicals (i.e. estradiols and their derivatives, antibiotic residues) can exert their undesirable effects on the receiving water bodies even at sub-microgram per liter concentrations [62]. Presence of antibiotics and their residues in natural waterways also promotes the development of undesired antibiotic resistance mechanisms

Micropollutant removed	Removal efficiency (%) compared to control experiments	Power output of MFC or power consumption/current input to the BES	Reference
Carbamazepine	22–34	15–16 mW/m^2 in CW-MFC	[64]
Diclofenac	52–57	161 mW/m^2 in CW - MEC	[64]
Ibuprofen	35–39	67 mW/m^2 in CW - MEC	[64]
Naproxen	30–40	36 mW/m^2 in CW - MEC	[64]
Chloramphenicol	85–90	Cathode potential maintained at – 1.5 V in the two-chamber MEC system	[65]
B-lactam antibiotics containing pharmaceutical wastewater	Up-to 86% organics removal	A hybrid UASB – MEC system where the cathode is poised at – 0.5 V	[66]
17β-Estradiol removal in pharmaceutical wastewater	Up-to 99.2% removal of 17β-Estradiol content	Application of current at a current density of 20 mA/cm^2 to the MEC	[67]

Table 5.
The utility of BES and affiliated hybrid systems for removal of micropollutants.

within bacteria found in natural aquatic ecosystems. It has been established in multiple studies that such micropollutants can easily pass through conventional wastewater treatment systems such as activated sludge systems, anaerobic digester systems and membrane bioreactors completely unchanged, back into the consumers of recycled water [63]. It is therefore clear that novel technologies are required to tackle such emerging pollutants such as various types of micropollutants.

The removal of some micropollutants has been demonstrated using BES in several previous studies. In a study conducted by [64], it was demonstrated that constructed wetland-MEC hybrid system operated at pilot-scale could successfully remove several micropollutants individually and their mixtures. The removal of four pharmaceutical compounds carbamazepine (CBZ), diclofenac (DCF), ibuprofen (IBU) and naproxen (NPX) were monitored in a constructed wetland – BES hybrid environment and the removal of all of these micropollutant compounds was observed (**Table 5**) [64].

Other studies conducted in this area also indicate similar results where micropollutant compounds such as chloramphenicol [65], β-lactam antibiotics removal from pharmaceutical wastewater [66] and 17β-Estradiol removal [67] were reported in earlier studies.

9. Conclusion

From the above analysis, it is evident that BES excels at removing certain types of refractory xenobiotic compounds more efficiently than other conventional treatment technologies. Certain types of xenobiotic pollutants such as azo dyes that resist biotransformation and subsequent degradation in conventional treatment systems can be dealt with relatively easily using BES systems such as MFCs and MEC systems or with hybrid systems combining BES with conventional treatment systems such as activated sludge systems and UASB reactors. The current level of scientific knowledge on the utility of this technology in biodegradation processes is insufficient but it is currently evolving into a better level of understanding where it could potentially be applied in pilot or industrial scales in order to deal with these refractory organic waste types in a more meaningful manner.

Conflict of interest

All authors declare no conflict of interest.

Author details

Eustace Fernando[1*], Godfrey Kyazze[2], Ahmed Ahsan[1] and Pavithra Fernando[1]

1 Department of Biological Sciences, Faculty of Applied Sciences, Rajarata University of Sri Lanka, Mihintale, Sri Lanka

2 School of Life Sciences, University of Westminster, London, UK

*Address all correspondence to: eustace6192@as.rjt.ac.lk

References

[1] Santoro, C., Arbizzani, C., Erable, B., Ieropoulos, I., 2017. Microbial fuel cells: From fundamentals to applications. A review. Journal of power sources 356, 225-244.

[2] Alipanahi, R., Rahimnejad, M., Najafpour, G., 2019. Improvement of sediment microbial fuel cell performances by design and application of power management systems. International journal of hydrogen energy 44, 16965-16975.

[3] Flimban, S.G.A., Ismail, I.M.I., Kim, T., Oh, S.-E., 2019. Overview of recent advancements in the microbial fuel cell from fundamentals to applications: design, major elements, and scalability. Energies 12, 3390.

[4] Santoro, C., Winfield, J., Theodosiou, P., Ieropoulos, I., 2019. Supercapacitive paper based microbial fuel cell: high current/power production within a low cost design. Bioresource technology reports 7, 100297.

[5] Fang, Z., Song, H.-L., Cang, N., Li, X.-N., 2013. Performance of microbial fuel cell coupled constructed wetland system for decolorization of azo dye and bioelectricity generation. Bioresource technology 144, 165-171.

[6] Fernando, E., Keshavarz, T., Kyazze, G., 2014. Complete degradation of the azo dye Acid Orange-7 and bioelectricity generation in an integrated microbial fuel cell, aerobic two-stage bioreactor system in continuous flow mode at ambient temperature. Bioresource Technology 156, 155-162. https://doi.org/10.1016/j.biortech.2014.01.036

[7] Yang, G., Wang, J., Zhang, H., Jia, H., Zhang, Y., Gao, F., 2019. Applying bio-electric field of microbial fuel cell-upflow anaerobic sludge blanket reactor catalyzed blast furnace dusting ash for promoting anaerobic digestion. Water research 149, 215-224.

[8] Fernando, E.Y., Keshavarz, T., Kyazze, G., 2019. The use of bioelectrochemical systems in environmental remediation of xenobiotics: a review. Journal of Chemical Technology & Biotechnology 94, 2070-2080.

[9] Logan, B.E., Regan, J.M., 2006. Electricity-producing bacterial communities in microbial fuel cells. TRENDS in Microbiology 14, 512-518.

[10] Frijters, C., Vos, R.H., Scheffer, G., Mulder, R., 2006. Decolorizing and detoxifying textile wastewater, containing both soluble and insoluble dyes, in a full scale combined anaerobic/aerobic system. Water research 40, 1249-1257.

[11] Fernando, E., Keshavarz, T., Kyazze, G., Fonseka, K., 2016. Treatment of colour industry wastewaters with concomitant bioelectricity production in a sequential stacked mono-chamber microbial fuel cells–aerobic system. Environmental Technology 37, 255-264. https://doi.org/10.1080/09593330.2015.1068378

[12] Dissanayake, M., Liyanage, N., Herath, C., Rathnayake, S., Fernando, E.Y., 2021. Mineralization of persistent azo dye pollutants by a microaerophilic tropical lake sediment mixed bacterial consortium. Environmental Advances 100038.

[13] Fernando, E., Keshavarz, T., Kyazze, G., 2012. Enhanced bio-decolourisation of acid orange 7 by Shewanella oneidensis through co-metabolism in a microbial fuel cell. International Biodeterioration & Biodegradation 72, 1-9. https://doi.org/10.1016/j.ibiod.2012.04.010

[14] Cardenas-Robles, A., Martinez, E., Rendon-Alcantar, I., Frontana, C., Gonzalez-Gutierrez, L., 2013. Development of an activated carbon-packed microbial bioelectrochemical system for azo dye degradation. Bioresource technology 127, 37-43.

[15] Xu, H., Quan, X., Chen, L., 2019. A novel combination of bioelectrochemical system with peroxymonosulfate oxidation for enhanced azo dye degradation and MnFe2O4 catalyst regeneration. Chemosphere 217, 800-807.

[16] Oon, Y.-L., Ong, S.-A., Ho, L.-N., Wong, Y.-S., Dahalan, F.A., Oon, Y.-S., Teoh, T.-P., Lehl, H.K., Thung, W.-E., 2020. Constructed wetland–microbial fuel cell for azo dyes degradation and energy recovery: Influence of molecular structure, kinetics, mechanisms and degradation pathways. Science of The Total Environment 720, 137370.

[17] Fang, Z., Song, H., Yu, R., Li, X., 2016. A microbial fuel cell-coupled constructed wetland promotes degradation of azo dye decolorization products. Ecological Engineering 94, 455-463.

[18] Zou, H., Wang, Y., 2017. Azo dyes wastewater treatment and simultaneous electricity generation in a novel process of electrolysis cell combined with microbial fuel cell. Bioresource technology 235, 167-175.

[19] Niu, C.-G., Wang, Y., Zhang, X.-G., Zeng, G.-M., Huang, D.-W., Ruan, M., Li, X.-W., 2012. Decolorization of an azo dye Orange G in microbial fuel cells using Fe (II)-EDTA catalyzed persulfate. Bioresource technology 126, 101-106.

[20] Mishra, S., Nayak, J.K., Maiti, A., 2020. Bacteria-mediated bio-degradation of reactive azo dyes coupled with bio-energy generation from model wastewater. Clean Technologies and Environmental Policy 1-17.

[21] Ilamathi, R., Sheela, A.M., Gandhi, N.N., 2019. Comparative evaluation of Pseudomonas species in single chamber microbial fuel cell with manganese coated cathode for reactive azo dye removal. International Biodeterioration & Biodegradation 144, 104744.

[22] Han, H.-X., Shi, C., Yuan, L., Sheng, G.-P., 2017. Enhancement of methyl orange degradation and power generation in a photoelectrocatalytic microbial fuel cell. Applied Energy 204, 382-389. 10.1016/j.apenergy.2017.07.032

[23] Gambino, E., Toscanesi, M., Del Prete, F., Flagiello, F., Falcucci, G., Minutillo, M., Trifuoggi, M., Guida, M., Nastro, R.A., Jannelli, E., 2017. Polycyclic Aromatic Hydrocarbons (PAHs) Degradation and Detoxification of Water Environment in Single-chamber Air-cathode Microbial Fuel Cells (MFCs). Fuel Cells 17, 618-626.

[24] Kronenberg, M., Trably, E., Bernet, N., Patureau, D., 2017. Biodegradation of polycyclic aromatic hydrocarbons: Using microbial bioelectrochemical systems to overcome an impasse. Environmental Pollution 231, 509-523.

[25] Jones, K.C., De Voogt, P., 1999. Persistent organic pollutants (POPs): state of the science. Environmental pollution 100, 209-221.

[26] Chandrasekhar, K., Mohan, S.V., 2012. Bio-electrochemical remediation of real field petroleum sludge as an electron donor with simultaneous power generation facilitates biotransformation of PAH: effect of substrate concentration. Bioresource Technology 110, 517-525.

[27] Hu, Y., Duan, C., Fu, D., Wu, X., Yan, K., Fernando, E., Karunarathna, S.C., Promputtha, I., Mortimer, P.E., Xu, J., 2020. Structure of Bacterial Communities in Phosphorus-Enriched Rhizosphere Soils. Applied Sciences 10, 6387.

[28] Adelaja, O., Keshavarz, T., Kyazze, G., 2015. The effect of salinity, redox mediators and temperature on anaerobic biodegradation of petroleum hydrocarbons in microbial fuel cells. Journal of hazardous materials 283, 211-217.

[29] Li, X., Wang, X., Zhang, Y., Zhao, Q., Yu, B., Li, Y., Zhou, Q., 2016. Salinity and conductivity amendment of soil enhanced the bioelectrochemical degradation of petroleum hydrocarbons. Scientific reports 6, 1-11.

[30] Li, X., Wang, X., Zhang, Y., Cheng, L., Liu, J., Li, F., Gao, B., Zhou, Q., 2014. Extended petroleum hydrocarbon bioremediation in saline soil using Pt-free multianodes microbial fuel cells. Rsc Advances 4, 59803-59808.

[31] Zhang, Y., Wang, X., Li, X., Cheng, L., Wan, L., Zhou, Q., 2015. Horizontal arrangement of anodes of microbial fuel cells enhances remediation of petroleum hydrocarbon-contaminated soil. Environmental Science and Pollution Research 22, 2335-2341.

[32] Yeruva, D.K., Jukuri, S., Velvizhi, G., Kumar, A.N., Swamy, Y. V, Mohan, S.V., 2015. Integrating sequencing batch reactor with bio-electrochemical treatment for augmenting remediation efficiency of complex petrochemical wastewater. Bioresource technology 188, 33-42.

[33] Yang, Y., Lu, Z., Lin, X., Xia, C., Sun, G., Lian, Y., Xu, M., 2015. Enhancing the bioremediation by harvesting electricity from the heavily contaminated sediments. Bioresource technology 179, 615-618.

[34] Yan, Z., Jiang, H., Cai, H., Zhou, Y., Krumholz, L.R., 2015. Complex interactions between the macrophyte Acorus calamus and microbial fuel cells during pyrene and benzo [a] pyrene degradation in sediments. Scientific reports 5, 1-11.

[35] Nsibande, S.A., Forbes, P.B.C., 2016. Fluorescence detection of pesticides using quantum dot materials–a review. Analytica Chimica Acta 945, 9-22.

[36] Verma, A., Prakash, S., 2018. Haematotoxicity of Phorate, an Organophosphorous pesticide on a Freshwater Fish, *Channa punctatus* (Bloch).

[37] Arya, S., Sudhakar, P., Dwivedi, N., 2021. Pesticides and Its Impact on Biodiversity and Environment.

[38] Cao, X., Song, H., Yu, C., Li, X., 2015. Simultaneous degradation of toxic refractory organic pesticide and bioelectricity generation using a soil microbial fuel cell. Bioresource Technology 189, 87-93.

[39] Quejigo, J.R., Domínguez-Garay, A., Dörfler, U., Schroll, R., Esteve-Núñez, A., 2018. Anodic shifting of the microbial community profile to enhance oxidative metabolism in soil. Soil Biology and Biochemistry 116, 131-138.

[40] Li, Yue, Li, X., Sun, Y., Zhao, X., Li, Yongtao, 2018. Cathodic microbial community adaptation to the removal of chlorinated herbicide in soil microbial fuel cells. Environmental Science and Pollution Research 25, 16900-16912.

[41] Li, X., Li, Yue, Zhang, X., Zhao, X., Chen, X., Li, Yongtao, 2020. The metolachlor degradation kinetics and bacterial community evolution in the soil bioelectrochemical remediation. Chemosphere 248, 125915.

[42] Khan, N., Khan, M.D., Nizami, A.-S., Rehan, M., Shaida, A., Ahmad, A., Khan, M.Z., 2018. Energy generation through bioelectrochemical degradation of pentachlorophenol in microbial fuel cell. RSC advances 8, 20726-20736.

[43] Carboneras, M.B., Rodrigo, M.A., Canizares, P., Villasenor, J., Fernandez-Morales, F.J., 2020. Removal of

oxyfluorfen from polluted effluents by combined bio-electro processes. Chemosphere 240, 124912.

[44] Zhang, Q., Zhang, L., Wang, H., Jiang, Q., Zhu, X., 2018. Simultaneous efficient removal of oxyfluorfen with electricity generation in a microbial fuel cell and its microbial community analysis. Bioresource technology 250, 658-665.

[45] Cao, X., Yu, C., Wang, H., Zhou, F., Li, X., 2017. Simultaneous degradation of refractory organic pesticide and bioelectricity generation in a soil microbial fuel cell with different conditions. Environmental technology 38, 1043-1050.

[46] Gomes, H.I., Dias-Ferreira, C., Ottosen, L.M., Ribeiro, A.B., 2015. Electroremediation of PCB contaminated soil combined with iron nanoparticles: effect of the soil type. Chemosphere 131, 157-163.

[47] Du, R.X., Fan, Z.X., Guo, D.F., Cai, L.J., Ding, H.F., Bi, Y.P., 2008. Analysis on the existence level of polychlorinated biphenyl in water environment in China. China Safety Science Journal 18, 16-21.

[48] Fitzgerald, E.F., Belanger, E.E., Gomez, M.I., Hwang, S., Jansing, R.L., Hicks, H.E., 2007. Environmental exposures to polychlorinated biphenyls (PCBs) among older residents of upper Hudson River communities. Environmental research 104, 352-360.

[49] Ramamoorthy, Sub, Ramamoorthy, Sita, 1997. Chlorinated organic compounds in the environment: Regulatory and monitoring assessment. CRC press.

[50] Carpenter, D.O., 2006. Polychlorinated biphenyls (PCBs): routes of exposure and effects on human health. Reviews on environmental health 21, 1.

[51] DeCastro, B.R., Korrick, S.A., Spengler, J.D., Soto, A.M., 2006. Estrogenic activity of polychlorinated biphenyls present in human tissue and the environment. Environmental science & technology 40, 2819-2825.

[52] Chen, S.-J., Tian, M., Zheng, J., Zhu, Z.-C., Luo, Y., Luo, X.-J., Mai, B.-X., 2014. Elevated levels of polychlorinated biphenyls in plants, air, and soils at an e-waste site in Southern China and enantioselective biotransformation of chiral PCBs in plants. Environmental science & technology 48, 3847-3855.

[53] Margesin, R., Schinner, F., 2001. Bioremediation (natural attenuation and biostimulation) of diesel-oil-contaminated soil in an alpine glacier skiing area. Applied and environmental microbiology 67, 3127-3133.

[54] Yan, Z., Song, N., Cai, H., Tay, J.-H., Jiang, H., 2012. Enhanced degradation of phenanthrene and pyrene in freshwater sediments by combined employment of sediment microbial fuel cell and amorphous ferric hydroxide. Journal of hazardous materials 199, 217-225.

[55] Xu, X., Zhao, Q.L., Wu, M.S., 2015. Improved biodegradation of total organic carbon and polychlorinated biphenyls for electricity generation by sediment microbial fuel cell and surfactant addition. RSC advances 5, 62534-62538.

[56] Wu, M., Xu, X., Lu, K., Li, X., 2019. Effects of the presence of nanoscale zero-valent iron on the degradation of polychlorinated biphenyls and total organic carbon by sediment microbial fuel cell. Science of the Total Environment 656, 39-44.

[57] Rismani-Yazdi, H., Christy, A.D., Dehority, B.A., Morrison, M., Yu, Z., Tuovinen, O.H., 2007. Electricity generation from cellulose by rumen microorganisms in microbial fuel cells. Biotechnology and bioengineering 97, 1398-1407.

[58] Ren, Z., Steinberg, L.M., Regan, J.M., 2008. Electricity production and microbial biofilm characterization in cellulose-fed microbial fuel cells. Water Science and Technology 58, 617-622.

[59] Sharma, R.K., Mukhopadhyay, D., Gupta, P., 2019. Microbial fuel cell-mediated lignin depolymerization: a sustainable approach. Journal of Chemical Technology & Biotechnology 94, 927-932.

[60] Kim, B.H., Chang, I.S., Gadd, G.M., 2007. Challenges in microbial fuel cell development and operation. Applied microbiology and biotechnology 76, 485-494.

[61] Ki, D., Popat, S.C., Rittmann, B.E., Torres, C.I., 2017. H2O2 production in microbial electrochemical cells fed with primary sludge. Environmental science & technology 51, 6139-6145.

[62] Das, S., Ray, N.M., Wan, J., Khan, A., Chakraborty, T., Ray, M.B., 2017. Micropollutants in wastewater: fate and removal processes. Physico-chemical wastewater treatment and resource recovery 3, 75-117.

[63] Komolafe, O., Mrozik, W., Dolfing, J., Acharya, K., Vassalle, L., Mota, C.R., Davenport, R., 2021. Fate of four Different Classes of Chemicals Under Aaerobic and Anaaerobic Conditions in Biological Wastewater Treatment. Frontiers in Environmental Science.

[64] Hartl, M., García-Galán, M.J., Matamoros, V., Fernández-Gatell, M., Rousseau, D.P.L., Du Laing, G., Garfí, M., Puigagut, J., 2021. Constructed wetlands operated as bioelectrochemical systems for the removal of organic micropollutants. Chemosphere 271, 129593.

[65] Guo, N., Wang, Y., Yan, L., Wang, X., Wang, M., Xu, H., Wang, S., 2017. Effect of bio-electrochemical system on the fate and proliferation of chloramphenicol resistance genes during the treatment of chloramphenicol wastewater. Water research 117, 95-101.

[66] Hu, D., Min, H., Chen, Z., Zhao, Y., Cui, Y., Zou, X., Wu, P., Ge, H., Luo, K., Zhang, L., 2019. Performance improvement and model of a bio-electrochemical system built-in up-flow anaerobic sludge blanket for treating β-lactams pharmaceutical wastewater under different hydraulic retention time. Water research 164, 114915.

[67] Hua, M., He, H., Fu, G., Han, F., 2019. 17β-Estradiol removal by electrochemical technology in the presence of electrochemically active bacteria in aerobic aquatic environments. Environmental Engineering Science 36, 316-325.

Chapter 9

Critical Studies on the Kinetics, Isotherms and Activation Energy of Sorption Phenomenon for Optimized Kenaf Shive Sorbent in Crude Oil/Seawater System

Zwahruddeen Muhammad Salisu, Diya'uddeen Bashir Hasan, Yahaya Gambo Liman, Suleiman Mohammed Awal, Ukanah Suleiman Pendo, Mohammed Kabir Yakubu and Ishiaku Semo Umaru

Abstract

The secondary effect discovery of synthetic sorbents opened another research direction for many field of studies. However, the sorption parameters of lignocellulosic sorbents are rarely reported most importantly, kenaf shive. This paper centered at the sorption behavior of optimized kenaf shive sorbents using Response Surface Methodology (RSM) via surface deposit technique. Five-level Central Composite Design (CCD) experimental matrix was used to analyze the effect of particle sizes (125–1000μm), stirring time (5–30 min) and methyltrimethoxysilane (MTMS) concentration (5–20% v/v) as individual and combined variables process in the developed sorbents. The unmodified shive was compared with the modified, and it reveals a positive shift in the sorption capability. Instrumental analysis such as FTIR (Fourier Transform Infra-Red), XRD (X-ray Diffraction), DT-TGA (Differential Thermal-Thermogravimetric analysis) and BET (Brunaure-Emmett–Teller) were carried out on the optimized sorbent and the results were in conformity with the sorption results. The sorption behavior deployed fits the pseudo-first-order and Langmuir isotherm with regression coefficient $R^2 = 0.9496$ and $R^2 = 0.9400$. The sorption property was found to be spontaneous and exothermic, however, the activation energy studies shows physic-sorption phenomenon with $25.3 kJmol^{-1}$ and $R^2 = 0.9360$.

Keywords: lignocellulosic, design matrix, sorbent, isotherm, physic-sorption

1. Introduction

The release of organic hydrocarbon to the sea or land is termed as crude oil spillage. The broadly speaking, global energy sources are categorized into: fossil fuel, nuclear fuel, non-renewable and renewable energy resource, hitherto, crude oil is one of the common and important energy sources for transportation amongst the fossils in the planet [1–3]. Oil spill is majorly as result environmental issues associated to exploration, transportation and refining [4–7]. Oil spills are usually transported by wind, current, temperature, weathering and salinity increases the transportation consequently, accumulate on sea surfaces or sediment at the debris [8, 9]. This effect of oil spill is not restricted to human body directly alone, but could affect the plants within the community and water sources [10, 11]. Furthermore, this menace affect aquatics consequently, affect the hygiene of the communities' citizenry through the inhalation toxicants [12–14].

This menace (crude oil spill) pause a challenge upon researchers to get solution for its containment and recovery. Containment and mechanical recovery; bumming; bioremediation; chemical dispersant and the use of sorbent were approaches employed to combat this disaster [15–17]. The concept of sorbent recovery came up in the last couple of decades. The sorbent source could be synthetic such as: polyethylene, polypropylene, polyurethane; natural inorganic such as: clay, perlite and graphite; natural organic such as: kenaf bast fibers, Sawdust and kenaf shive/core fibers [18–23]. The sorbent technique are examples for physical methods however, biological methods using microorganisms and chemical methods using in-situ bumming and dispersants are also feasible and practicable. Neither the biological

Figure 1.
Kenaf plant: (a) a standing Kenaf plant in field (b) cross-section of kenaf stalk (c) chopped kenaf shive (core) fibers and baled bast kenaf fibers.

nor chemical methods are viable [23]. The industrially acceptable physical methods for crude oil recovery is by using synthetic materials which are now considered hazardous owing to their non-biodegradability and are capital intensive [15, 24–26].

For sustainable and renewable-of course-cheap sorbents, bio-mops would need to be considered first. Kenaf plants acclimatized to different climatic changes and the bast is largely used in paper pulping, agro-packages etc. therefore, if such plant shive is modified for oil spill remediation is a long-way in research (see **Figure 1**). The preeminent properties imbibe by kenaf shive are: low cost, high efficiency and biodegradable properties of natural sorbents gained a high exploration. A high number of natural organic oil sorbents were reported, namely: wood chips, sugarcane bagasse, cotton and jute [27–29]. Jute plant having many a common properties with kenaf plant deems to be investigated. Jute and kenaf constitute of cellulose, hemicellulose (82–85%) and lignin [14, 30].

This research paper critically study the sorption behavior of this novel, ultralight, robust and facile sorbent in dense crude oil/seawater system with reference to temperature, oil concentration and sorption time.

2. Experimental

2.1 Materials

The used chemicals are analytical grades, without further purification and dried Kenaf stalks were obtained from National Research Institute for Chemical Technology (NARICT), Zaria. The crude oil and seawater samples used for the sorption test were obtained from Petroleum Research Laboratory, Warri, Delta state, Nigeria. The raw crude oil was kept at room temperature and sea water was stored below 0°C in refrigerator.

2.2 Fabrication and hydrophobia coating of kenaf shive sorbent

The sorbents were fabricated as reported by Salisu et al., [15]. For brevity, a pulverized Kenaf shive was dispersed in (2 wt%) sodium hydroxide/urea solution (1.9 wt%/10 wt%) and stirred for 6mins using mechanical stirrer to achieve homogeneity in the dispersion. Aftermath, the sample was gelated by placing in a refrigerator for more than 24 hrs. Then, the mixture was thawed at room temperature after frozen, followed by immersion into ethanol (99 vol %) for coagulation. The beaker in which the preparation takes place was use as mold to control the specimen thickness. It is imperative to know that no cross linker was used which makes the particles a bit loose. Coagulation was directly carried out by immersing the gel in DI water for 2 days. Freeze-drying was carried out on the sample for 2 days at approximately −60°C after pre-freezing the sample at −18°C for 12 h.

The afore-fabricated aerogel of kenaf shive was coated using chemical vapor deposition (CVD) technique for silation, i.e. methyltrimethoxysilane (MTMS). Then the resulted sample was capped and heated in an oven at 70°C for 2 hrs for a completed silanation reaction. Thereafter, the coated sample was placed in a vacuum oven to remove the excess coating reagent at approximately small pressure.

2.3 Characterizations

Infrared spectra of the sorbent in KBr pellets was analyzed and scanned from 4000–400 cm^{-1} using Shimadzu FTIR-8400S. The test was carried out on the raw (unmodified) and modified optimized unextracted sorbent that bears the highest oil sorption to confirm the modifications by taking the advantage of the unique vibration/stretching property for each functional group. The sorbent structure was determine using Shimadzu XRD 6000 (Tokyo, Japan) with CuKα radiation ($\lambda = 1.542\,\text{Å}$) operated at 30 kV and 30 mA whereby the ground sorbent was scanned at rate of $0.05°/\text{min}$ at angle range of $3° \geq 2\Theta \leq 90°$ The generated raw data were used to replot the diffractogram aided by **Origin Pro 9.0 16Bit, Figure 2**. Surface area was determined using Bmnauer, Emmette and Teller (BET) technique by (Quantachrome Instruments, Model Nova 1000e series, USA), however, the heat properties was not set aside but determine using DTA-TGA60 Shimadzu, Japan.

2.4 Adsorbability measurement

Oil adsorption capability for both preliminaries and the optimized extracted as well as unextracted sorbent of the modified kenaf shive fibers was investigated. According to ASTM F-726-12, the adsorption capacity formula is expressed as follows [14, 31]:

$$S_{\text{w}} = \frac{s_{wt} - s_o}{s_o} \tag{1}$$

Where; *Sw* is the sorption rate (g (liquid)/g (sorbent)), *So* is the quality of the shive fiber before sorption, and Swt is the quality of the kenaf shive fiber after sorption. 1 g of raw and modified shive fibers was immersed into a beaker, and measurements was recorded after every 5 min. According to ASTM –726–12, the test measures the rapid adsorption capacity (15 min soaking) and 24 h adsorption capacity. The sea water used for this test is a natural seawater not simulated.

2.5 Batch experiments

Equal mixture of 15 mL petroleum ether and 1 mL of 1 + 1 sulfuric acid were shaken in a reparatory funnel for 15 mins. The lower aqueous organic layer was

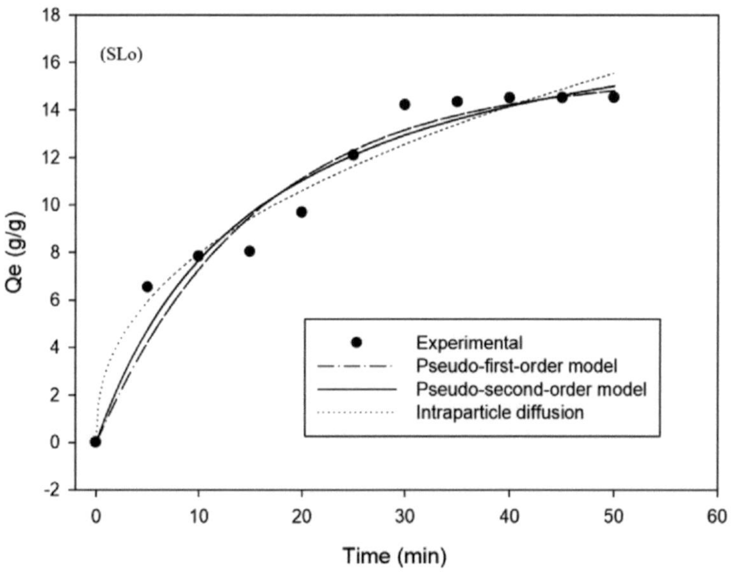

Figure 2.
Kinetics of crude oil sorption on silane optimized kenaf shive sorbent.

released after settling for about 10 min. The organic layer was poured into a beaker containing 1.2 g of drying agent (anhydrous sodium sulfate), then the mixture was drain into glass funnel. Consequently, the solution was filtered into the colorimeter coupled with 25 mL of petroleum ether (this was repeated with the same quantity of petroleum ether). The residual oil concentration was determined by filtering the sorbent and analyzed using UV–Vis spectroscopy.

Adsorption kinetics were performed by immersing 1 g of developed sorbent a mixture of oil/sea water at room temperature. Samples and crude oil concentration were, respectively, weighed and measured at different time interval, between 1–90 min.

Isotherm studies was carried out at room temperature (298 K) by varying the initial concentrations 5–30g/L at interval of 5 g/L using the aforementioned procedure.

The adsorption thermodynamics and activation energies (Ea) were determined via the batch experiments at different temperatures (298, 303, 313 and 323 K).

The crude adsorption capacity at equilibrium (Q) is calculated by the following formula:

$$Q = \frac{(C_o - C_e)V}{s} \tag{2}$$

Where, *Co* and *Ce* are, respectively, the initial and equilibrium concentrations of crude oil (g/L) at any time t. *V* is the volume of the solution (L), and *S* is the mass of the adsorbent (g).

2.6 Adsorption kinetics

2.6.1 Pseudo first-order model

The pseudo-first-order model is represented by the following equation [14]:

$$\frac{dQ_t}{dt} = K_1(Q_e - Q_t) \tag{3}$$

When boundary conditions are reached, $t = 0,\ Q = 0$ and $t = t,\ Q = Qt$, the equation can change to:

$$\ln(Q_e - Q_t) = lnQ_e - K_1 t \tag{4}$$

this is simplified as:

$$Q_t = Q_e\left(1 - e^{-K_1 t}\right) \tag{5}$$

Where, k_1 is the pseudo first-order rate constant; Qe and Qt are the adsorption capacities of the adsorbent at equilibrium.

2.6.2 Pseudo second-order model

The pseudo second-order model is represented as follows [14, 32]:

$$\frac{dQt}{dt} = k_2(Q_e - Q_t)^2 \tag{6}$$

The linearized-integrated form of the equation is:

$$Q_t = \frac{k_2 Q_{e^2} t}{1 + k_2 Q_e t} \qquad (7)$$

where k_2 is the pseudo second-order rate constant.

2.6.3 Intraparticle diffusion model

The intraparticle diffusion model can be used to analyze the removal of pollutants by an absorbent during a diffusion process. This is expressed as the following equation [33]:

$$Q_t = k_p t^{0.5} + C \qquad (8)$$

where k_p is the intraparticle diffusion rate constant; and C is a constant related to the bounding layer thickness.

2.7 Adsorption isotherm

2.7.1 Langmuir isotherm model

The Langmuir isotherm model assumes that adsorption occurs at a specific uniform location on the adsorbent surface. According to this model, the adsorbent forms a molecular monolayer.

The equation is as follows [14, 33]:

$$Q_e = \frac{k_1 Q_0 C_e}{1 + k_1 C_e} \qquad (9)$$

where Qo is the maximum adsorption capacity of the adsorbent (g/g); and K_1 is the Langmuir constant of equilibrium adsorption.

2.7.2 Freundlich isotherm model

The Freundlich isotherm model assumes that multilayer adsorption takes place at heterogeneous surfaces with different adsorption energies and characteristics. Here, the adsorption of the surface is calculated by the following equation:

$$Q_e = k_2 C_e^{1/n} \qquad (10)$$

where K2 (mg/g)(L/mg)1/n is the Freundlich constant; and n is the adsorption intensity.

2.8 Adsorption thermodynamics

The adsorption thermodynamics of the crude oil adsorption process need to be further investigated. Various thermodynamic parameters such as enthalpy (ΔH), entropy (ΔS), and Gibbs free energy (ΔG) can be obtained by isothermal adsorption studies [34]. ΔG of adsorption can be represented by the classical Van't Hoff equation:

$$\Delta G = RT \ln Ko \qquad (11)$$

where K_0 can be calculated by the following equation:

$$K_0 = \text{Qe}/\text{Ce}$$

The apparent enthalpy (ΔH) of adsorption and the entropy (ΔS) are calculated as follows:

$$\ln\left(\frac{Q_e}{c_e}\right) = \frac{\Delta S}{T} - \frac{\Delta H}{RT} \tag{12}$$

where ΔG is in (kJ/mol); ΔH is in (kJ/mol); ΔS is in (kJ/(molK)); R is the universal gas constant (8.314 J/mol); T is the adsorption temperature (K).

2.9 Activation energy

The activation energy can be determined from the change of the absorption rate constant, k with temperature, T(K) using the Arrhenius equation [32].

$$lnk \quad = lnA - \frac{E_a}{RT} \tag{13}$$

Where A is the pre-exponential factor obtained from the intercept plot of lnk (kinetic rate constant of the best fitted model) versus 1/T and R is the gas constant (8.314 J/mo1K). By plotting ln [k] against 1/T, Ea can be calculated from the slope.

3. Results and discussions

3.1 The crude oil sample was characterized using rheometer instrument

In **Table 1**, the physical properties of the used crude oil was expatiated. Hence, viscosity and density played a vital role in adsorption. However, different crude oil have unique physical properties and were recorded, thus results the yardstick for differentiation. This is insight of the heavy crude oil. This has some difficulties in penetrating through the sorbent than the medium or light crudes [11].

3.2 Structural characterization

The models were found to be statistically significant ($p < 0.05$) and therefore included in the models, analysis of variance (ANOVA) was performed. Based on the results, it is seen that the p-values for both responses in all the **Table 2**, were both less than 0.05 (< 0.0001), this indicates that is significance and both could be used for response prediction with Regression coefficient R^2, adjusted R^2 and predicted R^2 were used to evaluate the quality of the developed equation see my paper [15].

In **Figure 3** is the FTIR spectra indicating peaks of the raw (unmodified) and, modified (optimized sorbent) kenaf shive. The results are fully discussed in [5], for succinctness is not discussed herein, however, the most important functional

Sample	K. Viscosity (m^2/s)	Speed (m/s^2)	Torgue (Nm)	Temp. (°C)	Density (g/cm^3)
Crude oil	1.33	30.00	0.10	24.5	0.8965
	0.67	60.00	0.00	24.5	

Table 1.
Specifications of crude oil samples.

Run no.	Experimental Design				Results	
					Experimental	Predicted
	Resident time (min)-A	Particle size (μm)-B	Pamcle MTMS conc. (%)-C	Density (g cm^{-3})	% swelling	% swelling
1	5.00	1000.00	20.00	0.075	417.10	410.7694
2	17.50	562.50	12.50	0.068	652.30	642.9804
3	17.50	1000.00	12.50	0.088	460.00	502.6942
4	17.50	562.50	20.00	0.106	778.59	881.8884
5	5.00	125.00	5.00	0.140	234.00	250.9699
6	30.50	1000.00	5.00	0.120	283.20	364.7302
7	17.50	125.00	12.50	0.058	585.60	575.1541
8	17.50	562.50	5.00	0.072	654.30	583.2501
9	5.00	562.50	12.50	0.078	181.70	353.5144
10	5.00	125.00	20.00	0.071	654.30	564.7088
11	17.50	562.50	12.50	0.102	651.60	642.9804
12	17.50	562.50	12.50	0.074	659.31	642.9804
13	30.00	125.00	5.00	0.085	290.30	322.1401
14	30.00	125.00	20.00	0.083	654.30	705.5277
15	17.50	562.50	12.50	0.079	650.57	642.9804
16	30.00	562.50	12.50	0.074	664.20	524.6339
17	17.50	562.50	12.50	0.083	654.30	642.9804
18	5.00	1000.00	5.00	0.075	222.60	163.31
19	17.50	562.50	12.50	0.083	654.30	642.9804
20	30.00	1000.00	20.00	0.073	673.30	648.2679

Table 2.
Design matrix for crude oil silane (SL) modified sorbents.

groups are in **Table 3**. These functional group signifies the occurrence of the reaction between the silane compound and cellulosic materials of the shives. The Brunure-Emmitte-Teller (BET) results confirms the findings. is briefly discussed here. The BET results indicates an increase in surface area from 100 to 301.1 m^2/g. This attribute to the high crude oil sorption of the optimized sorbent was observed than in the unmodified shive based on the investigated variables (see **Table 4**). Couple with the cementing materials effect which was vividly shown in **Figure 3**. The cementing material decrease the oil sorption in the sorbent because it is less porous compare to the pulverized shives [15].

The DT-TGA spectra in **Figure 4** indicates the heat behavior and state transition of the optimized sorbent. The TG thermogram indicates four decomposition and weight loss labeled W, X, Y and Z at corresponding temperatures of 185, 355, 415 and 475°C respectively. The weight loss 10% at W was as a result of dehydration and pyrolysis in the sample via endothermic heat exchange. This phenomenon was proved by DT thermogram. The second stage exothermic heat was observed resulting to weight losses at X, Y and Z corresponding to 25, 10, 50% respectively leaving 5% residue, these indicate the optimized sorbent's degradation. This attribute indicates the optimized sorbent is highly organic and decomposability consequently, eco-friendly.

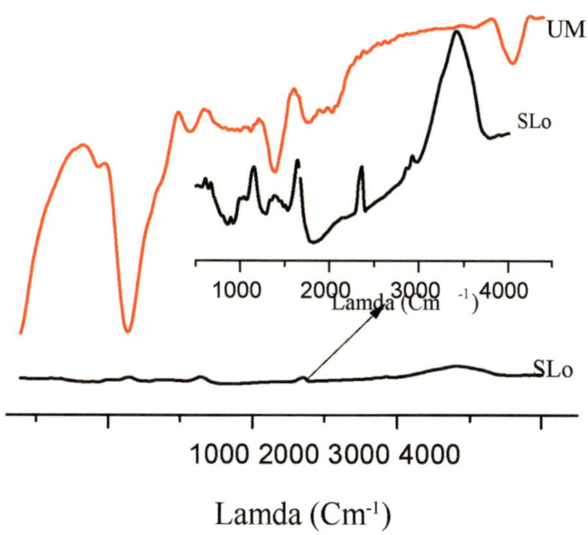

Figure 3.
FTIR photogram unmodified (UM) and optimized silane (SLo) kenaf shive sorbent.

Adsorbents	Maximum Sorption Capacity (g/g)	References
Crosslinked-l-Octene/styrene/DVB terpolymer	40	
Carbon fiber aerogel	115	
Graphene coated melamine sponge	165	
Silanized melamine sponge	163	
Polypropylene	15	[32]
Banana skins	5–7	[4]
Silanized cellulose aerogel from paper waste	24.4	[11]
Acrylic acid modified kenaf shive	7	[5]
Styrene modified kenaf shive	8.03	[14]
Silanized kenaf shive sorbent	12.02	[15]

Table 3.
Comparative adsorption capacities of different sorbents for crude oil.

A brief comparative sorption capacity of different crude oil sorbents is shown, **Table 3** and **Figure 5**. This is to buttress that the synthesized sorbent is within the sorption capacity of different sources. The obtained sorbent has crude oil sorption higher than that of acrylic acid and styrene modified kenaf shive as well as the sorbents obtained from banana skin.

3.3 Adsorption kinetics

Adsorption kinetics curve for the modified and optimized Kenaf Shive sorbent was exemplified in **Figure 2**. The relationship for the adsorption per unit time was tested in oil–water system. The slope at each point indicates the instantaneous sorption capacity. The adsorption capability increases rapidly at the initial stage i.e. 0–5 min . A slow increase in adsorption was observed up to 30 min, after, the curve flattens indicating

Wave number (cm^{-1})	Vibration	Structure
786	Vs (SiC)	–Si – C
2922	$\delta s(CH_3)$	$-CH_3$
3324	δs (OH)	–OH
2051	$\delta as(CH_2)$	$-CH_2$
1673	$\delta as(CO)$	–C = O
1591	$\delta s(CC)$	$-CH = CH_2$
1151	$Vs(CH_3)$	$-CH_3$
1021	δs(SiOSi)	–Si–O–Si

Vs: symmetrical vibration, δs: symmetrical stretching, δas: Asymmetrical stretching.

Table 4.
Functional group assignment of the optimized silane kenaf shive sorbent for FTIR spectra.

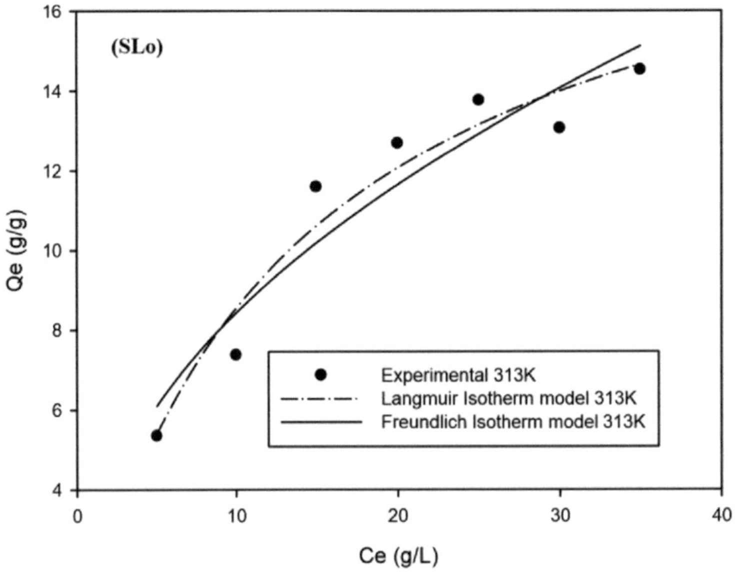

Figure 4.
Isotherm of crude oil sorption on modified and optimized kenaf shive sorbent (STo).

equilibrium adsorption [35]. This phenomenon was attributed to the increase in pore size of the optimized sorbent which was justified by the BET results analysis. Hence, the used oil is hydrophobic and viscose which made it slightly soluble in water, then couple with hydrophobic nature of the modifier leads to the high adsorption capability. The diffusion becomes slow when the pore sizes reduce this contributes to the slowness and little increase in sorption capacity after 30 min [14].

This study shows that out of the three (pseudo-first-order, pseudo-second-order and intraparticle diffusion) kinetic models used, the behavior that best fits the sorption capacity of this modified and optimized sorbent is pseudo-first-order. This was proven by correlation coefficient (R^2) of the three said models (**Figure 2**). The R^2 of pseudo-first-order is 0.950 with sorption capacity 12.020 g/g. The corresponding R^2 and sorption capacities were shown in **Table 5**. Despite the high adsorption shown in pseudo-second-order yet is less assured based on the recorded R^2 value [36].

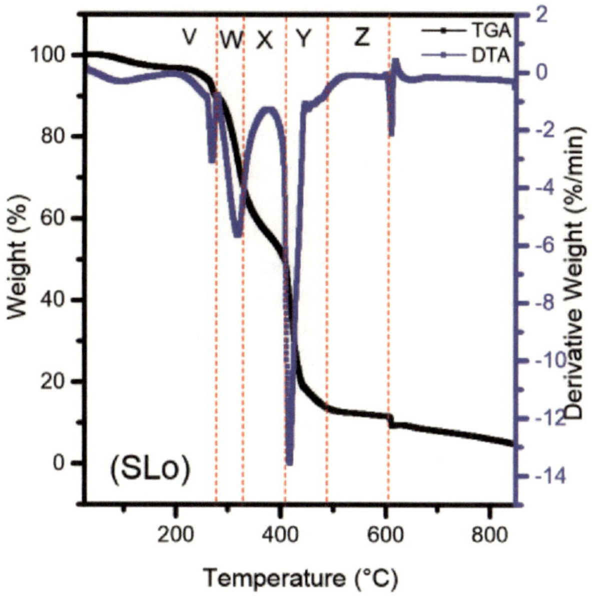

Figure 5.
DT-TGA plots showing the thermal effect on the optimized modified silane sorbent.

Kinetic Model	Parameters	Value
	Q_e	12.020
Pseudo-first-order	K_1	0.013
	R^2	0.950
	Q_e	15.040
Pseudo-second-order	K_2	0.611
	R^2	0.843
	K_3	0.145
Intraparticle diffusion	C	8.717
	R^2	0.720

Table 5.
Kinetic parameters for modified/optimized kenaf shive sorption in oil/water system.

3.4 Adsorption isotherm

Sorption isotherms describe the equilibrium existence between the liquid and solid phase, however, shows the interrelation between solute and sorbent. It is therefore, important in the sorbent optimization. Besides, it also gives the capacity of the sorbent and the equilibrium relationships between sorbent and sorbate. In other words, the ratio between the quantity sorbed and the remaining in solution at fixed temperature at equilibrium. In this study the data are fitted into prominent models; Langmuir and Freundlich isotherms. These isotherm models were depicted in **Figure 4** whose constant values express the affinity of sorbate to surface of sorbent.

The Langmuir Isotherm model was developed to describe a monolayer sorption onto a solid surface of specific finite number of identical binding sites. This model shows the equilibrium distribution of sorbate onto solid or liquid sorbents with the assumption monolayer formation on homogenous energy surface. The sorption mechanisms in this model involve three steps: the diffusion of ions residue to the extemal surface of sorbent; the diffusion into the pores of sorbent; and the sorption of the residue on the intemal surface of sorbent.

Initial concentration and contact time are the basic factors that affect the first part of this model and the final part is considered as rate determining step that is relatively quick process. Linearized form of Langmuir equation was used in this studies.

The Freundlich isotherm model is applied in the intensity estimation of sorbent towards sorbate. One major characteristic of the Freundlich isotherm, though not based on a theoretical background, is its ability to give a good representation of equilibrium data over a restricted range of concentration. The model assumes that the removal of crude oil molecules occurs on a heterogeneous sorbent surface and can be applied to multilayer sorption. The equilibrium data were treated with the linearized Freundlich isotherm equations.

The mathematical model for the adsorption isotherm for modified and optimized kenaf shive sorbent in an oil/water mixture at 313 K is presented. The results are shown in **Figure 4** and **Table 6**. Comparison of the R^2 values (**Table 6**) reveals that the Langmuir model is the best fitting to explain the adsorption of crude oil from the optimized kenaf shive sorbent (SLo).

3.5 Thermodynamic studies

The thermodynamic parameters, values ΔG and ΔH can be calculated by plotting *ln* (Qe/Ce) versus $1/t$ (**Figure 6** and **Table 7**). The ΔG values of the developed sorbent ranges between approximately-1.9 to 2.8 kJ/mol at temperatures 303, 313, 323, 33 K, indicates that in the adsorption process, crude oil molecules are relatively spontaneous for the mixture on to the surface of the optimized silane sorbent. This appeared for the sorbents having a negative ΔGs, however, for positive ΔG appeared implied nonspontaneous sorption process. It also observed that as the temperature increases ΔG reduces, in other words is inversely related with temperature. Consequently, higher temperatures leads to weaker driving force of adsorption, in addition, lead to more difficult sorption of the oil [14]. If $\Delta S < 0$, then the oil molecules movement in the developed sorbent is said to be limited and show a level of orderliness as well as decrease in randomness at the solid-mixture interface during the adsorption of crude oil/seawater system due to the highly ordered crude oil molecules in the hydrophobic layer of the sorbents at adsorption equilibrium. In other words, negative ΔS (entropy) shows an associated mechanism of the reaction and is enthalpy driven [33]. The negative enthalpy (ΔH) attributed to the exothermic behavior of the sorption phenomenon [14, 36].

Isotherm Model	Isotherm Constants	Temperature (313 K)
Langmuir	Q_o	12.60
	K_1	0.030
	R^2	0.940
Freundlich	n	0.600
	K_2	0.180
	R^2	0.840

Table 6.
Thermodynamic parameters for the sorption of crude oil onto optimized kenaf shive sorbent.

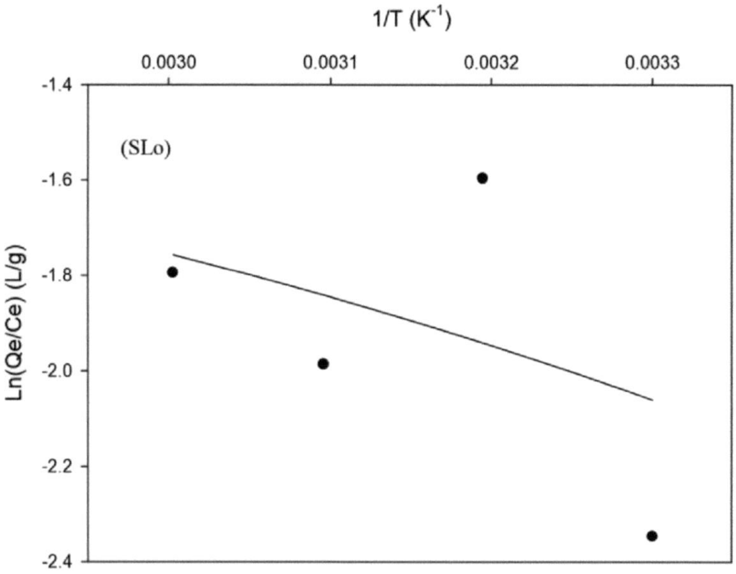

Figure 6.
Plot of $\ln\left(Q_e/C_e\right)$ *against* $\frac{1}{T}$ *for crude oil adsorption of optimized kenaf shive sorbent for thermodynamics parameters.*

T (K)	$\Delta G\left(\mathrm{kJ.mol^{-1}}\right)$	$\Delta H\left(\mathrm{kJ.mol^{-1}}\right)$	$\Delta S\left(J.\mathrm{mol}^{-1}.K^{-1}\right)$	Ea $\left(\mathrm{kJ.mol^{-1}}\right)$	R^2
303	−1.90				
313	−2.20	−7.04	−29.50	25.30	0.9360
323	−2.50				
333	−2.80				

Table 7.
Thermodynamics parameters for crude oil sorption on optimized kenaf shive sorbent.

3.6 Activation energy

Activation energy, Ea is an important thermodynamic parameter which must be overcome by a sorbate before sorption interaction occur with the functional groups of the sorbent surface.

The activation energy can be determined from the change of the absorption rate constant, k with temperature, T(K) using the Arrhenius equation [14]:

$$lnK \; = lnA - \frac{E_a}{RT}$$

Where A is the pre-exponential factor and R is the gas constant $(8.314\ \mathrm{J/mol\,K})$. By plotting $\ln\left[k_1\right]$ against 1/T, E_a and lnA can be calculated respectively, from the slope and intercept. The pseudo-first-order constant was used in the activation energy manipulation because the kinetic equation that best fitted the kinetic models is the pseudo-first-order.

In this studies, the best kinetic model of each sorbent was used at different temperatures of 303, 313, 323 and 333 K. The natural logarithms of the absorption rate constants, k_1 was plotted against the 1/T. In a nut shell, the sorbents that were best

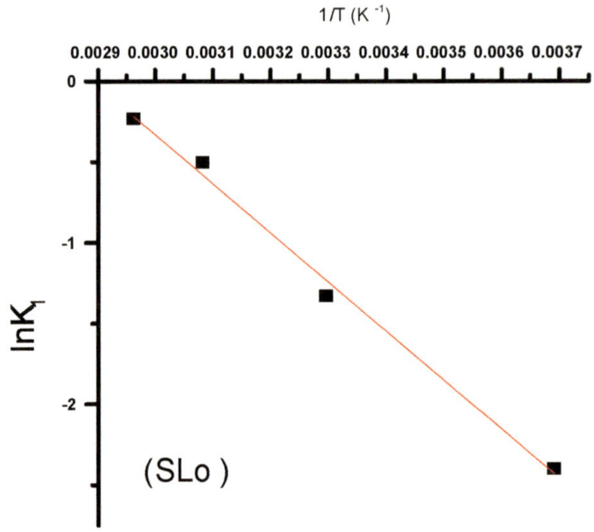

Figure 7.
Plot of $\ln k_1$ *against* $\frac{1}{T}$ *for crude oil adsorption of optimized silane kenaf shive sorbent for activation energy parameters.*

fitted with say, Pseudo-first-order, the rate constant k, was determined at four different temperatures. However, such rates were plotted against the corresponding 1/T.

Plots of Ink_1 versus 1/T are presented in (**Figure** 7), the activation energy value is presented in **Table** 7.

Generally speaking, the developed sorbent has lower activation energy because is between 5-50 kJ/mol [14]. Pseudo-second-order model has higher binding energy than those of the pseudo first-order model. This is because the corresponding models used for the absorption process controlled by chemisorption, which involves higher forces than in physic-sorption. Moreover, the physisorption phenomenon that was observed by the sorbents/mixture interface is an isosteric heat behavior of its enthalpy (ΔH) [15].

4. Conclusion

A silanized kenaf shive sorbents were feasible via surface deposit technique. Effect of some important parameters were studied and optimized using Response Surface Methodology that increase the sorption capability to $> 12g/g$. Containment of this menace using this agro-based waste with no/and or little economic value make an economic sense besides its eco-friendliness. To ascertain the feasibility of this facile and robust sorbent, analytical tests were carried out on the optimized sorbent such as: FTIR, BET, DT-TGA and XRD which show a backing information to this great achievement. Of course, in order to complete the studies enthertor, a critical study on sorption phenomenon were undertook such as: kinetics, isotherms and thermodynamics which respectively, reveals the fitness of pseudo-first-order, Langmuir and physic-sorption of the developed sorbent with an exothermic reaction process. From all the obtained results, show that sorbent from kenaf shive serves as an alterative for crude oil containment and recovery with economic value.

Author details

Zwahruddeen Muhammad Salisu[1*], Diya'uddeen Bashir Hasan[3], Yahaya Gambo Liman[1], Suleiman Mohammed Awal[2], Ukanah Suleiman Pendo[1], Mohammed Kabir Yakubu[2] and Ishiaku Semo Umaru[2]

1 National Research Institute for Chemical Technology (NARICT), Zaria, Nigeria

2 Department of Polymer and Textile Engineering, Ahmadu Bello University (ABU), Zaria, Nigeria

3 Center for Energy Research and Training (CERT), Ahmadu Bello University (ABU), Zaria, Nigeria

*Address all correspondence to: zwahruddeen.ms@narict.gov.ng; ruddeen.zms@gmail.com

References

[1] Lu, Y., and Yuan, W., 2017. Superhydrophobic/Superoleophilic and Reinforced Ethyl Cellulose Sponges for Oil.Water Separation: Synergistic Strategies of Crosslinking, Carbon Nanotube Composite and Nanosilica Modification. ACS Appl. Mater. Interfaces, DOI: 10.1021/acsami.7b09160.

[2] Ha HT, Son LT, Viet NTB, Dung NT, Khoi NV, et al. (2016) Oil Sorbents based on Methacrylic Acid-Grafted Polypropylene Fibers: Synthesis and Characterization. *J Chem Eng Process Technol* 7: 292. doi: 10.4172/2157-7048.1000292

[3] Zhang, C.; Chong, D.; Zhang, H.; Peng, S.; Xin, W.; Hu, Y. Regeneration of mesoporous silica aerogel for hydrocarbon adsorption and recovery. Mar. Pollut. Bull. **2017**, 122, 129-138.

[4] Bhairavi D.; Mika S.; Simo K. 2018. A review of bio-based materials for oil spill treatment. Journal of Water Research 135 (2018) 262-277

[5] Salisu, Z.; Umaru, IS.; Danladi, A.; Yakubu, MK.; Diya'uddeen, BH. 2019. Optimisation of crude oil adsorbent developed from a modified styrene kenaf shive. *Journal of Materials Science and Chemical Engineering, 2019*, 7, *38-51*, https://doi.org/10.4236/msce.2019.72004. *ISSN Online: 2327-6053, ISSN Print: 2327-6045*

[6] Ajay, K.; Amit, K.; Gaurav, S.; Ala'a, H.; Al-M, M.N.; Ayman, A.G.; Florian, J. S. Quatemary magnetic BiOCl/g-C3N4/Cu2O/Fe3O4 nano-junction for visible light and solar powered degradation of sulfamethoxazole from aqueous environment. Chem. Eng. J. 2018, 334, 462-478.

[7] Naushad, M.; Ahamad, T.; Al-Maswari, B.M.; Alqadami, A.A.; Alshehri, S.M. Nickel ferrite bearing nitrogen-doped mesoporous carbon as efficient adsorbent for the removal of highly toxic metal ion from aqueous medium. Chem. Eng. J. **2017**, 330, 1351-1360.

[8] Liu, H., Geng, B., Chen, Y., Wang, H., 2017. Review on the aerogel-type oil sorbents derived from nanocellulose. ACS *Sutainable Chem. Eng*. 5, 49-66

[9] NOAA, 2017. Spill Containment Methods. Office of Response and Restoration [Online] Available at: h ttps://response.restoration.noaa.gov/oil-and-chemicalspills/oil-spills/spill-containment-methods.html.

[10] Singh, V.; Kendall, R. J.; Hake, K.; Ramkumar, S.; 2014 "Crude Oil Sorption by Raw Cotton" *IndEng*. Chem. Res. 2013, 52, 6277.

[11] Son, T. Nguyen; Jingduo, Feng; Nhat, T. Le Ai; T. T. Le; Nguyen, Hoang; Vincent, B. C. Tan; Hai, M. Duong; 2013, "Cellulose Aerogel from Paper Waste for Crude Oil Spill Cleaning". Industrial and Engineering Research, dx.doi.org/10.1021/ie4032567

[12] Chang, S.E., Stone, J., Demes, K., Piscitelli, M., 2014. Consequences of oil spills: a review and framework for informing planning. Ecol. Soc. 19 (2), 26.

[13] ITOPF, 2014. ITOPF Response Techniques [Online] Available at: http://www.itopf. com/knowledge-resources/documents-guides/response-techniques/.

[14] Salisu, Z.; Umaru, IS.; Danladi, A.; Yakubu, MK.; Diya'uddeen, BH. 2019. Recovery of crude oil from aqueous medium by optimised styrene/kenaf shive graft-based sorbent via regeneration method: study of the equilibrium, kinetics and activation energy. *World Journal Innovative Research* (*WJIR*). ISSN: 2454-8236,

Volume-6, Issue-2, February 2019 Pages 27-34

[15] Salisu, Z.M., Yakubu, M.K., Diya'uddeen B.H., Ishiaku, S.U., Abdullahi, D. 2019c. Development Of Kenaf Shive Bio-Mop Via Surface Deposit Technique For Water Remediation From Crude Oil Spill Contamination. *J Results in Engineering*, Elsevier, dx.doi.org/10

[16] Junaid Saleem, Muhammad Adil Riaz, Gordon McKaya. 2018. Oil sorbents from plastic wastes and polymers: A review. Journal of Hazardous Materials 341 (2018) 424-437. www.elsevier.com/locate/jhazmat

[17] Si, Y., Guo, Z. 2015. Superwetting materials of oil water emulsion separation, Chem. Lett. 44 (2015) 874-883, http://dx.doi.org/10.1246/c 1.150223.

[18] Ge, J., Zhao, H., Zhu, H., Huang, J., Shi, L., Yu, S. 2016. Advanced sorbents for oil-spill cleanup: recent advances and future perspectives. *Advances Material* (2016)10459–10490, http://dx.doi.org/10.1002/adma.20160812.

[19] Al-Majed, A.A., Adebayo, A.R., Hossain, M.E. 2014. A novel technology for sustainable oil spills control, Environ. Eng. Manage. J. 13 (2014) 265-274

[20] Yu, L., Hao, G., Xiao, L., Yin, Q., Xia, M., Jiang, W. 2017. Robust magnetic polystyrene foam for high efficiency and removal oil from water surface, Sep. Purl Technol. 173 (2017) 121-128, http://dx.doi.org/10.1016/j.seppur.2016.09.022.

[21] Shah DU, Porter D, Vollrath F. Can silk become an effective reinforcing fibre? A property comparison with flax and glass reinforced composites. Compos Sci Technol 2014; 101: 173-183.

[22] Mustafa A, Bin Abdollah MF, Shuhimi FF, Ismail N, Amiruddin H, Umehara N. 2015. Selection and verification of kenaf fibres as an altemative friction material using Weighted Decision Matrix method. Mater Des 2015;67:577-582.

[23] Pickering, K.L., Aman Efendy, M. G., Le, T.M. 2016. A review recent developments in natural fibre composites and their mechanical performance. *Composites*: Part A 83 (2016) 98-112

[24] Wu, C.J.; Li, Y.F.;Woon,W.Y.; Sheng, Y.J.; Tsao, H.K. Contact Angle Hysteresis on Graphene Surfaces and Hysteresis-free Behavior on Oil-infused Graphite Surfaces. Appl. Surf. Sci. **2016**, 385, 153-161.

[25] Almasian, A.; Jalali, M.L.; Chizari Fard, Gh.; Maleknia, L. Surfactant grafted PDA-PAN nanofiber: optimization of synthesis, characterization and oil absorption property. Chem. Eng. J. **2017**, 326, 1232-1241.

[26] Cai, S. Kimura, M. Wada, S. Kuga, L. Zhang, 2008. Cellulose aerogels from aqueous alkali hydroxide-urea solution, ChemSusChem 1 (2008) 149-154

[27] Arfaoui, MA.; Dolez, PI.; Dube, M.; David, E. 2017. Development and characterization of hydrophobic treatment jute fibres based on zinc oxide nanoparticles and a fatty acid. Appl. Surface Sci.397, 19-29.

[28] Wang, A., et al., 2017a. The enhanced stability and biodegradation of dispersed crude oil droplets by Xanthan Gum as an additive of chemical dispersant. *Mar. Pollut*. Bull. 118, 275-280.

[29] Wang, Z.; Ma, H.; Chu, B.; Hsiao, B. S. Super-hydrophobic modification of porous natural polymer "luffa sponge" for oil absorption. Polymer **2017b**, 126, 470-476.

[30] Chen, Y.; Zhang, D., 2014. Adsorption kinetics, isotherm and thermodynamics studies of flavones from Vaccinium bracteatum Thunb leaves on NKA-2 resin, Chem. Eng. J. 254 (2014) 579-585.

[31] Babatope A.O., Funmilayo O. 2017. Comparative adsorption of crude oil using mango (*mangnifera indica*) shell and mango shell activated carbon. doi. org/10.4491/eer.2017.011, *University Lagos, Lagos, Nigeria*

[32] Jingduo Feng, Son T. Nguyen, Zeng Fan, Hai M. Duong. 2015. Advanced fabrication and oil absorption properties of super-hydrophobic recycled cellulose aerogels. Chemical Engineering Journal 270 (2015) 168-175.

[33] Sagnik Chakraborty, Shamik Chowdhury Papita Das Saha. 2012. Insight into biosorption equilibrium, kinetics and thermodynamics crystal violet onto *Ananas comosus* (pineapple) leafpowder. *Appl Water Sci* (2012) 2: 135-141, DOI 10.1007/sl320l-Ol2-0030-9

[34] Boparai, H.K., Joseph, M., O'Carroll, D.M., 2011. Kinetics and thermodynamics of cadmium ion removal by adsorption onto nano zerovalent iron particles, J. Hazard. Mater. 186 (2011) 458-465.

[35] Saha, P., Chowdhury, S., Insight into Adsorption Thermodynamics, in: M. Tadashi (Ed.), Thermodynamics, InTech, 2011, p. 450.

[36] Iman Mobasherpour, Esmail Salahi, Mohsen Ebrahimi. 2011. Thermodynamics and kinetics of adsorption of Cu(II) from aqueous solutions onto multi-walled carbon nanotubes. Joumal of Saudi Chemical Society2011, 586, 302-314.

Chapter 10

Differential Impact of the Prior Mix by Stirring in the Biodegradation of Sunflower Oil

Pedro Eulogio Cisterna Osorio and Barbara Faundez-Miño

Abstract

Fats and oils present in wastewater are usually eliminated by physical and biological processes. In this experience, the fatty wastewaters are treated biologically, and it assesses the impact of the mix in the fats and oils biodegradation and carried out the experiments in a laboratory scale unit. The biodegradation of fats and oils was analysed in two sceneries, with mix previous by mechanical agitation and without mix. Key parameters were monitored, such as the concentration of fats and oils in the influents and effluents, mass loading, and the efficiency of biodegradation. The mass loading range was similar in both sceneries. In the experimental activated sludge plant without mix, the biodegradation of fats and oils reached levels in the range of 28 to 42.5%. For the wastewater treatment plant with a previous mix by mechanical agitation, the levels of biodegradation of fats and oils ranged from 64 to 75%. Therefore, considering the efficiency of the biodegradation of fats and oils in both sceneries, the results indicated that the level mix is a high incidence.

Keywords: biodegradation, fats and oils, activated sludge

1. Introduction

Nowadays, the growing sensitivity with respect to environmental protection has led to increasing regulatory pressure on wastewater treatment, imposing severe limitations on pollutant concentrations before discharging them to the environment. In this context, one of the major challenges is represented by the biological treatment of oily wastewater [1]. Domestic and industrial wastewaters contain fats and oils. The fraction of lipids in urban wastewater is 30–40% of the chemical oxygen demand (COD) [2].

The biodegradation of lipids in activated sludge processes is not well known. The literature states that these can be treated by biological treatment, which eventually causes foam formation composed of filamentous bacteria and flocs that affect biodegradation [3]. Considering generic information, fats and oils are classified as slowly biodegradable substances. As for the biodegradation process, bacteria initially save these substances in their cytoplasm and later through the enzymatic process; it starts hydrolysis to produce an assimilable substrate that can be biodegraded [4].

There are three types of reactions catalysed by microbial enzymes: oxidative, hydrolytic, and synthesis. Hydrolytic enzymes are used to hydrolyse insoluble complex compounds, such as fats and oils, on simple components that pass through the cellular membrane by diffusion. These enzymes (i.e., oxidoreductases) act outside the cell wall [5]. The extracellular enzyme called lipases is the most common, releasing fatty acids as a consequence of enzymatic action [6]. Lipases break down molecules into simpler components, which appear as end products or intermediates that are consumed by microorganisms. If the solid substrate is sufficiently porous, the enzyme can diffuse and biodegradation can take place inside it; for low porosity material, such as oils where enzymes cannot diffuse, the reaction takes place on the outer surface of the particle [7].

Furthermore, a wider range of microorganisms biodegrades fatty acids from other microorganisms that do not produce extracellular lipolytic enzymes [8]. Wastewater with high lipid concentration inhibits the activity of microorganisms in biological treatment systems, such as active sludge and methane fermentation. To reduce such inhibitory effects, microorganisms capable of effectively degrading edible oils can be selected from different environmental sources [9]. There are many researches that study about elimination and biodegradation of lipids by biological treatment [10]. Biodegradation of fats and oils and other substrates, not soluble in water, is one of the greatest problems for the biological treatment of wastewater [11].

A wide variety of organic compounds such as carbon and energy are used as a source by microorganisms. When substrates possess low or zero solubility, the use of biosurfactants is recommended [12]. One of the main characteristics of emulsified mixtures is the presence of at least one hydrophilic polar liquid and at least one lipophilic; the simplest and most frequent case is when liquids are oil and water [13]. Depending on the emulsification process, the diameter of droplets in the internal phase ranges from 0.1 μm to 0.1 mm. Usually, there is a wide size distribution of bubbles; a narrow distribution can be considered when the ratio between the smallest and largest size droplets is 1:10. Emulsions of this class are normally thermodynamically unstable, so it is known the tendency to reduce the interface area between aqueous and oily phase, causing the oil droplets to coalesce. Coalescence of oil droplets can be reduced or even eliminated through stabilization mechanisms [13].

The unstable nature of an emulsion is because contact between oil and water molecules is not energetically favourable. Therefore, surfactants (emulsifiers) are added to the emulsion to improve the system stability; molecules inside are adsorbed to the surface of oil droplets during homogenization, providing a protective membrane that prevents flocculation and coalescence [14].

In a reactor, energy is delivered to the system, by mechanical agitation or electric fields, and the efficient contact of the phases increases the interfacial area and the transfer of matter. Mechanical agitation is simpler and of greater operational variety. Agitation speed is an important factor in the industrial application for the mixing efficiency to increase productivity [15]. The use of electric fields is more energetically efficient since the electric forces are applied on the interface of the fluids, unlike the mechanical agitation that delivers the energy to the bosom of the liquid and only a part of it is transferred to the interface [16]. A third technique to achieve an adequate mixture is ultrasound, which has mechanical and/or chemical effects, the first is due to the implosion of microbubbles, generating highly reactive free radicals, and the mechanical is caused by wave shocks during symmetric cavitation [17].

In biological treatment by activated sludge, the magnitude of the contact area of phases, water and oil is very important; therefore, a significant interfacial area is required. The interfacial area can be expanded by delivering energy to the system through mechanical stirring or an electric field. The increment of interfacial surface between the aqueous phase and the oily phase is often implemented by mechanical stirring [18].

It is known that the smaller the size of the emulsions, the more the biodegradation process is favoured, since the interface area increases and, therefore, the possibility that lipases in an aqueous medium can carry out the oil hydrolysis. As the stirring speed increases, the average droplet size decreases steadily [19]. It was observed that the enzymatic pre-hydrolysis under the influence of ultrasound drastically reduces the reaction time from 24 h to 40 min as compared to conventional stirring with improved yield [20].

Electrical potentials are also applied to reduce emulsion sizes and thereby increase the homogeneity of the oil and water system. There are studies, in which the following bubble sizes between 1 and 36 μm were obtained for a stirred system with two electrodes [21]. When electric fields are applied to increase the degree of dispersion of the discontinuous phase in the continuous one, the bubbles acquire a surface charge, which generates a self-rejection that leads to a reduction in the sizes of the bubbles [22].

The fatty acids produced during the course of the reaction act as surfactants, stabilizing the emulsions [18]. Biosurfactants are surface-active molecules that are produced by a wide range of microbes including bacteria, fungi, and yeast. They have several advantages over the chemical surfactants such as higher biodegradability, lower toxicity, better environmental compatibility, high selectivity, higher foaming, and specific activity under extreme conditions such as temperature, pH, and salinity [23].

Synthetic surfactants increase the desorption and solubilization of nonpolar compounds in soils and aquifer materials, but they cause environmental problems during their production, and they are resistant to biodegradation and can be toxic when they accumulate in the ecosystem [24].

Mechanical agitation is susceptible to improvement through the application of a non-stationary behaviour that consists of the displacement of the rotating propeller from top to bottom, which cancels the existence of segregated regions that are not affected by stable agitation due to the position of the propeller in the pond, and another mechanism is counter-rotation [25].

Mixing and corresponding dispersion is achieved in the aeration tank used; there are two important factors that largely determine the emulsion level: bubble size and distribution, and the fraction of the dispersed phase. The average bubble size is between 150 μm and250 μm [16], which can be obtained by means of a suitable booster and fine bubble diffusers. If there is a suitable enzyme concentration and the optimal interfacial area between phases: aqueous and oily, the mass transfer is solved and the hydrolysis stage starts [26]. Further studies on the effects of agitation speed from 0 rpm (static) to 200 rpm on tannase production showed that increasing agitation speed caused the fungal pellets to decrease in size but to increase in number per unit volume, increasing the interface area [27].

This work analyses and evaluates the differential impact caused by mixing by stirring on the behaviour of the biological treatment system for activated sludge on a laboratory scale when treating an influent containing fats and oils. In this experience, we work with vegetable oil.

2. Materials and methods

2.1 The choice of sunflower oil is explained based on the following criteria

- It is the most frequent oil used in Chile;
- It is well standardized, and
- It is a highly accessible product.

The chemical composition of sunflower oil is shown in **Table 1**.

Fat	Position	16:0	18	18:1(9)	18:2(9,12)	18:3(9,12,15)
Sunflower oil girasol	1	10.6	3.3	16.6	69.5	—
	2	1.3	1.1	21.5	76.0	—
	3	9.7	9.2	27.6	53.5	—

Table 1.
Fatty acids that make up sunflower oil and specific stereo analysis [18]. Results in % moles.

2.2 Physicochemical parameters and analytical methods

2.2.1 Chemical oxygen demand (COD)

The potassium dichromate method was used to evaluate COD levels. The method used is a variation of the standard method [28]; however, it maintains its basis. The variation used has the advantage it uses a significantly smaller sample and reagents. The sample is chemically oxidized through the action of potassium dichromate at 150°C for 2 h. Silver sulphate is used as a catalyst and mercury sulphate is used to avoid possible interferences with chloride. Afterward, determination by spectrophotometry at 600 nm is performed. Equipment and instruments are used to determine the various parameters to characterize the wastewater used.

2.2.2 Fats and oils

Determination of the fats and oils was used in Gravimetric Assay Soxhlet method. This method quantifies substances with similar characteristics, based on its common solubility in appropriate solvent, 213E method [29].

Total suspended solids (TSS) are determined by filtering a known volume of the sample on Whatman (Whatman plc., Maidstone, UK) 4.7 cm GF/C glass fibre filters and then drying it at 103–105°C. The difference in weight of the filter before and after filtration is used to estimate the TSS, 209C method [29].

Volatile suspended solids (VSS): The volatile suspended solids are determined by weight loss after calcination at 550°C, 208E method [29].

2.3 Continuous equipment

In the current investigation, an activated sludge plant at the laboratory scale is used to conduct biodegradability tests in wastewater with oils and fats. To meet these objectives, experimental work is required, with such parameter information describing the process dynamics regarding the aeration and sedimentation tanks regarding fat and oil content of the wastewater. For this purpose, BIOCONTROL-MARK 2 equipment was used. The details of the equipment are shown in **Figure 1**.

This experimental equipment consists essentially of the following parts:

- Control unit

Composed of the main switch, air cylinder, it is complemented with a flow meter and a flow regulation system. Additionally, the wastewater feed pump, that includes a flow rate regulation system, a timer for intermittent operations, and an ON–OFF switch allowing sludge recycling from the sedimentation tank to the aeration tank.

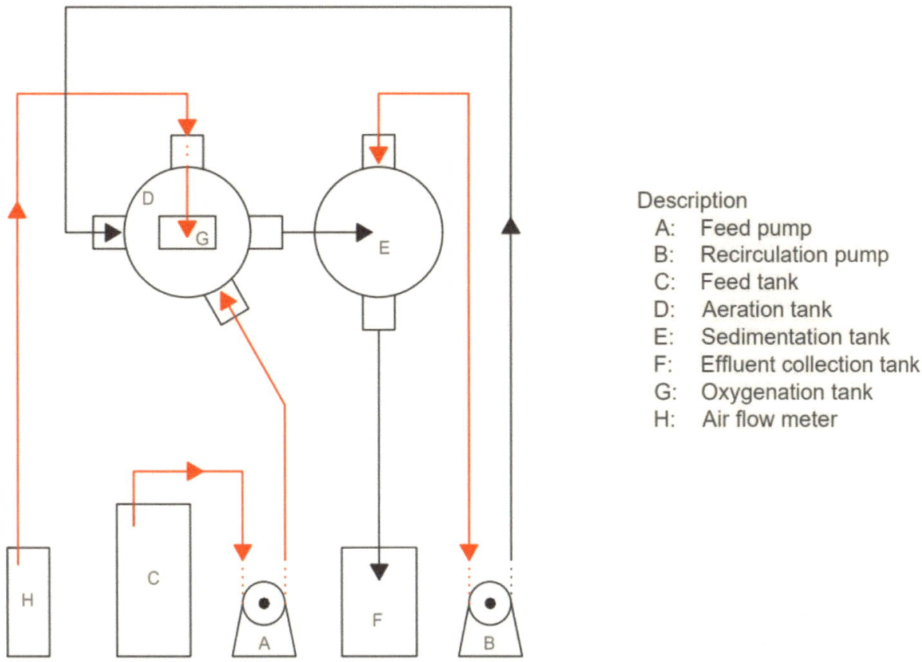

Figure 1.
Experimental equipment diagram.

- Aeration tank

It consists of a transparent Plexiglas® (Vittadini Riferimenti, Milan, Italy) cylinder with a height of 38 cm and a diameter of 20 cm, which has outlets at various heights associated with different volumes (7, 8, 9, and 10 l). There are two separated inlets allowing recirculation of sludge from the top. The influent to be treated is placed at the bottom. In addition, the system has two ceramic diffusers placed in the bottom in a way that they can disperse the air in tiny bubbles [30].

- Sedimentation tank

This consists of a transparent Plexiglas® (Vittadini Riferimenti, Milan, Italy) cylinder, where its lower part is cone-shaped to make sludge sedimentation and thickening easier.

The mixed liquor is fed from the aeration tank, which has an outlet in its upper part. This flow escapes by overflow, when it arrives in the sedimentation tank. The solid phase decantation gives a method to downward flow. Decanted sludge is separated and recirculated at the bottom through the pump towards the aeration tank. Treated water also uses the overflow mechanism to be evacuated to the storage tank.

2.4 Experimental methodology

a. Feed preparation

The treatment system was initially fed with synthetic wastewater prepared in the laboratory, according to strong urban wastewater typical characteristics of [31]. This wastewater has an approximate BOD of 400 mg/l, with the corresponding proportions of nitrogen and phosphorus in a relation to BOD: N: P = 100:5:1. Approximately 400 mg of saccharose, 20 mg of phosphate hydrogen of potassium, and 100 mg of ammonium chloride were added per litre of water. Measurements begin when sunflower oil is added, the concentration of this substrate gradually increases. Feed was prepared daily and nitrogen and phosphorus increased according to the organic input coming from fats and oils.

b. Operating modes

The synthetic wastewater was poured into a storage pond of approximately 50 l, where the stirring unit has been installed to disperse oil or fat. Through a peristaltic pump, controlled by the control unit, it drives the feed to the aeration tank. Oxygen feed and recirculation flow are controlled by the control unit. Process effluent is collected in a 30-l volume tank, where the samples are taken to be processed. The flow of synthetic wastewater is 25 l/day.

2.5 Mass balance: biodegradability determination

Experimental method is established in this protocol, which allows to carry out the material balance of fats and oils. From the process diagram, **Figure 2** shows flows and concentrations of inlet and outlet streams in different stages of the activated sludge process are indicated. To carry out material balance and estimate biodegraded oil in the activated sludge process, the oil in the feed tank must be estimated.

This experience works with influents that are biologically treated by active sludge, containing only vegetable oil. This retained fraction is an indicator of the mixing level reached in the feed tank, which corresponds to not emulsified oil, and therefore, it is not part of the influent entering the aeration tank; obviously, the size of this fraction indicates the mixing level of the system quantitatively.

Where:

F0, C0: Flow and concentration oil entering the mixing tank.

F1, C1: Flow and feed oil concentration to the aeration tank.

F2, C2: Flow and concentration of fats and oil leaving the aeration tank = (Flow and concentration of fats and oils entering the secondary settler).

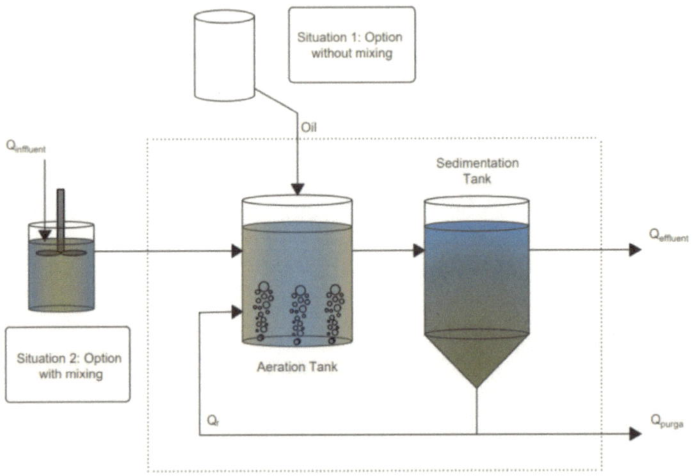

Figure 2.
Active sludge plant process diagram.

Operation days for balance sheet	Oil mass fed (g)	Retained oil mass (g)	Oil mass in effluent (g)	Accumulated oil mass (g)	Biodegraded oil mass (g)	Degradation efficiency (%)
1 a 8	79	12,3	5,4	6,5	54,8	75
9 a 16	120	28	8	19	65	71
17 a 23	157	70	10	21	56	64

Table 2.
Mass balance in active sludge for oily influent with prior stirring.

F3, C3: Flow and concentration of fats and oil from purified effluent.
F4, C4: Flow and fats and oil concentration in recirculation flow.
M1: Oil mass contained in the mixing liquor of the aeration tank.
M2: Oil mass floating in the upper part of sedimentation tank and oil mass at the bottom of sedimentation tank by biomass attached.

Mass balance and biodegradability determination of fats and oils in bench-scale activated sludge reactor.

In this experiment made in equipment in **Figure 1**, water and oil were mechanically stirred and mixed in the feed tank. Part of aggregated oil was accumulated in the feed tank, so the oil fraction not going to the aeration tank was known. The influent containing sunflower oil and concentration was gradually increased.

Biodegraded sunflower oil mass was determined from the mass balance equation.

Sunflower oil is part of the influent fed to the treatment system.

Initially, it is assumed the system has no oil.

Results of matter balance are shown in **Table 2**.

From material balance, the mass of biodegraded sunflower oil is obtained, which is part of the influent fed to the system.

For the aeration tank:

$$\frac{\Delta M_1}{\Delta t} = F_1 \cdot C_1 - F_2 \cdot C_2 - r_A \cdot V + F_4 \cdot C_4 \tag{1}$$

For the secondary sedimentation tank:

$$\frac{\Delta M_2}{\Delta t} = F_2 \cdot C_2 - F_3 \cdot C_3 - F_4 \cdot C_4 \tag{2}$$

It must be noticed that, $F_6 = 0$.

Where M_1 corresponds to the oils and fats in the aerator tank.

M2: Corresponds to oils and fats in the sedimentation tank.

V: Is the volume reactor.

r_A; Is the vegetable oil biodegradation rate.

The balance for the system as a whole is as follows:

From this expression we have: $r_A \cdot V$, corresponds to vegetable oil that disappears per unit of time, it means the oil is clearly biodegraded by microorganisms, such that:

$$r_A \cdot V = F_1 \cdot C_1 - F_3 \cdot C_3 - \frac{\Delta M_1}{\Delta t} - \frac{\Delta M_2}{\Delta t} \tag{3}$$

An important part of the oil floats and therefore does not biodegrade and passes to a secondary sedimentation pond in which it accumulates. For reasons of technical feasibility, the mass balance must be kept integral for some time. Fats and oils accumulated as a result of separation by flotation are measured, allowing to determine the oil biodegradation level in influent.

The equation for integral material balance for a certain period of time is as follows:

$$M_1 + M_2 = F_1 \cdot C_1 \cdot \Delta_t - F_3 \cdot C_3 \cdot \Delta_t - r_A \cdot V \cdot \Delta_t \quad (4)$$

Where is Δt the time elapsed.

Every day is identified by subscript as shown in the following examples:

Day 1:

$$M_{11} + M_{21} = F_{11} \cdot C_{11} \cdot \Delta_t - F_{31} \cdot C_{31} \cdot \Delta_t - r_A \cdot V \cdot \Delta_t \quad (5)$$

Day 2:

$$M_{1i} + M_{2i} = F_{1i} \cdot C_{1i} \cdot \Delta_t - F_{3i} \cdot C_{3i} \cdot \Delta_t - r_A \cdot V \cdot \Delta_t \quad (6)$$

Then, the mass balance for a given number of days of operation is as follows:

$$r_A \cdot V \cdot \Delta_t = F_{1i} \cdot \Delta_t \cdot \sum(C_{1i}) - F_{3i} \cdot \Delta_t \cdot \sum(C_{3i}) - \sum(M_{1i} + M_{2i}) \quad (7)$$

Where:

$$M_1 = \sum M_{1i}\, M_2 = \sum M_{2i} \quad (8)$$

Then, biodegradability is

$$B = r_A \cdot V \cdot \frac{\Delta_t}{F_{1i}} \cdot \Delta_t \cdot \sum(C_{1i}) \quad (9)$$

From the material balance calculation, the percentage of biodegraded oil of the influent is estimated. The flows and concentrations of the activated sludge process are indicated below.

The feed flow and the concentration of fats and oils (F0, C0) is known, and is set according to the conditions previously defined for the experimental phase.

F1, C1: The flow is determined according to the experimental conditions and regarding the concentration of fats and oils C1, and it is determined by the Soxhlet gravimetric method.

F3, C3: This flow is equal to inflow and, therefore, is given by predetermined conditions of the experiment, and the concentration of fats and oils, C3, is obtained by the Soxhlet method.

M1: This estimates the oil mass accumulated in a given period of time, and it is necessary to remove supernatant fats and oils, dry and weigh them.

M2: The mass of sunflower oil and derivatives accumulated in the secondary sedimentation tank in a determined period of time is determined. The upper or floating phase of the secondary settler is poured into an auxiliary tank, which is oil and its derivatives in humid conditions. Then, the remaining water is evaporated through a heat source and oil mass is obtained by gravimetric. Oil and derivatives accumulated in biomass in secondary settler are added to this mass, which is determined by the Soxhlet method for a 100-ml sample. With the concentration obtained, the oil retained in settled solids is estimated.

2.6 Incidence of agitation and biodegradation of sunflowers oil

In this experience, an influent that only has sunflower oil has been fed to the biological treatment system by active sludge, it is the only carbon source available for microorganisms under two sceneries, with and without previous agitation. Different operational parameters are monitored providing information on system behaviour. The oil biodegradation percentage is estimated by mass balance.

3. Material balance results: without agitation

Biodegradability is calculated from the mass balance in the system and concentrations and quantities of oil obtained. In this case, feeding is not subjected to a previous agitation, and this condition is achieved by introducing oil directly to the aeration tank by dripping, and the total volume entered into the system; it is measured every 24 h.

In the present experiment (**Table 3**), influent fed to active sludge system contains only oil and without stirring, and **Table 3** found that oil biodegradation levels range between 28.1 and 42.5%, which is considerable in spite of not being high. This is an interesting result because it opens the viability of biologically degrading substances, such as fats and oils, which evidently presents important comparative advantages when compared with the removal of fats or oils *via* flotation.

The above is explained by the lack of stirring, and it is worth noting that experience with olive oil production wastewater, a culture medium, with a concentration of 5% was used. Effects of agitation speeds were studied on the growth of species Scenedesmus microalgae, in a photobioreactor. A maximum specific growth rate of 0.031 1/h was obtained, using a speed of rotary impeller around 350 rpm, higher than that found in the absence of agitation 0.024 h^{-1} [32].

On the other hand, the synthesis of bio-lubricant from the effluent of palm oil plant, which is based on enzymatic hydrolysis. Effects of essential parameters were examined, which include stirring speed. The optimal hydrolysis rate (0.1639 mg/s.l) is achieved at 650 rpm, at 40°C, pH 7.0, 20 U ml of loading enzyme [33].

This biodegradation range achieved in this experience, although it is not an optimal result, it shows that oils and fats can be eliminated from influent through biodegradation process; it is also important to note this level of biodegradation can be increased with some modification processes.

The results for the oil removed by biodegradation are due to the fact that a considerable percentage of sunflower oil that floats in the aeration tank reaches the secondary settler, which is due to the very low solubility of substrate in water, which originates a two-phase system. This is increased by the oil tendency to float in water, which is favoured by the flow of air driven from the bottom of the aerobic tank.

Operation days for balance sheet	Oil mass fed (g)	Oil mass in effluent (g)	Accumulated oil mass (g)	Biodegraded oil mass (g)	Degradation efficiency (%)
1 a 22	160	11	81	68	42,5
23 a 28	95	14	52,5	28,5	30,0
29 a 33	90	5,22	59,5	25,28	28,1

Table 3.
Matter balance in active sludge for oily influent without previous stirring.

Figure 3.
Oil and grease accumulated in settling tank.

An increase in emulsification constitutes the vehicle that would generate more favourable conditions for the microbiological attack, since the hydrolysis of oil requires an oil-water interfacial area, which is favoured with the increase in the number of emulsions and with the decrease in their size. One of the causes of the results obtained with respect to biodegradability is the type of hydrodynamics of process reactor, since given the lower oil density in relation to the water and the injection of air from the bottom of the aeration tank, it generates conditions for that the oil floats and, therefore, does not remain in the treatment system during appropriate residence time. There is a part of the oil that is not biodegraded and is separated through flotation (**Figure 3**).

Among the already mentioned characteristics of oils and fats is their low solubility in water and their marked tendency to buoyancy, which is enhanced when the aeration tank is fed with air; as the oil floats, the coalescence process is favoured, reversing the emulsification achieved through aeration by the agitation caused a consequence of air injection into the aerobic pond.

It is noteworthy that the system worked for 50 days and the only carbon source was vegetable oil; with regard to toxicity, some authors [34] claim an inherent toxicity of fatty acids, since they are potential inhibitors of metabolic processes of cells due to their surfactant properties, and for this reason, it is considered they do not remain as such in cells; they are immediately converted into thioesters of their coenzyme A.

As already stated, the low solubility of fats and oils makes it difficult to disperse and distribute this substrate properly in the medium that supports microorganisms in water. In view of the aforesaid, the adequate size of the emulsions is not achieved, so contact surface with the mixing liquor, and therefore with bacterial flocs is limited by this operational factor.

Fats and oils are substances that do not favour the development and growth of bacterial colonies, due to the already explained, but when biomass is acclimatized for a considerable time to the new type of influent, it is found the biomass is adapted.

This is explained from the mutations caused by changes or chemical or physical agents, which change the DNA imparting new characteristics to the cell, thereby

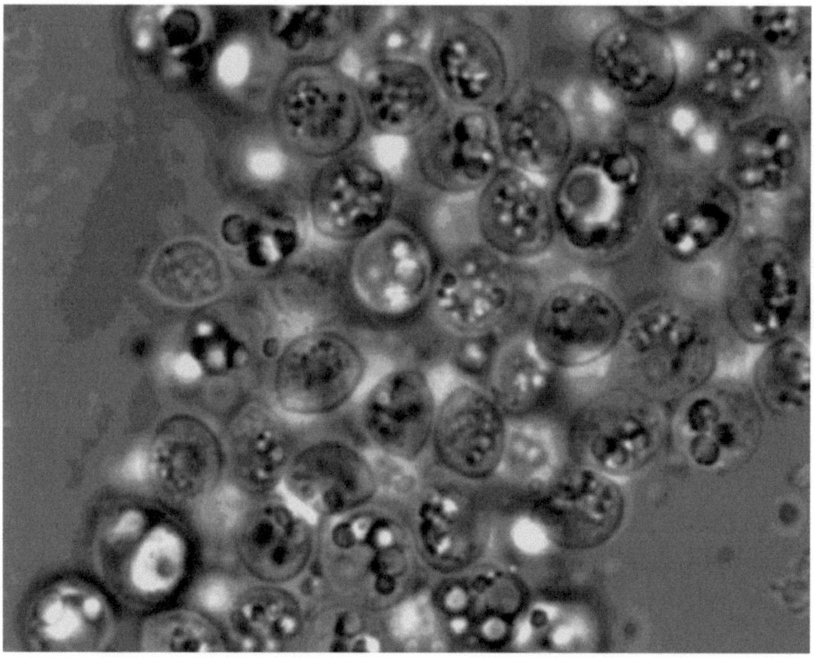

Figure 4.
Acclimatization of bacteria to oil particles.

allowing the cell to degrade a xenobiotic substance or facilitate biodegradation. Spontaneous mutations occur in one out of every 1,000,000 cells; however, the DNA molecule is capable of replicating [35]. **Figure 4** shows the environmental condition of the biomass in an oily medium.

The existence of an oily environment for the fat and oil-degrading biomass is an important factor and has to do with the above-mentioned. In kinetic monitoring experiences of the aerobic biodegradable processes of dairy fats and oils, we worked with native biomass from a lagoon that treated dairy water and commercial bioaugmentation inoculum, and the removal percentage was from 78 to 91% and from 82 to 95%, respectively. However, for high substrate concentrations, the native inoculum is more mineralizing than the commercial one [36].

- Oil concentration ratio in influent and effluent

The oil concentration in the effluent is not correlated with oil concentration in influent (**Figure** 5), which is mainly explained by the fact that an important part of the oil is retained in the sedimentation tank, instead of leaving through the effluent, given its buoyancy. There is no evidence that behaviour regarding the elimination of fats and oils in urban active sludge treatment plants is related to the concentration of fats and oils in influents in concentration ranges, which are the discharged influents usually present [37].

As the important part of oil floats is that it does not enter the interior mixed liquor of the active sludge system, this breaks the relationship that should exist between the concentrations of oil at the outlet and in the inlet, unlike other soluble substrates such as saccharose.

Stirring is the driving force that promotes the oil emulsification in the water, which depends on the flow and air pressure in the active sludge system.
The emulsified oil will be the most susceptible to biodegrade; it must be considered that the airflow also favours flotation, so both effects are opposed.

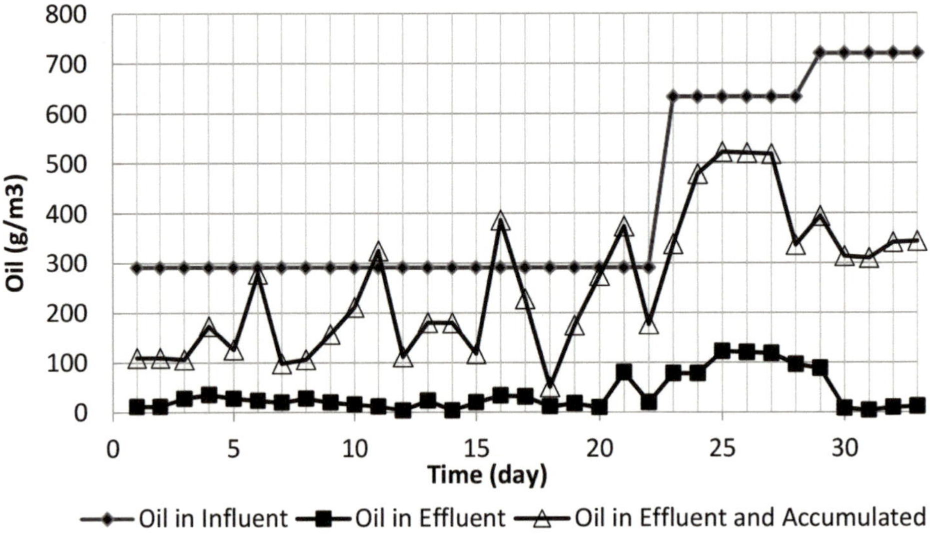

Figure 5.
Performance of oil concentration in influent, effluent, and effluent the most accumulated effluent, without previous stirring.

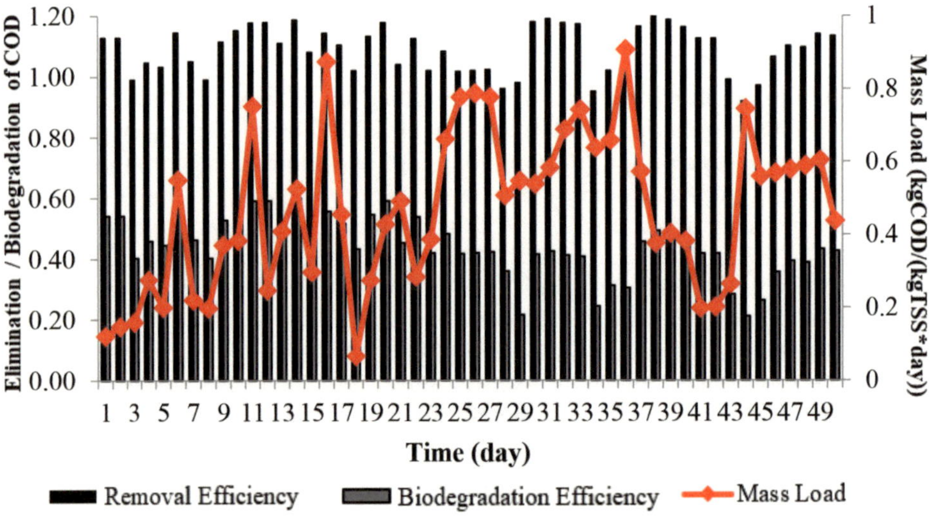

Figure 6.
Performance of COD removal and biodegradation efficiency and mass loading for sunflower oil feed without prior agitation.

Biomass constitutes an element of resistance from a hydrodynamic point of view to the flotation of the oil since these are particles that impede the flow of the air and on the other hand are a source of oil consumption, since they use it as a carbon source. The floating of the oil in the aeration tank is reduced according to the biomass concentration since it opposes resistance to the airflow.

Figure 6 shows that in this experience, removal and biodegradation efficiency do not depend on the mass load, and this indicates the blown air enhances flotation more than emulsification, given the initial system, condition, influencing without previous stirring.

The biodegradation efficiency is low, less than 40%, unlike the elimination that exceeds the average of 85%, which indicates that the flotation phenomenon prevails, it is worth noting that despite this pessimistic condition of mixture and size of emulsions, biodegradation level over 30% is also achieved.

4. Mass balance and biodegradability analysis

In this case, mass balance presents a difference with respect to the balance made for the case where the oil is directly added to the aeration tank, since in this experience the mixture of water and oil is previously prepared in the feed tank; part of the added oil remains accumulated in tank, so it is pertinent to estimate this fraction of oil does not enter the treatment system, as shown in **Figure 2**. In the following experiment, the mixture of water with oil is subjected to the previous stirring to evaluate the mixing effect on biodegradation and oil elimination, since a greater mixture stimulates the development of smaller emulsions and precisely treats to evaluate how this variable affects the biodegradation of fats and oils.

This experience worked with an influent that is subjected to previous mechanical stirring to evaluate the influence of and stirring on biodegradation of oil, since when stirring is carried out before entering the aerator tank, the level of emulsification of oil particles in the water is increased, and with this, the contact area between water-oil and oil-microorganisms is increased, which is closely related to the mass transfer phenomena that involve biodegradation of sunflower oil in active sludge treatment system.

Influents with concentrations of 300, 600, and 900 mg/l are prepared in a feeding tank, for a period of approximately one week, to measure the biodegradability levels achieved, and therefore compare the behaviour of system with results obtained where influent is not previously agitated. The operating conditions regarding airflow, residence time, recirculation ratio, and COD: N: P ratio at the feed level remain unchanged.

Biodegradation levels obtained improve considerably when a mixture of water and oil is subjected to mechanical stirring before being fed to an aeration tank; the agitation carried out is with a propeller-type mechanical stirrer. It should be mentioned that this type of stirring is limited which is significantly reflected in the oil that does not enter the system since it is not emulsified, which gives rise to the fraction of the oil that is retained.

Biodegradation values between 64 and 75% obtained with stirring are similar to those corresponding to aerobic biodegradation experience that eliminates fats and oils from the dairy industry, which uses a mixture of isolated and selected native bacteria as biomass, reaching 72% biodegradation efficiency [38].

Since the only substrate is oil, this system can be assimilated to the proposal for the treatment of fats and oils that is contemplated in the wastewater treatment project of the Los Angeles commune developed by DEGREMONT, which consists of treating the fats and oils together with the sludge in an aerobic digester; therefore, the only available substrate is the fats and oils collected in the primary treatment and that was dosed based on criteria of optimal distribution and mixing [39].

Recent researches have studied the behaviour and performance of aerobic thermophile bacteria for wastewater with a high oily organic content, verifying that between 55 and 58°C, the maximum growth rate is achieved [40]. Now, this is important given that at higher temperatures an increase in the dissolution of fats and oils is achieved, and as it is an aerobic digester, it is more feasible to reach temperatures in this range.

However, analysis of the first-order kinetic model constants showed that alteration in rotor speed resulted in an increase in the values of the kinetic constants (for instance, from 0.57 h^{-1} at 50 rpm to 0.84 h^{-1} at 75 rpm) [41].

It is important to note that the oil that enters the system is the one that has been emulsified and therefore the size of the oil droplets that enter the system reaches a comparatively much smaller size than in the previous case (without mixing), which allows a considerably greater interfacial area between the mixed liquor and the oil.

The bacterial mass is suspended in water, and therefore, contact level between microorganisms and oil droplets is considerably increased, which stimulates the production of the lipase by bacteria. Then, increasing the oil-water interfacial area, where occurs oil hydrolysis, the one that is increased and therefore the amount of fatty acids and glycerol, compounds that bacterial mass will biodegrade them; therefore, the increase of interfacial area allows the substantial increase in biodegradation levels.

4.1 COD elimination

According to observed, it is confirmed that an important part of organic matter corresponding to vegetable oil is accumulated in secondary settler and that the accumulated oil mass is clearly correlated with oil concentration in influent.

The COD values inform us that the sum of oil biodegradation and flotation phenomena makes it possible to achieve global elimination of vegetable oil, which is considerable. On the other hand, it is observed the biodegradation of oil does not depend on the mass load.

Considering the mass load values, it means an initial work with extended aeration and final stage in conventional type regime, which is consistent with theory and experience of wastewater treatment, since differences between both regimes are not observed in effluent quality.

In any case, the oil removal is mainly due to biodegradation of vegetable oil around 70%, while 20% corresponds to flotation (**Figure** 7).

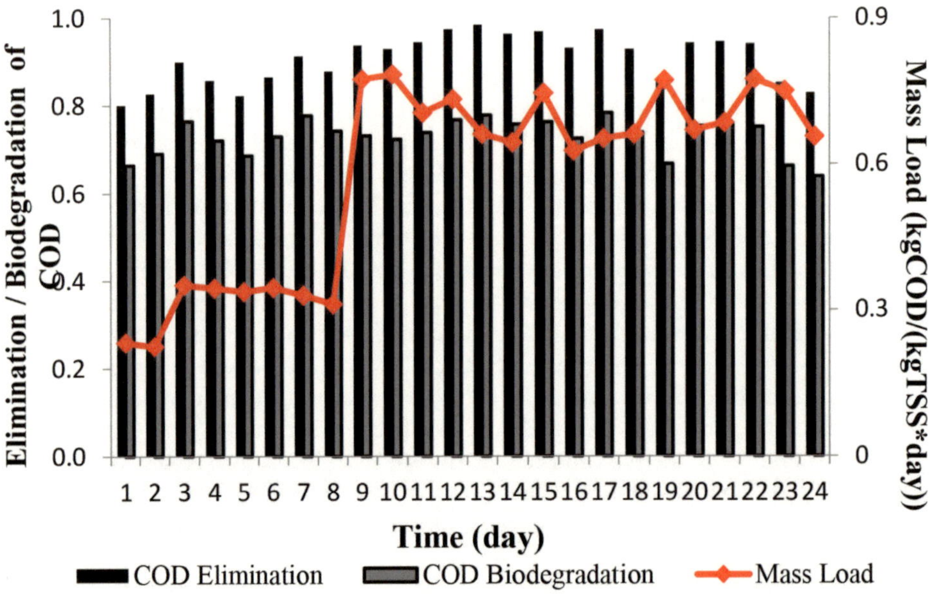

Figure 7.
Performance of COD removal efficiency and mass loading for sunflower oil feed with the previous stirring.

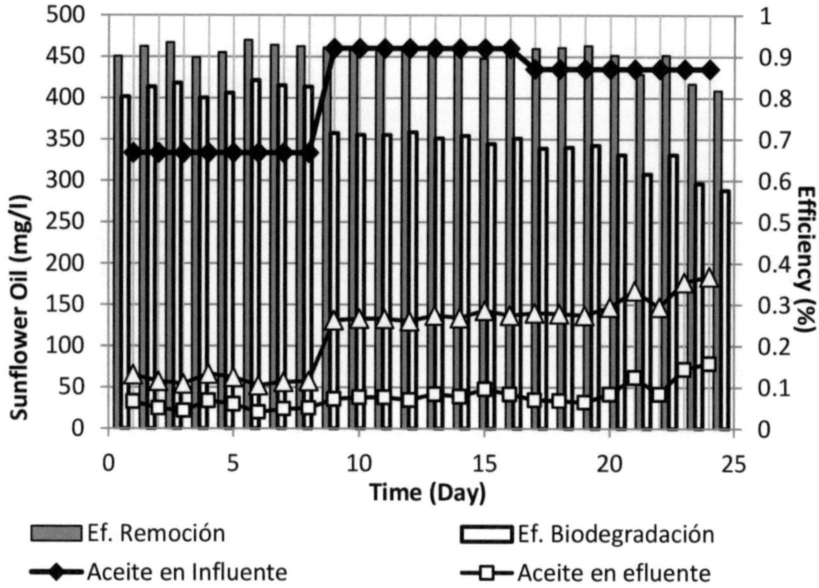

Figure 8.
The behaviour of sunflower oil concentration in influent, effluent, and effluent plus accumulated effluent and oil removal efficiency with prior agitation.

4.2 Removal of vegetable oil

The elimination of oil by biodegradation has a huge advantage over the physical and physicochemical processes currently used in the elimination of this substrate, since they generate a greasy residue that requires a final disposal and tends to accumulate, unlike the elimination by aerobic biodegradation that biologically oxidizes fats and oils to CO2, incorporate them into the carbon cycle.

From **Figure 8**, it can be seen that the concentration of sunflower oil of the influent is practically constant during a determined amount of approximately 8 days, three periods are distinguished, since the feeding during each period has different concentrations of vegetable oil in the influent. Initially, the oil concentration is 220 mg/l, later it increases to a value close to 470 mg/l and for the last days of operation, which corresponds to a concentration of vegetable oil in the influent close to 500 mg/l, this corresponds to a COD concentration above 1100 mg/l, and there is a slight increase in the COD of the effluent.

Regarding the elimination of oil, elimination levels that are around 90% are reached, where a considerable percentage of this elimination corresponds to oil that is biodegraded and the remainder is separated by flotation, as shown in **Figure 8**.

The amount of accumulated oil increases with the concentration of oil in the influent. It should be noted that the increase in oil concentration in the influent causes a decrease in the biodegradation performance of the activated sludge system, which is corroborated with the results of the material balance, such that the biodegradability of the oil decreases by 75% for vegetable oil concentrations of 220 mg/l and up to 64% when the concentration increases to 500 mg/l.

5. Conclusions

For influents with concentrations of fats and oils that range between 200 and 800 mg/l and that are not subjected to a previous mixing, the elimination by biodegradation of the same reaches 42.5 for the concentrations of smaller magnitude and for the concentrations of the highest rank decreases to 28%.

For influents with the concentrations of fats and oils ranging between 330 and 465 mg/l and that are subjected to a previous mixing, their elimination by biodegradation ranges from 64 to 75%, which has as a consequence a considerable reduction of greasy residues that accumulate and take up space to be disposed of.

From the results, it is concluded that the previous mixing is a relevant factor to increase the elimination by biodegradation of fats and oils in an oily influent.

The global elimination that includes biodegradation and flotation exceeds 80% at all events.

The biodegradation efficiency of sunflower oil increases through greater agitation, which is a contribution from the environmental point of view, since fats and oils are eliminated, transforming them into CO2 by the biological route and thus incorporating these residues into the cycle of carbon.

Author details

Pedro Eulogio Cisterna Osorio* and Barbara Faundez-Miño
Department of Civil and Environmental Engineering, University of Bio-Bio, Concepcion, Chile

*Address all correspondence to: pcisterna@ubiobio.cl

References

[1] Capodici M et al. Treatment of oily wastewater with membrane bioreactor systems. Water. 2017;**9**(6):412

[2] Dueholm TE, Andreasen KH, Nielsen PH. Transformation of lipids in activated sludge. Water Science and Technology. 2001;**43**(1):165-172

[3] Chipasa KB, Mędrzycka K. Behavior of lipids in biological wastewater treatment processes. Journal of Industrial Microbiology and Biotechnology. 2006;**33**(8):635-645

[4] Ronzano E, Dapena J. Tratamiento Biológico De Las Aguas Residuales. 2nd ed. Madrid,Spain: Diaz de Santos; 2002

[5] Rittmann B, McCarty P. *Biotecnología del Medio Ambiente*. Madrid, Spain: McGraw-Hill; 2001

[6] Kurashige J, Matsuzaki N, Makabe K. Modification of fats and oils by lipases. Journal of Dispersion Science and Technology. 1989;**10**(4–5):531-559

[7] Moo-Young M, Blanch HW. Fenómenos de transporte y diseño de biorreactores. En: Biotecnología Básica. Editorial Acribia, S.A.: Zaragoza, España; 1991

[8] Ratledge C. Microbial oxidations of fatty alcohols and fatty acids: biodegradation and biotransformations of oils and fats. Journal of Chemical Technology and Biotechnology. 1986, 1992;**55.4**:399-400

[9] Sugimori D, Utsue T. A study of the efficiency of edible oils degraded in alkaline conditions by Pseudomonas aeruginosa SS-219 and Acinetobacter sp. SS-192 bacteria isolated from Japanese soil. World Journal of Microbiology and Biotechnology. 2012;**28**(3):841-848

[10] Chipasa KB, Mdrzycka K. Characterization of the fate of lipids in activated sludge. Journal of Environmental Sciences. 2008;**20**(5): 536-542

[11] Volkering F, Breure A áM, Van Andel JG. Effect of micro-organisms on the bioavailability and biodegradation of crystalline naphthalene. Applied Microbiology and Biotechnology. 1993; **40**(4):535-540

[12] Donaji JI, Medina Moreno SA, Jorge Noel GR. Propiedades, aplicaciones y producción de biotensoactivos: una revisión. Revista internacional de contaminación ambiental. 2010;**26**(1): 65-84

[13] Schubert H, Armruster H. Principles of formation and stability of emulsions. International Chemical Engineering. 1992;**32**:14-27

[14] Capek I. Degradation of kinetically-stable o/w emulsions. Advances in Colloid and Interface Science. 2004;**107** (2–3):125-155

[15] Rahim A, Noraida S, et al. Effect of agitation speed for enzymatic hydrolysis of tapioca slurry using encapsulated enzymes in an enzyme bioreactor. International Journal of Chemical Engineering and Applications. 2015; **6**(1):38

[16] Tsouris C et al. Comparison of liquid-liquid dispersions formed by a stirred tank and electrostatic spraying. Chemical Engineering Communications. 1997;**160**(1):175-197

[17] Hagenson LC, Doraiswamy LK. Comparison of the effects of ultrasound and mechanical agitation on a reacting solid-liquid system. Chemical Engineering Science. 1998;**53**(1):131-148

[18] Weatherley LR, Rooney DW, Niekerk MV. Clean synthesis of fatty acids in an intensive lipase-catalysed

bioreactor. Journal of Chemical Technology & Biotechnology: International Research in Process, Environmental and Clean Technology. 1997;**68**(4):437-441

[19] Ruiz-Márquez D et al. Oil-in-water food emulsions stabilized by tuna proteins. Grasas y Aceites. 2010;**61.4**: 352-360

[20] Adulkar TV, Rathod VK. Ultrasound assisted enzymatic pre-treatment of high fat content dairy wastewater. Ultrasonics Sonochemistry. 2014;**21**(3):1083-1089

[21] Scott TC, Sisson WG. Droplet size characteristics and energy input requirements of emulsions formed using high-intensity-pulsed electric fields. Separation Science and Technology. 1988;**23**(12–13):1541-1550

[22] Slaughter JC, Weatherley LR, Wilkinson A. Electrically enhanced enzymic hydrolysis of vegetable oils using lipase from Candida rugosa. Enzyme and Microbial Technology. 1993;**15**(4):293-296

[23] Roy A. Review on the biosurfactants: properties, types and its applications. Journal of Fundamentals of Renewable Energy and Applications. 2017;**8**:1-14

[24] Park TJ et al. Petrochemical wastewater treatment with aerated submerged fixed-film reactor (ASFFR) under high organic loading rate. Water Science and Technology. 1996;**34**(10): 9-16

[25] Nomura T, Uchida T, Takahashi K. Enhancement of mixing by unsteady agitation of an impeller in an agitated vessel. Journal of Chemical Engineering of Japan. 1997;**30**(5): 875-879

[26] Albasi C et al. Enzymatic hydrolysis of sunflower oil: Characterisation of interface. Journal of Chemical Technology & Biotechnology: International Research in Process, Environmental and Clean Technology. 1997;**69.3**:329-336

[27] Purwanto LA, Ibrahim D, Sudrajat H. Effect of agitation speed on morphological changes in Aspergillus niger hyphae during production of tannase. World Journal of Chemical. 2009;**4**(1):34-38

[28] Crespi M, Huertas JA. Determinación simplificada de la demanda química de oxígeno por el método deldicromato. Water Technology. 1984;***13***:35-40

[29] Díaz de Santos SA. Métodos normalizados para el análisis de aguas totales y residuales. Madrid. Comité editorial conjunto: APHA, AWWA and WPCF 3; 1992

[30] Vittadini G. Catálogo de Información de Equipamiento de Biocontrol. Milan, Italy: Vittadini Riferiment; 1991

[31] Tchobanoglous G, Burton F, Stensel H. *Wastewater Engineering, Treatment, Disposal and Reuse*. 4th ed. New York, NY, USA: Metcalf & Eddy, Inc.; McGraw-Hill; 2003

[32] Hodaifa G et al. Influence of hydrodynamic stress in the growth of Scenedesmus obliquus using a culture medium based on olive-mill wastewater. Chemical Engineering and Processing: Process Intensification. 2010;**49**(11): 1161-1168

[33] Syaima MTS et al. The synthesis of bio-lubricant based oil by hydrolysis and non-catalytic of palm oil mill effluent (POME) using lipase. Renewable and Sustainable Energy Reviews. 2015;**44**: 669-675

[34] Ratledge C. Bioquımica del crecimiento y metabolismo.

In: Bu'Lock J, Kristiansen B, editors. Biotecnologıa básica. Editorial Acribia, Zaragoza 1991. pp. 11-55

[35] Bitton G. Wastewater Microbiology. New York: John Wiley & Sons; 1994

[36] Loperena L et al. Kinetic properties of a commercial and a native inoculum for aerobic milk fat degradation. Bioresource Technology. 2006;**97**(16): 2160-2165

[37] Young JC. Removal of grease and oil by biological treatment processes. Journal - Water Pollution Control Federation. 1979;**51**(8):2071-2087

[38] Loperena L, Ferrari MD, Díaz AL, Guzmán I, Pérez LV, Carvallo F, et al. Isolation and selection of native microorganisms for the aerobic treatment of simulated dairy wastewaters. Bioresource Technology. 2009;**100**:1762-1766

[39] Cisterna-Osorio P, Arancibia-Avila P. Comparison of biodegradation of fats and oils by activated sludge on experimental and real scales. Water. 2019;**11**(6):1286

[40] Sürücü G. Growth requirements of thermophilic aerobic microorganisms in mixed cultures for the treatment of strong wastes. Water Science and Technology. 1999;**40**(1):53-60

[41] Michelan R et al. Effect of impeller type and mechanical agitation on the mass transfer and power consumption aspects of ASBR operation treating synthetic wastewater. Journal of Environmental Management. 2009; **90**(3):1357-1364

Section 2

Inorganic Pollutants

Chapter 11

Conventional and Contemporary Techniques for Removal of Heavy Metals from Soil

Vaishali Arora and Babita Khosla

Abstract

One of the most important components of the natural environment is soil. Soil is a non-renewable natural resources on which the whole human society is dependent for various goods and services. The intensive, and unsustainable anthropogenic practices along with the rapid growth of the human population have led to continuous expansion and concern for the degradation of soil. The agricultural soil is exposed to a plethora of contaminants, the most significant contaminant among them is heavy metals. The major sources of heavy metal contamination are associated with agriculture, industries, and mining. The increase of heavy metal contents in the soil system affects all organisms via biomagnification. In this chapter, we will review various conventional and contemporary physical or chemical and biological techniques for remediation of contaminated soil. The advanced solution for degraded soil is integrating innovative technologies that will provide profitable and sustainable land-use strategies.

Keywords: Metal toxicity, Soil Pollution, *In-situ*, *Ex-situ* remediation, Bioremediation

1. Introduction

Soil is the uppermost layer of Earth's crust, which is produced at the rate of a few centimeters per thousand years by the continuous transformation of solid crust material. According to FAO, the soil consists of mineral particles, organic matter, water, air, and living organisms [1]. It is one of the most essential, complex, and non-renewable natural resources. It provides humanity with a wide range of ecological, economical, and cultural services. These include provisional services: food, fiber, raw materials; regulating services: mitigation against flood, drought, carbon storage, support hydrological and nutrient cycle, recycling of wastes; cultural services: recreational, esthetic, heritage values, and cultural identity [2]. According to McBratney, 2017 soil provides around US$ 11.4 trillion of ecosystem services [3].

Soil conditions underpin food security, habitat for various organisms, bio-economies, and above-ground biodiversity. It is the major variable in regulating the climate, hydrological, and nutrient cycles. However, anthropogenic activities including industrialization and urbanization have polluted the environment extremely and deteriorating the quality of life for all living organisms. There is enormous pressure on this finite, non-renewable natural resource. Further, inappropriate land-use management severely impacts the functions of soil, which is amplified by climate change. These stresses lead to degradation processes of soil like erosion, contamination, and degradation [4].

1.1 Soil pollution

In the era of the Anthropocene, the imprudent discharge of waste, and chemicals in the ecosystem has led to the increase of concentration of contaminants to critical levels. According to FAO. "Soil pollution" refers to the presence of a chemical or substance out of place and/or present at a higher than the normal concentration that has adverse effects on any non-targeted organism [1]. Although there is the contribution of contaminants through natural sources like, volcanic, seepage from parental rock, biogenic, and forest emissions, the widespread soil contamination and degradation are caused by anthropogenic activities. The rapid and injudicious industrialization, intensive agricultural practices, faulty mining practices and waste disposals are the major causes of heavy metal contamination of soil.

The pollutants introduced in soil by anthropogenic activities can arise from a plethora of sources. These might be discrete point sources or diffuse sources. The emission of heavy metals from point sources includes thermal power plants, coal mines, gold mines, smelting, electroplating, textiles, leather, and e-waste processing; and non-point sources include soil erosion, agricultural run-off, vehicular emissions, ash fallout, combustion of fuel, acid deposition, mining tailings, heavy metal mining and smelting, mismanaged radionuclides waste, and open freight storage (**Figure 1**).

One of the major concerns is the contamination of heavy metals in agricultural soil. It has increased tremendously in the soil system since the last decade. Although most of the heavy metals exist geologically, the emission of them in the ecosystem through anthropogenic sources like increased chemical discharge through the indiscriminate usage of pesticides and fertilizers into the agricultural soil has led to the accumulation of heavy metals concentration to dangerous levels. As soil holds the largest terrestrial pool for carbon, thus degrading soil will only worsen the phenomenon of climate change. The conditions of soil also underpin various Sustainable Development Goals (SDGs) set by the United Nations (**Figure 2**).

In view of these facts, strategies for remediation of contaminated soil must be implemented. Various remediation techniques have been developed to solve or minimize the influences of contamination. These technologies include physical, chemical, and biological methods.

1.2 Heavy metals

Heavy metals and metalloids are generally referred to as a group of elements that have densities >5 g cm^{-3}. These include lead (Pb), chromium (Cr), arsenic (As), zinc (Zn), cadmium (Cd), copper (Cu), mercury (Hg), and nickel (Ni). They are naturally occurring elements, whose natural concentration in the soil ecosystem is primarily dependent on parent rock material [5]. Some heavy metals, like Zn, Cu, Fe, Ni, Mn, Mo, and Cr are essential for the functioning of structural and biochemical processes in living organisms and are required in trace concentrations, hence called micronutrients. They can cause harmful effects to plants if absorbed in higher concentrations. While non-essential heavy metals, including Pb, Hg, As have unknown biological functions but are used for various processes in modern industrial applications. The non-essential heavy metals are toxic to plants even at low concentrations [6]. However, the emission rate of pollutants through anthropogenic sources has increased the concentration of heavy metals in soil to hazardous amounts.

Heavy metals speciation plays an important role in their long-lasting presence in the environment, as mobile forms are easily leachable thus making them to spread ubiquitously in different media, and the bioavailable heavy metals are easily

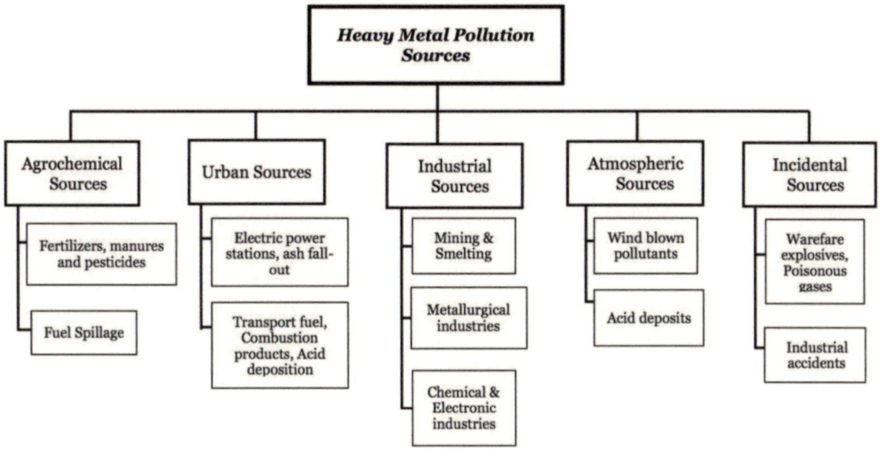

Figure 1.
Various anthropogenic sources of soil pollution.

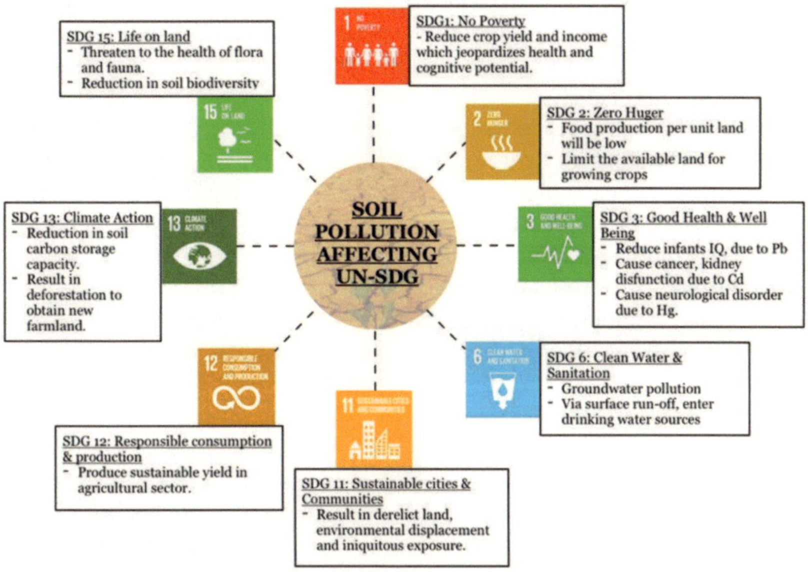

Figure 2.
The negative impacts of soil pollution on SDGs.

absorbed by living organisms. They are non-biodegradable and non-thermodegradable so their accumulation in living organisms can cause biomagnification of heavy metals, that is they can affect organisms throughout all levels of the food chain. Particularly humans, as they are at the top of the food chain. The physicochemical properties of soil, like pH, cation exchange capacity and soil texture, also play a key role in the accumulation and availability of heavy metals [7]. Once heavy metals are exposed to humans, via inhalation, ingestion, or absorbed through the skin, they can accumulate in vital organs such as the kidney, brain, liver where they can be a threat to the health of humans [8]. Heavy metal contamination in agricultural soils may cause disturbance in the structure of soil, interfere with plant growth, and be harmful for human health via entering the food chain [9], posing health problems for all living organisms [10]. Furthermore, degradation of agricultural soil will impact crop yield and will put the most vulnerable people at higher risk of economic loss and malnutrition [4].

Soil Pollution by heavy metals is now a global concern. Europe has been found with 2.8 million sites that are potentially contaminated with heavy metal soil pollution, in China 19% of agricultural soil contain harmful pollutants exceeding the standards of environmental quality [11]. In India, heavy metal pollution in soil cover is approximately 80% by anthropogenic origin in the states of Maharashtra, Gujarat, and Telangana [12]. Therefore, the studies on agricultural soils which are contaminated with heavy metals are of much concern, especially due to two reasons. Firstly, ingestion is the main source of heavy metal exposure to humans and the agricultural food chain is the primary source of various food products for humans [8]. Secondly, densely accumulated heavy metals in agricultural soil can percolate through pore spaces and enter groundwater systems, consequently deteriorating the groundwater quality [13].

2. Remediation techniques

The comprehensive objective of any soil remediation approach is to create a final solution that is protective of human health and the environment. The remediation strategies should incorporate reduction of metal bioavailability and the reduction should be demonstrated for a long term, only if the reduction of heavy metal is equated to reduced risk [14].

A successful process of remediation includes the following steps: 1) Technology pre-screening and treatability study scoping; 2) Remedial investigation of the contaminated site; 3) Feasibility study of pre-screened remediation technology; 4) Determination of best remediation method; 5) Design and implementation of remediation practices; 6) Evaluation and monitoring of remediation process; 7) Depletion in concentration and/or removal of toxic metal [15].

Various remediation techniques applied to soil can be employed via *ex-situ* or *in-situ* methodologies. Although the *ex-situ* methodology of soil remediation is less expensive, fast, and easier to apply, it generates a significant amount of waste product that must be treated before storing or releasing it in the landfill sites. While *in-situ* remediation methodology involves low land disturbance, applicable to a broad range of inorganic pollutants, lesser in cost, and reduced risk of spreading contamination. Broadly various remediation techniques known for improving the quality of contaminated soil are studied under three categories of their application:

- Physical Remediation Techniques
- Chemical Remediation techniques
- Biological Remediation Techniques

2.1 Physical remediation techniques

The remediation techniques that are applied through physical amendments to the soil are incorporated under this category. The physical techniques of remediation include the capping of contaminated sediments, washing, and excavation of soil.

2.1.1 Capping

It is a non-intrusive and cost-effective method for remediating contaminated sediment. The technique is utilized to decrease the solubility, mobility and transfer rate of heavy metals in the sediment [16]. It is usually applied in sub-aqueous conditions. Sandy material and apatite are usually tiered in specific proportions, which are placed on the contaminated sediment like a cap. The cap is usually composed of a, (i) stabilizing base layer which supports the added weight of cap; (ii) an isolation base layer, it isolates the contaminants from the sediment; (iii) a filter layer for hydraulic protection for the base layer; (iv) an armor layer, it inhibits erosion for the protection of filter and base layer. Capping can be performed in two ways, Passively (inactive) or Reactively (active). The former methodology includes a cap composed of clean and neutral material which provides a physical barrier between the environment and contaminated sediment. However, passive methods have been observed to cause leaks of toxic metals. The latter methodology includes the cap with reactive material which can reduce the mobility, toxicity, and bioavailability of contaminants in sediments. This technique is not appropriate for shallow water or marshes or water bodies with large water flows as the capping material can be washed away [17]. Below is a graphical representation of the capping methodology (**Figure 3**) [18].

2.1.2 Washing of soil

Sediment washing is a simpler technique that is performed *ex-situ*. In this technique, a solution is utilized to wash the contaminated sediment for the transfer of pollutants from sediment to an aqueous solution. This is achieved by mixing the soil with an aqueous solution of alkalis, acids, and surfactants [14]. Washing includes (i) excavation of highly contaminated sediment from the bulk soil; (ii) washing of sediment is processed with the help of aqueous mixtures; (iii) the solubilized contaminants are removed from aqueous solution through various chemical processes. For performing this method more efficiently additives are added to the aqueous solution, depending upon the physicochemical nature of contaminated sediment. These additives should have high treatment efficiency and environmental compatibility. Common additives used are inorganic acids (sulfuric acid, nitric acid), organic acid (oxalic acid, ascorbic acids), and surfactants (sophorolipids

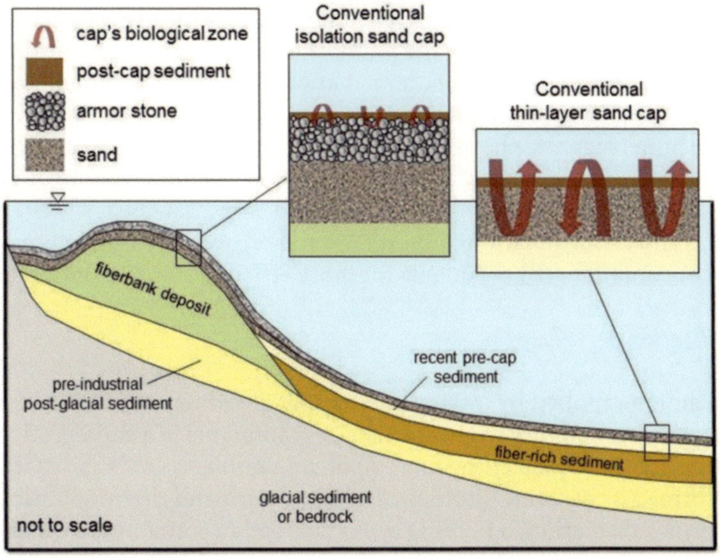

Figure 3.
Capping technique for isolating contaminated sediment [18].

and rhamnolipids) [16]. EDTA has been reported as the additive for the removal of heavy metals, due to its versatile chelating nature, however, the toxic effect of EDTA on the environment and its low biodegradability has reduced its application widely [19]. After washing, the sediment is considered contaminant depleted instead of contaminant free. Therefore, to make this technique successful, the number of contaminants treated should be quantified to be equivalent to the site-specific action limit. This technique is suitable for the contaminants which are weakly associated with sediments, and in coarse-grained sediments [14].

2.1.3 Excavation of soil

This technique includes physical removal of majorly contaminated soil from the bulk soil. There are several ways to perform this technique. It can be divided into three methodologies (i) substitution of polluted sediment by removing the soil and putting it in another soil. This method is more suitable for land contaminated in small areas; (ii) the deep excavation of contaminated sediment for natural degradation of heavy metals; (iii) importing new soil and mixed with contaminated soil for dilution of heavy metals. This technique is expensive and is efficiently applicable only on land with small areas of contamination [20].

2.2 Chemical remediation techniques

This technique includes the utilization of chemical reagents, reactions, and principles for the removal of contaminants. Major methodologies used under this technique are solidification, immobilization, vitrification, and electro kinetics.

2.2.1 Immobilization

This methodology is used to stabilize heavy metals, can be applied *ex-situ* and *in-situ*. It often uses organic and inorganic reagents for the reduction of heavy metals mobility, toxicity, and bioavailability in the soil. The primary objective of this technique is to alter the bioavailable phases of metals into more geo-chemically stable phases, with the immobilization of chemicals. It is achieved through combined mechanisms of adsorption, complexation, and precipitation. The stabilizing effect of amendments is dependent upon the physical, chemical, and biological characteristics of sediment, heavy metal type, remediation time, remediation method, and evaluation method. The most common inorganic reagents used for immobilization are silico-calcium reagents, phosphates, iron-containing materials, aluminum salts, and mineral-based amendments. Organic reagents for immobilization of heavy metals include manure, biochar, biosolids, bark, wood chips, sawdust, sewage sludge, and turf. A complex formulation of inorganic and organic amendments can also be applied to the contaminated sediments for more efficient stabilization [21].

2.2.2 Solidification

It is a technique applied by mixing contaminated sediments with materials that impart physical stability to encapsulate contaminants in a solid product. Solidification is the physical encapsulation of contaminants in a solid matrix, which are formed by cement, bitumen, asphalt, fly ash and thermoplastic binders. During *In-Situ* remediation, a binding agent is added to contaminated sediment which is followed by an auger spin mixing to transform the soil into a solid matrix [15]. The stabilization of heavy metals includes chemical reactions which reduce their mobility in the environment. The entrapped toxic metals are not leachable as the solid block is impermeable to water. A mixture of various salts can be used for the solidification or stabilization of contaminants in soil *ex-situ* or *in-situ*. Several economically effective and environmentally friendly waste resources have been reported for their application in contaminated sediment. These waste resources can also improve the quality of polluted soil, such as lime-based agents, calcined oyster shells, eggshells, waste mussel shells, and calcined cockle shells [20]. However, the process does not extract the pollutant. So, over the long term, if the integrity of solid matrix is deteriorated due to natural weathering or any uncontrolled physical disaster the contaminants which are trapped can mobilize into the environment. Therefore, this methodology is applied as a last option for remediation of soil. This technology is dependent on the concentration of contaminants present in the sediment, amount of water, and ambient temperature. These factors affect the binding reaction of contaminants to the solid material, it inhibits the binding and decreases the stability of the solid matrix [14].

2.2.3 Vitrification

This methodology of remediation is a type of stabilization/solidification technique. It requires high thermal energy in contaminated soil, at least 1400°C - 2000°C, for the removal of organic or volatile substances. It is achieved by mixing the contaminated sediments with glass-forming precursors, heating the mixture till its liquid solution is formed. The steam produced by introducing high thermal energy and the products of pyrolysis are collected from exhaust gas [21]. On the cooling of this solution, an amorphous homogenous glass is obtained. The contaminants can be stabilized by two ways of interactions with solid glass matrix, that is chemical bonding and encapsulation. For *in-situ* remediation, electrodes can be inserted directly into the contaminated sediments. This technique is efficient but expensive and complex to perform [20].

2.2.4 Electrokinetic remediation

In this technique, the electric field is applied to the wet contaminated sediments for the movement of ionized metals towards the cathode or anode. The pollutants are migrated towards electrodes through electro-migration (charged chemical movements), electro-osmotic flow (fluid movements), electrophoresis (charged particle movements), and electrolysis (chemical reaction due to electric field) procedures [21]. On the completion of the remediation process, the contaminant concentrated electrodes can be treated through several techniques for treating the heavy metals. This technique performs more efficiently in fine-grained clayey soil, where heavy metals are present as soluble ions, because of high electric conductivity and strong electric field [16]. To enhance the efficiency of this technique application of chelating agents can be performed, such as EDTA, nitrilinoacetic acid, succinic acid, citric acid. A schematic representation of this technique has been represented in (**Figure 4**).

2.3 Biological remediation

Biological remediation or bioremediation is a technique of transforming the heavy metals present in the contaminated soil, into a less toxic element. This technique uses biological phenomena that are intrinsic to plants and microorganisms, for the destruction, removal, or immobilization of hazardous contaminants from the polluted environment. Bioremediation is an eco-friendly and economically effective technique for heavy metal removal compared with the conventional chemical and physical methods, which are usually expensive and ineffective especially for sediments contaminated

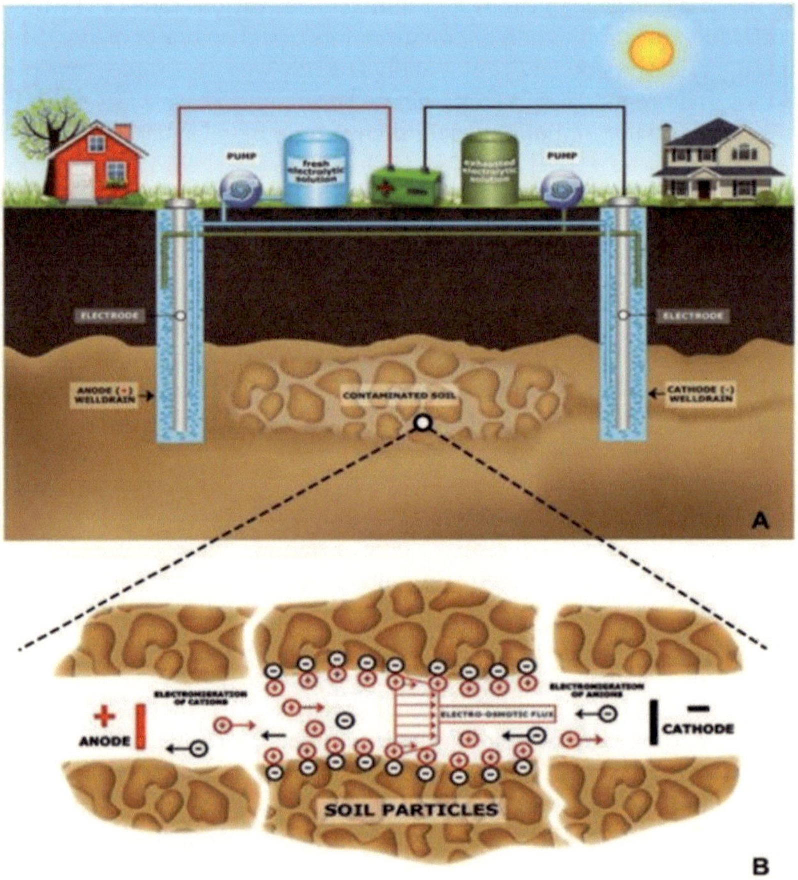

Figure 4.
A schematic representation of in-situ *electrokinetic installation; B schematic representation of detailed electrokinetic remediation technique [22].*

with low metal concentrations, in addition to producing significant amounts of toxic sludge [20]. The main objective of the bioremediation technique is to stimulate a favorable condition for microflora or plants at the contaminated site by providing suitable growth conditions. So, they can grow at their full potential and produce enzymes as secondary metabolites for immobilizing the toxic metals. During the bioremediation process of the contaminant, chemical bonds are broken, and energy is released, which is further utilized by the microorganisms for their growth. Various investigations show that the total transformation percentage of various heavy metals by microbes are Cr (27%), Co (20%), Cd (31%), Pb (22%) [23]. Bioremediation technology is aided with several methodologies, such as bioventing, bioleaching, and land farming, bioreactor, composting, and bioaugmentation, rhizo-filtration, and biostimulation. Therefore diverse metabolic activity inherent to microbes can be exploited for degradation, removal, or transformation of heavy metals in contaminated soil [24]. Mostly bioremediation can be performed by utilizing microorganisms (algae, fungi, and bacteria), and plants (phytoremediation), or with the combinations of both.

2.3.1 Phytoremediation

This technique involves the use of various native, imported, or genetically modified plant species for the reduction, and removal of contaminants from soil, sludge, wastewater, sediments, and groundwater. This technique is best applicable when the contaminants are present around the rhizosphere and in a wide area of land. The basic principle in phytoremediation involves the disintegration through secondary metabolites or absorption by roots, and storing them in leaves of plants, of contaminants present in soil [20]. Hyperaccumulation and hyper tolerance are very important characteristic for a plant for their utilization in phytoremediation. Phytoremediation technique includes phytoextraction, Phytofiltration, Phytostabilization, Phytovolatilization, and Phytodegradation [19].

Phytoextraction/Photoabsorption/Phytosequestration/Phytoaccumulation refers to a biochemical process where the assimilation of heavy metal contaminants from the sediment or water is processed through roots and translocated to any harvestable part of the plant, based on the mechanism of hyperaccumulation (**Figure 5**). Hyperaccumulators can concentrate 100 to 1000 times higher than those found in non-hyperaccumulators without suffering any apparent phytotoxic effect. This method includes three steps (i) cultivation of suitable plant species in the contaminated land; (ii) harvesting of biomass concentrated with metal; (iii) post-harvest treatment for obtaining economic value [25]. The most used hyperaccumulators are from the family *Fabaceae, Brassicaceae, Lamiaceae, Cryophylaceae, Violaceae, Asteraceae, Cyperaceae, and Poaceae* [24].

Phytofiltration is the cleanup method for a contaminated environment with the use of plant roots. It could be performed in three forms of rhizofiltration (plant roots), blastofiltration (seedlings), caulofiltration (excited plant shoots) [19].

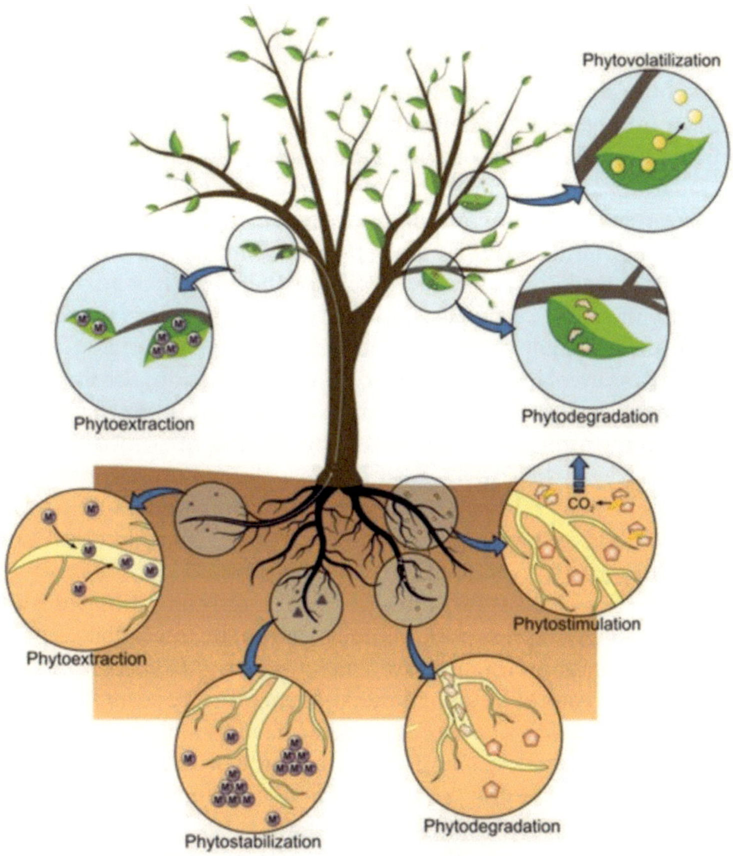

Figure 5.
Schematic representation of several strategies involved in phytoremediation technique [26].

Phytostimulation enhances the conditions of the rhizosphere for the efficient growth of microbes. It is performed for the removal of organic pollutants in the sediment.

Phytostabilization aims to the reduction of mobility and bioavailability of heavy metals in the environment by stabilizing the contaminants in the rhizosphere of plant species. It is performed by reducing the accessibility and mobility of heavy metals through precipitation, root sorption, metal valence reduction, and complexation. The efficiency of this technique can be enhanced by changing the pH and organic matter content in the sediment [25].

Phytodegradation is a technique utilized for degrading organic matter into non-hazardous chemicals through secondary metabolites or enzymes secreted by plants. Enzymes like nitroreductase and dehalogenases are used by plants for the degradation of organic matter. These enzymes are used only in optimal conditions (temperature, pH). This process can be performed more efficiently with the introduction of microorganisms in the contaminated soil, this technique is called Rhizodegradation [26].

Rhizofiltration is the process in which plants absorb and precipitate organic and inorganic contaminants through roots from contaminated wastewater, groundwater, and surface water. Major characteristic features of plants are hypoxia tolerant, and large absorption surface area for a suitable application of this technique. Terrestrial plants are more efficient for this purpose than aquatic plants [27].

2.3.2 Microbial remediation

Microorganisms can absorb or adsorb the heavy metals present in the soil to transform its chemical nature and reduce its mobility, bioavailability, and solubility. This remediation technique by microbes can be carried out in two ways, through mobilization or immobilization. These processes are accomplished by mechanisms, like bio-precipitation, biosorption, bioaccumulation, bio-assimilation, bioleaching, biodegradation, and biotransformation (**Figure 6**). Commonly microbial species used for remediation methodology are *Bacillus, Arthrobacter, Pseudomonas, Enterobacter, Aspergillus, Penicillium, Rhizopus, Rhodotorula, Candida utilis* [23].

Biosorption is a mechanism where microbes either absorb or adsorb the inorganic contaminants on the cell surface or into the cell. While adsorption is performed on the surface of the cell, absorption involves an entire volume of material. Several mechanisms involved in biosorption are precipitation, the formation of stable complexes with organic ligands, and redox reaction. The process of adsorption involves forming a complex of the heavy metals and functional groups on the cell surface, from where they can be absorbed into the cell. Adsorption is executed by binding heavy metals to the cell surface through electrostatic interaction, complexation, and ion exchange. According to Jin et al. [28], microbes perform adsorption predominantly in comparison to absorption.

Bioleaching is the mobilization of heavy metals from contaminated soil through biological dissolution, complexation, or bio-oxidation by microbial activity. The best-known microbes for bioleaching are *Thiobacillus* and *Leptospirillum ferrooxidans*. Various mechanisms of microbial metabolism produce several secretions, like low molecular organic acids. These organic acids have shown to effectively dissolve heavy metals and soil particles containing toxic heavy metals [28].

Bioaccumulation includes the agglomeration of contaminants into the microbe where it is concentrated, where metal is sequestered.

Bio-assimilation of heavy metals includes the active transport of microbial cell's siderophore for the chelation of toxic metals. Siderophores are biomolecules that are produced when microbes are present in iron-deficient media/environment. These

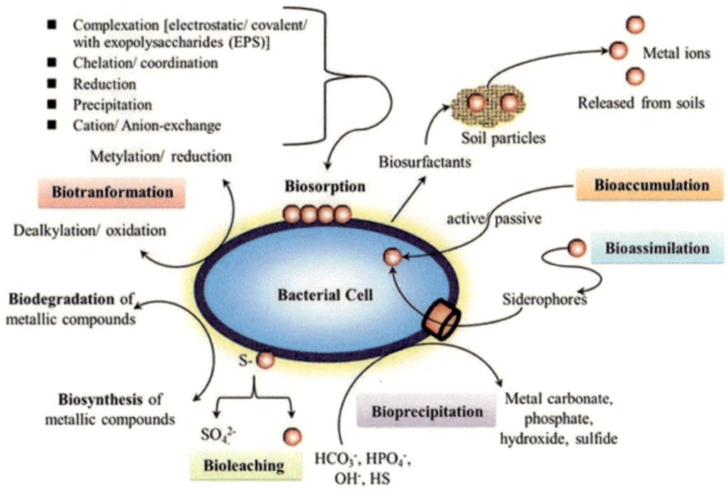

Figure 6.
Schematic representation of various mechanisms involved in microbial remediation of heavy metal contaminated soil [3].

Methodology	Remediation Technique	Applicability	Advantages	Limitations
Physical Remediation	Surface Capping	*In Situ*, in areas with excessive heavy metal pollution	Applicability is unchallenging, low operating cost, high security	Limited to small land areas, and applicable at specific geographic locations, deprivation of land
	Landfilling	*Ex Situ*, applicable to areas with high metal pollution	Immediate restoration, high security	High capital cost, supplementary land is required for storing of the unproductive sediment
	Encapsulation	*In Situ*, applicable to areas with high heavy metal pollution	Isolation of heavy metal from contaminated sediment is effective, installation can be done quickly	Limited to small scale and shallow land areas, costly,
	Soil Washing	*Ex Situ*, applicable to soil with moderate to high heavy metal pollution	Efficiency is high, immediate remediation can be observed, cost-effective, removal of heavy metals are absolute	Effectiveness varies with the variation in physicochemical nature of soil, drastic soil disturbance has been observed
	Excavation of Soil	*Ex Situ*, applicable to areas with high heavy metal contamination	Removal of heavy metal is effective, Less time is required for completion of process	Production of harmful waste products which can have negative impact on soil, costly

Methodology	Remediation Technique	Applicability	Advantages	Limitations
Chemical Remediation	Stabilization	*In Situ*, applicable to areas with high heavy metal contamination	Affordable, easy to applicability, instantaneous effect on contaminated soil, covers a broad-spectrum of inorganic pollutants	Specific to different metals, temporary effectiveness, constant monitoring is required, remnants of contaminants will still be present in the soil
	Solidification	*In Situ* and *Ex Situ*, applicable to areas with high heavy metal contamination	Implementation is quick, high efficacy	High capital cost, treated land loses important ecological functions
	Vitrification	*In Situ* and *Ex Situ*, applicable to areas with high heavy metal contamination	High efficiency, easy to install, applicable to various contaminants	High capital cost due to energy requirement, limited to a small scale areas, treated land loses its environmental function
	Electrokinetics	*Ex Situ*, fine soil, applicable to soil with moderate to high heavy metal pollution	Application is easy, economically effective, deterioration of soil functions are minimum	Time-consuming, low efficiency, best for fine-textured soil with low permeability, pH of soil has to be controlled
Bioremediation	Phytoremediation	*In Situ*, applicable to soil with low to moderate heavy metal pollution	More public acceptance, economically effective, easy to apply	Limited to shallow land, time-consuming, restricted to specific metals, effectiveness depends on the growth conditions, and bioavailability of heavy metals.
	Contaminant transformation with the help of microbes	*In Situ*, applicable to soil with low to moderate-heavy metal pollution	Easy to implement, economical, disturbance to soil is low, remediation is less time consuming	Depends on microbes, soil, metal type, and plant, low efficacy

Table 1.
Mechanisms, advantages and disadvantages of the available remediation techniques for heavy metal contaminated soil [19].

biomolecules are specifically iron (Fe III) chelators which are finally transported into microbes by various uptake proteins. Many reports have suggested that if siderophores are bonded with other metals, they can still be recognized by uptake protein for its transportation into the microbial cell [16, 24].

Bioprecipitation is a method that uses the mechanism of immobilization for the reduction of mobility and bioavailability of heavy metals in soil. It involves converting soluble heavy metals into insoluble hydroxides, carbonates, sulfides, and phosphates.

Biotransformation changes the chemical nature of heavy metals, altering their toxicity, mobility, and bioavailability. This methodology includes methylation, reduction, dealkylation, and oxidation of heavy metals for altering their soluble form into an insoluble form [16].

The applicability of these individual techniques in any specific soil remediation project is determined primarily by contamination site geography, characteristics of contaminants, the goal of remediation, cost-effectiveness, financial budget, readiness in implementing the technique, the time provided, and public acceptability (**Table 1**). Integration of more than one technique has been experimentally proved to be more efficient, such as application of chemical remediation in highly heavy metal contaminated sediment, which can be followed by phytoremediation for further removal of remaining contaminants [15].

3. Conclusion

Over-exploitation of natural resources, land mismanagement, industrialization, and urbanization has led to the discharge of heavy metal through anthropogenic activities. The contamination of soil by heavy metals is of great concern because of its potential impact on human, animal, and plant health. Therefore, effectual technologies of remediation are necessary. Although the traditional physical and chemical methods for cleanup of sediment contaminated with high concentrations of heavy metal are low in cost, but simultaneously can modify soil properties and native microflora and can also produce secondary pollutants in the soil. By comparison, bioremediation is a better alternative to solve this issue. It is environmentally friendly, cost-effective, does not impact the natural microflora of soil, and the use of nature-based products enhances the attainment of UN Sustainable Goals. However, various aspects of bioremediation make the method moderately debilitated, such as longer time is required for transforming the heavy metals. Integration of various techniques can help in achieving a more efficient result for remediating the contaminated soil. Furthermore, the screening of various native plants for remediation of polluted soil with toxic heavy metals as well as advancement in the application of biotechnological approaches has offered various modified plants for phytoremediation.

Author details

Vaishali Arora and Babita Khosla*
Department of Environmental Sciences, Maharshi Dayanand University,
Rohtak, India

*Address all correspondence to: babitakhosla@gmail.com

References

[1] FAO: Eugenio NR, McLaughlin M, Pennock D. A soil pollution: A hidden reality. Rome; 2018:142 pp. ISBN: 978-92-5-130505-8

[2] Dominati E, Mackay A, Green S, Patterson M. A soil change-based methodology for the quantification and valuation of ecosystem services from agroecosystems: A case study of pastoral agriculture in New Zealand. 2014; 100:119-129. DOI: 10.1016/j.ecolecon. 2014.02.008

[3] McBratney AB, Morgan CLS, Jarrett LE. The Value of Soil's Contributions to Ecosystem Services. In: Field DJ, Morgan CLS, McBratney AB, editors. Global Soil Security. Progress in Soil Science. Cham: Springer; 2017. p. 227-235. DOI: 10.1007/978-3-319-43394-3_20

[4] European Commission: Perez AP, Eugenio NR. Status of local soil contamination in Europe: Revision of the indicator "Progress in the management contaminated site in Europe". Luxembourg: Joint Research Centre (EUR), European Commission; 2018. Report No.: JRC 107508. Contract No.: EUR 29124 EN. DOI: 10.2760/093804.

[5] Rai PK, Lee SS, Zhang M, Tsang YF, Kim K. Heavy metals in food crops: Health risks, fate, mechanisms, and management. Environ Int. 2019; 125:365-385. DOI: 10.1016/j.envint.2019. 01.067

[6] Zwolak A, Sarzyńska M, Szpyrka E, Stawarczyk K. Sources of soil pollution by heavy metals and their accumulation in vegetables: A review. Water Air Soil Pollution. 2019; 230:164. DOI: 10.1007/s11270-019-4221-y

[7] Vareda JP, Valente AJM, Duraes L. Assessment of heavy metal pollution from anthropogenic activities and remediation strategies: A review. Journal of Environmental Management. 2019; 246:101-118. DOI: 10.1016/j.jenvman. 2019.05.126

[8] Wu W, Wu P, Yang F, Sun D, Zhang D-X, Zhou Y-K. Assessment of heavy metal pollution and human health risks in urban soils around an electronics manufacturing facility. Science of the Total Environment. 2018; 630:53-61. DOI: 10.1016/j.scitotenv.2018.02.183

[9] Esmaeilzadeh M, Jaafari J, Mohammadi AA, Panahandeh M, Javid A, Javan S. Investigation of the extent of contamination of heavy metals in agricultural soil using statistical analyses and contamination indices. Human and Ecological Risk Assessment: An International Journal. 2019;25(5): 1125-1136. DOI: 10.1080/10807039.2018. 1460798

[10] Adriano, Domy C. Bioavailability of Trace Metals. In: Adriano, Domy C, editors. Trace elements in terrestrial environments. New York: Springer;2001. Pp. 61-89. DOI: 10.1007/978-0-387-21510-5

[11] Hou D, O'Connor D, Igalavithana DA, Alessi SD, Luo J, Tsang WCD. Metal contamination and bioremediation of agricultural soils for food safety and sustainability. Nat Rev Earth Environ. 2020;1: 366-381. DOI: 10.1038/s43017-020-0061-y

[12] Kumar V, Sharma A, Kaur P, Sidhu GPS, Bali Shreeya A, Bhardwaj A. Pollution Assessment of heavy metals in soil of India and ecological risk assessment: A-state-of-the-art. Chemosphere.2019;216:449-462. DOI: 10.1016/j.chemosphere.2018.10.066

[13] Adimalla N, Qian H, Wang H. Assessment of heavy metal contamination in agricultural soil lands in northern Telangana, India: an

approach of spatial distribution and multivariate statistical analysis. Environ Monit Assess. 2019; 191:246. DOI: 10.1007/s10661-019-7408-1

[14] Wauna AR, Felix EO. Heavy metals in contaminated soil: A review of sources, chemistry, risks, and best available strategies for remediation. ISRN Ecology. 2011; 2011: 1-20. Article ID 402647. DOI: 10.5402/2011/402647

[15] Liu L, Li W, Song W, Guo M. Remediation techniques for heavy metal-contaminated soils: Principle and applicability. Science of the Total Environment. 2018; 633:206-219. DOI: 10.1016/j.scitotenv.2018.03.161

[16] Peng W, Li X, Xiao S, Fan W. Review of remediation technologies for sediments contaminated by heavy metals. J Soils Sediments. 2018; 1701-1719. DOI: 10.1007/s11368-018-1921-7

[17] Vandenbossche M, Jimenez M, Casetta M, & Traisnel M. Remediation of Heavy Metals by Biomolecules: A Review. Critical Reviews in Environmental Science and Technology. 2014;45(15):1644-1704. DOI: 10.1080/10643389.2014.966425

[18] FIBREM: Remediation of Sweden's fibre bank sediment-Planning ahead [Internet]. Available from: https://www.geo.uu.se/digitalAssets/606/c_606274-l_3-k_fibrem-2-copy.jpg

[19] Li C, Zhou K, Qin W, Tian C, Qi M, Yan X, Han W. A review of heavy metals contamination in soil: Effets, sources, and remediation techniques. Soil and Sediments Contamination: An International Journal. 2019;28(4):380-394. DOI: 10.1080/15320383.2019.1592108.

[20] Dhaliwal SS, Singh J, Taneja PK, Mandal A. Remediation techniques for removal of heavy metals from the soil contaminated through different sources: a Review. Environ Sci Pollut Res. 2019; 27:1319-1333. DOI: 10.1007/s11356-019-06967-1

[21] Lwin CS, Seo B, Kim H, Owens G, Kim K. Application of soil amendments to contaminated soils for heavy metal immobilization and improved soil quality: a critical review. Soil Science and Plant Nutrition. 2018;64(2):156-167. DOI: 10.1080/00380768.2018.1440938

[22] Rosestolato D, Bagtain R, Ferro S. Electrokinetic remediation of soils polluted by heavy metals (mercury in particular). Chemical Engineering Journal. 2015; 264:16-23. DOI: 10.1016/j.cej.2014.11.074.

[23] Pratush A Kumar, A Hu Z. Adverse effect of heavy metals (As, Pb, Hg, and Cr) on health and their bioremediation strategies: A review. Int Microbiol. 2018; 21:97-106. DOI: 10.1007/s10123-018-0012-3

[24] Gupta S, Singh D. Role of genetically modified microorganisms in heavy metal bioremediation. In: Kumar R, Sharma A, Ahluwalia S, editors. Advances in Environmental Biotechnology. Singapore: Springer; 2017. p. 197-214. DOI: 10.1007/978-981-10-4041-2_12

[25] Ojuederie BO, Babalola OO. Microbial and plant assisted remediation of heavy metal polluted environment: A review. Int. J. Environ Res. Public Health. 2017; 14(12):1504. DOI: 10.3390/ijerph14121504

[26] Favas JCP, Pratas J, Varun M, D'Souza R, Paul SM. Phytoremediation of soils contaminated with metals and metalloids at mining areas: Potential of native flora. In: Hernandez-Soriano MC, editor. Environmental risk assessment of soil contamination. InTechOpen; 2014. DOI: 10.5772/57469.

[27] Awa SH, Hadibarat T, Removal of heavy metals in contaminated soil by

phytoremediation mechanism: a review. Water Air Soil Pollut. 2020;231(47):1-15. DOI: 10.1007/s11270-020-4426-0

[28] Jin Y, Luan Y, Ning Y, Wang L. Effects and mechanisms of microbial remediation of heavy metals in soil: A review. Appl Sc. 2018; 8:1336. DOI: 10.3390/app8081336

Chapter 12

Phytoremediation of Arsenic Contaminated Water Using Aquatic, Semi-Aquatic and Submerged Weeds

Dibakar Roy, Dasari Sreekanth, Deepak Pawar, Himanshu Mahawar and Kamal K. Barman

Abstract

Arsenic (As) is the one the most toxic element present in earth which poses a serious threat to the environment and human health. Arsenic contamination of drinking water in South and Southeast Asia reported one of the most threatening problems that causes serious health hazard of millions of people of India and Bangladesh. Further, use of arsenic contaminated ground water for irrigation purpose causes entry of arsenic in food crops, especially in Rice and other vegetable crops. Currently various chemical technologies utilized for As removal from contaminated water like adsorption and co-precipitation using salts, activated charcoal, ion exchange, membrane filtration etc. are very costly and cannot be used for large scale for drinking and agriculture use. In contrast, phytoremediation utilizes green plats to remove pollutants from contaminated water using various mechanisms such as rhizofiltration, phytoextraction, phytostabilization, phytodegrartion and phytovolatilization. A large numbers of terrestrial and aquatic weed flora have been identified so far having hyper metal, metalloid and organic pollutant removal capacity. Among the terrestrial weed flora *Arundo donax, Typha latifolia, Typha angustifolia, Vetivaria zizinoids* etc. are the hyper As accumulator. Similarly *Eicchornea crassipes* (Water hyacinth), *Pistia stratiotes* (water lettuce), *Lemna minor* (duck weed), *Hyrdilla verticillata, Ceratophyllum demersum, Spirodella polyrhiza*, *Azola, Wolfia* spp., etc. are also capable to extract higher amount of arsenic from contaminated water. These weed flora having As tolerance mechanism in their system and thus remediate As contaminated water vis-à-vis continue their life cycle. In this chapter we will discuss about As extraction potential of various aquatic and semi aquatic weeds from contaminated water, their tolerance mechanism, future scope and their application in future world mitigating As contamination in water resources.

Keywords: Arsenic, Phytoremediation, Weed

1. Introduction

Arsenic (As) is the one the most toxic element present in earth which poses a serious health hazard to animal and human health. Generally arsenic is present in the earth crust in the form minerals, especially associated with iron pyrite and zinc ores. Arsenic contamination occurs through both by natural as well as anthropogenic processes [1]. Unlike other toxic heavy metals (Cadmium, mercury and chromium) arsenic contamination in environment predominately occurs through natural biogeochemical process [2] and some manmade activities play important role (triggering the process) in that process. Anthropogenic activities such as coal mining and burning smelting of As containing metal ores and other industrial activities are also responsible for distribution of arsenic in the environment [3]. Arsenic contamination of drinking water in South and Southeast Asia reported one of the most threatening problems that causes serious health hazard of millions of people of India and Bangladesh [4]. The source of As contamination in water in those countries were due to two different natural processes; oxidation of arsenopyrite minerals lies below ground water table due to water mining process and reduction of As containing iron hydroxides [5]. Arsenic exists in the nature in −3, 0, +3 and + 5 oxidation states and environmental forms include arsenious acids, arsenic acids, arsenites, arsenates, methylarsenic acid, dimethylarsinic acid, arsine, etc. Two inorganic forms are very common in natural waters: arsenite (AsO_3^{3-}) and arsenate (AsO_4^{3-}), referred to as arsenic (III) and arsenic (V). Pentavalent (+5) or arsenate species are AsO_4^{3-}, $HAsO_4^{2-}$, $H_2AsO_4^-$ while trivalent (+3) arsenites include $As(OH)_3$, $As(OH)_4^-$, AsO_2OH^{2-} and AsO_3^{3-}. The solubility of inorganic species depends on pH and redox potential of the environment and arsenite (As^{3+}) is the most soluble form inorganic As. Pentavalent species or arsenate (As^{5+}) predominate in oxygen rich aerobic environments, where as trivalent arsenites (As^{3+}) dominant in moderately reducing anaerobic environments such as groundwater [4].

Arsenic concentration in drinking water reported more than 50 μg L^{-1} in many areas in the world [6], whereas maximum permissible limit set by World Health Organization (WHO) is 10 μg L^{-1}. The use of arsenic contaminated ground water for irrigation purpose causes build up of As in soil and leads to entry of As in food crops, especially in rice and vegetables [7, 8]. This causes serious health hazard, in those As containing areas. In Southeast Asian countries like Bangladesh, Eastern parts of India (West Bengal and Bihar) and Vietnam, rice is consumed as major staple food and is very efficient in As translocation in grains [9]. Thus rice crop play a major pathway for As entry in human body living in those contaminated areas apart from drinking water. Thus remediation of arsenic contaminated water is important for environmental point of view. Various technologies are for remediation of arsenic contaminated water like ion exchange, electro dialysis, membrane filtration, adsorption and coagulation-flocculation generates lot of arsenic enriched waste. That waste material generally dumped or disposed in nearby surroundings, from where arsenic can also come back to soil and water by leaching thus making system susceptible to arsenic contamination. Along with above mentioned problem, huge cost is involved in this existing arsenic remediation technology. That necessitates finding out an alternate low cost technology which can take care of arsenic contaminated water.

Phytoremediation is an alternate and low cost technology that utilizes green plant to extract arsenic from water and store it vegetative cells. Phytoremediation process includes phytoextraction, phytostabilization, phytovolatilization, phytotransformation, and rhizofiltration [10]. Researchers find out that plants uptake arsenic by roots through phosphate uptake pathway and transfer it their above ground parts (shoot and leave). But how much amount of arsenic translocated from source (water) to sink (plant parts) depends on phytoremediation efficiency of the plant concern. However, more than 90% of total arsenic accumulated into the plant is stored in roots.

The plants utilized for phytoremediation have some criteria like (1) plant have higher specific growth rate under contaminated environment, (2) higher translocation capability of the toxic element concerned [11]. Metal translocation capability depends on factors like (1) bio concentration factor (BCF) and (2) translocation factor (TF). Plants having BCF >1 are ideal for Phytoremediation. Chinese brake fern (*Pteris vittata*) is the most promising plant for phytoremediation of arsenic from contaminated soil [12]. For instance, plants species like water hyacinth (*Eichhornia crassipes*), duck weed (*Lemma minor, Spirodela polyrhiza* and *Wolfia globosa*), water lettuce (*Pistia stratiotes*) and fern (*Azolla pinnata*) have been successfully utilized for arsenic removal from water purpose by many researchers [13–15]. Among the semi aquatic weeds. Apart from these free floating aquatic weed flora such as *Arundo donax*, *Vetivaria* sp., and *Alternanthera philoxeroides* had been successfully utilized for remediation of As contaminated water [16, 17]. In this chapter we are going to discuss about arsenic removal potential of various aquatic and semi aquatic weeds along with their future use for phytoremediation purpose.

2. Phytoremediation pathways

The terminology "Phytoremediation" consists of two words, "Phyto" means "green plants" and "remediation" means "curative measures or restoration". The word "phytoremediation" was first given by Chaney [18]. In phytoremediation process, generally green plants are used which uptake toxic chemical substances (such as heavy metals and metalloids, pesticide residues etc.) from contaminated sites (soil and water) by various mechanisms and remove them from environment. Various crop and weed plants are found to be suitable for phytoremediation purpose. But research results indicated that weed flora had higher phytoremediation potential than cultivated crops (Example- Brassica sp). There are various pathways of phytoremediation process such as, rhizofiltration, phytoaccumulation or phytoextraction, phytostabilization, phytodegradation or phytotransformation and phytovolatilization etc.

- Rhizofiltration: Plants uptake toxics substances by their roots through adsorption or absorption process and sequester in their root system. Aquatic plants mainly exhibited this process.

- Phytoaccumulation or phytoextraction: Plants uptake toxic substances by their root system and translocated to other plant parts such as stem and leave or other modified plant parts. This mechanism mainly exhibited this process are suitable for remediation of contaminated soil.

- Phytostabilization: In this process, plants restrict movement of toxic substances in soil or water, thus reduced their availability to plants. In this method, plants do not uptake toxic substances from environment. Rather, plants secrets some root exudates or photochemicals which form stable chemical bond with toxic substances and increases its stability in environments.

- Phytodegradation or Phytotransformation: In this process, plants uptake toxic substance from soil or water and degrade these primary toxic substances into nontoxic forms. A large number of metabolic and physiological factors are involved in this process.

- Phytovolatilization: Plant uptake toxic substances by their root system and translocated to their aerial plant parts especially in leaves; and release toxic substances in the form of vapor which may not be toxic as their primary source.

Apart from this there are some other terminologies often used in phytoremediation process are bioconcentration factor (BCF) and translocation factor (TF).

BCF = toxic substance uptake by plant/toxic substance present in environment (soil or water).

TF = toxic substance present in shoot or stem/toxic substance present in roots or.

Toxic substance present in leaves/Toxic present in shoot or stem.

For, Hyper accumulator plants both BCF and TF is >1 is desired. In other words, plants suitable for phytoremediation, BCF >1 is always desirable. But for aquatic weeds, as their dominant pathways is rhizofiltration; their toxic substances BCF >1 but TF for root to shoot or shoot to leaves is <1.

3. Potential of various aquatic plants for phytoremediation

3.1 Phytoremediation by free floating aquatic weeds

Eichhornia crassipes: *Eichhornia crassipes* is commonly known as water hyacinth, a free-floating perennial aquatic plant native to tropical and sub-tropical South America, and is now wide spread in all tropic climates. The genus Eichhornia comprises seven species of water hyacinth among which *E. crassipes* is the most common and have been reported to grow very first. However, its enormous biomass production rate, high tolerance to pollution and absorption capacity of heavy-metal and nutrient qualify it for use in wastewater treatment [19].

The capability of removing arsenic from contaminated water was earlier observed by Misbahuddin and Fariduddin [20] and they observed that water hyacinth can removes arsenic from water within 3–6 hr. exposure time. Amount of arsenic removed depends on number of the plant used, exposure time, presence of air and sunlight. They concluded that whole plants were more effective than fibruous roots alone. It was observed that dried roots of water hyacinth can rapidly reduces As content in contaminated water within below WHO recommended critical level (<10 μg Lg^{-1}) [21]. A fine powder was prepared from dried roots of water hyacinth plants (obtained from Dhaka, Bangladesh) removed more than 93% arsenite and 95% of arsenate from a solution containing As @ 200 μg L^{-1} within 1 hr. exposure time [21]. Higher biomass production ability of water hyacinth allow it to remove As at higher rate (600 mg As ha^{-1} day^{-1}) and greater efficiency (17%) compared to lower biomass producing aquatic macrophytes such as lesser duck weed (*Lemna minor*) which removed As at lower rate (140 mg As ha^{-1} day^{-1}) and lesser efficiency (5%); though there was no difference in bioaccumulation capacity [13]. Similarly better As extraction capacity of water hyacinth (80%) compared to *Lemna minor* and Spirodella Polyrhiza from tropical coalmine effluent was also been reported [22] from India. Unlike lower biomass producing aquatic macrophytes, water hyacinth poses better As extraction ability compared to higher biomass producing vetivar grass [23]. Not only higher biomass, higher reproduction ability also plays an important role in As phytoremediation by water hyacinth. Water hyacinth was a suitable phytoremediation agent when As present in contaminated water at lower concentrations. When As was provided at lower concentrations @ 1 and 2 mg L^{-1}, water hyacinth removed 90 and 65% of total As from contaminated solutions (1 and 2 mg L^{-1} respectively) provided respectively within 7 days [24] and maximum As stored in roots. Water hyacinth can extract higher amount As from contaminated water but their presence in water bodies reduces dissolved oxygen content (DOC), which makes its application for a larger water bodies a problematic pathway which needs to be taken care.

Pistia stratiotes: Pistia stratiotes is commonly called as water lettuce There are many previous studies indicated that Pistia stratiotes capable of removing toxic heavy metals from contaminated water [25–27], but there were few studies was done on As uptake by water lettuce. Earlier a field study carried out using *P. stratiotes* and results showed that *Pistia stratiotes* can remove As from contaminated water, along with higher bioconcentration factor (BCF) for root (8632) vis-à-vis lower BCF for leaf (2342) [28]. In a laboratory study it was demonstrated that maximum As removal efficiency of *P. stratiotes* was found at pH 6.5 and Pistia removed 87.5% of the metalloid provided in the solution [29]. From Laboratory study it was revealed that *P. stratiotes* can accumulate As efficiently when As was provided at lower concentrations, though total As uptake was increased with increase in As concentration in the solution [30]. Arsenite accumulation in *P. Stratiotes* was found more in root and less in leaves like water hyacinth. Arsenic accumulation in roots and leaves were respectively 1120 and 31.60 $\mu g\ g^{-1}$ DW respectively when 10 μM As (As^{3+}) solutions are employed [31]. When higher concentration of As solutions used (>20 μM), As toxicity symptoms like chlorosis, suppressed growth, lower photosynthetic rate, suppressed enzymatic activities and increased cell damage were observed in *P. stratiotes* [30, 31].

Lemna, Spirodella and Wolfia: Weeds belongs to Lemna, Spirodella and Wolfia are generally known as Duckweeds. Duckweeds are small free-floating aquatic weed plants which generally found in water bodies, mainly comprises of four genera, *Lemma, Spirodela, Wolfia*, and *Wolfiella,* and of 34 species. Among these Lemna, Spirodela, and Wolfia have been widely reported to accumulate arsenic from contaminated water [13, 32–34]. Research studies indicated that, total As accumulation in *Lemna gibba* was more in field condition compared to laboratory conditions due to higher exposure time in field condition [32]. However higher accumulation of As in plant parts is not always correlated with bio-concentration factor (BCF). It was found that total As accumulation plant parts may be higher in field condition, but higher BCF was obtained at laboratory conditions [32] due to better availability of external nutrients.

However nutrients like phosphate addition may suppressed As uptake by duckweeds as both phosphorus and arsenic belongs same group-V(b) element family in periodic table [33]. In most of the phytoremediation study carried out in laboratory condition, As is provided either in the form of arsenite (As^{3+}) or arsenate (As^{5+}). But some studies included dimethyl arsenic acid (DMAA), an organic form of arsenic for evaluation of As phytoremediation potential of duckweed species. In a lab study, *Spirodela polyrhiza* was exposed to two forms of As species, arsenate and DMAA with concentrations ranged from 1, 2, and 4 μM and their interaction with phosphate (100 to 500 μM) was studied [33]. Results obtained showed that arsenate uptake was affected by higher phosphate concentrations whereas DMAA uptake was not influenced by phosphate concentration indicating that *Spirodela polyrhiza* had separate mechanisms for DMAA uptake. Duckweeds showed contrasting As uptake behavior when provided in two separate inorganic forms (As^{5+} vs. As^{3+}) and maximum As uptake was reported with arsenite form (As^{3+}) [34]. Spirodela polyrhiza extracted 17408 and 8674 $\mu g\ g^{-1}$ As (dry weight basis) respectively from solutions containing As in the form of As^{3+} and As^{5+} (64 μM As each) respectively within 6 days [34]. Maximum amount of As extracted by duckweeds is still questionable and it is varied with As exposure time, concentrations of As in contaminated solution, and research type (laboratory vs. field study). *Spirodela polyrhiza* reported to uptake 400 $mg\ kg^{-1}$ As (dw basis) without showing any toxicity symptoms, but can accumulate up to 900 $mg\ kg^{-1}$ As (dw basis) when subjected to 320 $\mu M\ ml^{-1}$ As containing solutions [35]. Under natural condition, *Lemna minor* was found to accumulate 430 $mg\ kg^{-1}$ As (dry weight basis) under As contaminated environment [36]. There are few studies on As uptake by Wolfia globosa (rootless duckweed). Wolfia globosa had been reported to extract more than 1000 mg kg^{-1} (frond dry weight basis) from contaminated water [37]. Like other duckweeds, Wolfia globosa also uptake more arsenite form compared to arsenate form [37]. Later studies confirmed that Wolfia globosa produced phytochelatins which played an important role minimizing toxic effects of As in their body parts [38]. These above cited studies showed that Lemna minor, *Spirodela polyrhiza* and Wolfia globosa are suitable for phytoremediation of As from contaminated water.

Salvinia: Salvinia is a floating fern belongs to genus salviniaceae, commonly called as butterfly fern. The genus salviniaceae contains 12 different species, out of them only 3 had been investigated for As phytoremediation were namely *Salvinia molesta*, *Salvinia minima* and *Salvinia natans* [39–41]. *Salvinia minima* have been reported as an efficient scavenger of Pb (34 mg g^{-1} dw) and less efficient remover of As (0.05 mg g^{-1}) from contaminated medium and uptake of both Pb and As increased with exposure time duration and concentration of the element in the medium concerned [40]. The plant showed toxicity symptoms when As^{3+} concentration was more than 100 μM and tolerates up to 300 μM. Addition of phosphate in solution, reduced As uptake of as occurred in other aquatic weed plant also been recorded in their study. Similarly negative impact of phosphate and iron on As uptake by Salvinia natans was observed [41]. Phosphate addition reduced As uptake when provided in the from arsenate (As^{5+}), in contrast no impact when As was provided in the form of DMAA. Like other aquatic weeds (Eicchornia, Pistia and Spirodela), *Salvinia molesta* also showed As toxicity upon exposure to higher concentration. To counter As stress, antioxidant enzyme activities and reactive oxygen species (ROS) were increased in floating leaves [39]. These studies indicated that Salvinia can play an important role for As phytoremediation as it had own defense mechanism.

Azolla: Azolla is a small, free floating aquatic fern commonly found in paddy fields, ponds, river and lakes. There are numerous studies carried out globally showed that Azolla can remediate heavy metal toxicity from contaminated water [42–44]. But studies on As phytoremediation capability of azolla were scarce. In As contaminated area of Bangladesh, Mahmud et al. [45] evaluated 49 different plant species for As uptake and BCF; found that *Azolla pinnata* along with

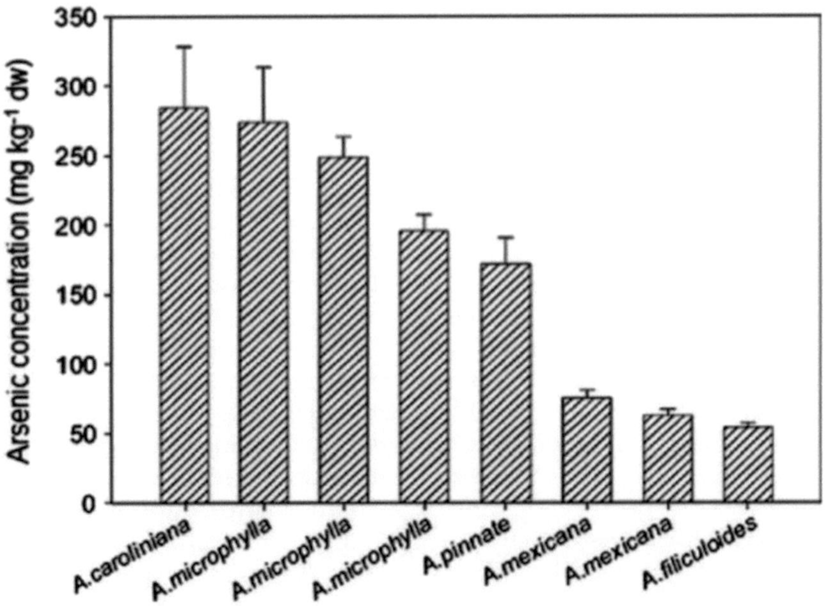

Figure 1.
Arsenic uptake pattern in different Azolla sp. (adapted from Zhang et al., 2009).

Eichhornia crassipes and *Spirodella polyrhiza* showed higher BCF and TF in paddy field. Among 49 plant species, *Azolla pinnata* showed highest BCF 10.92 indicated its suitability to reduce As uptake by paddy plants in field condition. A study using Azolla conducted in China using 50 different strains of Azolla spp. based on their uptake and speciation [46]. As uptake was ranged from 29 to 397 mg kg^{-1}; *A. caroliniana* accumulated maximum As followed by *A. macrophylla* and minimum accumulation was associated with *A. filiculoides* when all strains were grow in 50 μM As^{5+} solution for 10 days (**Figure 1**). Arsenic speciation in followed in the order of arsenate (As^{5+}) > arsenite (As^{3+}) > DMAA and MMAA accounting 50–60, 25–40 and 1–5% of total arsenic in *A. caroliniana* respectively. In contrast, asrenite (As^{3+}) was dominant As species in *A. filiculoides* governs 55–69% of total As [46]. Another study was conducted on phytoremediation of As by *A. caroliniana* wild using various As concentrations (0, 0.25, 0.5, 1.0 and 1.5 mg L^{-1}) and impact of As exposure on plant enzymatic properties were investigated [47]. Maximum As uptake (386 mg kg^{-1}) was reported at highest As concentration (1.5 mg kg^{-1}). It was observed that peroxidases, glutathione reductase, catalase and superoxide dismutase activities were enhanced at lower As doses and reduced at higher doses. In exposure to higher As concentration, thiol content and anthocyanin production were increased and correlated with higher As uptake.

3.2 Phytoremediation of arsenic by semi aquatic weeds

Some semi aquatic weed such as *Alternathera philoxeroides*, *Arundo donax*, *Vetivaria Zizinoids*, *Typha latifolia*, *Phragmites* spp. and *Canna* spp. had been widely reported to accumulate As in their body parts from contaminated soils and water [16, 17, 48–51]. *Alternanthera philoxeroides* had potential to extract As from contaminated water and stored in root system [52, 53]. Reports from previous studies indicates that As accumulation in *A. philoxeroides* followed in the order of root > stem > leaf and average BCF for root ranged from 106 to 191, when exposed to various doses of As containing solutions (1, 2 and 5 mg kg^{-1}) under laboratory condition [52]. Under natural condition, *Alternanthera philoxeroides* observed to uptake 12.94 mg kg^{-1} total As dw from pulp paper industry water with average BCF- 3.58 and TF-0.51 [53]. Higher BCF under laboratory condition observed due to used of higher As containing solution and availability of external nutrients for weed plants which may trigger As uptake through phosphate uptake pathway.

Arundo donax is a perennial semi aquatic weed mostly found in submerged condition offer a tremendous potential to uptake As from contaminated water. Earlier research work showed that Arundo donax can grow efficiently up to 50–600 μg L^{-1} As concentration without showing any toxicity symptom and maximum As uptake, BCF (15), TF (4.93) were recorded at 600 μg L^{-1} [16]. Toxicity symptoms appeared when plants were exposed to solutions containing 1000 μg L^{-1} As [16]. Further, combined use of plant growth promoting rhizobacteria (PGPR) such as Stenotrophomonas maltophilia and Agrobacterium sp. increased bioaccumulation of As in roots of Arundo donax plant upon exposure to higher concentration As (20 mg kg^{-1}) and enhanced overall phytoremediation efficiency of Arundo donax in presence of PGPR bacteria [51]. The As accumulation in *Phragnites austratlis* followed in the order of roots > rhizomes > leaves and maximum total As uptake was registered 32.5 mg kg^{-1} [54]· *V. zizinoids*, another semi aquatic weed reported to be capable of extracting As from contaminated water [17, 55]. In a hydroponic study (21 days), root to shoot As uptake it was increased with increase in As concentrations by *V. zizinoids* can uptake [17]. The BCF and TF for As were 10 and 0.86 indicates that *V. zizinoids* was an As hyper accumulator and stored higher proportion of As in their root system. Combined use of arbuscular mycorrhizal fungi (Glomus spp.) enhanced As uptake capability and growth of vetivar grass *(Chrysopogon zizanioides)* [55]. *Typha latifolia* also had the potential to uptake higher proportion of As from contaminated environment (soil), but most of the studies conducted using *Typha latifolia* were focused in soil. Most of the studies showed that semi aquatic weeds store more As in their root system and lower in upper vegetative parts. Higher plant vigor, higher As extraction capacity and perennial nature make them suitable phytoremediation agent for constructed wetland system. Combined use of submerged weeds like Hydrilla, Ceratophyllum, Potamogeton along with semi aquatic weeds (*Arundo donax, Vetivaria zizinoids, Phragmites* spp. and Typha sp.) and PGPR like VAM, As oxidizing bacteria may be highly useful to treat and remediate As contaminated water in constructed wetland system. Semi aquatic weeds are highly efficient when As present in higher concentrations and when As concentration in the system become lower submerged weeds come to play their role, as they are highly efficient As remover at lower concentrations. Again use PGPR will increase overall phytoremediation efficiency. Future research may be undertaken in these aspects for better information and output.

3.3 Phytoremediation by submerged aquatic weeds

Among the submerged aquatic weeds *Hydrilla verticillata*, *Ceratophyllum dermersum, Potamageton crispus, Valisnaria natans, Eleocharis acicularis* and *Elodea Canadensis* widely reported by many researchers to extract As from contaminated water. Studies conducted in laboratory and field conditions indicated that *Hydrilla verticillata*, and *Ceratophyllum demersum* can uptake higher proportion of As from contaminated water depending on exposure time and concentration of metalloid [22, 56, 57]. Unlike *Spirodela polyrhiza*, *Hydrilla verticillata* also uptake

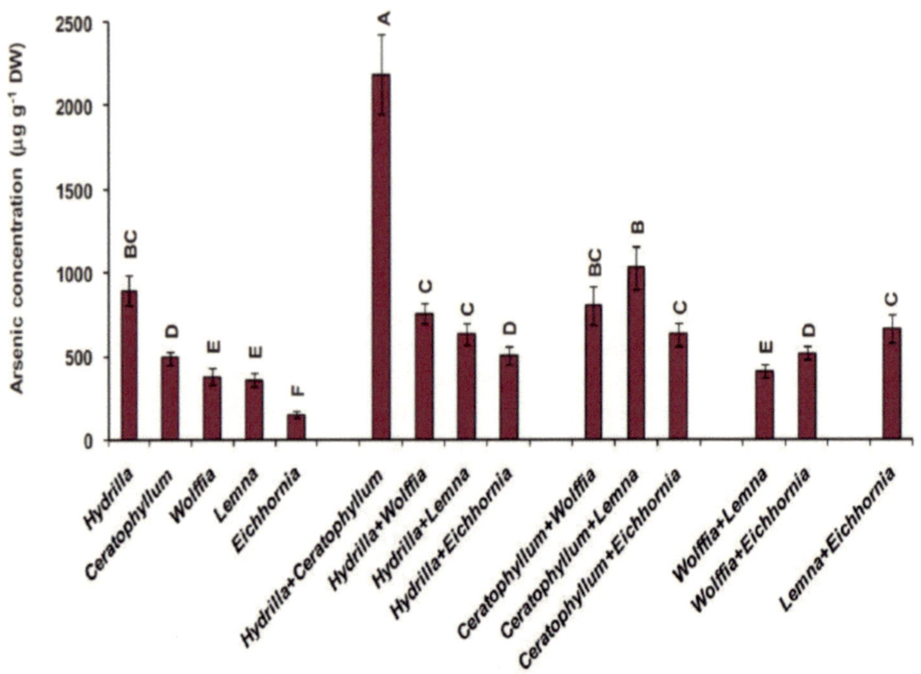

Figure 2.
Arsenic uptake comparison between various weed plants (Hydrilla, Ceratophyllum, Wolffia, Lemna and Eicchornia) grown in singly or in various combinations upon arsenic exposure for 30 days (adapted from Srivastava et al., 2014).

more arsenite (As^{3+}) form rather arsenate (As^{5+}) form [50]. Under simulated field condition (aquatic pond) Hydrilla verticillata alone removed sum total 8546 µg (348 µg g^{-1}) of As from contaminated water (As concentration 1500 µg L^{-1}) which is 72% of the total arsenic supplied [56]. Ceratophyllum demersum reported to accumulate 76 µg g^{-1} in 4 days and and As accumulation further increased to 201 µg g^{-1} in 7 days when exposed to 50 µM As solutions [22]. Maximum As accumulation by Ceratophyllum was recorded 525 µg g^{-1} dw when subjected to with 250 µM As^{5+} solution for 7 days [22]. Uptake of As by *Ceratophyllum demersum* depends on species of As present ($As3^{+}$ vs. As^{5+}) and pH of the medium. Maximum uptake of As^{3+} by Ceratophyllum was reported at pH 6.5 [58]. This variation in selective uptake of As species largely depends on uptake pathways and plant metabolism.

In natural conditions, submerged weeds grow in water bodies in association with floating macrophytes. Use of Combinations of submerged and floating weeds found more effective for phytoremediation purpose than submerged and floating weeds alone. Research work carried out using Hydrilla, Certophyllum, lemna and Wolfia at various combinations showed that Ceratophyllum + lemna combination (3326 µg) combination removed maximum total As followed by *Hydrilla + Wolfia* (1896 µg) (**Figure 2**). When the contribution of single plant considered, contribution of Hydrilla is more than 50% [56]. Arsenic phytoextraction potential of five different submerged weeds namely *Ceratophyllum dermersum*, *Potamageton crispus, Myriophyllum spicatum, Hydrilla verticillata* and *Vallisneria natans* were compared

Name of the plants	Key findings	Reference
Eichhornia crassipes	Removed 600 mg As ha^{-1} day^{-1} within 21 days with 18% removal efficiency when As was applied @ 0.15 mg L^{-1}	[13]
Lemna minor	Removal rate 140 mg As ha^{-1} day^{-1} within 21 days with 5% removal efficiency when As was applied @ 0.15 mg L^{-1}	[13]
	Removed relatively higher As^{3+} (17408 µg g^{-1}) and lower As^{5+} (8674 µg g^{-1}As) from As containing solutions (64 µM As each)	[35]
Pistia stratiotes	Accumulates 1120 µg g^{-1} As in *roots* and 31.60 µg g^{-1} As in leaves (dry weight basis) from 10 µM As containing solution	[31]
Salvinia natans	Accumulates 50 µg g^{-1} As in roots	[41]
Hydrilla verticillata	Removed sum total 8546 µg (348 µg g^{-1}) of As from contaminated water (As concentration 1500 µg L^{-1})	[49]
Ceratophyllum demersum	Accumulates 525 µg g^{-1} (dry weight baisis) from 250 µM As^{5+} solution for 7 days	[22]
Potamogeton crispus	Accumulates 1000 mg kg^{-1} (dry weight basis) from As contaminated environment	[52]
Myriophyllum spicatum	Accumulates 1000 mg kg^{-1} (dry weight basis) from As contaminated environment	[52]
Vallisnaria natans	Accumulates 1000 mg kg^{-1} (dry weight basis) from As contaminated environment	[52]
Alternanthera philoxeroides	Extract 12.94 mg kg^{-1} total As (dry weight basis) from pulp paper industry effluents	[61]
Arundo donax	Accumulates As at the rate 9 mg kg^{-1} with TF = 4.93 and BF = 15.00 for the arsenic containing solution 600 µg L^{-1}.	[16]
Phragnites australis	Accumulates 32.5 mg kg^{-1} As in root	[62]

Table 1.
Phytoremediation ability of various aquatic and semi aquatic weeds.

under natural As contaminated environement [59]. Results showed that all plants accumulated more 1000 mg kg^{-1} dw As; highest and lowest As accumulation and BCF were associated with *Vallisneria natans* (BCF- 361) and *Ceratophyllum dermersum* (BCF- 221) [59]. Similarly ability of potamogeton spp., Myriophyllum spp. and Valisnaria app to uptake As from contaminated water were also been reported by many authors [60–62]. Arsenic uptake by various types of aquatic, semi-aquatic and submerged weeds has been outlined in **Table 1**.

4. Mechanisms of arsenic uptake and detoxification in aquatic weeds

4.1 Mechanisms of arsenic uptake in aquatic macrophytes

Three pathways for arsenic uptake in marine macrophytes have been described – (i) active uptake through phosphate uptake transporters, (ii) passive uptake through aquaglyceroporins, and (iii) physicochemical adsorption on root surfaces. Plants mainly uptake As(V) through phosphate uptake transporters [63, 64]. As(III), DMAA and MMAA gets into the plants by passive mechanism through the aquaglyceroporin channels [64].

4.1.1 Active uptake through phosphate uptake transporters

As(V) and phosphate are chemical analogs, and compete for uptake carriers in the plasmalemma [65]. As a result, as the phosphate content rises, more As (V) is required to be desorbed in the solution. Mkandawire and Dudel. [32] and Rahman et al. [33] showed that As (V) is taken up by aquatic plants through the phosphate uptake pathway, it competes with phosphate for uptake in tissues of *L. gibba* L. and *S. polyrhiza* L.

4.1.2 Passive uptake through aquaporins/aquaglyceroporins

Physiological studies indicate that these arsenic species are transported in rice through aquaporins /aquaglyceroporins via passive uptake mechanisms [66, 67]. Molecular studies revealed that Nodulin26-like intrinsic membrane proteins (NIPs), one of the major subfamilies of aquaporins transporters that promote the transport of neutral molecules like water, glycerol, and urea, are responsible for transporting As(III) into rice roots [68]. Aquaporins and aquaglyceroporins are two of three subfamilies of water channel proteins (WCPs), the transmembrane proteins that have a specific three-dimensional structure with a pore that permeates water molecules [69], which are permeable to water, glycerol, and/or other small, neutral molecules. Glycerol and As(III) compete for uptake in rice (*Oryza sativa* L.), indicating that this arsenic species is carried via the plasma membrane by aquaporins/ aquaglyceroporins [67].

4.1.3 Physicochemical adsorption on root surfaces

Arsenic is adsorbing and accumulating on the surfaces of aquatic plants due to suspended iron oxides (Fe-plaque). Robinson et al. [70] discovered a strong association between arsenic and iron concentrations in aquatic plants, which is believed to be due to arsenic adsorption on plant surfaces' iron oxides. Rahman et al. [14] investigated arsenic species adsorption on precipitated iron oxides on *S. polyrhiza* L. roots/fronds and revealed a strong association between arsenic and iron concentrations in tissues when the plant was exposed to As (V). There was no association between arsenic and iron in plant tissue when *S. polyrhiza* L. was exposed to As (III), DMAA, and MMAA.As (V) is primarily adsorbed on precipitated iron oxides on the roots of aquatic plants and deposited by a physicochemical adsorption process, according to the findings.

4.2 Arsenic metabolism and detoxification in aquatic macrophytes

Arsenic occurs primarily as As (V) in an oxic environment and as As (III) in a reduced environment [64]. In plants, As (V) and phosphate share the same transporter, while As(III) enters plant cells through NIPs'aquaporins [57, 64]. Because of their distinct molecular properties, these two types of arsenic elicit different biochemical responses in aquatic plants [71]. As (V) has no affinity for thiol ligands, while As(III) has a strong affinity for peptides with sulfhydryl (-SH) groups, such as glutathione (GSH) and phytochelatins (PCs) [64, 72]. Even though plants had been exposed to As, arsenic speciation in plant tissues indicates that arsenic is primarily present in the As(III) oxidation state (V).This suggests that As(V) is effectively reduced to As(III) in plant cells after uptake, and that most plants have high As(V) reduction competence [64]. The reduction of As(V) to As(III) is mediated by GSH [73] and by enzyme [74], which is thought to be a detoxification mechanism of the plants. As(V) and As(III) have been shown to generate reactive oxygen species (ROS) within cells when they are taken up [75], and plants counteract the generation of ROS by various enzymes and cellular compounds [76]. The GSH can act as an antioxidant and is required for the synthesis of Phytochelatins which are required for metalloid chelation [71].

The mechanism of arsenic accumulation and detoxification was studied by many others in aquatic plant *H. verticillata* [57, 71]. In the presence of As (III) or As(V), *H. verticillata* enhanced the biosynthesis of thiols such as PCs, and increased antioxidant enzyme activity. Although the levels of thiolic compounds such as NP-SH, cysteine, GSH, and oxidized glutathione (GSSG) were significantly enhanced in *H. verticillata* upon exposure to both As(III) and As(V), As(III) was found to enhance the activities of cysteine synthase and c-glutamylcysteine synthetase and the amount of cysteine and GSH to higher levels than As(V) . The analysis of PCs indicates that the accumulation of PC1 and PC2 in *H. verticillata* was enhanced with the increase of both As(III) and As(V) concentrations [71]. Thus, during As (III) and As(V) stress, phytochelatins and antioxidant systems in *H. verticillata* react differently, which is considered to be the plant's detoxification mechanism.

5. Biotechnological interventions for phytoremediation

Plants have been utilized for phytoremediation of toxic metals and metalloids, however due to heavy metal phytotoxicity to plants; this process has been slow and largely rendered ineffective [77]. Natural heavy metal hyperaccumulators are also available, however, they are limited to specific geo-climatic conditions and also lack the crucial biomass required for efficient phytoremediation. Phytoremediation has a lot of potential using genetic engineering technologies to improve plant tolerance and heavy metal accumulation. Furthermore, various new studies using omics technologies such as genomics, transcriptomics, proteomics, and metabolomics to elucidate the genetic determinants and pathways involved in heavy metal and metalloid tolerance in plants have been identified. Presently there are three main biotechnological approaches for the phytoremediation of heavy metals and metalloids are currently being used to engineer plants for phytoremediation of heavy metals and metalloids: (1) manipulating metal/metalloid

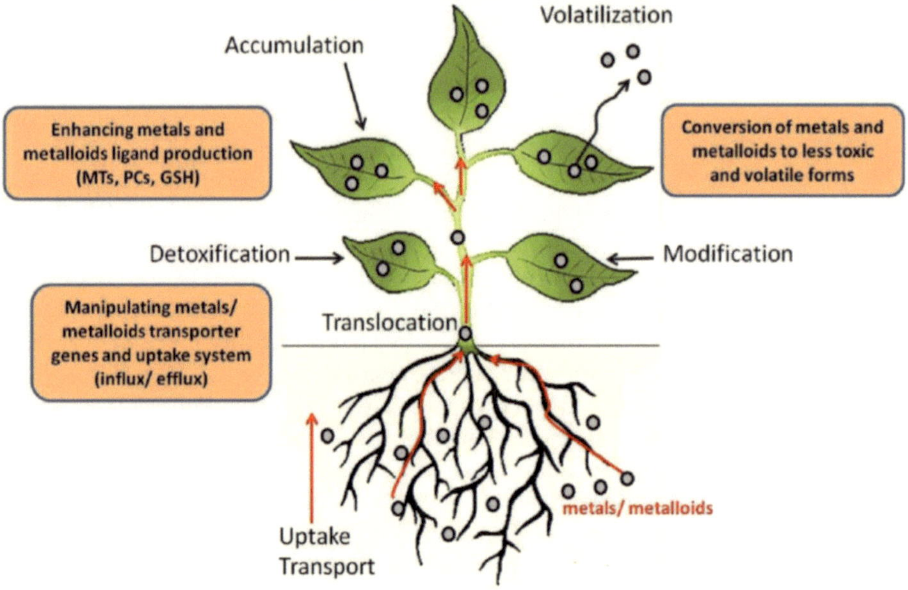

Figure 3.
Potential biotechnological strategies for phytoremediation. Heavy/toxic metals can be mobilized and transported (influx) into roots through plasma membrane transporters. They can then be transported (efflux) out of the roots into the xylem and translocated into the shoots. At this stage, plant tolerance to toxic elements may be enhanced through manipulation of influx/efflux transporters or by increasing the levels of ligands/chelators. Volatilization of the toxic elements can be achieved through enzymes that modify these toxic elements. Chelators or efflux transporters can also be used to export the toxic elements out of the cytosol and into vacuoles or the cell wall. Adapted from Dhankher et al. (2011).

transporter genes and uptake systems; (2) enhancing metal and metalloid ligand production; (3) conversion of metals and metalloids to less toxic and volatile forms [78] (**Figure 3**).

5.1 Manipulating metal/metalloid transporter genes and uptake system

Enhanced heavy metal tolerance and bioaccumulation has been attained in different plant species by genetic manipulation of metal transporter genes. For example, the overexpression of full length *NtCBP4* (plasma membrane channel protein) in *Nicotiana tabacum* showed Pb^{2+} hypersensitivity and enhanced accumulation of Pb^{2+} in the genetically manipulated plants. However, the overexpression of a truncated version of *NtCBP4* generated by deletion of its C-terminal, calmodulin- binding domain and part of the putative cyclic nucleotide- binding domain showed improved tolerance to Pb^{2+} and less accumulation of Pb^{2+} [79]. *Nicotiana tabacum* plants expressing *CAX2* (calcium exchanger 2) gene accumulated more Ca^{2+}, Mn^{2+} and Cd^{2+} and also showed enhanced tolerant to elevated Mn^{2+}. It was also observed that overexpression of *CAX2* gene in *Nicotiana tabacum* increased Mn^{2+} and Cd^{2+} transport in the root tonoplast vesicles in the transgenic plants [80]. Moreover, T-DNA mutants of the *Arabidopsis CNGC1* (cyclic nucleotide- gated ion channel 1) gene, that encodes a homologous protein to *NtCBP4*, also showed Pb^{2+} hypersensitivity and enhanced accumulation of Pb^{2+} in the genetically manipulated plants. These findings suggest that *NtCBP4* and *AtCNGC1* play an important role in the transport pathway of Pb^{2+} [79, 81]. The overexpression of yeast *YCF1* (Yeast Cadmium Factor 1) gene in *Arabidopsis thaliana* resulted in enhanced accumulated higher amounts and tolerance to Pb^{2+} and Cd^{2+} metals in plants [82].

Recent research findings have revealed arsenite is transported in plants by proteins belonging to the aquaporins [83, 84]. It is observed that in efficient arsenic hyperaccumulators such as *Pteris vittata* has highly well-organized system of arsenic translocation from root to shoot tissues [85, 86], However, most non-hyperaccumulators show low mobility rate compared to *P. vittata*, also variable Arsenic mobility rate is observed among different plant species, suggesting that it is controlled by genes. Arsenic loading to the xylem is a critical stage in arsenic translocation from root to shoot, however it is a poorly known mechanism. Ma et al. [87, 88] has identified and characterized *Lsi2* gene encoding an efflux protein, plays an important role in loading arsenite into the xylem. Mutation in *Lsi2* gene caused about 50% reduction in arsenic accumulation in the shoot. The *Lsi2* gene is a homolog of the *E. coli* ArsB gene, an As (III)/H^+ exchanger that confers bacterial arsenite tolerance [89].

Genome-wide gene expression analysis in *Oryza sativa* roots treated with different heavy metals and metalloids; As(V), Cr(VI), Pb, and Cd, showed numerous differentially expressed genes as well as unique genes. Various genes belonging to different transporter families were identified [90]. Recently Wang et al. [91], has identified genes for Cu tolerance in the *Paeonia ostii* with the help of *de novo* transcriptome sequencing approach. Such genes may further be transferred to crop plants for enhancing heavy metal tolerance. Therefore, strategies of developing transgenic plants for arsenic (As) phytoremediation include enhancing plant uptake for phytoextraction, decreasing plant uptake, improving the plants' tolerance to As contamination, and increased methylation for enhanced food safety.

5.2 Enhancing metals and metalloids ligand production

Complexation of Arsenic with phytochelatins (PCs), or metallothionein (MTs) or glutathione (GSH) is an proficient way to detoxify As(III), since these complexes are sequestered in the vacuoles, this process is catalyzed by the homologs of multidrug resistance proteins (MRPs) [92, 93]. Enhancing the accumulation or synthesis of PCs and/or GSH and/or MTs may be one way to increase phytoremediation of arsenic. The overexpression of *PCS* in *Brassia juncea* enhanced its tolerance to arsenic but no significant increase arsenic accumulation was observed, this may be due to the fact that PC synthesis is also limited by the production of GSH [94]. The overexpression of *AtPCS1* and *GSH1*genes, that encode g-glutamylcysteine synthetase (g-ECS), the rate-limiting step in GSH biosynthesis, individually in *Arabidopsis thaliana* increased both arsenic tolerance and as well as accumulation [95].

Arsenic (As) tolerance in plants can also be increased by modifying GSH and PCs. Dhankher et al. [96] transferred and co-expressed two bacterial genes, *E. coli* arsenate reductase (*arsC*) and γ-glutamylcysteine synthetase (*γ-ECS*), in *Arabidopsis thaliana*, the transgenic plants grown in the presence of 125 μM sodium arsenate accumulated threefold more arsenic in the aboveground biomass and showed almost 17-fold higher biomass than wild type WT plants. The overexpression of AtPCS1 under constitutive promoter in *A. thaliana* enhanced tolerance to arsenate but failed to enhance arsenic accumulation [97]. These studies showed that manipulation of genes for increasing the production of metal chelation agents hold great potential for improving heavy metal and metalloid tolerance and accumulation in plants.

The *de novo* transcriptome sequencing analysis in *Raphanus sativus L.* roots under cadmium stress was carried out to discover differentially expressed genes and microRNAs (miRNAs) involved in Cd-responsive regulatory pathways. Various candidate genes encoding PCs, GSHs, and MTs; and other genes belonging to zinc iron permease (ZIPs) and ABC transporters were identified [98]. Likewise, in *de novo* transcriptome analysis in radish roots under chromium stress, showed that

1561 unigenes down-regulated and 1424 unigenes were up-regulated, various transcription factors such as Chromium stress-responsive genes involved in chelate compounds, signal transduction and antioxidant biosynthesis were discovered [99]. Such candidate genes can further be transferred into the crop plants to enhance heavy metal tolerance as well as accumulation.

5.3 Conversion of metals and metalloids to less toxic and volatile forms

There are several reports for developing phytoremediation strategies for heavy metals with the help of biotechnological interventions by conversion of these metals to less toxic and volatile forms. It is observed that many organisms, including bacteria, fungi, and animals, methylate arsenic. Methylated arsenic have been discovered in several plant species, including rice grain [100, 101], and suggest that this is the process is a result of endogenous methylation by the plants themselves. The final product of this pathway is the gas trimethylarsine (TMAs(III)), that can be volatilized from the plant. Qin et al. [102] have cloned a gene encoding an As(III)-S-adenosylmethionine methyltransferase (arsM) from the soil bacterium *Rhodopseudomonas palustris*. Expression of the *arsM* gene in an arsenic-sensitive strain of *E. coli* that resulted in the biosynthesis of several methylated forms of arsenic, including volatile TMAs(III) and conferred arsenic tolerance in the plants. These findings show that the expression of the single methyltransferase (*arsM*) gene is sufficient to produce both volatilization and tolerance to arsenic (As). A gene for an ArsM homolog in a primitive plant, the eukaryotic alga Cyanidioschyzon merolae has been idenfied [103]. Cells expressing *CmArsM* methylates As(III), as like the purified enzyme. In a rice microarray study, a putative gene annotated as a methyltransferase was found to be upregulated upon exposure to arsenate in the growth solution [104]. These findings indicate the possibility of engineering arsenic volatilization for the phytoremediation of arsenic-contaminated water and soil and also to improve the safety of the food supply.

6. Conclusions

Contamination of soils and water by arsenic is one the serious threat for food security and human health in throughout the world. Some severe skin and other diseases occur due to continuous consumption of As contaminated foods and water. This necessitates a suitable technology to handle arsenic contaminated water carefully, so that above mentions points can be satisfied. Phytoremediation of arsenic contaminated water by aquatic and semi aquatic weeds offers low cost, economically feasible and eco-friendly technology to remove arsenic from contaminated water for long term. Some weeds have tremendous potential to accumulate higher amount of arsenic in their plant parts such as Eichhornia crassipes, Hydrilla verticillata, Spirodella polyrhiza, Arundo donax and Vetivaria spp. More specifically semi aquatic weeds like Arundo donax and Vetivaria sp. (perennial) can be used with in combination with Eicchornia, Spirodella and Hydrilla to remove arsenic more efficiently from treatment tanks or constructed wetland system. Although management of plant biomass will be another concern for disposal, but these plant materials can be used for making fiber (water hyacinth), handcraft items (Arundo and Typha stems) and biofuel purpose. Moreover, with advancement of molecular genetics in future As tolerance genes can be transferred to food crops (specially rice) which can store huge amount of As in their roots or very low transfer coefficient from root to grain so that transgenic rice crops will able to grow using As contaminated water and contribute in food security in upcoming days.

Author details

Dibakar Roy*, Dasari Sreekanth, Deepak Pawar, Himanshu Mahawar and Kamal K. Barman
ICAR-Directorate of Weed Research, Jabalpur, India

*Address all correspondence to: dibakar499176@gmail.com

References

[1] Shukla A, Srivastava S. A review of phytoremediation prospects for arsenic contaminated water and soil. Phytomanagement of polluted sites. 2019 Jan 1:243-54.

[2] Bhattacharyya R, Chatterjee D, Nath B, Jana J, Jacks G, Vahter M. High arsenic groundwater: mobilization, metabolism and mitigation–an overview in the Bengal Delta Plain. Molecular and cellular biochemistry. 2003 Nov; 253(1):347-55.

[3] Shaji E, Santosh M, Sarath KV, Prakash P, Deepchand V, Divya BV. Arsenic contamination of groundwater: A global synopsis with focus on the Indian Peninsula. Geoscience Frontiers. 2020 Oct 1.

[4] Srivastava S, Suprasanna P, D'souza SF. Mechanisms of arsenic tolerance and detoxification in plants and their application in transgenic technology: a critical appraisal. International journal of phytoremediation. 2012 May 1;14(5):506-17.

[5] Saha D, Sahu S. A decade of investigations on groundwater arsenic contamination in Middle Ganga Plain, India. Environmental geochemistry and health. 2016 Apr 1;38(2):315-37.

[6] Mukherjee AB, Bhattacharya P, Jacks G, Banerjee DM, Ramanathan AL, Chandan M, Chandrashekharam D, Debashis C, Naidu R. Groundwater arsenic contamination in India: extent and severity. CSIRO Publishing; 2006.

[7] Tuli R, Chakrabarty D, Trivedi PK, Tripathi RD. Recent advances in arsenic accumulation and metabolism in rice. Molecular Breeding. 2010 Aug; 26(2):307-23.

[8] Spognardi S, Bravo I, Beni C, Menegoni P, Pietrelli L, Papetti P. Arsenic accumulation in edible vegetables and health risk reduction by groundwater treatment using an adsorption process. Environmental Science and Pollution Research. 2019 Nov; 26(31):32505-16.

[9] Awasthi S, Chauhan R, Srivastava S, Tripathi RD. The journey of arsenic from soil to grain in rice. Frontiers in Plant Science. 2017 Jun 20; 8:1007.

[10] Rahman MA, Hasegawa H. Aquatic arsenic: phytoremediation using floating macrophytes. Chemosphere. 2011 Apr 1;83(5):633-46.

[11] Nazir A, Malik RN, Ajaib M, Khan N, Siddiqui MF. Hyperaccumulators of heavy metals of industrial areas of Islamabad and Rawalpindi. Pak J Bot. 2011 Aug 1;43(4):1925-33.

[12] Ma LQ, Komar KM, Tu C, Zhang W, Cai Y, Kennelley ED. A fern that hyperaccumulates arsenic. Nature. 2001 Feb;409(6820):579-.579.

[13] Alvarado S, Guédez M, Lué-Merú MP, Nelson G, Alvaro A, Jesús AC, Gyula Z. Arsenic removal from waters by bioremediation with the aquatic plants Water Hyacinth (*Eichhornia crassipes*) and Lesser Duckweed (*Lemna minor*). Bioresource technology. 2008 Nov 1;99(17):8436-40.

[14] Rahman MA, Hasegawa H, Ueda K, Maki T, Rahman MM. Arsenic uptake by aquatic macrophyte *Spirodela polyrhiza* L.: Interactions with phosphate and iron. Journal of hazardous materials. 2008 Dec 30; 160(2-3):356-61.

[15] Roy C, Jahan M, Rahman S. Characterization and treatment of textile wastewater by aquatic plants (macrophytes) and algae. European Journal of Sustainable Development Research. 2018; 2(3):29.

[16] Mirza N, Mahmood Q, Pervez A, Ahmad R, Farooq R, Shah MM, Azim MR. Phytoremediation potential of *Arundo donax* in arsenic-contaminated synthetic wastewater. Bioresource technology. 2010 Aug 1;101(15):5815-9.

[17] Mirza N, Mubarak H, Chai LY, Yong W, Khan MJ, Khan QU, Hashmi MZ, Farooq U, Sarwar R, Yang ZH. The potential use of *Vetiveria zizanioides* for the phytoremediation of antimony, arsenic and their co-contamination. Bulletin of environmental contamination and toxicology. 2017 Oct;99(4):511-7.

[18] Chaney, R.L., 1983. Plant uptake of inorganic waste constituents. In: Parr, J.F.E.A. (Ed.), Land Treatment of Hazardous Wastes. Noyes Data Corp, Park Ridge, NJ, pp. 50-76.

[19] Ebel M, Evangelou MW, Schaeffer A. Cyanide phytoremediation by water hyacinths (*Eichhornia crassipes*). Chemosphere. 2007 Jan 1;66(5):816-23.

[20] Misbahuddin, M.I.R. and Fariduddin, A.T.M., 2002. Water hyacinth removes arsenic from arsenic-contaminated drinking water. Archives of Environmental Health: An International Journal, 57(6), pp.516-518.

[21] Al Rmalli SW, Harrington CF, Ayub M, Haris PI. A biomaterial based approach for arsenic removal from water. Journal of environmental monitoring. 2005;7(4):279-82.

[22] Mishra, V.K., Upadhyay, A.R., Pathak, V. and Tripathi, B.D., 2008. Phytoremediation of mercury and arsenic from tropical opencast coalmine effluent through naturally occurring aquatic macrophytes. Water, air, and soil pollution, 192(1), pp.303-314.

[23] Taleei MM, Ghomi NK, Jozi SA. Arsenic removal of contaminated soils by phytoremediation of vetiver grass, chara algae and water hyacinth. Bulletin of environmental contamination and toxicology. 2019 Jan 15;102(1):134-9.

[24] Islam A, Saha PK, Iqbal M, Islam MN, Nayeem M. Removal of arsenic by water hyacinth from arsenic contaminated water. Water Int. 2016;1(2):36-41.

[25] Rodrigues AC, do Amaral Sobrinho NM, dos Santos FS, dos Santos AM, Pereira AC, Lima ES. Biosorption of toxic metals by water lettuce (*Pistia stratiotes*) biomass. Water, Air, & Soil Pollution. 2017 Apr 1;228(4):156.

[26] Zhou YQ, Li SY, Shi YD, Lv W, Shen TB, Huang QL, Li YK, Wu ZL. Phytoremediation of Chromium and Lead Using Water Lettuce *Pistia stratiotes* L. InApplied Mechanics and Materials 2013 (Vol. 401, pp. 2071-2075). Trans Tech Publications Ltd.

[27] Odjegba VJ, Fasidi IO. Accumulation of trace elements by *Pistia stratiotes*: implications for phytoremediation. Ecotoxicology. 2004 Oct;13(7):637-46.

[28] Lee CK, Low KS, Hew NS. Accumulation of arsenic by aquatic plants. Science of the total environment. 1991 Apr 15;103(2-3):215-27.

[29] Basu A, Kumar S, Mukherjee S. Arsenic reduction from aqueous environment by water lettuce (*Pistia stratiotes* L.). Indian Journal of Environmental Health. 2003 Apr 1;45(2):143-50.

[30] Farnese FD, Oliveira JA, Lima FS, Leão GA, Gusman GS, Silva LC. Evaluation of the potential of *Pistia stratiotes* L.(water lettuce) for bioindication and phytoremediation of aquatic environments contaminated with arsenic. Brazilian Journal of Biology. 2014 Aug;74(3):S108-12.

[31] De Campos FV, de Oliveira JA, da Silva AA, Ribeiro C, dos Santos Farnese F. Phytoremediation of arsenite-contaminated environments: is *Pistia stratiotes* L. a useful tool?. Ecological Indicators. 2019 Sep 1;104:794-801.

[32] Mkandawire M, Dudel EG. Accumulation of arsenic in *Lemna gibba* L. (duckweed) in tailing waters of two abandoned uranium mining sites in Saxony, Germany. Science of the Total Environment. 2005 Jan 5;336(1-3):81-9.

[33] Rahman MA, Hasegawa H, Ueda K, Maki T, Okumura C, Rahman MM. Arsenic accumulation in duckweed (*Spirodela polyrhiza* L.): a good option for phytoremediation. Chemosphere. 2007 Sep 1;69(3):493-9.

[34] Duman F, Ozturk F, Aydin Z. Biological responses of duckweed (*Lemna minor* L.) exposed to the inorganic arsenic species As (III) and As (V): effects of concentration and duration of exposure. Ecotoxicology. 2010 Jun;19(5):983-93.

[35] Zhang X, Hu Y, Liu Y, Chen B. Arsenic uptake, accumulation and phytofiltration by duckweed (*Spirodela polyrhiza* L.). Journal of environmental sciences. 2011 Apr 1;23(4):601-6.

[36] Favas PJ, Pratas J, Prasad MN. Accumulation of arsenic by aquatic plants in large-scale field conditions: opportunities for phytoremediation and bioindication. Science of the total Environment. 2012 Sep 1;433:390-7.

[37] Zhang X, Zhao FJ, Huang Q, Williams PN, Sun GX, Zhu YG. Arsenic uptake and speciation in the rootless duckweed *Wolffia globosa*. New Phytologist. 2009 Apr;182(2):421-8.

[38] Zhang X, Uroic MK, Xie WY, Zhu YG, Chen BD, McGrath SP, Feldmann J, Zhao FJ. Phytochelatins play a key role in arsenic accumulation and tolerance in the aquatic macrophyte *Wolffia globosa*. Environmental Pollution. 2012 Jun 1;165:18-24.

[39] Da Silva AA, Oliveira JA, Campos FV, Ribeiro C, Farnese FD. Role of glutathione in tolerance to arsenite in *Salvinia molesta*, an aquatic fern. Acta Botanica Brasilica. 2017 Dec;31(4):657-64.

[40] Hoffmann T, Kutter C, Santamaria J. Capacity of *Salvinia minima* Baker to tolerate and accumulate As and Pb. Engineering in Life Sciences. 2004 Feb 5;4(1):61-5.

[41] Rahman MA, Hasegawa H, Ueda K, Maki T, Rahman MM. Influence of phosphate and iron ions in selective uptake of arsenic species by water fern (*Salvinia natans* L.). Chemical Engineering Journal. 2008 Dec 15;145(2):179-84.

[42] Sood A, Uniyal PL, Prasanna R, Ahluwalia AS. Phytoremediation potential of aquatic macrophyte, Azolla. Ambio. 2012 Mar;41(2):122-37.

[43] Pandey VC. Phytoremediation of heavy metals from fly ash pond by *Azolla caroliniana*. Ecotoxicology and Environmental Safety. 2012 Aug 1;82:8-12.

[44] Rai PK. Phytoremediation of Hg and Cd from industrial effluents using an aquatic free floating macrophyte *Azolla pinnata*. International journal of phytoremediation. 2008 Jul 23;10(5):430-9.

[45] Mahmud R, Inoue N, Kasajima SY, Shaheen R. Assessment of potential indigenous plant species for the phytoremediation of arsenic-contaminated areas of Bangladesh. International Journal of Phytoremediation. 2008 Apr 3;10(2):119-32.

[46] Rofkar JR, Dwyer DF, Bobak DM. Uptake and toxicity of arsenic, copper,

and silicon in *Azolla caroliniana* and *Lemna minor*. International journal of phytoremediation. 2014 Feb 1;16(2):155-66.

[47] Zhang X, Lin AJ, Zhao FJ, Xu GZ, Duan GL, Zhu YG. Arsenic accumulation by the aquatic fern Azolla: comparison of arsenate uptake, speciation and efflux by *A. caroliniana* and *A. filiculoides*. Environmental Pollution. 2008 Dec 1;156(3):1149-55.

[48] Srivastava S, Sounderajan S, Udas A, Suprasanna P. Effect of combinations of aquatic plants (*Hydrilla*, *Ceratophyllum*, *Eichhornia*, *Lemna* and *Wolffia*) on arsenic removal in field conditions. Ecological engineering. 2014 Dec 1;73:297-301.

[49] Srivastava S, Mishra S, Dwivedi S, Tripathi RD. Role of thiol metabolism in arsenic detoxification in *Hydrilla verticillata* (Lf) Royle. Water, Air, & Soil Pollution. 2010 Oct;212(1):155-65.

[50] Khang HV, Hatayama M, Inoue C. Arsenic accumulation by aquatic macrophyte coontail (*Ceratophyllum demersum* L.) exposed to arsenite, and the effect of iron on the uptake of arsenite and arsenate. Environmental and experimental botany. 2012 Nov 1;83:47-52.

[51] Chen G, Liu X, Brookes PC, Xu J. Opportunities for phytoremediation and bioindication of arsenic contaminated water using a submerged aquatic plant: *Vallisneria natans* (Lour.) Hara. International journal of phytoremediation. 2015 Mar 4;17(3):249-55.

[52] Norouznia H, Hamidian AH. Phytoremediation efficiency of pondweed (*Potamogeton crispus*) in removing heavy metals (Cu, Cr, Pb, As and Cd) from water of Anzali wetland. International Journal of Aquatic Biology. 2014 Sep 10;2(4):206-14.

[53] Krayem M, Baydoun M, Deluchat V, Lenain JF, Kazpard V, Labrousse P. Absorption and translocation of copper and arsenic in an aquatic macrophyte *Myriophyllum alterniflorum* DC. in oligotrophic and eutrophic conditions. Environmental Science and Pollution Research. 2016 Jun;23(11):11129-36.

[54] Li B, Gu B, Yang Z, Zhang T. The role of submerged macrophytes in phytoremediation of arsenic from contaminated water: A case study on *Vallisneria natans* (Lour.) Hara. Ecotoxicology and environmental safety. 2018 Dec 15;165:224-31.

[55] Datta R, Quispe MA, Sarkar D. Greenhouse study on the phytoremediation potential of vetiver grass, *Chrysopogon zizanioides* L., in arsenic-contaminated soils. Bulletin of environmental contamination and toxicology. 2011 Jan 1;86(1):124-8.

[56] Jomjun N, Siripen T, Maliwan S, Jintapat N, Prasak T, Somporn C, Petch P. Phytoremediation of arsenic in submerged soil by wetland plants. International journal of phytoremediation. 2010 Nov 18;13(1):35-46.

[57] Raj A, Jamil S, Srivastava PK, Tripathi RD, Sharma YK, Singh N. Feasibility Study of *Phragmites karka* and *Christella dentata* Grown in West Bengal as Arsenic Accumulator. International journal of phytoremediation. 2015 Sep 2;17(9):869-78.

[58] Guarino F, Miranda A, Castiglione S, Cicatelli A. Arsenic phytovolatilization and epigenetic modifications in *Arundo donax* L. assisted by a PGPR consortium. Chemosphere. 2020 Jul 1;251:126310.

[59] Simmons ZD, Suleiman AA, Theegala CS. Phytoremediation of arsenic and lead using alligator weed (*Alternathera philoxeroides*).

Transactions of the ASABE. 2007;50(5):1895-900.

[60] Sharma P, Tripathi S, Chandra R. Highly efficient phytoremediation potential of metal and metalloids from the pulp paper industry waste employing *Eclipta alba* (L) and *Alternanthera philoxeroide* (L): Biosorption and pollution reduction. Bioresource Technology. 2021 Jan 1;319:124147.

[61] Ghassemzadeh F, Yousefzadeh H, Arbab-Zavar MH. Arsenic phytoremediation by *Phragmites australis*: green technology. International journal of environmental studies. 2008 Aug 1;65(4):587-94.

[62] Caporale AG, Sarkar D, Datta R, Punamiya P, Violante A. Effect of arbuscular mycorrhizal fungi (*Glomus* spp.) on growth and arsenic uptake of vetiver grass (*Chrysopogon zizanioides* L.) from contaminated soil and water systems. Journal of soil science and plant nutrition. 2014 Dec;14(4):955-72.

[63] Tripathi RD, Srivastava S, Mishra S, Singh N, Tuli R, Gupta DK, Maathuis FJ. Arsenic hazards: strategies for tolerance and remediation by plants. Trends in biotechnology. 2007 Apr 1;25(4):158-65.

[64] Zhao FJ, Ma JF, Meharg AA, McGrath SP. Arsenic uptake and metabolism in plants. New Phytologist. 2009 Mar;181(4):777-94.

[65] Mkandawire M, Lyubun YV, Kosterin PV, Dudel EG. Toxicity of arsenic species to *Lemna gibba* L. and the influence of phosphate on arsenic bioavailability. Environmental Toxicology: An International Journal. 2004 Feb;19(1):26-34.

[66] Abedin MJ, Feldmann J, Meharg AA. Uptake kinetics of arsenic species in rice plants. Plant physiology. 2002 Mar 1;128(3):1120-8.

[67] Meharg AA, Jardine L. Arsenite transport into paddy rice (*Oryza sativa*) roots. New phytologist. 2003 Jan;157(1):39-44.

[68] Ma JF, Yamaji N, Mitani N, Xu XY, Su YH, McGrath SP, Zhao FJ. Transporters of arsenite in rice and their role in arsenic accumulation in rice grain. Proceedings of the National Academy of Sciences. 2008 Jul 22;105(29):9931-5.

[69] Benga G. Water channel proteins (later called aquaporins) and relatives: past, present, and future. IUBMB life. 2009 Feb;61(2):112-33.

[70] Robinson B, Kim N, Marchetti M, Moni C, Schroeter L, van den Dijssel C, Milne G, Clothier B. Arsenic hyperaccumulation by aquatic macrophytes in the Taupo Volcanic Zone, New Zealand. Environmental and Experimental Botany. 2006 Dec 1;58(1-3):206-15.

[71] Srivastava S, Mishra S, Tripathi RD, Dwivedi S, Trivedi PK, Tandon PK. Phytochelatins and antioxidant systems respond differentially during arsenite and arsenate stress in *Hydrilla verticillata* (Lf) Royle. Environmental science & technology. 2007 Apr 15;41(8):2930-6.

[72] Raab A, Ferreira K, Meharg AA, Feldmann J. Can arsenic–phytochelatin complex formation be used as an indicator for toxicity in *Helianthus annuus*?. Journal of Experimental Botany. 2007 Apr 1;58(6):1333-8.

[73] Delnomdedieu M, Basti MM, Otvos JD, Thomas DJ. Reduction and binding of arsenate and dimethylarsinate by glutathione: a magnetic resonance study. Chemico-biological interactions. 1994 Feb 1;90(2):139-55.

[74] Bleeker PM, Hakvoort HW, Bliek M, Souer E, Schat H. Enhanced arsenate

reduction by a CDC25-like tyrosine phosphatase explains increased phytochelatin accumulation in arsenate-tolerant *Holcus lanatus*. The Plant Journal. 2006 Mar;45(6):917-29.

[75] Meharg AA, Hartley-Whitaker J. Arsenic uptake and metabolism in arsenic resistant and nonresistant plant species. New Phytologist. 2002 Apr;154(1):29-43.

[76] Mittler R. Oxidative stress, antioxidants and stress tolerance. Trends in plant science. 2002 Sep 1;7(9):405-10.

[77] Dhankher OP, Pilon-Smits EA, Meagher RB, Doty S. Biotechnological approaches for phytoremediation. InPlant biotechnology and agriculture 2012 Jan 1 (pp. 309-328). Academic Press.

[78] Kotrba P, Najmanova J, Macek T, Ruml T, Mackova M. Genetically modified plants in phytoremediation of heavy metal and metalloid soil and sediment pollution. Biotechnology advances. 2009 Nov 1;27(6):799-810.

[79] Sunkar R, Kaplan B, Bouché N, Arazi T, Dolev D, Talke IN, Maathuis FJ, Sanders D, Bouchez D, Fromm H. Expression of a truncated tobacco NtCBP4 channel in transgenic plants and disruption of the homologous Arabidopsis CNGC1 gene confer Pb2+ tolerance. The Plant Journal. 2000 Nov;24(4):533-42.

[80] Hirschi KD, Korenkov VD, Wilganowski NL, Wagner GJ. Expression of Arabidopsis CAX2 in tobacco. Altered metal accumulation and increased manganese tolerance. Plant physiology. 2000 Sep 1;124(1):125-34.

[81] Zeng H, Xu L, Singh A, Wang H, Du L, Poovaiah BW. Involvement of calmodulin and calmodulin-like proteins in plant responses to abiotic stresses. Frontiers in plant science. 2015 Aug 11;6:600.

[82] Song WY, Sohn EJ, Martinoia E, Lee YJ, Yang YY, Jasinski M, Forestier C, Hwang I, Lee Y. Engineering tolerance and accumulation of lead and cadmium in transgenic plants. Nature biotechnology. 2003 Aug;21(8):914-9.

[83] Bienert GP, Thorsen M, Schüssler MD, Nilsson HR, Wagner A, Tamás MJ, Jahn TP. A subgroup of plant aquaporins facilitate the bi-directional diffusion of As (OH) 3 and Sb (OH) 3 across membranes. BMC biology. 2008 Dec;6(1):1-5.

[84] Mosa KA, Kumar K, Chhikara S, Mcdermott J, Liu Z, Musante C, White JC, Dhankher OP. Members of rice plasma membrane intrinsic proteins subfamily are involved in arsenite permeability and tolerance in plants. Transgenic research. 2012 Dec 1;21(6):1265-77.

[85] Xu XY, McGrath SP, Zhao FJ. Rapid reduction of arsenate in the medium mediated by plant roots. New Phytologist. 2007 Nov;176(3):590-9.

[86] Duan GL, Zhu YG, Tong YP, Cai C, Kneer R. Characterization of arsenate reductase in the extract of roots and fronds of Chinese brake fern, an arsenic hyperaccumulator. Plant Physiology. 2005 May 1;138(1):461-9.

[87] Ma JF, Tamai K, Ichii M, Wu GF. A rice mutant defective in Si uptake. Plant Physiology. 2002 Dec 1;130(4):2111-7.

[88] Ma JF, Tamai K, Yamaji N, Mitani N, Konishi S, Katsuhara M, Ishiguro M, Murata Y, Yano M. A silicon transporter in rice. Nature. 2006 Mar;440(7084):688-91.

[89] Meng YL, Liu Z, Rosen BP. As (III) and Sb (III) uptake by GlpF and efflux by ArsB in *Escherichia coli*. Journal of Biological Chemistry. 2004 Apr 30;279(18):18334-41.

[90] Dubey S, Shri M, Misra P, Lakhwani D, Bag SK, Asif MH, Trivedi PK, Tripathi RD, Chakrabarty D. Heavy metals induce oxidative stress and genome-wide modulation in transcriptome of rice root. Functional & integrative genomics. 2014 Jun;14(2):401-17.

[91] Wang Y, Dong C, Xue Z, Jin Q, Xu Y. De novo transcriptome sequencing and discovery of genes related to copper tolerance in Paeonia ostii. Gene. 2016 Jan 15;576(1):126-35.

[92] Lu YP, Li ZS, Rea PA. AtMRP1 gene of Arabidopsis encodes a glutathione S-conjugate pump: isolation and functional definition of a plant ATP-binding cassette transporter gene. Proceedings of the National Academy of Sciences. 1997 Jul 22;94(15):8243-8.

[93] Tommasini R, Vogt E, Fromenteau M, Hörtensteiner S, Matile P, Amrhein N, Martinoia E. An ABC-transporter of *Arabidopsis thaliana* has both glutathione-conjugate and chlorophyll catabolite transport activity. The Plant Journal. 1998 Mar;13(6):773-80.

[94] Gasic K, Korban SS. Transgenic Indian mustard (*Brassica juncea*) plants expressing an Arabidopsis phytochelatin synthase (*AtPCS1*) exhibit enhanced As and Cd tolerance. Plant molecular biology. 2007 Jul;64(4):361-9.

[95] Guo J, Dai X, Xu W, Ma M. Overexpressing *GSH1* and *AsPCS1* simultaneously increases the tolerance and accumulation of cadmium and arsenic in *Arabidopsis thaliana*. Chemosphere. 2008 Jul 1;72(7):1020-6.

[96] Dhankher OP, Li Y, Rosen BP, Shi J, Salt D, Senecoff JF, Sashti NA, Meagher RB. Engineering tolerance and hyperaccumulation of arsenic in plants by combining arsenate reductase and γ-glutamylcysteine synthetase expression. Nature biotechnology. 2002 Nov;20(11):1140-5.

[97] Li Y, Dhankher OP, Carreira L, Lee D, Chen A, Schroeder JI, Balish RS, Meagher RB. Overexpression of phytochelatin synthase in Arabidopsis leads to enhanced arsenic tolerance and cadmium hypersensitivity. Plant and Cell Physiology. 2004 Dec 15;45(12):1787-97.

[98] Xu L, Wang Y, Liu W, Wang J, Zhu X, Zhang K, Yu R, Wang R, Xie Y, Zhang W, Gong Y. De novo sequencing of root transcriptome reveals complex cadmium-responsive regulatory networks in radish (*Raphanus sativus* L.). Plant Science. 2015 Jul 1;236:313-23.

[99] Xie Y, Ye S, Wang Y, Xu L, Zhu X, Yang J, Feng H, Yu R, Karanja B, Gong Y, Liu L. Transcriptome-based gene profiling provides novel insights into the characteristics of radish root response to Cr stress with next-generation sequencing. Frontiers in plant science. 2015 Mar 31;6:202.

[100] Williams PN, Price AH, Raab A, Hossain SA, Feldmann J, Meharg AA. Variation in arsenic speciation and concentration in paddy rice related to dietary exposure. Environmental science & technology. 2005 Aug 1;39(15):5531-40.

[101] Zhu YG, Sun GX, Lei M, Teng M, Liu YX, Chen NC, Wang LH, Carey AM, Deacon C, Raab A, Meharg AA. High percentage inorganic arsenic content of mining impacted and nonimpacted Chinese rice. Environmental science & technology. 2008 Jul 1;42(13):5008-13.

[102] Qin J, Rosen BP, Zhang Y, Wang G, Franke S, Rensing C. Arsenic detoxification and evolution of trimethylarsine gas by a microbial arsenite S-adenosylmethionine methyltransferase. Proceedings of the National Academy of Sciences. 2006 Feb 14;103(7):2075-80.

[103] Qin J, Lehr CR, Yuan C, Le XC, McDermott TR, Rosen BP. Biotransformation of arsenic by a Yellowstone thermoacidophilic eukaryotic alga. Proceedings of the National Academy of Sciences. 2009 Mar 31;106 (13):5213-7.

[104] Norton GJ, Lou-Hing DE, Meharg AA, Price AH. Rice–arsenate interactions in hydroponics: whole genome transcriptional analysis. Journal of experimental botany. 2008 May 1;59(8):2267-76.

Chapter 13

A Review on the Resistance and Accumulation of Heavy Metals by Different Microbial Strains

Madhuri Girdhar, Zeba Tabassum, Kopal Singh and Anand Mohan

Abstract

Heavy metals accumulated the earth crust and causes extreme pollution. Accumulation of rich concentrations of heavy metals in environments can cause various human diseases which risks health and high ecological issues. Mercury, arsenic, lead, silver, cadmium, chromium, etc. are some heavy metals harmful to organisms at even very low concentration. Heavy metal pollution is increasing day by day due to industrialization, urbanization, mining, volcanic eruptions, weathering of rocks, etc. Different microbial strains have developed very efficient and unique mechanisms for tolerating heavy metals in polluted sites with eco-friendly techniques. Heavy metals are group of metals with density more than 5 g/ cm^3. Microorganisms are generally present in contaminated sites of heavy metals and they develop new strategies which are metabolism dependent or independent to tackle with the adverse effects of heavy metals. Bacteria, Algae, Fungi, Cyanobacteria uses in bioremediation technique and acts a biosorbent. Removal of heavy metal from contaminated sites using microbial strains is cheaper alternative. Mostly species involved in bioremediation include Enterobacter and *Pseudomonas* species and some of bacillus species too in bacteria. *Aspergillus* and *Penicillin* species used in heavy metal resistance in fungi. Various species of the brown algae and Cyanobacteria shows resistance in algae.

Keywords: heavy metal resistance, bioremediation, biosorption, bioleaching, plasmid

1. Introduction

Air, water and soil which are the essential elements of life are contaminated rapidly due to increasing population, urbanization, mining activities and industrialization [1]. Heavy metals toxicity is causing problem to humans, animals, aquatic animals, plants and even microbes too.

Various methods are introduced to remove the heavy metal pollution like chemical techniques such as chemical precipitation, oxidation or reduction method, electrochemical treatment. Physical techniques such as ion exchange, evaporation, filtration, membrane technology, reverse osmosis. Biological techniques like microorganisms such as bacteria, fungi, algae, cynobacteria, lichens, etc.

Heavy metals damage cell membranes, alter functioning of enzymes, inhibit protein synthesis, denature protein and damage the structure of DNA. Toxicity is mainly created by the dislocation of essential metals from their real binding sites or ligand interactions [2]. Bioremediation is cost-effective, safe and eco-friendly; can be virtually restored a result to the heavy metal pollution issue as it is natural process. Biological methods are best to control short term or long term environmental pollution. Various heavy metals are accumulated with the help of bacteria, fungi, cyanobacteria, lichens, etc. and helps in bioremediation and used as bio-indicators. They are not harmful human heath as well as ecosystem. Such organisms are used for indication and controlling heavy metal pollution. Mostly genes encoded by heavy metal resistant bacteria are located on plasmids. Biosorption is environmentally safe and low cost methodology of removing metals from the ecosystem. Various analysis were observed throughout previous 5 decades provided quantity of data regarding differing kinds of biosorbents and their mechanism of absorption of heavy metal. Additional research is to explore new biosorbents from surroundings [3].

Since last few years, various physical and chemical methods are used to remove heavy metals but it is expensive, needs laboratory and inefficient. According to various studies bioremediation and biosorption techniques are much more beneficial, cheap, non-toxic, natural process.

Minimum inhibitory concentration (MIC) is the lowest concentration at which the isolate or antimicrobial agent is completely suppressed is recorded. Microorganisms correspond to heavy metals using various defense systems, such as exclusion, compartmentalization [4], complex formation and synthesis of binding proteins, such as metallothioneins [5].

Bioremediation strategies have been proposed as an attractive alternative owing to their low cost and high efficiency [6].

Different methods are used to study characterization of heavy metals on microbes by 16S RNA sequence, biodegradability test, siderophore assay, biochemical test, morphological test, antibiotic resistance, nucleotide sequencing, etc. Microbial pigmentation and enzymatic activities like catalase, gelatin hydrolysis, oxidase, nitrate reductase, were characteristics selected to examine their outcomes.

Bioremediation is of two types: in-situ bioremediation and ex-situ bioremediation. In-situ bioremediation process is mainly used due to its ability in decreasing disturbance of ecosystem at the heavy metal polluted sites whereas ex-situ bioremediation, it takes place inside bioreactors, bio-piles and land farming. In-situ bioremediation is much more efficient and eco-friendly (**Figure 1**).

Metal microbe interactions developed by microbial cells are bio-transformation, bio-leaching, bio-degradation, bio-mineralization, bio-adsorption and bio-accumulation in bioremediation method.

Biofilm used as efficient bioremediation tool and stabilization too. Even at harmful conditions, they show high resistance towards heavy metals. With the help of genetic engineering one can insert desired characters like ability to resist heavy metals, tolerate metal stress, etc. For example: engineered *Chlamydomonas reinhardtii* shows increased resistance to cadmium toxicity. *Corynebacterium glutamicum* was genetically modified using ars (operon) to accumulate arsenic polluted sites. Biofilm combines or work with biosorbent or any exopolymeric substance which consist of surfactants or emulsifier properties. The study was conducted on *Rhodotorula mucilaginosa* shows efficiency in heavy metal removal and develops 91.7–95.4% biofilm cells. Biosurfactants studied were surfactin, rhamnolipid and sophorolipid for removal of several heavy metals.

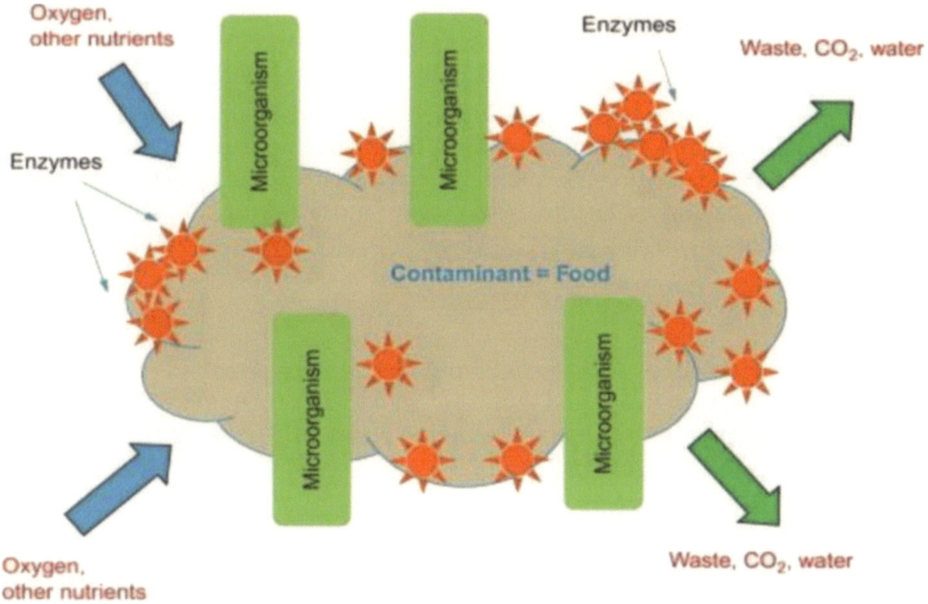

Figure 1.
Bioremediation (enzyme-catalyzed destruction) of contaminants. The use of power ultrasound in biofuel production, bioremediation and other applications [7].

The aim of the review is to study the source of the heavy metals on earth, consequences of the heavy metals on plants as well as on animals, various isolated microbial strains from bacteria, fungi and algae tolerance towards heavy metals and to study mechanism adapted by strain to accumulate heavy metals.

Future approaches in bioremediation are genetic modification of microbes or genetic engineered microbes, genetic technologies and forms specificity using biofilm by optimization process and immobilization process can be attained, biofilm mediated remediation, formation of microbial fuel cell (MFC), use of nano-particles with algae and bacteria, gene transfer within biofilm, transgenic cynobacteria, modify gene or enzyme in microbes. In Rhizo-remediation technique, *rhizosphere* bacteria and *mycorrhizae* combine for uptake.

2. Source of the heavy metals

High amount of heavy metals in the soil, water and air arise from various sources, which consist of natural sources include natural emission, atmospheric decomposition, sea salt spray, forest fires, rock weathering, biogenic means and wind borne soil particles and artificial sources such as mining activities, agricultural waste, domestic effluents, smelters, sewage sludge irrigation, improper stacking of the industrial solid waste, the excess utilization of pesticides, insecticides and fertilizers, etc. [8, 9].

2.1 Lead in environment

Lead (Pb) is unnecessary metal on the crust. It is a important contaminant that is present in the soil, water and air as a dangerous waste. It is extremely injurious to the human, animals, plants and even microbes too. The crucial sources of lead metal are children toys, drinking water, dust, petroleum, electronic industries, water pipes, battery, pottery, paint, stained glass, cosmetics and biocide preparation [10, 11].

2.2 Arsenic in environment

Arsenic (As) is non-essential metal. Arsenic is also present in pyrotechnics, in bronzing and hardening other metals. Arsenic is originated from the weathering of rocks and mineral, volcanic eruptions, fossil fuels, agricultural products, preservatives, medicinal products and industrial activities. Herbicides, pesticides, insecticides, fungicides and fertilizers also contribute to arsenic contamination and extremely deadly and carcinogenic [12] (**Figure 2**).

2.3 Mercury in environment

Natural activities like volcanoes and forest fire release mercury in environment. The burning of coal, oil, wood and mining of gold releases mercury in the environment. It affects immune system as well as nervous system. Methyl-mercury damages the developing embryos too [14, 15].

2.4 Chromium in environment

Chromium is released to environment by combustion processes and from metal industries and chemical manufacturing industries as waste. Chromium 4 is most dangerous form and may lead health issues like allergy, nose irritations, skin rashes, liver damage, kidney damage and even death [16, 17].

2.5 Cadmium in environment

Cadmium is also a non-essential member and highly dangerous to mankind. Cadmium is used in semiconductors, nickel-cadmium batteries, electroplating, municipal wastes such as plastics, PVC manufacturing, alloys, overuse of fertilizers rich in phosphate and control rod for nuclear reactors. Soils and water pollution by cadmium produced by the mining sites and smelting industries, sewage sludge

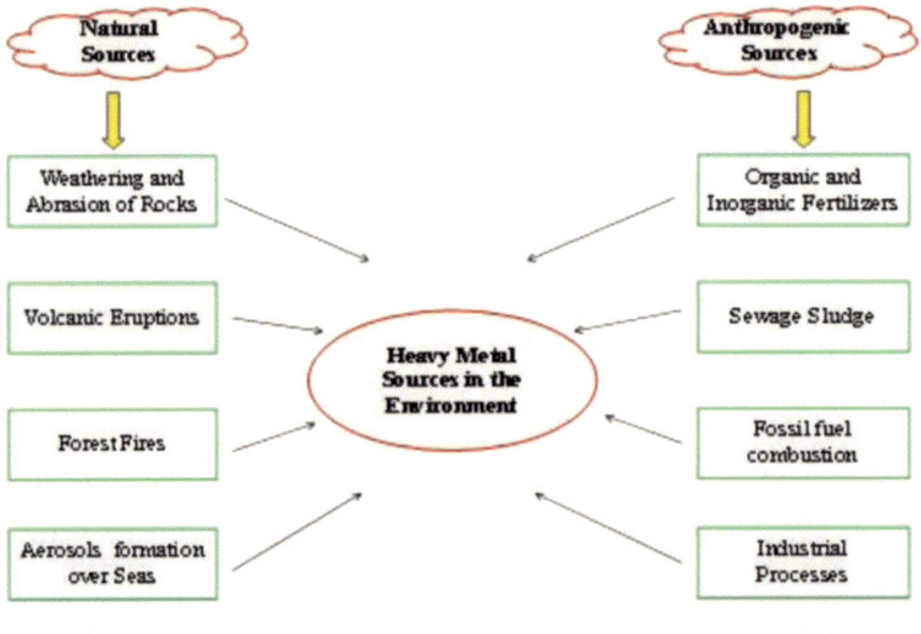

Figure 2.
Heavy metal sources in the environment [13].

application and burning of fossil fuels like coal, petroleum, etc. Chronic exposure of cadmium in human has many harmful effects such as high blood pressure and destroys to different organs such as lung, liver kidney and testes in males [18, 19].

2.6 Copper in environment

Copper a transition metal and also an essential element for living organisms including humans and other animals at low concentrations. Copper is released in ecosystem through decaying of vegetation, forest fire, sea-sprays, wind-blown dust. Copper is utilized as the alloy in the manufacture of wire, pipe, and various metal products. Copper are majorly used in agriculture to treat plant diseases, like mildew, or for water treatment and as preservatives, leather and fabrics. Intake of excessive amount of copper, it can cause nausea, vomiting, stomach cramps, diarrhea and can destroy liver and kidney and even lead to death [20, 21].

2.7 Zinc in environment

Zinc (Zn) is also a transition metal and zinc is utilized in galvanizing and alloying and also in the manufacture of electric goods, dying, insecticides, pesticides and cosmetics. Mining activities, smelting of metals and production of steel and other waste can release zinc into the environment. It may cause health issues in living organisms such as dehydration, nausea, electrolyte imbalance, vomiting, abdominal pain, dizziness, acute renal failure, muscular incardination and damage of hepatic parenchyma [22].

2.8 Manganese in environment

Manganese is released from sewage sludge, combustion of fossil fuels, mining processes, etc. it can cause toxicity in plants and causes swelling of cell walls, brown spots on leaves, etc. [23].

2.9 Iron in environment

The major sources of iron are metal refining, sewage, dust from iron mining, iron and steel industry. Iron sulphate is utilized in fertilizer and herbicide [24].

2.10 Other heavy metals in environment

Thallium is present in insecticides, metal alloys and fire cracker. Phosphorus is found in insecticides such as organophosphate for example: malathion [25].

The environmental factors plays very crucial role in biosorption of heavy metals and these factors are pH, temperature, biomass concentration, metal ion concentration. Algae, fungi and bacteria acts as biosorbents and helps in mechanism of biosorption [26].

3. Adverse effects of heavy metals

Heavy metal pollution is causing severe health effects in human body as well as animals and plants too. Heavy metals are also effected the growth of microbes which are used in treatment or accumulation of heavy metals by damaging their DNA. Heavy metals can cause skin allergies, cancer, effect major organs like kidney, liver, brain, lung, etc., and enter in blood stream and even death too in animals and humans. Retarded growth and development, bad shoot induction and root formation, less nutrient and mineral content and can even cause death in plants [27].

3.1 Adverse affects of heavy metals on humans

Heavy metals like lead, chromium, nickel, mercury, cadmium, arsenic, etc. may destroy and alter functioning of various prime organs such as the liver, lungs, kidney, brain, heart and even blood also. Heavy metal infectivity may be either quick (within few hours/days) or long term (within months). Prolonged exposure of few toxic heavy metals at even less concentration can cause cancer or even death too. Heavy metals may cause various severe health risk and diseases [28].

Heavy metals can affect human body by lead is carring to liver and kidney by red blood cells. Cadmium binds to blood cells, liver and kidney tissues. Arsenic is accumulated in blood, kidney, heart, muscle, lung liver and also in nails, hair, etc.

The effect of toxicity depends on the exposure route and chemical nature of particular heavy metal like lipid solubility, volatility, etc.

Some heavy metals like arsenic, lead, mercury, nickel, cadmium, etc. have carcinogenic effect. Some heavy metals like lead, manganese, etc. may induce neurotoxicity [29].

Heavy metals function as a pseudo element of the body while they can interrupt with metabolic processes. Few metals, like aluminum may be separated through excretory activities, and few metals get absorbed in the body and even in food chain, showing long term exposure. Heavy metal toxicity depends upon the absorbed amount, the path of exposure and time of exposure. This may lead to several health risks and can also result in huge loss due to oxidative stress induced by free radical formation [30].

Arsenic is most harmful heavy metal which is highly toxic and carcinogenic. It mainly affects endocrine system, lungs, kidney, pulmonary, nervous system and skin. It causes skin cancer, respiratory cancer, perforation of nasal septum, dermatomes, etc. ingestion in gastrointestinal tract results in vomiting, disturbance in circulation, damage nervous system and led to death. Other consequences are high blood pressure, heart attacks, decrease in production of blood cells, enlargement of liver, change in skin color, loss of sensation in limbs. Exposure of arsenic through air can cause lung cancer and bladder cancer [31].

Cadmium is another dangerous heavy metal and it targets renal region, bones, testes, cardiovascular, skeletal system and pulmonary organ. It causes proteinuria, glucosuria, osteomalacia, emphysemia, aminoaciduria, etc. It may damage kidney and lung [19].

Chromium damages the organ such as lungs, kidney, pancreas, testes, liver, pulmonary region of body. It causes problems like ulcer, perforation of nasal septum, respiratory track cancer [17].

Lead is also very toxic even in less amount and targets multiple organs such as spleen, bones, the nervous system, hemotopoietic system, cardiovascular, gastrointestinal, renal region and reproduction system too. It causes issues like anemia, central nervous system disorders, peripheral neuropathy, encephalopathy [32].

Manganese is required in small concentration in body but in excessive damages nervous system and led to central and peripheral neuropathies and brain damage [23].

Nickel damages pulmonary system and skin too. It results high chances of lung cancer, nose cancer, larynx cancer and prostate cancer and skin allergy or skin rashes. It also shows symptom like sickness, dizziness, birth defects, asthma, chronic bronchitis, lung embolism, heart disorders [19].

Zinc may cause nausea, vomiting, illness, anemia, stomach cramps, damage to nervous system and skin irritation. It causes skin allergy, dermatitis, brain disorder.

Increased amount of zinc effects pancreas, disturbs the metabolism of protein and amino acids in body and arteriosclerosis too [33].

Cobalt can cause vomiting, nausea, loss of appetite and may affect on lungs causing asthma, pneumonia and wheezing when exposed with cobalt metal and may develop various allergies or skin rashes. Mainly it is dangerous for heart muscle and causes heart muscle disease known as cardiomyopathy and shows rapid increase in count of red blood cells after long time exposure [34].

Copper damages liver, brain, cornea, lungs, immune system including blood cells. It causes gastrointestinal symptoms such as vomiting, nausea, abdominal pain and even lead to liver and kidney damage, genetic disorders, reproductive or developmental effects, delayed growth, prolonged bone formation and less body weights [35].

Tin effect both nervous system and pulmonary system. Exposure may lead to skin and eye irritation or respiratory tract problems. It causes pneumoconiosis, central nervous system disorders, visual defects, changes in EEG too [36]. Phosphorus symptom caused by exposure of phosphorus on human health includes sweating, headache, vomiting, abdominal cramps, weakness, ptosis, miosis, and severe issues are sensorimotor, polyneuropathy, atrophy and even led to respiratory paralysis [37].

The consequences of thallium exposure include blood vomiting, nausea, abdomen pain, eye disorder, mental retardation, hair loss and severe issues are cardiac failure, brain disorder and even coma too [25].

Mercury attacks the nervous system and renal region and may cause proteinuria. Inhalation of mercury may cause headache, memory loss, insomnia, tremors, neuromuscular and thyroid damage. It damages the chromosome structure and DNA. Effects on reproductive system by low sperm count, birth defects and even miscarriages too. During pregnancy, it may pass through placental barrier to embryo or baby for exposure [38].

The major organs targeted by these heavy metal mercury and lead causes neurotoxicity (brain), arsenic lead to hepatotoxicity (liver), cadmium causes nephrotoxicity (kidney)/pulmonotoxicity (lungs) and zinc mainly induce hematoxicity (blood).

The heavy metals interrupt in metabolic processes in two ways [39]:

- They are absorbed and thereby disturb role in major organs and glands such as the heart, liver, brain, kidneys, bone, etc.

- They displace the important nutritional minerals from their real place hindering their biological function. Consumption of foods, beverages, skin exposure, and the inhaled air are ways through which these contaminants can be present in body. It is unfeasible to reside in heavy metal free surrounding.

Various heavy metals produce ROS and damages DNA of the cell and disrupt reproduction cycle. Arsenic damages kidney and liver and may cause abdominal cramping, etc.

3.2 Adverse effects of heavy metals on marine animals

Heavy metals present in water by industrial effluent or agricultural waste like fertilizers, pesticides, etc. and deposited in water bodies and settle down and can present on surface with help of aquatic plants and aquatic macrophytes. Heavy metals stimulate the production of reactive oxygen species (ROS) that can damage aquatic organisms.

Several heavy metals accumulate in various major organs of the fish causing mortality. Firstly it affects the circulatory system by entering in blood and alters the components of blood. It makes the fish anemic and weak.

Huge amount of heavy metal shows inhibitory effects on the growth and development of aquatic organisms like fishes, phytoplankton and zooplankton. Heavy metals may cause disruption in respiration, damage respiratory track which leads to suffocation, reduces the sperm count, egg production and short life span. Heavy metals can disturb oxygen level, reduction of developmental growth or give rise to developmental anomalies, byssus formation and reproduction too. In juvenile phase shows high mortality and in adults decreased breeding ability. Heavy metal shows changes in structure and organs and may exhibit functional changes and transform metabolic pathways. Results of a research [40] showed that ten different fish species had the highest concentration of heavy metals is in liver and kidney.

The fishes like *Labeo rohita* and aquatic organism are eaten by humans as rich protein sources and heavy metal pollution may cause health risk in humans too through aquatic species. Cadmium can be bioaccumulated in mussels, oysters, shrimps, lobsters and fishes too.

Mercury in fish muscles occur as Methyl mercury which is formed in aquatic sediments. Movement of heavy metals in fish takes place through the blood where the ions are generally attached to proteins. There are five potential routes for the contaminants to enter an aquatic organism. The pathways are through the food, non-food particles, gills, the skin and oral consumption of water. Once the contaminants are accumulated, they are carried by the blood to the liver for modification and storage. If contaminants are altered by the liver, they can be stored or excreted in the bile produced in liver or reversed back into the blood stream for elimination by the gills or kidneys or stored in fat which is a hepatic tissue.

3.3 Adverse affects of heavy metals on plants

Plants require various heavy metals for their growth and excessive amount of heavy metals can damage cell structure, inhibition of major enzymes, inhibit the photosynthesis process and growth of plants, altered water balance, nutrient assimilation and can even cause plant death [41].

Heavy metal give rise to chlorosis, slow and poor plant growth, yield depression and even less nutrient absorption, disorders in plant metabolic processes and decreased potential to fixate molecular nitrogen in legumes of plants.

Seed germination was gradually retarded in the presence of large amount of lead. It can be due to long term incubation of the seeds and have resulted to compensate the toxic outcomes of lead by various mechanisms such as leaching, chelation, metal binding or absorption by microorganisms [42].

Replacing of major essential nutrients at cation exchange sites reveals indirect toxic effects on plant development. Enzyme metabolism is extremely crucial for growth and development of plants and heavy metals effect enzymes to inhibit many other major metabolisms in plants.

Heavy metals may lead to loss of fertility of soil by reduction in decomposition of organic matter by depletion of various microbes present inside the soil [43].

Copper is required as micronutrients in plants and helps in synthesis of ATP and assimilation of carbon dioxide. Excessive copper may exhibit oxidative stress and decreases growth of root.

Zinc required as micronutrient for synthesis of chlorophyll in plants. It retards growth of plants and nutrient level. It causes manganese and copper deficiency in shoot region.

Cadmium results in inhibition of growth and development, browning of roots tips and even death too.

Mercury can effects whole food chain and induces ROS and oxidative stress too. It causes depletion of germination in seeds, height of plant reduced flowering and fruit production, retarded growth and development.

Chromium induces the oxidative stress and degrades photosynthesis pigments in plants [30].

Lead degrades the development of roots and arsenic effects yield of crop and chlorosis, plant height and decreases ability of seed for germination [44].

Nickel is important and considered as macronutrient in plants but present in excessive amount can inhibit root growth, short shoot yield, etc. [45].

Enzymes and co-enzymes both are made up various elements such as cobalt. High concentration of cobalt may cause depletion in nutrients like proteins, amino acids, carbohydrates, etc. Also exhibit retarded plant growth and development.

Photosynthesis is prime phenomena in plants and it requires iron element. The excessive concentration of iron can inhibit photosynthesis itself [24].

Plants experience oxidative stress upon exposure to heavy metals that leads to cellular damage and disrupt of cellular ionic homeostasis. To decrease the detrimental outcomes of heavy metal exposure and their absorption, plants have participated in detoxification processes highly based on chelation and sub-cellular compartmentalization. A primary class of heavy metal chelator known in plants is phytochelatins (PCs), are produced by non-translation from reduced glutathione (GSH) in a transpeptidation reaction catalyzed by the enzyme phytochelatin synthase (PCS) [39].

The various biosorption techniques adopted by the plants such as phytoextraction, phytoextraction, rhizofiltration, phytovolatilisation and many others.

4. Bioremediation of heavy metals by microbial strains

Various microbial strains can accumulate the toxicity of heavy metals from bacteria, fungi, algae and helps in bioremediation and biosorption [46]. Bacterial strains show five different mechanisms in resistance to heavy metals. These mechanisms are by inhibiting the entrance of metals into the cell. The cell wall, membrane and capsule prohibit entry of metal ions inside the cellular body. Carbonyl group in polysaccharides of bacterial capsule accumulates the ions of heavy metals. Ions of metal like zinc, lead, and copper resulted resistance by *Pseudomonas aeruginous* biofilm [47].

In bacteria, active transport illustrate largest group of heavy metal resistance. Active transport remove metal ions from cell membrane and it can be placed on either on plasmid or on chromosomes [48, 49].

In intracellular sequestration, combination of metal ions to form large ion is done by several compounds inside cytoplasm of cell. Example; *P. putida* shows potential of intracellular sequestration of metal ions such as zinc, candium and copper [50].

In extracellular sequestration, metal ions are collected by periplasm or outer membrane of cells as insoluble compounds [51].

Condensation of metal ions was done by the bacterial strains. Strains decreasing chromate, vanadate and moyhybadate were observed from surroundings. Metal ions were utilized as electron donors for generating energy by bacterial isolates. Example: *S. aureus* strain for resistance of arsenic (As^{5+}/As^{3+}) [52], *Klebsiella pneumonia* for resistance of mercury (Hg^{2+}/Hg) [53].

4.1 Tolerance against heavy metals in bacteria

There are various processes of heavy metal resistance like extracellular barrier, extracellular sequestration, and active transport of metal ions (efflux), intracellular sequestration, and reduction of metal ions by microbial cells.

B. subtilis revealed the excessive potential to remove the amount of the cadmium.

Bacteria resistant to mercury are *Alcaligenes faecalis*, *Bacillus pumilus*, *Pseudomonas aeruginosa*, and *Brevibacterium iodinium* for the eradication of cadmium and lead metals.

59 isolated actinobacteria have shown resistance to the five heavy metals. Using molecular identification 16S rRNA, 27 strains were found to classified in the *Streptomyces* and *Amycolatopsis* genera [54].

Three strains were identified up to genus level based on their morphological, cultural, physiological and biochemical characteristics as *Gemella* sp., *Micrococcus* sp. and *Hafnia* sp. Among these three isolates, *Gemella* sp. and *Micrococcus* sp. exhibited the resistance towards lead, chromium and cadmium metals whereas *Hafnia* sp. exhibited reactivity to cadmium (Cd). All strains revealed dissimilar MICs against the heavy metals at different concentrations using Atomic Absorption Spectrophotometer [55].

Bacterial cell wall experiencing the metal ion is the primary constituent of biosorption. The metal ions get connected to the various functional groups such as (amine, carboxyl, hydroxyl, phosphate, sulfate, amines) exist on the cell wall of the microbe. The metal uptake mechanism involves binding of metal ions to reactive groups lies on cell wall followed by internalization of metal ions inside cell protoplast. Some metal in more amount are accumulated by Gram positive strains due to presence of glycoproteins in their cell wall. Fewer metal absorption by Gram negative strains is reported due to phospholipids and LPS in their cell wall.

4.1.1 Arsenic resistant bacteria

Gram positive and gram negative bacterial strains have been investigated in the absorption of heavy metals.

Arsenic resistant bacteria species are *Enterobacter* sp. and *Klebsiella pneumoniae* based on phylogenetic analysis of 16S rDNA sequence [56].

The *Enterobacter* sp. (MNZ1), *K. pneumoniae* 1 (MNZ4) and *Klebsiella pneumonia* 2 (MNZ6) species shows resistance towards arsenic and survive in the presence of high level of arsenic [57].

10 isolates of rhizobacteria out of which some were Gram-positive bacteria *Arthrobacter globiformis*, *Bacillus megaterium*, *Bacillus cereus*, *B. pumilus*, and *Staphylococcus lentus*), and few were Gram-negative bacteria (*Enterobacter asburiae* and *Rhizobium radiobacter*). *R. radiobacter* exhibited the highest MIC of greater than 1500 ppm of the arsenic metal [58].

Aeromonas, *Exiguobacterium*, *Acinetobacter*, *Bacillus* and *Pseudomonas* are isolates of bacteria that can tolerate high levels of arsenic species [59].

Acidithiobacillus, *Deinococcus, Bacillus*, *Desulfitobacterium* and *Pseudomonas* show resistance against arsenic [60] (**Table 1**).

4.1.2 Cadmium resistant bacteria

Cadmium resistant bacterium, *Salmonella enterica 43C* is isolated from industrial effluent was characterized on the basis of biochemical and 16S rRNA ribotyping [62].

S. no.	Microorganisms	Accumulation of heavy metals in ppm	References
1.	*Pseudomonas aeruginosa*	1596.6	[61]
2.	*Bervibacillius choshinensis*	1011.18	[61]
3.	*R.radiobacter*	1500	[58]

Table 1.
Arsenic removal by bacterial strains.

S. no.	Microorganism	Accumulation of heavy metals in ppm	References
1.	*Pseudomonas aeruginosa*	2200 mg/L	[64]
2.	*Alcaligenes eutrophus*	320 ppm	[65]
3.	*Staphylococcus xylosus*	278 mg/g	[66]
4.	*Rhodotorula* sp. Y11	11.38 mg/g	[67]

Table 2.
Removal of heavy metal by cadmium resistant bacteria.

The efflux processes involves cadA and cadB gene method, and encodes several efflux pump proteins and various functional groups like amine, carboxyl, phosphate and hydroxyl ease cadmium binding sites to bacterial surface such as chemisorption. The membrane impermeablility is regulated by enzymes used in detoxifying the cadmium metal [63]. Various processes on the basis of morphological, biochemical characteristics, 16S rDNA gene sequencing and phylogeny analysis exhibited that the strain RZCd1 was recognized as *Pseudomonas* sp. M3. In log phase, industrial strains revealed more than 70% of the cadmium accumulation [57] (**Table 2**).

4.1.3 Mercury resistant bacteria

With the help of 16S rRNA gene sequence, *Vibrio fluvialis CASKS5* strain was recognized. The mercury-absorption ability of *V. fluvialis* was examined at several amount of concentration and exhibit large MIC (Minimum Inhibitory Concentration) but low antibiotic resistance [68].

Staphylococcus, *Bacillus*, *Pseudomonas*, *Citrobacteria*, *Klebsiella*, and *Rhodococcus* are several species mainly used in bioremediation of mercury [69].

Highly mercury resistant bacteria strains were *Brevundimonas* sp. HgP1 and *Brevundimonas* sp. HgP2 with 16S rDNA from a gold mine situated in village Pongkor, West Java with high MIC of 575 ppm. The aim was to examine the effect of mercury on bacterial development and morphological changes of bacterial population. The development was observed by measuring optical density at 600 n [70].

Mercury-resistance in the bacteria isolates were classified into the various genera such as *Pseudomonas*, *Enterobacteriaceae*, *Proteus*, *Xanthomonas*, *Alteromona*, and *Aeromonas* [71].

Attachment to the cell membrane, influx and efflux adsorption, detoxification of toxic metals to less harmful form, the use of *metallothionein* protein were several processes for heavy metal resistance. Removal of the any ion can be decreased by efflux, an active extrution of the heavy-metal ion [72] (**Table 3**).

4.1.4 Lead resistant bacteria

Lead accumulation processes operated by the lead resistant bacteria isolates includes efflux mechanism, extracellular sequestration, biosorption, precipitation,

S. no.	Microorganism	Accumulation of heavy metals in ppm	References
1.	*Brevundimonas* sp.	575	[70]
2.	*Pseudomonas aeruginosa*	294.6	[61]
3.	*Bervibacillius choshinensis*	58.93	[61]

Table 3.
Removal of mercury by bacterial strains.

S. no.	Microorganism	Metal concentration in ppm	References
1.	*Pseudomonas aeruginosa*	625.8	[61]
2.	*Bervibacillius choshinensis*	625.8	[61]
3.	*Gemella* sp.	1350	[55]
4.	*Micrococcus* sp.	1100	[55]

Table 4.
Removal of Lead by bacterial strains.

alteration in cell morphology, enhanced siderophore production and intracellular lead bioaccumulation [73].

Four distinct bacteria were isolated with high levels of resistance to lead, each exhibited resistance to 2 mM lead on the minimal medium. Two were identified as Gram-positive genus *Corynebacterium* and two were the Gram-negative genus *Pseudomonas*. Three strains transferred no observable plasmid, indicating that the metal resistance is encoded by chromosomal [74] (**Table 4**).

Lead-resistant bacteria play an important role in the development of lead-exposed plants. The endophyte *Bacillus sp.* MN3-4 increases Pb(II) absorption in *Alnusfirma*, and *Pseudomonas fluorescens G10* and *Mycobacterium sp. G16* enhances plant development and growth and decreased Pb toxicity *in Brassica napus* [75].

4.1.5 Nickel resistant bacteria

The nickel-resistant bacteria were identified *as Shigella*, *Enterococci* and *Enterobacter*, but they were anaerobic, they only grew in the human samples from obese people and they tolerated a maximum concentration of 1 mM nickel [76].

Few strains *Cupriavidus sp.* ATHA3, *Klebsiellaoxytoca* ATHA6 and *Methylobacterium sp.* ATHA7 and their recognization was concluded on the basis of morphological, biochemical characteristics and 16SrDNA gene sequencing [77] (**Table 5**).

Alcaligenes eutrophus H16 and N9A strains and derivatives of strain CH34 lacking one or another of its natural metal resistance plasmid were used as recipients. Both of the plasmid, pTOM8 and pTOM9 of strain 31A conveyed resistance features which were expressed except *A. eutrophus* H16 [79].

S. no.	Microorganism	Metal concentration	References
1.	*Klebsiella oxytoca* strain ATHA6	83 mg/mL	[77]
2.	Enterobacter sp.	200 ppm	[78]

Table 5.
Removal of nickel by bacterial strains.

Nickel resistance isolates from bacteria isolated from New caledonia by DNA-DNA hybridization. The biotinylated probes of DNA were obtained from *Alcaligeneseutrophus CH34*, *Alcaligenes xylosoxidans 31A*, *Alcaligenes denitrificans 4a-2*, and *Klebsiella oxytoca CCUG 15788*. 9 probes were crossed with endonuclease-cleaved plasmid and all DNA samples from 56 nickel-resistant determinants. Few Caledonian isolates were recognized as *Acinetobacter*, *Pseudomonas mendocina*, *Comamonas*, *Hafniaalvei*, *Burkholderia*, *Arthrobacter aurescens*, and *Arthrobacter ramosus*isolates [80].

4.1.6 Copper resistant bacteria

Copper-resistant bacteria have been isolated from the different sources, but copper-resistant *Escherichia coli* strains were isolated from agricultural sewage and phytopathogenic *Pseudomonas* and *Xanthomonas* strains.

The *copA* gene was noticed in the copper resistant strains *Sphingomonas*, *Stenotrophomonas* and *Arthrobacter* isolated from the contaminated soil from agricultural fields [81] (**Table 6**).

Bacterial strains showed high level of removal of heavy metals, determinants like YJ3 and YJ7 maybe resistance to Cu and isolates like SWJ11, MT16, GZC24 and YAH27 may be resistance to heavy metals such as Cu, Pb, Cd, Ni and Zn. It has been observed that plant growth-promoting bacteria can enhance the development and heavy metal uptake of plants [83, 84].

Numerous bacterial species show resistance to heavy metal such as thallium, tungsten, uranium, plutonium, have been observed from sediment and water sample. *Pseudomonas aeruginosa* strains results in accumulation and resistance to these heavy metals. Plutonium is harmful for soil microorganism even at very low concentration and stops the growth of bacteria fungi present in soil and affects soil respiration [85].

4.2 Tolerance against heavy metals in fungi

Fungi are ubiquitous in nature and found in water and soil. Recent strains isolated from contaminated sites have shown exceptional potential to tolerate heavy metals [86].

Fungi show potential as biocatalysts to accumulate heavy metals and convert them into very less toxic metals. Fungi mostly use chelation method to upgrade the tolerance to harmful heavy metals.

Recent studies have concluded many fungal strains like Rhizopus *stolonifer* in tolerance to lead, cadmium, copper and zinc. *Pleurotus ostreatus* in strain is used in nickel resistance. *Aspergillus niger* lead to the removal of lead, zinc, iron by bioleaching process and *Aspergillus niger* lead to removal of Zinc, nickel, lead, cadmium, manganese by immobilized cells [87].

Fungus as biosorbents used in removal of heavy metal ions. Bioleaching involves use of heterotrophic fungi and their metabolic products for accumulation of heavy

S. no.	Microorganisms	Meta concentration in mg/L	References
1.	*Bacillus pumilus*	121.82	[82]
2.	*Staphylococcus pasteuri*	80	[82]
3.	*Agrobacterium tumefaciens*, strain CCNWRS33-2	300	[82]

Table 6.
Removal of copper by bacterial strains.

metals from solid waste. Bioleaching is alternative method to traditional methods and fungal strains such as *Aspergillus* and *Penicillin* are used. Micro colonial fungi (MCF) can be used as a aspect of future research in bioremediation field.

Fungi show two mechanisms for heavy metal tolerance:

a. Extracellular sequestration.

b. Intracellular sequestration.

Extracellular mechanism inhibits metal ions to entrance and intracellular mechanism decrease metal ions inside the cytosol. In extracellular system, fungal cells excrete the organic compound that does not belong to cell wall compounds to chelate metal ions.

In intercellular system, metal transport proteins show resistant by ejection of metal ions from inside the cytosol [88].

Fungi strains to tolerate heavy metals are *Aspergillus foetidus and Penicillin simplicissimum*.

4.2.1 Cadmium resistant fungi

Aspergillus versicolor, *Aspergillus fumigatus*, *Microsporum species*, *Cladosporium species*, *Paecilomyces species*, *Terichoderma* were investigated by results of Fazli et al. [89]. Biological mechanism of fungal isolate directly relies on resistance against cadmium metal. *Paecilomyces species* could accumulate 400 mg/L concentration of cadmium which is the highest MIC standard observed yet. Highly versatile fungus to cadmium stress was *Aspergillus versicolor* and most sensitive fungus species for inhibition of mycelia growth are *Microsporum species* and *Cladosporium s*pecies. Unique and advance technologies in bio treatment of heavy metals are metal uptake technique natively, utilizing combination of isolates and cell structures manipulation by autoclaving [90] (**Table 7**).

4.2.2 Lead resistant fungi

Penicillin oxalicumis species acts as a biosorbent and removes lead from aqueous solution. The isolates reveals uptake ability and tolerance to lead are *Aspergillus fumigatus*, *Penicllum simplicissimum* etc. Fungus biomass which is physically and chemically retreated again was a technique applied for biosorption of lead metal [94] (**Table 8**).

S. no.	Microorganism	Accumulation of heavy metals in ppm	Reference
1.	*Pencillium notatum*	500	[91]
2.	*Saccharmyces serviciae*	500	[91]
3.	*Penicillium verrucosum*	400	[91]
4.	*Penicillumfuniculosum*	500	[91]
5.	*Aspergillus niger*	400	[92]
6.	*T. ghaneuse*	1000	[55]
7.	*R. micosporus*	100	[55]
8.	*Trichodermabervicompactum QYCD-6*	150–200	[93]

Table 7.
Metal concentration of cadmium used in studying metal resistance in fungi.

S. no.	Microorganism	Accumulation of heavy metals in ppm	References
1.	*Aspergillus niger*	2000	[92]
2.	*F. meliae*	400	[55]
3.	*T. ghaneuse*	400	[55]
4.	*R. micosporus*	800	[55]
5.	*Pencilliumnotatum*	800	[91]
6.	*Saccharmyces serviciae*	700	[91]
7.	*Penicillium verrucosum*	700	[91]
8.	*Penicillum funiculosum*	800	[91]
9.	Trichodermabervicompactum QYCD-6	1600	[93]

Table 8.
Metal concentration of Lead used in studying metal resistance in fungi.

4.2.3 Mercury resistant fungi

Aspergillus niger and *Aspergillus flavus* used in bioremediation process in mercury contaminated soil. Both belongs to phylum Ascomycota and are soil fungi [95].

Fungal sensitivity against heavy metals alters the origination of fungal spores. Sporulation is a natural response created by fungi as metal avoidance strategy in heavy metal contaminated sites.

Formation of Metallothionein polypeptides reduce cytotoxicity and metabolize heavy metals in fungi. [96] (**Table 9**).

4.2.4 Nickel resistant fungi

Various fungi species such as *Aspergillus niger*, *Aspergillus giganteus*, *Penicillin vermiculatum*, *Gliocladium species*, *Beauvaria species*, *Trichodermaviride* and *Rhizopusstolonifera induces* shows sporulation due to increase in concentration of nickel in contaminated sites. Environmental factors like pH temperature organic matter and metal ions impacts toxicity of nickel. Alteration of magnesium transport minimizes nickel. Generation of chelating compounds like glutathione deactivates toxicity of nickel [97] (**Table 10**).

4.2.5 Arsenic resistant fungi

Bioaccumulation and biovolatilization through arsenic resistant species like *Penicillin sp.*, *Aspergillus sp.*, *Neosartorya sp.*, *Gliocladiumreseum* and the yeast *Candida humicola* in removal of arsenic have been studied [98–101].

Microbes involved in biochemical mechanisms to exploit arsenic oxy-anions either as an electron acceptor (arsenate) for anaerobic respiration or as an electron donor (arsenite) to support chemoautotrophic fixation of carbon dioxide into cell carbon [102].

S. no.	Microorganism	Accumulation of heavy metals in ppm	Reference
1.	*Aspergillus niger*	2000	[92]

Table 9.
Metal concentration of mercury used in metal resistance in fungi.

S. no.	Microorganism	Accumulation of heavy metals in ppm	References
1.	*Penicillum funiculosum*	400	[91]
2.	*Saccharmyces serviciae*	300	[91]
3.	*Penicillium verrucosum*	400	[91]
4.	*Pencillium notatum*	400	[91]
5.	*Aspergillus niger*	1000	[92]
6.	*Aspergillus foetidus*	500	[88]

Table 10.
Accumulation of nickel by fungal strains.

Two arsenic resistant fungi are *Fimetariella rabenhortii* and *Hormonema viticola* were isolated from contaminated soil. In fungi, Evaluation of plant growth promoting factors. Arsenic shows resistance by mediation of phosphate solubilization. *F. rabenhortii* and *H. viticola* had capacity to produce indole acetic acid and siderophores [103].

acrA biosensor strain is first fungal biosensor for arsenic detection. Using fungi as whole cell biosensors have various advantages [104].

A non-pathogenic strain *Aspergillus niger* is broadly used in Industrial applications. Presence of lead and zinc does not affect the fungal spore growth (**Table 11**).

4.2.6 Iron-resistant fungi

Iron is essential in low concentration but very harmful in high amount of concentration. The fungal strains useful in iron resistance are *Aspergillus niger* and *Aspergillus foetidus* and some *Penicillium species* too. Fungal strains have good ability for bio leaching process by interfering functional groups of enzymes [105] (**Table 12**).

4.2.7 Cobalt resistant fungi

Cobalt metal is found in state of cobaltite, linnaeite, smaltite, etc. Some fungal strains help in accumulation of cobalt are *Aspergillus niger*, *Aspergillus foetidus* and

S. no.	Microorganism	Accumulation of heavy metals in ppm	References
1.	*Aspergillus niger (arsenic III)*	1200	[92]
2.	*Aspergillus niger (arsenic IV)*	1000	[92]
3.	*T. ghaneuse*	800	[55]

Table 11.
Removal of arsenic by fungal strains.

S. no.	Microorganism	Metal concentration in ppm	References
1.	*Aspergillus niger*	2000	[88]
2.	*Aspergillus foetidus*	3500	[88]
3.	*Penicilium* sp.	8000	[88]
4.	*F.meliae*	800	[55]
5.	*T.ghaneuse*	500	[55]
6.	*R. micosporus*	800	[55]

Table 12.
Removal of Iron by fungal strains.

S. no.	Microorganisms	Metal concentration in ppm	References
1.	*Aspergillus niger*	1500	[88]
2.	*Aspergillus foetidus*	500	[88]
3.	*Penicillium* sp.	2500	[88]

Table 13.
Removal of cobalt by fungal strains.

Penicillium spp. The factors that improve the removal of cobalt were fungal biomass, incubation time, pH, temperature, concentration of metal ions [106] (**Table 13**).

4.3 Tolerance against heavy metals in algae

Metal detoxification or chelation is one more strategy defense for heavy metal resistance. Algae secrete chelating molecules in response to metal ions that successively bind to them resulting in the sequestration of complexed metals in cellular organelles. Most of the algae strains are rumored to accumulate elevated metal ion concentration in cellular organelles. Additionally, the appliance of this metal resistance in biogenesis of metal nano-particles and metal compound nano-particles has been investigated by [107].

Algae are aquatic plants which absence of true roots and stems. Even when less nutrition is provided still they can grow in large biomass. Large size, high sorption ability and no production of harmful components are responsible for good biosorbent material. Features required for binding algae surface to heavy metal ions are algae species, ionic charge of metal and chemical composition of metal ion solution. Amine, carboxyl, sulfate, phosphate, sulfhydryl, hydroxyl, imidazole groups are metal ion binding sites on algal surfaces [108].

Algae show various mechanism such as formation of proteins which binds with metals, changes in structure of cell membrane, complexation or elimination of ions. Heavy metals can be eliminated for contaminated sites by either living cells or dead cells by usage of inactive biomass. Mechanism of absorption of living cells is very much complex than intracellular uptake [109].

Two processes in algal biosorption are involved. 1. Ion exchange method where ions present on algal membrane Ca, Mg, K, Na. They are displaced by metal ions. 2. Complexation between metal ions and functional groups. The metal removal process of algae is similar to bacteria by bonding of metal ions with the membrane [110].

Cladophora species are best bio indicator and *scenedesmus species* results in stress tolerance and accumulation of heavy metal like copper and chromium. In brown algae, cell wall contains fucoidin and olginic acid which helps in accumulation of heavy metals too [111].

Three fresh water microalgal determinants *Phormidium ambiguum* (Cyanobacterium), *Pseudochlorococcum typicum* and *Scenedesmus quadricauda* var. *quadrispina* (Chlorophyta) were tried for resistance and absorption of mercury (Hg^{2+}), lead (Pb^{2+}) and cadmium (Cd^{2+}) in aqueous solution. Transmission electron microscopy (TEM) was examined to contemplate the connection between heavy metal ions and *P. typicum* cells. At ultrastructural level, electron thick layers were recognized on the algal cell membranes when exposed to Cd, Hg and Pb [110] (**Table 14**).

Bifurcaria bifurcate, *oocystis*, *Pithophora spp.*, *Sargassum* sp., *Sagassumtenerrimum*, *Fucusvesiculosus (brown algae)*, *Ascophyllumnodosum*are resistant to cadmium. *Pithophora* spp., *Sargassum* sp., *Spirogyra* sp., are resistant to chromium. *Calotropisprocera*, *Pithophora* spp., *Fucusvesiculosus* are species resistant

S. no.	Metal	Microrganism	Biosorption of metals	References
1.	Lead	*Ascophyllumnodosum*	370 mg/g	[112]
2.	Lead	*Nostoc* sp.	93.5 mg/g	[113]
3.	Lead	*Synechococcus* sp.	0.25 mg/g	[114]
4.	Lead	*Fucus vesiculosus*	370 mg/g	[112]
5.	Lead	Oedogonium sp.	145 mg/g	[113]
6.	Cadmium	*Chlorella* sp.	40 mg/g	[115]
7.	Cadmium	*Sargassum vulgare*	0.79 mmol/g	[116]
8.	Cadmium	*Sargassum natans*	135 mg/g	[117]
9.	Cadmium	*Chlorella sorokiniana*	43 mg/g	[118]
10.	Nickel	*Fucusvesiculosus*	40 mg/g	[112]
11.	Nickel	*Ascophyllum nodosum*	30 mg/g	[112]
12.	Nickel	*S. natans*	24.44 mg/g	[112]
13.	Zinc	*Cyanobium species*	0.125 mg/g	[114]
14.	Zinc	*Hydrodictyon reticulatum*	390 μg/g	[119]
15.	Zinc	*Rhizoclonium hieroglaphicum*	77.29 μg/g	[119]
16.	Zinc	*Fucus vesiculoses*	0.80 mmol/g	[112]
17.	Copper	*Cyanobium species*	0.212 mg/g	[114]
18.	Copper	*C. sorokiniana*	46.4 mg/g	[118]
19.	Copper	*Laminaria japonica*	1.59 mmol/g	[120]

Table 14.
Heavy metal shows biosorption potential in algal species.

to lead. *Cladophorafascicularis, Spirogyra hyaline, Sargassum sp.* are resistant to mercury metal and *Sargassum* sp., *Fucusvesiculosus*, *Ascophyllumnodosum* are resistant to nickel [121].

Red algae *Porphyra leucostica* was used to treatment heavy metal accumulation in wastewater and contaminated water sites by Ye et al. [122]. It was reported that this species are so efficient biosorbent.

Microalgae are also capable in utilizing the removal of heavy metals for water contaminated sites. Microalgae are unicellular organisms and also known as phytoplankton which are visible under microscope only and found in both fresh and marine water. Microalgae show positive responses in the resistance towards the heavy metals and convey better chances of bioremediation. Microalgae are also used as a bio-indicator to check or identify the effects of contaminants on ecosystem. Microalgae exhibit biosorption methods to accumulate heavy metals by showing extracellular mechanism and intracellular mechanism to deal with high toxic concentration. Microalgae mostly used to treat wastewater as it releases oxygen as a byproduct during process [123].

Bioremediation by Cynobacteria (Blue Green algae):

Cynobacteria is efficient tool for enhancing the productivity of crop, and plants, formation of bio fuel, rise in fertility of soil and bioremediation also. To explore multiple functional bioagents, genetically engineered cynobacteria should be introduced heavy metals like cadmium, lead, copper, cobalt, manganese were treated with different cynobacterial species such as *N. muscorum*, *A. subcylindrica*, *Nostoc*, *linckia*, *N. rivularis,* etc present in sewage and industrial waste water [124, 125].

Heavy metals develop oxidative stress by generation of reactive oxygen species (ROS) which is extremely toxic and damages the nucleic acid-DNA and RNA, protein and lipids also.

Cynobacteria acts as bioremediator because of their photoautotrophic nature and capability in nitrogen fixation. It is able to tertiary level of agro industrial effluents like oil refineries, paper and pharmaceutical industry. *Nostoc species* and *Microcystis species* accumulate wide range of organophosphate insecticides. As it is found in contaminated water sites and helps in high yield of plants and utilized for bioaccumulation. It can help to enhance the fertility of soil and useful as bio-fuels. It can be used as a good biofertilizers. Mechanism adopted by cynobacteria response to salanity result in bio-polymer production.

Cynobacteria develop bio-flocculants that shield there body mechanism from toxicity of heavy metas. Bio-flocculants are outlined by the presence of various negatively charged binding sides that permit cynobacteria in resistance of heavy metal from contaminated sites [126]. Cynobacteria have flourished numerous mechanisms for reducing the metal stress by intracellular metal sequestration, extracellular mechanism or binding of metals ions.

Metalithionein are metal binding proteins released by cynobacteria that support organism in metal sequestration of dangerous heavy metal ions.

Use of cynobacteria is much better than other microbes like bacteria fungi because of various other benefits like growth promoters, bio stabilizer, bio energy resource (bio-diesel), bio fertilizer, wasteland reclamation, carbon dioxide sequestration, methane oxidation.

Cynobacteria are very much efficient because of short generation time and helps in atmospheric nitrogen fixation.

Lichens in bioremediation:

Lichens are made by symbiotic association of fungi and algae in which both benefit each other. In wastewater remediation, lichens used as a biosorbents.

In heavy metal contamination, lichens can be used as bio-monitors too and the capability to accumulate heavy metal allows the monitoring ability. Lichen *Permelia perlata* shows the potential in biodegradation in contaminated sites.

Lichens adopt numerous processes for metal uptake such as extracellular uptake by ion exchange method intracellular removal and capturing of metal particles. The studies done by UK researchers on lichen results that lichen reproduces on land contaminated with uranium particles from mining activities and lichen converts uranium into dark particles. Endolithic lichen can be studied as a future approach in field of bioremediation [127].

5. Conclusion

Heavy metal pollution are very harmful for humans, animals, aquatic species and plants too and they were accumulated on earth crust by natural process as well as human activities such as industrialization, urbanization, mining and extraction, agricultural practices, etc. Bioremediation is the process which use either naturally occurring or deliberately introduced microorganisms to consume and break down environmental pollutants, in order to clean a polluted site. Various studies had been done and various strains were investigated are above mentioned. Bacteria, Fungi, Algae all are helpful in maintaining tolerance against heavy metals in different contaminated sites. There are several microbes present that provide heavy metal resistance through develop different method of resistance against different heavy metal. It can reduce heavy metals from environment to some extent. Further research area needs to be extended on the focus of gene transfer within bio-films for Bioremediation and use of genetic modified organisms. These strategies would facilitate the development of improved techniques for the bioremediation of heavy metals in the environment.

Acknowledgements

I would like to thank Lovely Professional University providing me the opportunity. I thank the anonymous referees for their useful suggestions.

Author contribution

This study was conducted in collaboration between all the authors. All authors read and approved the final manuscript.

Conflict of interest

The authors declare that they have no conflict of interest in the publication.

Author details

Madhuri Girdhar*, Zeba Tabassum, Kopal Singh and Anand Mohan
School of Bioengineering and Biosciences, Lovely Professional University, Phagwara, Punjab, India

*Address all correspondence to: madhurigirdhar007@gmail.com

References

[1] Chhikara S, Dhankhar R. Biosorption of Cr (VI) ions from electroplating industrial effluent using immobilized *Aspergillus niger* biomass. Journal of Environmental Biology. 2008;**29**(5): 773-778. Available from: https://pubmed.ncbi.nlm.nih.gov/19295081/

[2] Olaniran AO, Balgobind A, Pillay B. Bioavailability of heavy metals in soil: Impact on microbial biodegradation of organic compounds and possible improvement strategies. International Journal of Molecular Sciences. 2013;**14**(5):10197-10228. Available from: https://www.researchgate.net/publication/236911118_Bioavailability_of_Heavy_Metals_in_Soil_Impact_on_Microbial_Biodegradation_of_Organic_Compounds_and_Possible_Improvement_Strategies

[3] Igiri BE, Okoduwa SIR, Idoko GO, Akabuogu EP, Adeyi AO, Ejiogu IK. Toxicity and bioremediation of heavy metals contaminated ecosystem from tannery wastewater: A review. Journal of Toxicology. 2018;**2018**:16. DOI: 10.1155/2018/2568038

[4] Valls M, Atrian S, deLorenzo V, Fernández H, Luis Á. Engineering a mouse metallothionein on the cell surface of Ralstoniaeutropha CH34 for immobilization of heavy metals in soil. Nature Biotechnology. 2000;**18**(6):661-665. Available from: https://www.researchgate.net/publication/12481868_Engineering_a_mouse_metallothionein_on_the_cell_surface_of_Ralstonia_eutropha_CH34_for_immobilization_of_heavy_metals_in_soil

[5] Adams MW, Aveling R, Brockington D, Dickson B, Elliott J, Jon H, et al. Biodiversity conservation and the eradication of poverty. Science. 2004;**306**(5699):1146-1149. Available from: https://pubmed.ncbi.nlm.nih.gov/15539593/

[6] Mejare M, Bülow L. Metal-binding proteins and peptides in bioremediation and phytoremediation of heavy metals. Trends in Biotechnology. 2001;**19**(2): 67-73. Available from: https://pubmed.ncbi.nlm.nih.gov/11164556/

[7] Virkutyte J. The use of power ultrasound in biofuel production, bioremediation, and other applications. In: Gallego-Juárez JA, Graff KF, editors. Power Ultrasonic. Sawston, UK: Woodhead Publishing; 2015. pp. 1095-1112. Available from: https://www.sciencedirect.com/science/article/pii/B9781782420286000363?via%3Dihub

[8] Brad HB. Sources and origins of heavy metals. In: Brad HB, editor. Interface Science and Technology. Neubrucke, Germany: Elsevier; 2005. pp. 1-27. DOI: 10.1016/S1573-4285(05)80020-1

[9] Tchounwou PB, Yedjou CG, Patlolla AK, Sutton DJ. Heavy metal toxicity and the environment. Experientia Supplementum. 2012;**101**:133-164. DOI: 10.1007/978-3-7643-8340-4_6

[10] Zhang R, Wilson VL, Hou A, Meng G. Sources of lead pollution, its influence on public health and the countermeasures. International Journal of Health Animal Science & Food Safety. 2015;**2**:18-31

[11] Tiwari S, Tripathi IP, Tiwari HL. Effect of lead on environment. International Journal of Emerging Research in Management &Technology. 2013;**2**(6):1-5

[12] Chung JY, Yu SD, Hong YS. Environmental source of arsenic exposure. Journal of Preventive Medicine and Public Health. 2014;**47**(5):253-257. DOI: 10.3961/jpmph.14.036

[13] Sidhu GPS. Heavy metal toxicity in soils: Sources, remediation technologies

and challenges. Advances in Plants & Agriculture Research. 2016;**5**(1):445-446. DOI: 10.15406/apar.2016.05.00166

[14] Rice KM, Walker EM Jr, Wu M, Gillette C, Blough ER. Environmental mercury and its toxic effects. Journal of Preventive Medicine and Public Health. 2014;**47**(2):74-83. DOI: 10.3961/jpmph.2014.47.2.74

[15] Driscoll CT, Mason RP, Chan HM, Jacob DJ, Pirrone N. Mercury as a global pollutant: Sources, pathways, and effects. Environmental Science & Technology. 2013;**47**(10):4967-4983. Available from: https://www.ncbi.nlm.nih.gov/pmc/articles/PMC3701261/

[16] Ayele A, Godeto YG. Bioremediation of chromium by microorganisms and its mechanisms related to functional groups. Journal of Chemistry. 2021;**2021**:21. Available from: https://www.hindawi.com/journals/jchem/2021/7694157/

[17] Tumolo M, Ancona V, De Paola D, Losacco D, Campanale C, Massarelli C, et al. Chromium pollution in European water, sources, health risk, and remediation strategies: An overview. International Journal of Environmental Research and Public Health. 2020;**17**(15):5438. DOI: 10.3390/ijerph17155438

[18] Han JX, Qi S, Yu D. Review: Effect of environmental cadmium pollution on human health. Health. 2009;**01**(03):159-166. Available from: https://www.researchgate.net/publication/277653223_Review_Effect_of_environmental_cadmium_pollution_on_human_health

[19] Genchi G, Sinicropi MS, Lauria G, Carocci A, Catalano A. The effects of cadmium toxicity. International Journal of Environmental Research and Public Health. 2020;**17**(11):3782. DOI: 10.3390/ijerph17113782

[20] Rehman M, Liu L, Wang Q, Saleem MH, Bashir S, Ullah S, et al. Copper environmental toxicology, recent advances, and future outlook: A review. Environmental Science and Pollution Research. 2019;**26**:18003-18016. DOI: 10.1007/s11356-019-05073-6

[21] Shrivastava AK. A review on copper pollution and its removal from water bodies by pollution control technologies. Indian Journal of Environmental Protection. 2009;**29**(6):552-560. Available from: https://www.researchgate.net/publication/287550012_A_review_on_copper_pollution_and_its_removal_from_water_bodies_by_pollution_control_technologies

[22] Klimek B. Effect of long-term zinc pollution on soil microbial community resistance to repeated contamination. Bulletin of Environmental Contamination and Toxicology. 2012;**88**(4):617-622. DOI: 10.1007/s00128-012-0523-0

[23] Das AP, Ghosh P, Mohanty S, Sukla LB. Toxicological & Environmental Chemistry. 2015;**96**(7):981-997

[24] Rout GR, Sahoo S. Role of iron in plant growth and metabolism. Reviews in Agricultural Science. 2015;**3**:1-24. DOI: 10.7831/ras.3.1

[25] Peter ALJ, Viraraghavan T. Thallium: A review of public health and environmental concerns. Environment International. 2005;**31**(4):493-501. DOI: 10.1016/j.envint.2004.09.003

[26] Volesky B, Holan ZR. Biosorption of heavy metals. Biotechnology Progress. 1995;**11**(3):235-250. DOI: 10.1021/bp00033a001

[27] Balali-Mood M, Naseri K, Tahergorabi Z, Khazdair MR, Sadeghi M. Toxic mechanism of five heavy metals: Murcury, lead, chromium, cadmium, and arsenic. Frontiers in

Pharmacology. 2021;**12**:643972. DOI: 10.3389/fphar.2021.643972

[28] Lentini P, Zanoli L, Granata A, Signorelli SS, Castellino P, Dell'Aquila R. Kidney and heavy metals—the role of environmental exposure (review). Molecular Medicine Report. 2017;**15**(5): 3413-3419

[29] Sankhla MS, Sharma K, Kumar R. Heavy metal causing neurotoxicity in human health. International Journal of Innovative Research in Science, Engineering and Technology. 2017;**6**(5): 7721-7726

[30] Jaishankar M, Tseten T, Anbalagan N, Mathew BB, Beeregowda KN. Toxicity, mechanism and health effects of some heavy metals. Interdisciplinary Toxicology. 2014;**7**(2): 60-72. Available from: https://www.ncbi.nlm.nih.gov/pmc/articles/PMC4427717/

[31] Hong YS, Song KH, Chung JY. Health effects of chronic arsenic exposure. Journal of Preventive Medicine and Public Health. 2014;**47**(5):245-252. DOI: 10.3961/jpmph.14.035

[32] Wani AL, Ara A, Usmani JA. Lead toxicity: A review. Interdisciplinary Toxicology. 2015;**8**(2):55-64. DOI: 10.1515/intox-2015-0009

[33] Plum LM, Rink L, Haase H. The essential toxin: Impact of zinc on human health. International Journal of Environmental Research and Public Health. 2010;**7**(4):1342-1365. DOI: 10.3390/ijerph7041342

[34] Linna A, Uitti J, Oksa P, Toivio P, Virtanen V, Lindholm H, et al. Effects of occupational cobalt exposure on the heart in the production of cobalt and cobalt compounds: A 6-year follow-up. International Archives of Occupational and Environmental Health. 2020;**93**:365-374. Available from: https://pubmed.ncbi.nlm.nih.gov/31745627/

[35] Leyssens L, Vinck B, Straeten CVD, Wuyts F, Maes L. Cobalt toxicity in humans-A review of the potential sources and systemic health effect. Toxicology. 2017;**387**:43-56. DOI: 10.1016/j.tox.2017.05.015

[36] Cima F. Tin: Environmental pollution and health effects. In: Nriagu JO, editor. Encyclopedia of Environmental Health. 2011. pp. 351-359. DOI: 10.1016/B978-0-444-52272-6.00645-0

[37] Mallin MA, Cahoon LB. The hidden impacts of phosphorus pollution to streams and rivers. Bioscience. 2020;**70**(4):315-329. Available from: https://academic.oup.com/bioscience/article/70/4/315/5734751

[38] Amadi CN, Iqweze ZN, Orisakwe OE. Heavy metals in miscarriages and stillbirths in developing nations. Middle East Fertility Society Journal. 2017;**22**(2):91-100. Available from: https://www.sciencedirect.com/science/article/pii/S1110569017300377

[39] Singh R, Gautam N, Mishra A, Gupta R. Heavy metals and living systems: An overview. Indian Journal of Pharmacology. 2011;**43**(3):246-253. Available from: https://www.ncbi.nlm.nih.gov/pmc/articles/PMC3113373/

[40] Begum A, HariKrishna S, Khan I. Analysis of heavy metals in water, sediments and fish samples of Madivala Lakes of Bangalore, Karnataka. International Journal of ChemTech Research. 2009;**1**(2):245-249. Available from: https://www.researchgate.net/publication/237399772_Analysis_of_Heavy_metals_in_Water_Sediments_and_Fish_samples_of_Madivala_Lakes_of_Bangalore_Karnataka

[41] Arif N, Yadav V, Singh S, Singh S, Ahmad P, Mishra RK, et al. Influence of high and low levels of plant-beneficial heavy metal ions on plant growth and

development. Frontiers in Environmental Science. 2016;**4**(69):1-10. DOI: 10.3389/fenvs.2016.00069

[42] Sethy SK, Ghosh S. Effect of heavy metals on germination of seeds. Journal of Natural Science, Biology, and Medicine. 2013;**4**(2):272-275. DOI: 10.4103/0976-9668.116964

[43] Singh J, Kalamdhad AS. Effects of heavy metals on soil, plants, human health and aquatic life. International Journal of Research in Chemistry and Environment. 2011;**1**(2):15-21. Available from: https://www.researchgate.net/publication/265849316_Effects_of_Heavy_Metals_on_Soil_Plants_Human_Health_and_Aquatic_Life

[44] Shahid M, Khalid S, Abbas G, Shahid N, Nadeem M, Sabir M. Heavy metal stress and crop productivity. In: Hakeem K, editor. Crop Production and Global Environmental Issues. Springer: Cham; 2015. DOI: 10.1007/978-3-319-23162-4_1

[45] Ahmad MSA, Ashraf M. Essential roles and hazardous effects of nickel in plants. Reviews of Environmental Contamination and Toxicology. 2011;**214**:125-167. DOI: 10.1007/978-1-4614-0668-6_6

[46] Kapahi M, Sachdeva S. Bioremediation options for heavy metal pollution. Journal of Health & Pollution. 2019;**9**(24):191203. DOI: 10.5696/2156-9614-9.24.191203

[47] Selvi A, Tamil E, Anjugam R, Devi A, Madhan B, Kannappan S, et al. Isolation and characterization of bacteria from tannery effluent treatment plant and their tolerance to heavy metals and antibiotics. Asian Journal of Experimental Biological Sciences. 2012;**3**(1):34-41. Available from: https://www.ajebs.com/vol3(1)/6.pdf

[48] Bruins MR, Kapil S, Oehme FW. Microbial resistance to metals in the environment. Ecotoxicology and Environmental Safety. 2000;**45**(3):198-207. Available from: https://pubmed.ncbi.nlm.nih.gov/10702338/

[49] Silver S. Bacterial resistances to toxic metal ions—A review. Gene. 1996;**179**(1):9-19. Available from: https://pubmed.ncbi.nlm.nih.gov/8991852/

[50] Shamim S, Rehman A, Qazi MH. Cadmium-resistance mechanism in the bacteria Cupriavidus metallidurans CH34 and Pseudomonas putida mt2. Archives of Environmental Contamination and Toxicology. 2014;**67**(2):149-157. Available from: https://link.springer.com/article/10.1007%2Fs00244-014-0009-7

[51] Cha JS, Cooksey DA. Copper resistance in *Pseudomonas syringae* mediated by periplasmic and outer membrane proteins. Proceedings of the National Academy of Sciences. 1991;**88**(20):8915-8919. Available from: https://www.pnas.org/content/88/20/8915

[52] Ghosh S, Mohapatra B, Satyanarayana T, Sar P. Molecular and taxonomic characterization of arsenic (As) transforming Bacillus sp. strain IIIJ3-1 isolated from As-contaminated groundwater of Brahmaputra river basin, India. BMC Microbiology. 2020;**20**(1):1-20. Available from: https://pubmed.ncbi.nlm.nih.gov/32807097/

[53] Zeroual Y, Moutaouakkil A, Blaghen M. Volatilization of mercury by immobilized bacteria (*Klebsiella pneumoniae*) in different support by using fluidized bed bioreactor. Current Microbiology. 2001;**43**(5):322-327. Available from: https://link.springer.com/article/10.1007/s002840010310

[54] Soraia EB, Baz M, Barakate M, Hassani L, El Gharmali A, Imziln B. Resistance to and accumulation of heavy metals by Actinobacteria isolated from

abandoned mining areas. The Scientific World Journal. 2015;**2015**:14. DOI: 10.1155/2015/761834

[55] Oladipo OG, Awotoye OO, Olayinka A, Bezuidenhout CC, Maboeta MS. Heavy metal tolerance traits of filamentous fungi isolated from gold and gemstone mining sites. Brazilian Journal of Microbiology. 2017;**49**(1):29-37. Available from: https://www.researchgate.net/publication/319015738_Heavy_metal_tolerance_traits_of_filamentous_fungi_isolated_from_gold_and_gemstone_mining_sites

[56] Abbas SZ, Riaz M, Ramzan N, Zahid MT, Shakoori FR, Rafatullah M. Isolation and characterization of arsenic resistant bacteria from wastewater. Brazilian Journal of Microbiology. 2015;**45**(4):1309-1315 https://www.ncbi.nlm.nih.gov/pmc/articles/PMC4323304/

[57] Abbas SZ, Rafatullah M, Ismail N, Lalung J. Isolation, identification, and characterization of cadmium resistant Pseudomonas sp. M3 from industrial wastewater. Journal of Waste Management. 2014;**2014**:6. Available from: https://www.hindawi.com/journals/jwm/2014/160398/

[58] Titah HS, Abdullah SRS, Idris M, Anuar N, Basri H, Mukhlisin M, et al. Arsenic resistance and biosorption by isolated rhizobacteria from the roots of ludwigiaoctovalvis. International Journal of Microbiology. 2018;**2018**:10. DOI: 10.1155/2018/3101498

[59] Anderson CR, Cook GM. Isolation and characterization of arsenate-reducing bacteria from arsenic-contaminated sites in New Zealand. Current Microbiology. 2004;**48**(5):341-347. Available from: https://link.springer.com/article/10.1007/s00284-003-4205-3

[60] Suresh M, Sudhakar S, Tiwari KN. Prioritization of watersheds using morphometric parameters and assessment of surface water potential using remote sensing. Journal of the Indian Society of Remote Sensing. 2004;**32**:249-259. DOI: 10.1007/BF03030885

[61] Durve A, Naphade S, Bhot M, Varghese J, Chandra N. Characterisation of metal and xenobiotic resistance in bacteria isolated from textile effluent. Advances in Applied Science Research. 2012;**3**(5):2801-2806. Available from: https://www.researchgate.net/publication/260287429_Characterization_of_metal_and_xenobiotic_resistance_in_bacteria_isolated_from_textile_effluent

[62] Khan Z, Rehman A, Hussain SZ, Nisar MA, Zulfiqar S, Shakoori AR. Cadmium resistance and uptake by bacterium, Salmonella enterica 43C, isolated from industrial effluent. AMB Express. 2016;**6**(54):1-16. DOI: 10.1186/s13568-016-0225-9

[63] Silver S, Phung LT. Bacterial heavy metal resistance: New surprises. Annual Review of Microbiology. 1996;**50**(1):753-789. Available from: https://www.researchgate.net/publication/14302132_Silver_S_Phung_LT_Bacterial_heavy_metal_resistance_new_surprises_Annu_Rev_Microbiol_50_753-789

[64] Lin X, Mou R, Zhaoyun C, Xu P, Wu X, Zhu Z, et al. Characterization of cadmium-resistant bacteria and their potential for reducing accumulation of cadmium in rice grains. The Science of the Total Environment. 2016;**569-570**:97-104. Available from: https://pubmed.ncbi.nlm.nih.gov/27341110/

[65] Diels L. Accumulation and precipitation of Cd and Zn ions by *A. eutrophus* strains. In: Sally J, McCready RGL, Wichlacz PG, editors. CANMET Publisher; 1990. pp. 369-377

[66] Ziagova M, Dimitriadis G, Aslanidou D, Papaioannou X,

Tzannetaki EL, Liakopoulou-Kyriakides M. Comparative study of Cd (II) and Cr(VI) biosorption on *Staphylococcus xylosus* and Pseudomonas sp. in single and binary mixtures. Bioresource Technology. 2007;**98**(15): 2859-2865. Available from: https://pubmed.ncbi.nlm.nih.gov/17098422/

[67] Li Z, Hongli Y. Characterization of cadmium removal by Rhodotorula sp. Y11. Applied Microbiology and Biotechnology. 2006;**73**(2):458-463. Available from: https://pubmed.ncbi.nlm.nih.gov/16736089/

[68] Saranya K, Sundaramanickam A, Shekhar S, Swaminathan S, Balasubramanian T. Bioremediation of mercury by *Vibrio fluvialis* screened from industrial effluents. BioMed Research International. 2017;**2017**: 6509648. DOI: 10.1155/2017/6509648

[69] Keramati P, Hoodaji M, Tahmourespour A. Multimetal resistance study of bacteria highly resistant to Mercury isolated from dental clinic effluent. African Journal of Microbiology Research. 2011;**5**(7):831-837. Available from: https://www.researchgate.net/publication/215927616_Multimetal_resistance_study_of_bacteria_highly_resistant_to_Mercury_isolated_from_dental_clinic_effluen

[70] Irawati W, Paricia SY, Baskoro AH. A study on murcury-resistant bacteria isolated from a gold mine in Pongkor village, Bogor, Indonesia. Hayati Journal of Biosciences. 2012;**19**(4):197-200. Available from: https://cyberleninka.org/article/n/1274042/viewer

[71] De J, Sarkar A, Ramaiah N. Bioremediation of toxic substances by mercury resistant marine bacteria. Ecotoxicology. 2006;**15**(4):385-389. Available from: https://pubmed.ncbi.nlm.nih.gov/16673165/

[72] Wagner-Döbler I. Pilot plant for bioremediation of mercury-containing industrial wastewater. Applied Microbiology and Biotechnology. 2003;**62**(2-3):124-133. Available from: https://www.researchgate.net/publication/10770401_Pilot_Plant_for_Bioremediation_of_Mercury-Containing_Industrial_Wastewater

[73] Naik MM, Dubey SK. Lead resistant bacteria: Lead resistance mechanisms, their applications in lead bioremediation and biomonitoring. Ecotoxicology and Environmental Safety. 2013;**98**:1-7. Available from: https://pubmed.ncbi.nlm.nih.gov/24144999/

[74] Gummersheimer BS, Giblin T. Identification of lead resistant bacteria from a heavily contaminated site. Bios. 2003;**74**(2):48-54. Available from: http://www.jstor.org/stable/4608669

[75] Sheng XF, Xia JJ, Jiang CY, He LY, Qian M. Characterization of heavy metal-resistant endophytic bacteria from rape (*Brassica napus*) roots and their potential in promoting the growth and lead accumulation of rape. Environmental Pollution. 2008;**156**(3): 1164-1170. Available from: https://pubmed.ncbi.nlm.nih.gov/18490091/

[76] Lusi EA, Patrissi T, Guarascio P. Nickel-resistant bacteria isolated in human microbiome. New Microbes and New Infections. 2017;**19**:67-70. Available from: https://www.ncbi.nlm.nih.gov/pmc/articles/PMC5501881/

[77] Alboghobeish H, Tahmourespour A, Doudi M. The study of Nickel Resistant Bacteria (NiRB) isolated from wastewaters polluted with different industrial sources. Journal of Environmental Health Science and Engineering. 2014;**12**(1):44. Available from: https://www.ncbi.nlm.nih.gov/pmc/articles/PMC3931474/

[78] Nirbhavane HM, Bagde US. Study of lead and nickel resistance mechanism in Enterobacter species. Indian Journal of Microbiology Research. 2018;**5**(2):

229-235. DOI: 10.18231/2394-5478.2018.0048

[79] Van Houdt R, Mergeay M. Genomic context of metal response genes in Cupriavidus metallidurans with a focus on strain CH34. In: Metal Response in Cupriavidus metallidurans. Cham: Springer; 2015. pp. 21-44. Available from: https://publications.sckcen.be/portal/en/publications/genomic-context-of-metal-response-genes-in-cupriavidus-metallidurans-with-a-focus-on-strain-ch34(d7310a3a-5cd3-42c8-b9d5-6fe19d67a25b)/export.html

[80] Stoppel RD, Meyer M, Schlegel HG. The nickel resistance determinant cloned from the enterobacterium Klebsiella oxytoca: Conjugational transfer, expression, regulation and DNA homologies to various nickel-resistant bacteria. Biometals. 1995;**8**(1):70-79. DOI: 10.1007/BF00156161

[81] Altimira F, Yáñez C, Bravo G. Characterization of copper-resistant bacteria and bacterial communities from copper-polluted agricultural soils of central Chile. BMC Microbiology. 2012;**12**:193. DOI: 10.1186/1471-2180-12-193

[82] Andreazza R, Pieniz S, Okeke BC, Camargo FAO. Evaluation of copper resistant bacteria from vineyard soils and mining waste for copper biosorption. Brazilian Journal of Microbiology. 2011;**42**(1):66-74. DOI: 10.1590/S1517-83822011000100009

[83] Madhaiyan M, Poonguzhali S, Sa T. Metal tolerating methylotrophic bacteria reduces nickel and cadmium toxicity and promotes plant growth of tomato (*Lycopersicon esculentum* L.). Chemosphere. 2007;**69**(2):220-228. Available from: https://www.researchgate.net/publication/6320537_Metal_tolerating_methylotrophic_bacteria_reduces_nickel_and_cadmium_toxicity_and_promotes_plant_growth_of_tomato_Lycopersicon_esculentum_L

[84] Rajkumar M, Freitas H. Influence of metal resistant-plant growth-promoting bacteria on the growth of Ricinus communis in soil contaminated with heavy metals. Chemosphere. 2008;**71**(5):834-842. Available from: https://pubmed.ncbi.nlm.nih.gov/18164365/

[85] Abdelbary S, Elgamal MS, Farrag A. Trends in heavy metals tolerance and uptake by Pseudomonas aeruginosa. In: *Pseudomonas aeruginosa*—An Armory Within. London, UK: IntechOpen; 2019. DOI: 10.5772/intechopen.85875

[86] Iram S, Ahmad I, Javed B, Yaqoob S, Akhtar K, Kazmi MR, et al. Fungal tolerance to heavy metals. Pakistan Journal of Botany. 2009;**41**(5):2583-2594. Available from: https://www.researchgate.net/publication/265158665_Fungal_tolerance_to_heavy_metals_Pak_J_Bot

[87] Imran M, Ahmad I, Barasubiye T, Abulreesh HH, Samreen, Monjed MK, et al. Heavy metal tolerance among free-living fungi isolated from soil receiving long term application of wastewater. Journal of Pure and Applied Microbiology. 2020;**14**(1):157-170. Available from: https://microbiologyjournal.org/heavy-metal-tolerance-among-free-living-fungi-isolated-from-soil-receiving-long-term-application-of-wastewater/

[88] Anahid S, Yaghmaei S, Zahra N. Heavy metal tolerance of fungi. Scientia Iranica. 2011;**18**(3):502-508. Available from: https://www.researchgate.net/publication/257451542_Heavy_metal_tolerance_of_fungi

[89] Fazli MM, Soleimani N, Mehrasbi M. Highly cadmium tolerant fungi: Their tolerance and removal potential. Journal of Environmental Health Science & Engineering.

2015;**13**(19):1-9. DOI: 10.1186/s40201-015-0176-0

[90] Talukdar D, Sharma R, Jaglan S, Vats R, Kumar R, Mahnashi MH, et al. Identification and characterization of cadmium resistant fungus isolated from contaminated site and its potential for bioremediation. Environmental Technology & Innovation. 2020;**17** 100604. Available from: https://www.researchgate.net/publication/338445950_Identification_and_characterization_of_cadmium_resistant_fungus_isolated_from_contaminated_site_and_its_potential_for_bioremediation

[91] Joo JH, Hussein KA. Heavy metal tolerance of fungi isolated from contaminated soil. Korean Journal of Soil Science and Fertilizer. 2012;**45**(4):565-571. DOI: 10.7745/KJSSF.2012.45.4.565

[92] Acosta-Rodríguez I, Cárdenas-González JF, Pérez R, Oviedo JT, Martínez-Juárez VM. Bioremoval of different heavy metals by the resistant fungal strain *Aspergillus niger*. Bioinorganic Chemistry and Applications. 2018;**2018**:7. Available from: https://www.hindawi.com/journals/bca/2018/3457196/

[93] Zhang D, Yin C, Abbas N. Multiple heavy metal tolerance and removal by an earthworm gut fungus Trichoderma brevicompactum QYCD-6. Scientific Reports. 2020;**10**:6940. DOI: 10.1038/s41598-020-63813-y

[94] Iskandar NL, Zainudin NAIM, Tan SG. Tolerance and biosorption of copper (Cu) and lead (Pb) by filamentous fungi isolated from a freshwater ecosystem. Journal of Environmental Sciences. 2011;**23**(5):824-830. Available from: https://pubmed.ncbi.nlm.nih.gov/21790056/

[95] Hindersah R, Asda KR, Herdiyantoro D, Kamaluddin N. Isolation of mercury-resistant fungi from mercury-contaminated agricultural soil. Agriculture. 2018;**8**(3):33. Available from: https://www.researchgate.net/publication/323433651_Isolation_of_Mercury-Resistant_Fungi_from_Mercury-Contaminated_Agricultural_Soi

[96] Pietro-Souza W, de Campos Pereira F, Mello IS, Stachack FFF, Terezo AJ, da Cunha CN, et al. Mercury resistance and bioremediation mediated by endophytic fungi. Chemosphere. 2020;**240**:124874. Available from: https://pubmed.ncbi.nlm.nih.gov/31546184/

[97] Joho M, Inouhe M, Tohoyama H, Murayama T. Nickel resistance mechanisms in yeasts and other fungi. Journal of Industrial Microbiology. 1995;**14**(2):164-168. Available from: https://pubmed.ncbi.nlm.nih.gov/7766209/

[98] Visoottiviseth P, Nootra P. Selection of fungi capable of removing toxic arsenic compounds from liquid medium. Science Asia. 2001;**27**(2):83-92. Available from: https://www.researchgate.net/publication/255659242_Selection_of_Fungi_Capable_of_Removing_Toxic_Arsenic_Compounds_from_Liquid_Medium

[99] Cernansky S, Kolencik M, Sevc J, Urik M, Hiller E. Fungal volatilization of trivalent and pentavalent arsenic under laboratory condition. Bioresource Technology. 2009;**100**(2):1037-1040. Available from: https://pubmed.ncbi.nlm.nih.gov/18774290/

[100] Urik M, Čerňanský S, Ševc J, Simonovicova A, Littera P. Biovolatilization of arsenic by different fungal strains. Water, Air, and Soil Pollution. 2007;**186**:337-342. Available from: https://link.springer.com/article/10.1007/s11270-007-9489-7

[101] Srivastava PK, Vaish A, Dwivedi S, Chakrabarty D. Biological removal of

arsenic pollution by soil fungi. Science of the Total Environment. 2011;**409**(12):2430-2442. DOI: 10.1016/j.scitotenv.2011.03.002

[102] Wang S, Zhao X. On the potential of biological treatment for arsenic contaminated soils and groundwater. Journal of Environmental Management. 2009;**90**(8):2367-2376. DOI: 10.1016/j.jenvman.2009.02.001

[103] Soto J, Ortiz J, Herrera H, Fuentes A, Almonacid L, Charles TC, et al. Enhanced arsenic tolerance in *Triticum aestivum* inoculated with arsenic-resistant and plant growth promoter microorganisms from a heavy metal-polluted soil. Microorganisms. 2019;**7**(9):348. Available from: https://pubmed.ncbi.nlm.nih.gov/31547348/

[104] Choe SI, Gravelat FN, Al Abdallah Q, Lee MJ, Gibbs BF, Sheppard DC. Role of *Aspergillus niger* acrA in arsenic resistance and its use as the basis for an arsenic biosensor. Applied and Environmental Microbiology. 2012;**78**(11):3855-3863. Available from: https://www.ncbi.nlm.nih.gov/pmc/articles/PMC3346401/

[105] Balaban BG, Yılmaz Ü, Alkım C, Topaloğlu A, Kısakesen Hİ, Holyavkin C, et al. Evolutionary engineering of an iron-resistant Saccharomyces cerevisiae mutant and its physiological and molecular characterization. Microorganisms. 2020;**8**(1):43. Available from: https://www.ncbi.nlm.nih.gov/pmc/articles/PMC7023378/

[106] Cárdenas González JF, Rodríguez Pérez AS, Vargas Morales JM, Martínez Juárez VM, Rodríguez IA, Cuello CM, et al. Bioremoval of Cobalt (II) from aqueous solution by three different and resistant fungal biomasses. Bioinorganic Chemistry and Applications. 2019;**2019**: 1-8. DOI: 10.1155/2019/8757149

[107] Priyadarshini E, Priyadarshini SS, Pradhan N. Heavy metal resistance in algae and its application for metal nanoparticle synthesis. Applied Microbiology and Biotechnology. 2019;**103**(8):3297-3316. DOI: 10.1007/s00253-019-09685-3

[108] Wilde EW, Benemann JR. Bioremoval of heavy metals by the use of microalgae. Biotechnology Advances. 1993;**11**(4):781-812. Available from: https://www.sciencedirect.com/science/article/abs/pii/0734975093900036

[109] Mehta SK, Gaur JP. Use of algae for removing heavy metal ions from wastewater: Progress and prospects. Critical Reviews in Biotechnology. 2005;**25**(3):113-152. Available from: https://www.tandfonline.com/doi/abs/10.1080/07388550500248571?journalCode=ibty20

[110] Shanab S, Essa A, Shalaby E. Bioremoval capacity of three heavy metals by some microalgae species (Egyptian Isolates). Plant Signaling & Behavior. 2012;**7**(3):392-399. Available from: https://www.ncbi.nlm.nih.gov/pmc/articles/PMC3443921/

[111] Baumann HA, Morrison L, Stengel DB. Metal accumulation and toxicity measured by PAM—chlorophyll fluorescence in seven species of marine macroalgae. Ecotoxicology and Environmental Safety. 2009;**72**(4):1063-1075. Available from: https://www.researchgate.net/publication/222331193_Metal_accumulation_and_toxicity_measured_by_PAM-Chlorophyll_fluorescence_in_seven_species_of_marine_macroalgae

[112] Holan ZR, Volesky B. Biosorption of lead and nickel by biomass of marine algae. Biotechnology and Bioengineering. 1994;**43**(11):1001-1009. Available from: https://pubmed.ncbi.nlm.nih.gov/18615510/

[113] Gupta VK, Rastogi A. Biosorption of lead from aqueous solutions by green algae Spirogyra species: Kinetics and

equilibrium studies. Journal of Hazardous Materials. 2008;**152**(1):407-414. DOI: 10.1016/j.jhazmat.2007.07.028

[114] Shuhei O, Yamaguchi H, Vanel F, Fuchida S, Koshikawa H, Yamagishi T, et al. Differential heavy metal sensitivity in seven algal species from the NIES culture collection based on delayed fluorescence assays. Physiological Research. 2019;**68**(1):41-49. DOI: 10.1111/pre.12403

[115] Matsunaga T, Takeyama H, Nakao T, Yamazawa A. Screening of marine microalgae for bioremediation of cadmium-polluted seawater. Journal of Biotechnology. 1999;**70**(1-3):33-38. DOI: 10.1016/S0168-1656(99)00055-3

[116] Davis T, Volesky B, Mucci A. A review of the biochemistry of heavy metal biosorption by brown algae. Water Research. 2003;**37**(18):4311-4330. Available from: https://pubmed.ncbi.nlm.nih.gov/14511701/

[117] Holan ZR, Volesky B, Prasetyo I. Biosorption of cadmium by biomass of marine algae. Biotechnology and Bioengineering. 1993;**41**(8):819-825. Available from: https://pubmed.ncbi.nlm.nih.gov/18609626/

[118] Yoshida N, Ishii K, Okuno T. Purification and characterization of cadmium-binding protein from unicelluar alga Chlorella sorokinian. Current Microbiology. 2006;**52**(6): 460-463. DOI: 10.1007/s00284-005-0328-z

[119] Dwivedi S, Srivastava S, Mishra S, Kumar A, Tripathi RD, Rai UN, et al. Characterization of native microalgal strains for their chromium bioaccumulation potential: Phytoplankton response in polluted habitats. Journal of Hazardous Materials. 2010;**173**(1-3):95-101. Available from: https://www.sciencedirect.com/science/article/abs/pii/S0304389409013338

[120] Fourest E, Volesky B. Alginate properties and heavy metal biosorption by marine algae. Applied Biochemistry and Biotechnology. 1997;**67**:215-226. DOI: 10.1007/BF02788799

[121] Shamim S. Biosorption of Heavy Metals. London, UK: IntechOpen; 2007. Available from: https://www.intechopen.com/books/biosorption/biosorption-of-heavy-metals

[122] Ye J, Xiao H, Xiao B, Xu W, Gao L, Lin G. Bioremediation of heavy metal contaminated aqueous solution by using red algae *Porphyra leucosticta*. Water Science and Technology: A Journal of the International Association on Water Pollution Research. 2015;**72**(9):1662-1666. Available from: https://pubmed.ncbi.nlm.nih.gov/26524459/

[123] Kaplan D. Absorption and adsorption of heavy metals by microalgae. Handbook of Microalgal Culture: Applied Phycology and Biotechnology. 2013;7(2):602-611. Available from: https://onlinelibrary.wiley.com/doi/10.1002/9781118567166.ch32

[124] Zinicovscaia I, Cepoi L, editors. Cyanobacteria for Bioremediation of Wastewaters. Switzerland: Springer International Publishing; 2016. Available from: https://www.springer.com/gp/book/9783319267494

[125] Dubey SK, Dubey J, Mehra S, Tiwari P, Bishwas AJ. Potential use of cyanobacterial species in bioremediation of industrial effluents. African Journal of Biotechnology. 2011;**10**(7):1125-1132. Available from: https://www.researchgate.net/publication/230788895_Potential_use_of_cyanobacterial_species_in_bioremediation_of_industrial_effluents

[126] Bender J, Lee R, Phillips P. Uptake and transformation of metals and metalloids by microbial mats and their

use in bioremediation. Journal of Industrial Microbiology. 1995;**14**(2): 113-118. Available from: https://academic.oup.com/jimb/article/14/2/113/598845

[127] Saier MH Jr. Beneficial bacteria and bioremediation. Journal of Molecular Microbiology and Biotechnology. 2005;**9**(2):6. Available from: https://www.karger.com/Article/Abstract/88836

Chapter 14

Phytoremediation Potential of *Chrysopogon zizanioides* for Toxic Elements in Contaminated Matrices

Sebastián A. Ramos-Arcos, Edith G. González-Mondragón, Eduardo S. López-Hernández, Ana R. Rodríguez-Luna, Carlos M. Morales-Bautista, Selene Lagunas-Rivera and Sugey López-Martínez

Abstract

Many researchers have demonstrated the advantages of plants in the phytoremediation of soils and waters contaminated with heavy metals, herbicides, pesticides, leachates, etc. The unique morphological characteristics of *Chrysopogon zizanioides*, commonly known as vetiver, make it a hyperaccumulator of metals; its roots can store high concentrations of heavy metals such as As, Cd, Cr, Cu, Hg, Ni, Pb, Se, and Zn, and it has thus been successfully used in the field of environmental protection. This chapter presents the importance of vetiver, its characterization, and its potential use as phytoremediation potential for toxic elements in contaminated matrices.

Keywords: vetiver, leachate, phytoremediation, metals, hyperaccumulation

1. Introduction

Heavy metals are natural elements that have a high atomic weight and a density at least five times that of water, due to their high degree of toxicity, some such as Arsenic (As), Cadmium (Cd), Chromium (Cr), Lead (Pb), Copper (Cu), nickel (Ni), selenium (Se) Zinc (Zn) and Mercury (Hg) are considered harmful to health and the environment, raising concerns for setting up adequate prevention or restoration measures that reduce these risks. A special topic of global interest is the residual concentrations of heavy metals since some studies have shown that heavy metals, especially because are considered bioaccumulative in various matrices (range from ng kg^{-1} to less than 10 mg kg^{-1}) [1].

These last components are generated mainly by human activities such as mining, emissions, agriculture, and industrial waste; some studies mention that in high concentrations heavy metals such as Cd, Cr, and Pb can have potential toxic effects, for example, some studies they have observed that they could interact with the to the growth and general metabolism of humans and animals [2]. Also, it was reported that mean concentrations of heavy metals could affect the biodiversity through their bioaccumulation in different organisms, although it has been observed that this also depends on the type of ecosystem, the exposure time and other environmental factors [3]; such as, some reports suggest that the disposal mechanisms could also depend on the balance between sorption and desorption, as well as the natural dynamics of the soils on which they are deposited, the soil constituents (inorganic and organic), and the chemical nature of the soil. Compound [4].

Many studies have found that waste dumps are sources of heavy metals, most have reported As, Cd, Cr, Cu, Hg, Ni, Pb, and Zn, although the receptor organs are diverse, due to the conditions of storage, disposal, and importance in food production, the treatment of soils and aquifers contaminated by these compounds has gained interest in the last decade [5–7]. The disposition that each of them could have to the environment, can be measured in terms of leachates, whose composition varies from one site to another since they regularly are created by the biodegradation of waste, in some cases, depending on their diffusion capacity in the soil, they could pollute both groundwater and surface water [8]. Numerous studies have emphasized the importance of remediating these sites, mentioning that feasible and long-term alternatives must be created, especially, that guarantee low exposure of these pollutants in places that have a population immersed or that are destined for activities of the primary sector [5].

Heavy metal contamination and pesticides is a serious problem worldwide due to their toxicity, furthermore, assessing the impacts is very complex due to the fact that many species have cumulative and non-biodegradable properties, but cases have been reported, in which certain species of plants could be indicators of these pollutants [6, 8]; also, although some organisms usually transport or extract them from a matrix, they only transform it to other oxidation states in the soil, in terms of bioremediation, some technologies take advantage of this behavior to reduce their mobility and toxicity, however, if they are not remediated sites, metals can reach humans [9–11].

However, although it has been shown that these methods tend to control various types of organic or inorganic pollutants in the long term [12–14], some studies have warned about the risk factor of those plant species that tend to be hyperaccumulative and can also be a food source for some grazing or wild species (it has been reported that the concentration of Cd or Pb metals in hyperaccumulating plants is usually between 10 and 100 times higher than that of the soil) [15, 16].

Vetiver grass is a perennial herb of the Poaceae family, native to India. It is a plant that has been cultivated for many years in Asia, especially in India [17], can grow in a wide range of climatic conditions, and if planted correctly can be used anywhere in tropical, subtropical or Mediterranean climates [18].

Compiled by Méndez-Cano [19]; the plant vetiver is a perennial herb that forms dense clumps **Figure 1**; it has sterile inflorescences and seeds and reproduces vegetatively It can withstand extreme droughts due to the high salt content in the sap of its leaves, it can withstand extreme droughts due to the high salt content in the sap of its leaves and also flooding for long periods. It grows in a wide range of soils with different levels of fertility, it is tolerant to extreme climatic variations, such as prolonged droughts, floods and temperatures ranging from −9–55°C. It grows in soils, including rocky soils, and can also be grown in hydroponic conditions. It tolerates pH levels between 3.3 and 12.5, as well as saline, acidic, alkaline and sodic media with a high load of nutrients and heavy metals. It is classified as a C4 type plant due to its high atmospheric CO2 fixation capacity.

Recent research compares the variability in biomechanical properties of *Chrysopogon zizanioides*, including tensile strength, Young's modulus and strain at break, which have a direct implication to root reinforcement to slope [20], interesting studies reveal that biomass extracted from the roots of the species can be used as

Figure 1.
Chrysopogon zizanioides.

activated carbon. This work offers an innovative and environmentally safe approach to control porosity in biomass-derived activated carbon (BAC) materials for energy storage applications [21]. Natural fibers as compared to synthetic fibers are having higher strength, rigidity and also in supporting the structural load of matrix. Vetiver fiber is used as reinforcement for the polymer composites with polypropylene and polyethylene as matrix material [22].

Authors demonstrate the application of vetiver grass has been widely promoted in tropical regions as a cost-effective and environmental-friendly solution for slope stabilization and erosion control for many years. Despite its potential, vetiver grass utilization has not been widely accepted by disadvantaged agricultural communities at landslide hazard areas [23].

Also floating Hydroponic System (FHS) is a potential and cost-effective technique for wastewater treatment. Vetiver is a more efficacious material for phytoremediation due to its physiological and morphological properties [24].

Although there are reports of several species discovered with high potential for phytoremediation, vetiver is a grass species that meets all the criteria required to eliminate contaminants in water and soil, but are few reports of use [12]; an important point is that this plant can survive under hydroponic conditions, has been used for a long time in water and soil conservation [25–27], in the rehabilitation and restoration of landfills, as in the phytoremediation of leachates, it survives under hydroponic conditions [28]. Many species have been reported as metal phytoremediators but few have been reported to be able to adapt to extreme altitude, climate, variable pH, and exposure conditions in eutrophic systems; thus, it is of great importance to continue studying native species to identify potential alternative phytoremediators

[29]. For these reasons, in this study, we present a review of the importance of vetiver, its characterization, and its potential use as a remediation alternative.

2. General characteristics of vetiver (*Chrysopogon zizanioides*)

Vetiver belonging to the Poaceae family, native to India [19]. Is one of the few species of grass that meets all the criteria necessary to eliminate contaminants [28]. Regarding its morphological characteristics, it's a tall grass (1–2 m) with abundant vegetative growth, characterized by a massive, finely structured and deep root apparatus, capable of reaching 3–4 m deep in the first year [27]. For this trait, vetiver grass is well known for its effectiveness in controlling erosion and sediments [30].

It has long, narrow leaves that produce a thick growth barrier that cuts and separates runoff water. This type of growth also allows vetiver to act as an effective filter by trapping sediments and contaminants linked to them such as heavy metals and some pesticide residues [31]. One of the most useful physiological characteristics of vetiver is its high tolerance to high concentrations of heavy metals such as Al, B, Ba, Be, Co, Cr, Cu, Fe, Mg, Mn, Ni, Pb, S, Se, Tl, V, and Zn [12, 13, 19, 31, 32]. In **Figure 2**, shows a summary of the morphological characteristics of vetiver plant, and heavy metals to tolerant.

Some studies mention that the dense and finely structured root of this plant creates an ideal environment for microbiological processes in the rhizosphere, these characteristics also make vetiver a good alternative for stabilizing river banks and road embankments and preventing erosion [31]. However, the efficiency and cost-effectiveness in water and soil conservation, particularly in the treatment of wastewater, were only recognized in the decade of the 80s when its outstanding physiological and morphological characteristics were identified [31], but these distinctive features, make it an effective phytoremediator species for the treatment of various types of contaminants; also these attributes, together with its high biomass production, type of reproduction, and adaptations to climate changes, also make vetiver an ideal species for the phytoremediation not only of soil but also of artificial systems such as wetlands [12].

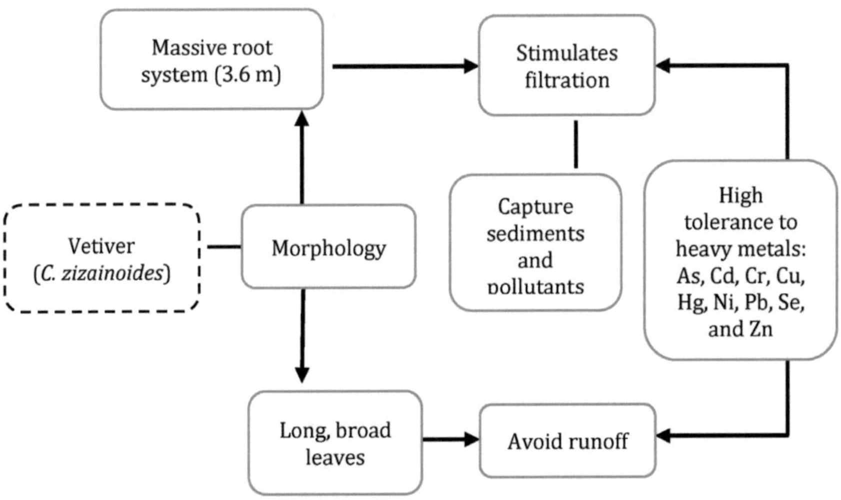

Figure 2.
Morphological characteristics of vetiver (C. zizainoides).

3. Methods for the characterization of heavy metals in phytoremediation

The different techniques applied for the characterization of heavy metals are presented in **Table 1**.

One of the techniques that can be used for the identification of heavy metals is Atomic Absorption Spectroscopy (AAS), this analytical technique is widely used to determine more than 70 elements in solution and in different matrices, in quantities as low as 10–14 g with reasonable selectivity, little manipulation, and minimum sample size. It can indirectly identify anions and organic compounds [33, 34]. This technique is older than ICP-OES (Inductively Coupled Plasma Optical Emission Spectroscopy), and various authors have reported studies comparative with other methods cited in environmental regulations [12, 28, 38], some mention that it makes it possible to quantitatively determine the chemical elements that constitute a material quickly, precisely, and accurately [7, 13, 39].

To convert solid and liquid samples into aqueous solutions for analysis with ICP-OES and AAS, it is necessary to eliminate all organic material to avoid interferences and obtain the analytes of interest at detectable concentrations [12, 38, 40–43]. Acid digestion is a necessary process in the identification of metals, which is done by acid decomposition at high temperatures [36] or using mixtures of HNO_3 and H_2O_2 [37].

Another method used is X-Ray Fluorescence spectroscopy (XRF). It can identify analytes or other components of interest and it is thus very useful for qualitative analysis. It is currently used in the fields of archeology, forensic sciences, medicine, geology, coatings, materials, electronics, pharmaceutics and environmental sciences, used this method to perform qualitative and quantitative analyses of heavy metals [35, 44].

Technique	Characteristics	References
Atomic Absorption Spectroscopy (AAS)	Identify at least 70 elements in quantities as low as 10–14 g, high selectivity.	[33, 34]
Inductively Coupled Plasma Optical Emission Spectroscopy (ICP-OES)	Quantify chemical elements that constitute a material quickly, accurately.	[7, 13, 35]
X-Ray Fluorescence spectroscopy (XRF)	Identify analytes or other components of interest, is very useful for qualitative analysis.	[36, 37]

Table 1.
Techniques for characterization of heavy metals.

4. Vetiver: potential use in phytoremediation

Several authors have achieved the mitigation of different types of heavy metals using vetiver grass (**Table 2**) and determined the amount of Cr absorbed from the residual sludge of a tanning facility and found a concentration of 596.92 mg kg^{-1} in the leaf tissue [38], others studies showed that the effect on vetiver of having a concentration of As of 225 mg kg^{-1} is a slight yellowing of the leaves and a small decrease in biomass [36]; these results confirm that this grass can survive successfully in soils moderately contaminated by As [31].

The capacity of vetiver to remove contaminants has also been tested using compost leachate, an experiment that was allowed to stand for 112 days without

Heavy metals	Origin	References
Cr	Residual sludge tannery	[39]
Cd, Cu, Fe, and Pb	Compost leach	[38]
As, Cd, Ni, Pb, and Zn	Ash remediation	[28]
Al, Cr, Cu, Fe, Mn, Ni, and Zn	Rehabilitation of iron mine	[13]
Cu, Fe, Mn, Pb, and Zn	Water polluted with heavy metals	[12]
Cr, Cu, Pb, and Zn	Soil contaminated with heavy metals	[38]
Cu, Fe, Mn, Pb, and Zn	Industrial waters	[24]
B	Industrial waters	[44]

Table 2.
Heavy metals absorbed for vetiver grass.

aeration, showed that the concentration of Cd, Cu, Fe, and Pb decreased after the treatment with vetiver, and therefore can be used to for the bio-purification of compost leachate [42]; others study evaluated the efficiency of vetiver in the absorption of metals based on the translocation and bioaccumulation factors, the results revealed that roots have a high uptake capacity for Cd, Pb, and Zn, however, there was a low translocation of metals such as Cd, As, Ni, and Pb towards the aerial part of the plant and accumulation of Zn in the roots was the highest at 100% [28].

However, some similar reports found a highs amount of Fe accumulated in the roots, despite this, the results show that vetiver is a good phythostabilizer and potential accumulator of heavy metals since in the roots they also found the presence of Al, Cu, Mn, Zn, Cr, and Ni, but in concentrations, inferiors to Fe [13]. In research similar, the absorbed metals were found to be in order Fe > Pb > Cu > Mn > Zn, the results also showed that as the length and density of the roots increases, so does the absorption of heavy metals, but suggest being careful if in the site intends to develop other species, due to the competition of Fe and its importance in the physiological processes of plants [12].

In 2007, a study assessed the efficiency of the vetiver grass in the phytoextraction of Cr, Cu, Pb, and Zn in order to establish whether this plant could be considered a good hyperaccumulator of those heavy metals. Phytoextraction experiments showed that vetiver was little efficient in the uptake of Cr and Cu (less than 0.1% in shoots and roots after 30 days for both metals), but highly efficient in the uptake of Pb and Zn (0.4% in shoots and 1% in roots for Pb and 1% in both shoots and roots for Zn, after 30 days), for these reasons, vetiver grass can be considered a good enough "hyperaccumulator" of Pb and Zn [41].

In 2013, other researchers measured the ability to remove heavy metals from industrial wastewater. Vetiver were grown on four samples of industrial wastewater taken from a milk factory, a battery manufacturing plant, an electric lamp plant, and an ink manufacturing plant, the results indicated that could tolerate and grow in wastewater [24].

On the other hand, some studies have evaluated the efficiency of vetiver in the treatment of leachates with the aim of reducing chemical oxygen demand, total suspended solids, total dissolved solids and total organic carbon in municipal landfill leachates. The results revealed a removal efficiency of approximately 90% [45]. A relevant study evaluated the differences in tolerance and accumulation of boron between reed (*Phragmites australis* L.), cattail (*T. latifolia* L.) and vetiver, these plants survived concentrations of B of up to 250, 500, and 750 mg L^{-1}, respectively, therefore, vetiver showed the highest tolerance to B [40]. Thus, the evidence described above confirms the phytoremediation potential of the vetiver grass, the findings of different studies have confirmed the potential of vetiver as a phytoremediation plant for use in the removal of heavy metals from contaminated soils [12, 46, 47] and in the rehabilitation of landfills [35]. Although it is not an aquatic plant, vetiver can grow and survive under hydroponic conditions [48] and can be used to remediate eutrophic waters, wastewater from pig farms [49], and waste leachates [50].

Further studies could focus on increasing the uptake of heavy metals using, for example, chelating agents [41] and explore the ability of vetiver to participate in the remediation of other pollutants such as endosulfan [49]. The dense growth of vetiver roots can prevent erosion and landslides and act as a natural barrier that could be used in landfill cells to prevent leachates from infiltrating the aquatic mantle, regardless of the impermeable barrier (geomembrane) that is commonly used in landfills.

5. Conclusions

The new trends in the restoration of degraded soils, wastewater and even leachates generated from urban waste include phyto-management as part of a Circular Economy model which is an attractive and viable alternative that is already being explored by different companies; it is based on the principles of preservation and optimization of natural resources, as well as improving the efficiency of production systems by eliminating or reducing environmental contaminants. Therefore, phytoremediation can be considered a circular economy strategy because it aims to reduce both the entry of materials and the production of waste.

In different matrices water, soil, air there are inorganic contaminants which include trace elements that are essential for the growth and development of plants, heavy metals and some non-metallic elements such as As and B are also included. Toxicity varies according to many factors, such as the chemical form of the elements, concentration, persistence among other factors, some compounds can be transformed to their less toxic forms such as Cr. *Chrysopogon zizanioides* is a hyper-accumulator species with a sometimes unpleasant appearance and its growth capacity makes it ideal for phytoremediation.

Based on group experience we know that this species can survive, tolerate, absorb and transform. Also based on the literature we know that there must be periods of acclimatization of the species for its transformation and or ideal absorption of the compounds. Due to previous knowledge about the phytoremediation process, which is an integral methodology where at the same time the species phytovolatilizes, rhizofiltration, phytodegrades. Due to the characteristics of the species, it can be a permeable membrane to prevent or sequester toxic elements to the water table, but thanks to the life cycle of this species it can absorb significantly contaminated by its modular growth. However, dead leaves may contain some compounds that cannot be degraded and these should be confined or incinerated to ensure that they do not return to the soil.

The essential oil extracted from vetiver roots can be used in the perfume industry, vetiver leaves can be used for roofing of rustic houses and the plant is already used as a fire barrier because it keeps growing even after being burned. In addition, the use of vetiver has the purpose of improving the management of degraded spaces and their restoration through innovative phytoremediation techniques. Vetiver could be used in many countries throughout the world due to its economical accessibility and ability to adapt to different climatic conditions, as well as its capacity to remove different types of pollutants as has already been evidenced.

Acknowledgements

Grateful for the postgraduate scholarship from the National Council of Science and Technology (CONACyT) -Mexico (430983), the Catedra-CONACyT 781 and DAIP-UGto Project # 811/2016. The authors also thank Promotora Ambiental (S.A. de C.V.) for the financial support provided and for the permission to use their facilities.

Conflicts of interest

The authors declare no conflict of interest.

Nomenclature

Al	aluminum
B	boron
Ba	barium
Be	beryllium
Cd	cadmium
Co	cobalt
Cr	chromium
Cu	copper
Fe	iron
Mg	magnesium
Mn	manganese
Ni	nickel
Pb	lead
S	sulfur
Se	selenium
Tl	thallium
V	vanadium
Zn	zinc

Author details

Sebastián A. Ramos-Arcos[1], Edith G. González-Mondragón[2],
Eduardo S. López-Hernández[1], Ana R. Rodríguez-Luna[1],
Carlos M. Morales-Bautista[1], Selene Lagunas-Rivera[3] and Sugey López-Martínez[1*]

1 University Juárez Autonomous of Tabasco, Villahermosa, Tabasco, México

2 Institute of Agroindustry, Technological University of the Mixteca, Huajuapan de León, Oaxaca, Mexico

3 CONACyt-Department of Chemistry, DCNyE, University of Guanajuato, Guanajuato, Mexico

*Address all correspondence to: sugey.lopez@ujat.mx

References

[1] Tchounwou, P. B.; Yedjou, C. G.; Patlolla, A. K.; & Sutton, D. J. Heavy metal toxicity and the environment. Molecular, clinical and environmental toxicology. 2012; 133-164. DOI: 10.1007/978-3-7643-8340-4_6

[2] Subhashini, V.; & Swamy, A. V. V. S. Phytoremediation of lead, cadmium and chromium contaminated soils using selected weed plants. Acta Biologica Indica. 2015; 4, 205-212.

[3] Kannan, N.; Thirunavukkarasu, N.; Suresh, A.; & Rajagopal, K. Analysis of heavy metals accumulation in mangroves and associated mangroves species of Ennore mangrove ecosystem, east coast India. Indian Journal of Science and Technology. 2016; 9, 1-11. DOI: 10.17485/ijst/2016/v9i46/101551

[4] Shaheen, S. M.; Tsadilas, C. D.; & Rinklebe, J. A review of the distribution coefficients of trace elements in soils: Influence of sorption system, element characteristics, and soil colloidal properties. Advances in Colloid and Interface Science. 2013; 201, 43-56. DOI: 10.1016/j.cis.2013.10.005

[5] Wuana, R. A.; & Okieimen, F. E. Heavy metals in contaminated soils: a review of sources, chemistry, risks and best available strategies for remediation. ISRN Ecology. 2011; 2011, 1-20. DOI: 10.5402/2011/402647

[6] Rajaganapathy, V.; Xavier, F.; Sreekumar, D.; & Mandal, P. K. Heavy metal contamination in soil, water and fodder and their presence in livestock and products: a review. Journal of Environmental Science and Technology. 2011; 4, 234-249.

[7] Xu, Y.; Wu, Y.; Han, J.; & Li, P. The current status of heavy metal in lake sediments from China: Pollution and ecological risk assessment. Ecology and evolution. 2017; 7, 5454-5466. DOI: 10.1002/ece3.3124

[8] Salem, Z. B.; Capelli, N.; Grisey, E.; Baurand, P. E.; Ayadi, H.; & Aleya, L. First evidence of fish genotoxicity induced by heavy metals from landfill leachates: the advantage of using the RAPD-PCR technique. Ecotoxicology and environmental safety. 2014; 101, 90-96. DOI: 10.1016/j.ecoenv.2013.12.014

[9] Hu, B.; Jia, X.; Hu, J.; Xu, D.; Xia, F.; & Li, Y. Assessment of heavy metal pollution and health risks in the soil-plant-human system in the Yangtze river delta, China. International Journal of Environmental Research and Public Health. 2017; 14, 1042. DOI: 10.3390/ijerph14091042

[10] Ahmadpour, P.; Ahmadpour, F.; Mahmud, T. M. M.; Abdu, A.; Soleimani, & M.; Tayefeh, F. H. Phytoremediation of heavy metals: A green technology. African Journal of Biotechnology. 2012; 11, 14036-14043. DOI: 10.1201/b16566-15

[11] Barbafieri, M.; Japenga, J.; Romkens, P.; Petruzzelli, G.; & Pedron, F. Protocols for applying phytotechnologies in metal-contaminated soils. In Plant-based remediation processes. Springer, Berlin Heidelberg. 2013; p. 19-37. ISBN 978-3-642-35564-6. DOI: 10.1007/978-3-642-35564-6_2

[12] Suelee, A. L.; Hasan S. N. M. S.; Kusin, F. M.; Yusuff, F. M.; & Ibrahim, Z. Z. Phytoremediation Potential of Vetiver Grass (*Vetiveria zizanioides*) for Treatment of Metal-Contaminated Water. Water, Air, & Soil Pollution. Ahmadpour 2017; 228, 158. DOI: 10.1007/s11270-017-3349-x

[13] Banerjee, R.; Goswami, P.; Pathak, K.; & Mukherjee, A. Vetiver grass: An environment clean-up tool for heavy metal contaminated iron ore mine-soil. Ecological Engineering. 2016; 90, 25-34. DOI: 10.1016/j.ecoleng.2016.01.027

[14] Valderrama, A.; Tapia, J.; Peñailillo, P.; & Carvajal, D. E. Water phytoremediation of cadmium and copper using *Azolla filiculoides* Lam. in a hydroponic system. Water and Environment Journal. 2013; 27, 293-300. DOI: 10.1111/wej.12015

[15] Zhou, C.; Huang, M.; Ren, H.; Yu, J.; Wu, J.; & Ma, X. Bioaccumulation and detoxification mechanisms for lead uptake identified in *Rhus chinensis* Mill. seedlings. Ecotoxicology and Environmental Safety. 2017; 142, 59-68. DOI: 10.1016/j.ecoenv.2017.03.052

[16] Khaokaew, S.; & Landrot, G. A field-scale study of cadmium phytoremediation in a contaminated agricultural soil at Mae Sot District, Tak Province, Thailand: (1) Determination of Cd-hyperaccumulating plants. Chemosphere. 2015; 138, 883-887. DOI: 10.1016/j.chemosphere.2014.09.108

[17] Truong, P., Tran Tan Van and Pinners. E. Vetiver System for the Prevention and Treatment of Contaminated Water and Land (Special Reference to Domestic and Municipal Wastewater Treatment in Australia). Extended Abstract. Ethiopian National Workshop. Addis Abba, March 2009.

[18] Truong, P., Van, T. T., & Pinners, E. Vetiver System Applications: A Technical Reference Manual. The Vetiver Network International, 2008; 107.

[19] Méndez-Cano M, López-Martínez S, Lagunas-Rivera S, González-Mondragón E.G., Rodríguez-Luna A.R., López – Hernández E. S. Evidence of phytoremediation potential in the poacea family and its efficiency characteristics family poaceae as a tool for remediation. Sylwan. 2019; 163:4, 1-20.

[20] Wu, Z., Leung, A. K., Boldrin, D., & Ganesan, S. P. Variability in root biomechanics of Chrysopogon zizanioides for soil eco-engineering solutions. Science of The Total Environment, 2021; 776, 145943.

[21] Vinayagam, M., Babu, R. S., Sivasamy, A., & de Barros, A. L. F. Biomass-derived porous activated carbon from Syzygium cumini fruit shells and Chrysopogon zizanioides roots for high-energy density symmetric supercapacitors. Biomass and Bioenergy, 2020; 143, 105838.

[22] Babji, R., Reddy, U., & Shakthivel, S. Characteristic Investigation and Comparison between Vetiver fiber-reinforced polypropylene and polyethylene with Coconut shell powder and Maleic anhydride as filler and coupling agents. Materials Today: Proceedings, 2020; 24, 2339-2351.

[23] Leknoi, U., & Likitlersuang, S. Good practice and lesson learned in promoting vetiver as solution for slope stabilisation and erosion control in Thailand. Land Use Policy, 2020; 99, 105008.

[24] Davamani, V., Parameswari, C. I., Arulmani, S., John, J. E., & Poornima, R. Hydroponic phytoremediation of paperboard mill wastewater by using vetiver (Chrysopogon zizanioides). Journal of Environmental Chemical Engineering, 2021; 105528.

[25] Darajeh, N.; Idris, A.; Truong, P.; Abdul Aziz, A.; Abu Bakar, R.; & Che Man, H. Phytoremediation potential of vetiver system technology for improving the quality of palm oil mill effluent. Advances in Materials Science and Engineering. 2014; 1-10. DOI: 10.1155/2014/683579

[26] Singh, V.; Thakur, L.; & Mondal, P. Removal of lead and chromium from synthetic wastewater using *Vetiveria zizanioides*. CLEAN–Soil, Air, Water. 2015; 43, 538-543. DOI: 10.1002/clen.201300578

[27] Truong, P.; & Danh, L. T. The vetiver system for improving water quality. The Vetiver Network International, San Antonio, TX, USA. 2015.

[28] Ghosh, M.; Paul, J.; Jana, A.; De, A.; & Mukherjee, A. Use of the grass, *Vetiveria zizanioides* (L.) Nash for detoxification and phytoremediation of soils contaminated with fly ash from thermal power plants. Ecological Engineering. 2015; 74, 258-265. DOI: 10.1016/j.ecoleng.2014.10.011

[29] Badejo, A. A.; Omole, D. O.; Ndambuki, J. M.; & Kupolati, W. K. Municipal wastewater treatment using sequential activated sludge reactor and vegetated submerged bed constructed wetland planted with *Vetiveria zizanioides*. Ecological Engineering. 2017; 99, 525-529. DOI: 10.1016/j.ecoleng.2016.11.012

[30] Fasani, E.; DalCorso, G.; Zerminiani, A.; Ferrarese, A.; Campostrini, P.; & Furini, A. Phytoremediatory efficiency of *Chrysopogon zizanioides* in the treatment of landfill leachate: a case study. Environmental Science and Pollution Research. 2019; 26(10), 10057-10069. DOI: 10.1007/s11356-019-04505-7

[31] Truong, P. Vetiver system for environmental protection. Vetican Consulting, Brisbane. 2019.

[32] Cruz López, C. A. D. L.; Ramos Arcos, S. A.; & López Martínez, S. Efecto de la adición de ácidos orgánicos sobre la bioacumulación de Plomo, Talio y Vanadio en *Chrysopogon zizanioides* creciendo sobre suelos contaminados de un relleno sanitario. Nova scientia. 2018; 10(21), 403-422. DOI: 10.21640/ns.v10i21.1582

[33] Ogunfowokan, A. O.; Adekunle, A. S.; Oyebode, B. A.; Oyekunle, J. A. O.; Komolafe, A. O.; & Omoniyi-Esan, G. O. Determination of heavy metals in urine of patients and tissue of corpses by atomic absorption spectroscopy. Chemistry Africa. 2019; 2(4), 699-712. DOI: 10.1007/s42250-019-00073-y

[34] Paul, V.; Pandey, R.; Ramesh, K. V.; & Meena, R. C. Atomic Absorption Spectroscopy (AAS) for Elemental Analysis of Plant Samples. Manual of ICAR Sponsored Training Programme for Technical Staff of ICAR Institutes on "Physiological Techniques to Analyze the Impact of Climate Change on Crop Plants". 2017; 84.

[35] Roongtanakiat, N.; Tangruangkiat, S.; Meesat, R. Utilization of vetiver grass (*Vetiveria zizanioides*) for removal of heavy metals from industrial wastewaters. Science Asia. 2007; 33, 397-403. DOI: 10.2306/scienceasia1513-1874.2007.33.397

[36] Datta, R.; Quispe, M. A.; Sarkar, D. Greenhouse study on the phytoremediation potential of vetiver grass, *Chrysopogon zizanioides* L., in arsenic-contaminated soils. Bulletin of environmental contamination and toxicology. 2011; 86, 124-128. DOI: 10.1007/s00128-010-0185-8

[37] Santos, H. M.; Coutinho, J. P.; Amorim, F. A. C.; Lôbo, I. P.; Moreira, L. S.; Nascimento, M. M.; & de Jesus, R. M. Microwave-assisted digestion using diluted HNO_3 and H_2O_2 for macro and microelements determination in guarana samples by ICP-OES. Food Chemistry. 2019; 273, 159-165. DOI: 10.1016/j.foodchem.2017.12.074

[38] Torres Rodríguez, D.; Cumana, A.; Torrealba, O.; & Posada, D. Uso del vetiver para la fitorremediación de cromo en lodos residuales de una tenería. Revista Mexicana de Ciencias Agrícolas. 2010; 1, 175-188.

[39] Cedeño Ochoa, C. J. Metales en agua por Plasma Acoplado por Inducción (Cd, Cr, Cu, Ni, Pb, Zn). Instituto de Hidrología, Meteorología y Estudios

Ambientales. Ministerio de Ambiente, Vivienda y Desarrollo Territorial-República de Colombia. 2006.

[40] Xin, J.; & Huang, B. Comparison of boron uptake, translocation, and accumulation in reed, cattail, and vetiver: an extremely boron-tolerant plant, vetiver. Plant and Soil. 2017; 416, 17-25. DOI: 10.1007/s11104-017-3186-0

[41] Antiochia, R.; Campanella, L.; Ghezzi, P.; & Movassaghi, K. The use of vetiver for remediation of heavy metal soil contamination. Analytical and bioanalytical chemistry. 2007, 388, 947-956. DOI: 10.1007/s00216-007-1268-1

[42] Ibezute, A. C.; Tawari-Fufeyin, P.; & Oghama, O. E. Analysis of pollution removal from compost leachate by vetiver grass (L.) Nash plant (*Vetiveria zizanioides*). Resources and Environment. 2014; 4, 268-273.

[43] Benavides Montoya, A. Problemática y alternativas tecnológicas para la remoción de arsénico en la obtención de agua potable. Curso impartido en el Centro de Investigación de Materiales Avanzados. México. 2013.

[44] Hansen, T. H.; Laursen, K. H.; Persson, D. P.; Pedas, P.; Husted, S.; & Schjoerring, J. K. Micro-scaled high-throughput digestion of plant tissue samples for multi-elemental analysis. Plant Methods. 2009; 5, 12. DOI: 10.1186/1746-4811-5-12

[45] Pazoki, M.; Abdoli, M. A.; Karbassi, A.; Mehrdadi, N.; & Yaghmaeian, K. Attenuation of municipal landfill leachate through land treatment. Journal of Environmental Health Science and Engineering. 2014; 12(1), 12. DOI: 10.1186/2052-336X-12-12

[46] Kočevar Glavač, N.; Djogo, S.; Ražić, S.; Kreft, S.; & Veber, M. Accumulation of heavy metals from soil in medicinal plants. Arhiv za Higijenu Rada i Toksikologiju. 2017; 68, 3, 236-244. DOI: 10.1515/aiht-2017-68-2990

[47] Borowczak, M.; & Hołtra, A. The content of heavy metals in soils and plants around the waste landfill in Siechnice (Poland). In E3S Web of Conferences 2017. Vol. 17. EDP Sciences. DOI: 10.1051/e3sconf/20171700009

[48] Xia, H.; Liu, S.; & Ao, H. A study on purification and uptake of garbage leachate by vetiver grass. In: Proc. of the 2nd International Conference on Vetiver, Thailand. 2000.

[49] Abaga, N. O. Z.; Dousset, S.; Munier-Lamy, C.; & Billet, D. Effectiveness of vetiver grass (*Vetiveria zizanioides* L. Nash) for phytoremediation of endosulfan in two cotton soils from Burkina Faso. International Journal of Phytoremediation. 2014; 16, 95-108. DOI: 10.1080/15226514.2012.759531

[50] Kong, X.; Lin, W.; Wang, B.; & Luo, F. Study on vetiver's purification for wastewater from pig farm. In Proc. Third International Vetiver Conference, Guangzhou, China. 2003.

Section 3

Bioremediation Technology

Chapter 15

Fungal Deterioration of Cultural Heritage Objects

Željko Savković, Miloš Stupar, Nikola Unković, Aleksandar Knežević, Jelena Vukojević and Milica Ljaljević Grbić

Abstract

Significant percent of world cultural heritage artifacts is threatened by fungal infestation. Fungi can deteriorate different substrates via various physical and chemical mechanisms. Hyphal growth and penetration into the substrate can cause symptoms like discoloration, biopitting, cracking, exfoliation and patina formation. On the other hand, chemical mechanisms include acid secretion, release of extracellular enzymes, pigment production, oxidation/reduction reactions and secondary mycogenic minerals formation. These processes can lead to serious, both esthetic and structural, alterations which may be irreversible and could permanently impair artworks. Proper isolation and identification of autochthonous isolates, as well as employment of different microscopic techniques and *in vitro* biodegradation tests are pivotal in understanding complex biodeterioration mechanisms caused by microorganisms, including fungal deteriogens. Biodeterioration and biodegradation studies require multidisciplinary approach and close collaboration of microbiologists, chemists, geologists and different personnel responsible for the safeguarding of cultural heritage monuments and artifacts, especially restorers and conservators.

Keywords: alterations, biodegradation, cultural heritage, fungi, multidisciplinary research

1. Introduction

Ars longa, vita brevis – states the ancient Roman proverb, emphasizing that human need for artistic expression is as old as the civilization itself. Unfortunately, extant artworks are only a fragment of humanity's creations throughout history. Along with artistic creation, there is a need for protection of the artwork from external, frequently damaging influences. Since works of art are an essential part of the cultural heritage legacy of every nation, they ought to be protected for future generations. Biodeterioration is defined as any undesired alteration of the property of the material which is caused by living organisms and cultural heritage objects are frequently prone to this process [1]. Mentioned alterations can be induced by both macroorganisms (plants and animals) and microorganisms (bacteria, algae and fungi). Inadequate storage and irregular maintenance of artifacts in archives, museums and depots oftentimes favorize microbial, especially fungal, proliferation [2]. Since fungi are ubiquitous organisms, with pronounced metabolic activities, they are capable of colonizing various types of microenvironments therefore constantly causing problems in cultural heritage collections around the world [3].

Fungal propagules - spores and mycelial fragments, are always present in the air, their concentrations being dependent on environmental factors [4, 5]. Namely, during their life cycle, fungi produce various types of sexual and asexual spores which are actively or passively released into the surrounding environment and dispersed by air currents to available substrates [3]. The successful colonization of available substrates requires propagules to be viable in addition to favorable growth conditions [4, 6]. It is known that due to their metabolic activities, numerous fungal species could cause both esthetic and physical damage to a variety of substrates, including stone, paint, paper, wood, textile and other materials of which cultural heritage artworks are made. Therefore, the application of adequate microscopic techniques, proper species identification and physiological characterization of autochthonous isolates are very important to appropriately assess potential threats to cultural heritage artworks, especially on those stored in inadequate conditions [3]. Consequently, biodeterioration and biodegradation studies require a multidisciplinary approach and a close collaboration of scientists (microbiologists, chemists, geologists etc.) and the specialists responsible for the safeguarding of cultural heritage objects, such as restorers and conservators. Therefore, this work addresses general mechanisms of biodeterioration caused by fungi and their role in the deterioration of different materials which constitute cultural heritage artworks.

2. Biodeterioration mechanisms

Fungi present on artworks can affect them in two ways – mechanically and chemically. The aforementioned processes, more often than not, are taking place simultaneously. Depending on the substrate's nature, exogenic and endogenic factors, the effect of one process can prove more prominent than the other [7, 8]. Notably, depending on its location, fungal colonizers can affect the substrate in two ways – from the surface to its interior and *vice versa* [7].

2.1 Physical processes

Physical processes are taking place under the influence of hyphal apical growth or by the formation of fruiting bodies on the surface and/or the inner layers of the colonized material. If the fungal growth is superficial, it results in the formation of spreading mycelium which covers the substrate and changes the original appearance, hence the esthetic value of the artifact [7]. Inner fungal growth might lead to further damage of the artworks and, especially if paintings are concerned, to the detachment of painted layers (exfoliation). Melanized micromycetes are well known inducers of mechanical deterioration, especially of stone substrates, since melanin provides mechanical rigidness to fungal structures, enhances the turgor pressure and facilitates hyphal penetration [8, 9]. In order to study mechanical deterioration, the application of different microscopic techniques is pivotal, especially *in situ* optical microscopy and scanning electron microscopy (**Figure 1**). The multimicroscopic approach is essential to ensure detailed information, not only about the deterioration status, but also to elucidate alterations that affect works of art, and to detect potential biodeterioration "culprits" [10].

2.2 Chemical processes

Mechanisms of chemical biodeterioration are much more complex and prominent than physical ones. Fungi can chemically alter the substrate via assimilation

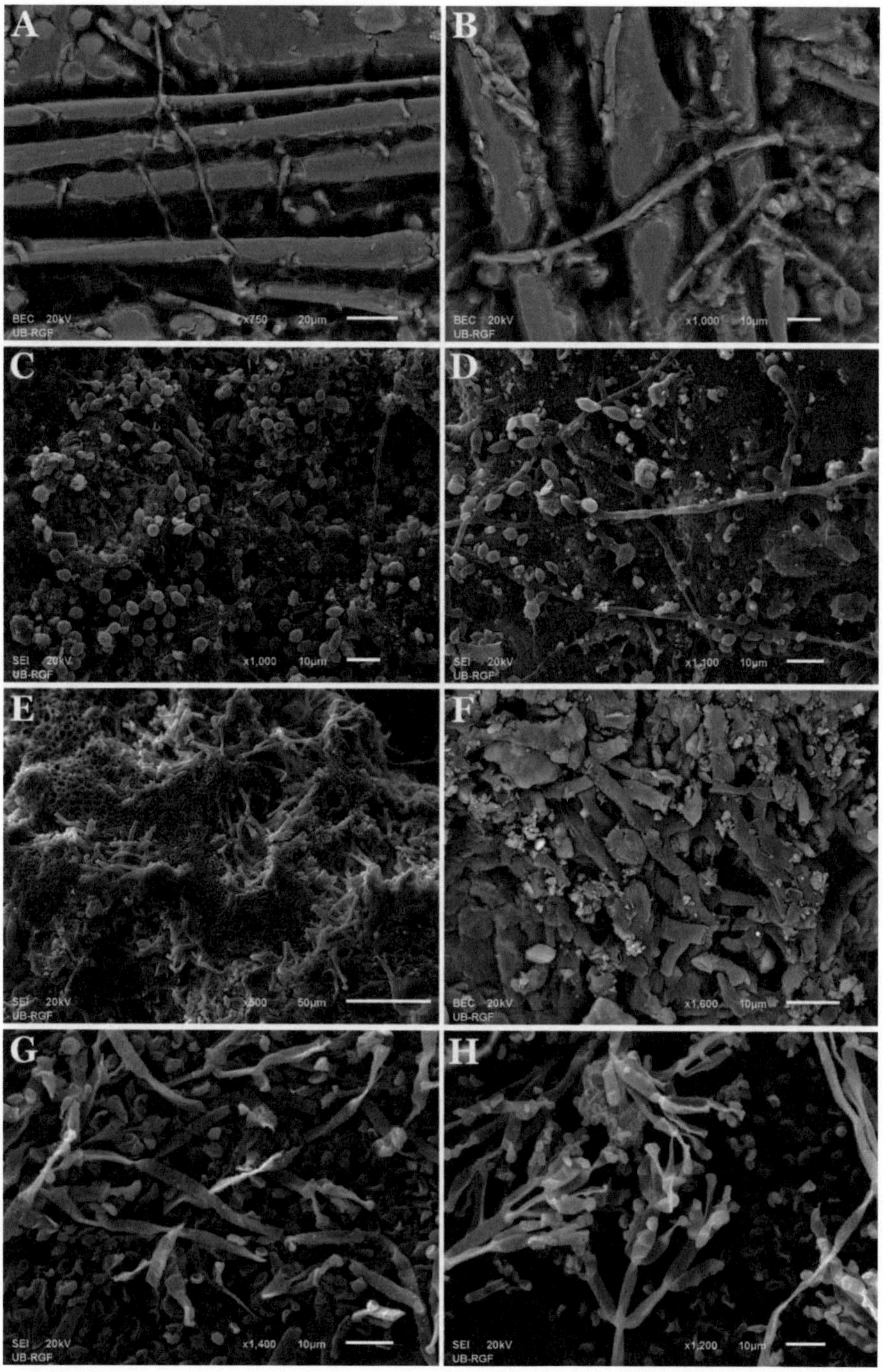

Figure 1.
Scanning electron micrographs depicting deteriorated surfaces of cultural heritage objects: A, B. deteriorated icon silk fibers with cracks and gaps formed by Chaetomium globosum hyphal growth; C, D. profusion of Cladosporium sp. hyphal network with conidial mass on deteriorated wall painting; E, F. Mycobiont of lichenized fungi from the surface of deteriorated limestone monument; G, H. anamorphic state of Xylariaceae infesting wooden iconostasis.

and dissimilation processes [11]. In case of the former, fungi utilize nutrients from the substrate by secreting various enzymes which catalyze the macromolecules' degradation. In contrast, dissimilation represents the production of various extracellular metabolites such as organic acids and pigments. These substances modify or damage the colonized substrate. Since hyphae have high surface to volume ratio, these metabolites can quickly diffuse between the cells as well as from the cell into the substrate [7, 8]. Nowadays, various microbiological, biochemical and petrographical tests are employed to study chemical biodeterioration. *In vitro* tests provide rapid, cost effective estimation of fungal degradation capacity, which helps in evaluating a potential risk to cultural heritage artifacts [3].

Acid production and acidolysis are the most studied biodeterioration mechanisms, particularly on inorganic materials [8, 12]. Due to their metabolic activities, fungi produce organic acids such as gluconic, citric, oxalic, malic, succinic, itaconic etc. [13]. Once the spore germination occurs, organic acids are produced by respiration in mitochondria as intermendiary products of the citric acid cycle. If the fungi grow on nutrient enriched substrates, these acids are formed in excess and excreted as secondary metabolites [6]. The secreted acids then react with different substances via cation solubilization and chelation reactions. The reaction of acids with different metals (i.e. K, Fe and Mn) results in the formation of organic salts and complex compounds [7]. It should be mentioned that many organic acids, especially the oxalic, are able to chelate different metals in the process called complexolysis. The oxalic acid is able to form complexes with diverse metals (Ca, Mg, Fe, Cu and others), consequently leading to secondary mycogenic minerals formation, calcium oxalate being the most well-known [14]. The aforementioned crystals are present in patinas on stone, frescoes, oil paintings, glass, wood and other materials [8]. It is ascertained that most of the fungi have the ability, in greater or lesser extent, to produce oxalic acid, and subsequently precipitate oxalates [3]. Furthermore, CO_2, as a product of respiration, in the conditions of increased humidity is transformed into carbonic acid, which then solubilizes calcium carbonate and magnesium carbonate present in limestone, mortar and gypsum. As a result, water soluble bicarbonates are formed. Additionally, increased H^+ concentration favorizes the colonization of acidofilic fungi, which further facilitates the biodeterioration process [7, 8].

Enzymes. Fungi are able to digest organic matter, altering and weakening those materials, by the action of extracellular hydrolytic enzymes, such as lignocellulases, proteases, lipases, pectinases, chitinases, etc. [15, 16]. Enzymes that convert large, complex and often water-insoluble compounds (cellulose, hemicellulose, lignin, proteins and lipids) into low-molecular-weight soluble compounds, play an important role in the biodeterioration and biodegradation processes [7]. Although filamentous fungi primarily use simple sugars as a carbon source, they can be producers of lignocellulolytic enzymes to depolymerize wood or cellulose material for nutritional purposes [17]. Cellulolytic enzyme complex, which is responsible for degradation of cellulose to glucose monomers, comprises of: endoglucanase (hydrolyzes β-1,4-glycosidic bonds within cellulose fibers), exoglucanase (hydrolyzes β-glycosidic bonds and remove cellobiose units from the free ends of chains) and β-glucosidase (hydrolyzes cellobiose and cellodextrin to glucose) [18, 19]. Hemicellulases hydrolyze hemicellulose, which is made up of hexoses (mannose, glucose, galactose) and pentoses (xylose, arabinose) to monomeric sugars and acetic acid. The complex of enzymes that hydrolyze hemicellulose consists of at least eight enzymes: endo-1,4-β-D xylanase, exo-1,4-β-D xylocuronidase, α-L arabinofuranosidase, endo-1,4-β- D mananase, β-mannosidase, acetyl-acid esterase, α-glucuronidase and α-galactosidase [19]. Lignin degradation, characteristic for white-rot fungi, is catalyzed by nonspecific polyphenol oxidases: manganese oxidizing peroxidases, lignin peroxidases, and laccase. This process involves breaking inter-monomer bonds, demethylation, hydroxylation, side chain modification, and aromatic ring cleavage [20, 21]. Some fungal species inhabit art objects that are substrates rich in fibrillar proteins (wool, parchment, leather, silk, etc.). Proteolytic enzymes (proteases) degrade various protein fibers such as collagen (wool), fibroin (silk) and keratin (parchment) [11]. Lipases catalyze the hydrolysis of triacylglycerols to glycerol and fatty acids. These enzymes might take part in the degradation of widely used painting constituents, linseed and shellac, derived primarily from unsaturated oleic, linoleic and linolenic acids [22].

Pigment production. Micromycetes produce pigments which vary in chemical composition and color and are species specific [7, 8]. They are present in hyphae, conidia, or are secreted into the substrate whilst their production is determined by the availability of nutrients and minerals, UV radiation, pH, temperature and other environmental factors [6]. Pigment secretion on/into the substrate leads to the appearance of different, frequently irreversible, colorations leading to the observable changes on cultural heritage objects. This diminishes the aesthetic value of the artwork and accelerates biodeterioration process [7, 11, 23]. Many fungi produce dark colored pigments – melanins, which are responsible for characteristic brownish color of mycelia and reproductive structures. These water insoluble, very stable and resistant, molecules are formed by oxidative polymerization of phenolic compounds (ortho-dihydroxy phenols) [24]. Dark colored melanins are characteristic for dematiaceous fungi (representatives of the former family Deamtiaceae) and are present in cell walls in both granular of amorphic form, while their amount increases with aging [6, 25]. On the other hand, the secretion of various water soluble exopigments into the substrate can also cause different aesthetic damage to artworks, especially on those made from organic materials. A vast number of different pigments are identified and grouped in three main families: the derivatives of toluquinone, of naphthoquinone and of beta-methyl-quinone [8].

3. Stone artifacts

A significant percentage of world cultural heritage objects is represented by stone artworks, such as architectural monuments, statues and tombstones, to name just a few [26]. Stone is considered an extreme environment for microbial growth and proliferation, mostly due to the intensive oscillations of diurnal and seasonal microclimatic parameters and low nutrient and water content [27]. Stone surfaces directly exposed to sunlight could achieve temperatures above 60°C, while being simultaneously susceptible to freezing–thawing cycles [28, 29]. Regardless, certain groups of microorganisms, the so called lithobionts, are able to colonize such environments. Primary colonizers are photoautothrophic organisms – cyanobacteria, algae and lichens, while hemolithotrophic and hemoorganotrophic bacteria and fungi are considered secondary. The latter are oligotrophic or poikilotrophic organisms which adapted to survive and grow in harsh or variable environmental conditions [29].

Microbial dwellers of stone surfaces frequently form biofilm, highly structurised microbial consortium embedded in a mutual extracellular matrix. Biofilm formation starts with unspecific, reversible interactions, followed by stable interactions which are initiated as a consequence of the formation of specific molecules and structures (lipopolysaccharides, membrane proteins, flagellae). After initial adhesion, extracellular polymeric substances are produced and excreted, enhancing the adhesion and cohesion of cells [8]. Evolutionary advantages of biofilm are to provide protection, resistance to physical and chemical stressors, metabolic cooperation and mutually regulated gene expression [30]. All groups of microbial dwellers of stone are characterized with a high phenotypic plasticity, which is reflected in polymorphism – the change of growth or sporulation forms with regard to external conditions. Therefore, micromycetes, which colonize this environment, are able to form different somatic and reproductive structures - sclerotia, chlamidospores, conidial clusters, perithecia and pycnidia [29, 31]. Moreover, stone interior is a specific microenvironment for the growth of certain microorganisms such are endolithic fungi. The microclimatic conditions in this environment are a little more favorable – water retention is higher and solar radiation and air current intensities are lower [27].

Among fungal colonizers of stone, micorcolonial fungi constitute a specific ecological group characterized by its slow growth and formation of irregular shaped cells, often packed in aggregates. They rarely form specialized reproductive structures and in turn, by active growth, form thick, pigmented cell walls which enable transition to the state of dormancy during prolonged, unfavorable environmental conditions. The ability to secrete extracellular polymeric substances and thick cell walls, enable water retention, nutrient absorption, desiccation reduction and cellular adhesion/cohesion. These organisms are able to survive for long periods without metabolic activities and their metabolic rates are low even during optimal environmental conditions [27].

Dematiaceous fungi are considered the most important agents of stone deterioration. All representatives intensively produce dark colored melanins which provide protection from excessive environmental radiation (UV radiation, x- and γ-rays) and chemical stressors. This group encompasses microcolonial fungi and black yeasts, species of genera: *Acrodictys*, *Aureobasidium*, *Capnobotryella*, *Coniosporium*, *Exophiala*, *Hormonema*, *Hortaea*, *Knuffia*, *Lichenothelia*, *Monodictys*, *Phaeococcus*, *Phaeococcomyces*, *Phaeosclera*, *Sarcinomyces* and *Trimmatostroma*, which are very important stone colonizers in arid and semiarid environments [27, 32]. Additionally, deatiaceous fungi include cosmopolitan filamentous species of genera *Alternaria*, *Cladosporium*, *Ulocladium*, *Epicoccum* etc. which are main colonizers of stone in more favorable conditions of temperate and humid environments [32]. They are especially important deteriogens of restored stone artifacts [27].

Fungi can deteriorate stone via physical and chemical mechanisms. Physical mechanisms include hyphal penetration of the rock surface which causes its fragmentation, while chemical ones include secretion of acidic metabolites and pigments and oxidation of mineral forming cations. Although many microorganisms are able to produce acids, fungi are considered as the most potent ones in nature that degrade rocks and minerals. The production of various acidic metabolites leads to the biocorrosion - dissolution of the mineral substrate, resulting in the formation of various secondary mycogenic minerals [26]. *In vitro* acid production and formation of calcium oxalate and calcium carbonate minerals have been reported by autochthonous isolates from limestone monuments such as ancient Roman stela and Portuguese king tomb [26, 33]. Synthesis and excretion of extracellular pigments mostly affect the stone aesthetically, although studies concerning pigment production on stone monuments are generally scarce [26]. Due to the mentioned processes, symptoms such as biopitting, biogenic patina and colorations can occur [11]. The growth of dematiaceous fungi results in the presence of dark stains while microcolonial fungi are the main culprits of biopitting phenomena on limestone and marble artworks, [32, 34]. Sanctuary of Delos in Greece is an example of mentioned biopitting phenomena [35].

Lichenized fungi are important colonizers of stone substrates. These organisms have a high tolerance to variations of environmental factors, especially temperature, insolation and water availability, which is responsible for their cosmopolitan distribution and ability to colonize extreme environments [36]. These organisms are poikilohydric, i.e. they are capable of enduring cycles of desiccation and rehydration due to their ability of lowering their metabolic rate and enter cryptobiosis under conditions of low water availability [37]. Endolithic lichens are of special importance to stone deterioration, since they are capable of the deepest penetration into the stone compared to other microorganisms [38]. Apart from the fact that hyphae of the mycobiont can penetrate the rock surface (**Figure 1E** and **F**), perithecia formation by some endolithic species can penetrate the surface from the inside out, which leads to biopitting [27]. Additionally, lichen growth could cause exfoliations, encrustations and disaggregation of the stone surface [11, 39]. Conversely, lichen morphology and its adhesive capability aren't always in correlation with its capacity to alter the substrate and physiological differences between the species are considered to be more significant [40, 41]. In fact, synthesis of different chemical deterioration agents is done by the mycobiont. Apart from carboxylic, lichens have the ability to produce lichen acids, semisoluble polyphenolic compounds which are able to form complexes with metal cations. The capability of lichens to absorb and maintain water enhances the duration of chemical reactions and therefore facilitates deterioration process [39, 41]. Lastly, some authors have reported the presence of orange-brownish patinas (*scialbatura*) on stone monuments made from limestone and marble. These colorations mainly consist of calcium carbonate and calcium oxalate minerals, sometimes intermixed with fragmented lichen thalli [42, 43]. Although this symptom is associated with deterioration, it is hypothesized that it may have a protective role to the monuments [27].

4. Wall paintings

Wall painting, as the pictorial technique, encompasses all painting techniques aimed at beautifying wall surfaces. There is no universally accepted definition of fresco, as well as consensus on what techniques can be included in this type of wall painting, however, the term *al fresco* generally refers to paintings made on a fresh lime mortar with mineral pigments mixed with water [44]. On the contrary, in *fresco a secco* or *al secco* technique painting is done on a dry plaster with paints prepared by mixing mineral pigments with various organic binders [32, 45]. Although hallmarked as an extreme type of habitat, painted layer and lime mortar are also considered to be very suitable and bioreceptive substrates for fungal growth. This is due to the mineral composition and porous nature of lime mortar, and the fact that organic and inorganic components of the painted layer represent a suitable niche for the development of a wide range of heterotrophic microorganisms [46].

Fungal infestation of wall paintings can occur from several sources including contaminated indoor air as the main, but also soil and plants of immediate vicinity, visitors, contaminated conservation tools, and indoor hotspots as secondary sources [7]. Whether a certain fungus will be able to colonize the painted layer or mortar depends on the ecological and physiological requirements of a given species. If the requirements are met the process is further controlled by three main factors: nutrient availability, relative humidity, and temperature [32]. The origin of nutrients in fresco painting is related to (1) additives (chaff, wheat paste, barley flakes, animal hair, hemp and flax fibers, egg whites, oils, fats) mixed with mineral and complex fillers of chopped straw and lime mortar; (2) additives used in the preparation of mortar (liquid resins, tar, polymer latex, emulsions, bitumen, milk, olive and linseed oil, lard, animal blood); (3) binders of plant and animal origin mixed with mineral pigments; (4) casein, paraloid mixtures, fixatives and consolidants based on polymer components (cellulose acetates, polyvinyl acetate, polymethyl acrylate, etc.) used in restoration works [32, 45, 47–50]. These organic components determine the richness of the fresco mycobiota. Since the composition of the painted layer and mortar is predominated by inorganic components, its mycobiota differs greatly from the fungal communities established on other painted works of art [45]. Furthermore, heterogeneously pigmented zones of the painted layer can be considered as selective substrates that condition the development of a specific mycobiota [51]. Using culture-dependent methods, the most commonly documented fungi on painted layer and mortar of wall paintings are Ascomycota of genera *Acremonium*, *Alternaria*, *Arthrinium*, *Aspergillus*, *Aureobasidium*, *Beauveria*, *Botrytis*, *Chaetomium*, *Chrysosporium*, *Cladosporium*, *Curvularia*, *Dreschlera*, *Engyodontium*, *Epicoccum*, *Eurotium*, *Exophiala*, *Fusarium*, *Geomyces*, *Gliomastix*, *Phoma*, *Penicillium*, *Scopulariopsis*, *Sepedonium*, *Sporotrichum*, *Stachybotrys*, *Stemphylium*, *Trichoderma*, *Trichotecium*, *Ulocladium* and *Verticillium* [7, 45, 52–54]. Contamination by fungi from phyla Basidiomycota is rare (e.g. *Coprinus* spp.), while Zygomycota of genera *Mucor* and *Rhizopus* are isolated frequently but are considered only surface contaminants [45, 55].

Species of the genera *Aspergillus*, *Aureobasidium*, *Alternaria*, *Cladosporium* and *Penicillium*, are frequently listed as the most common wall paintings contaminants, as well as the primary fresco deterioration agents in temperate climates [56]. Many *Cladosporium* species are recognized as the main biological agents in the process of biodeterioration of wall paintings since they are able to not only induce brown discolorations, but also penetrate through the entire painted layer all the way to the mortar support [57] (**Figure 1C** and **D**). Many species of the genus *Penicillium* are known to develop and intensely sporulate in a period of only few days to a few weeks on periodically moist fresco paintings [58]. Furthermore, in addition to dominant members of fungal community, species from less represented genera can also significantly contribute to the damage of wall paintings. For example, isolation of *Phoma* species from the surface of painted layer indicates that given wall paintings are in an active process of decay [59]. *Chaetomium*, *Aureobasidium* and *Epicoccum* species, due to strong proteolytic activity, degrade protein binders of the painted layer, which results in the lifting and separation of the painted layer from the support. Likewise, it has been recently contemplated that species of the genera *Mucor* and *Rhizopus* might be more involved in process of biodeterioration of wall paintings than originally considered, as it was shown in in vitro experiments that they are able to degrade protein binders and epoxy resins [60].

Mechanical (1) and chemical (2) activity of fungi directly results in damages to structural and esthetic integrity of fresco paintings. (1) Hyphal penetration, together with formation of fruiting bodies and various modifications of mycelium, increases internal pressure thereby forming new cracks in the painted layer and mortar, as well as expanding the existing ones. Damages caused by mechanical activity are considered by some to be of greater importance compared to changes induced by environmental factors and fungal chemical activity [61]. Furthermore, aside from mechanical activity, damages as the result of change in substrate properties can also incur due to utilizing fresco components as a source of nutrients for fungal growth (2′1) and/or due to secretion and interaction of fungal metabolites with organic and inorganic components of the painted layer and mortar (2′2) [7, 62]. (2′1) Extracellular enzymes break down complex organic components into simpler molecules enabling their absorption and easier penetration of hyphae into the substrate which results in cracking and peeling of the painted layer and mortar. The main enzymes involved in this process are β-glucosidase, phosphatase, lipase, arylsulfatase, esterase, protease and endo-N-acetyl-PD-glucosamidase [63, 64]. (2′2) Excreted organic acids chelate metal ions present in mineral pigments and mortar, resulting in the formation of mineral salts and complex compounds that increase pressure in pores, which leads to cracking, peeling, and loss of fragments of the painted layer and mortar [65, 66]. Additionally, salts stimulate formation of surface irregularities that serve as suitable sites for the settlement of heterotrophic microorganisms, thereby increasing the bioreceptivity of fresco painting [67]. In these circumstances, there is an uncontrolled biofilm development and acceleration of chemical dissimilation activity through oxidation, reduction and transformation of metal ions in pigments, primarily Fe and Mn, but also As, Pb, Cu, Zn and Hg, resulting in the alterations to the original color of the painted layer [53, 68]. Aside from organic acids, very stable and persistent fungal pigments (melanins, mycosporins, quinones, hydroxyanthraquinones and carotenoids) secreted onto the surface induce changes in the original coloration, which is process that depends on the chemical composition of the pigment, environmental conditions and interactions with substrate components [7, 32].

5. Canvas oil paintings

One of the best-known pictorial technique today, oil painting on canvas, emerged in the Middle Ages and has since been one of the most important art expressions, constituting outstanding works of art with important historic and cultural value [69]. Structurally speaking, these works of art are composed of the pictorial layer between the protective covering varnish and the ground (or preparatory layer) spread on a linen canvas. Compared to the other forms of artwork, oil paintings on canvas possibly provide the widest range of microhabitats and nutrients that may be exploited by a large variety of microbial species [70]. Materials that constitute the painting, i.e. the cellulose of the canvas support, organic adhesives (i.e. various animal, fish and plant glues) used in sizing the support, natural varnishes, and the oils used in binding the pigments (linseed, turpentine and other oils) are all composed of organic molecules of high nutritional content that are all easily degraded [32, 63]. These organic molecules encompass sugars, gums, and other polysaccharides, proteins and waxes, but also less chemically defined mixtures of biomolecules, such as egg yolk, bile, and urine, as well [45]. Organic glue pastes used to coat the back of paintings with linen canvas, i.e. "re-lining", may also represent a rich nutrient source [71]. Furthermore, dirt, dust and other environmental contaminants (volatile hydrocarbons released from machinery, respiration and cigarette smoke) deposited on the surface of the oil paintings provide nutrients as well [63].

Given the wide range of organic molecules that are present in oil paintings, many different microorganisms may grow provided inadequate storage and favorable environmental conditions, primarily high relative humidity and temperature, are met [45]. These specific environmental conditions may start and/or accelerate the microbial growth, which otherwise would persist on the obverse and the reverse side of the painting in a dormant metabolic state [70]. Among multitudes of different microorganisms, fungi are notorious for their ability to inhabit and decay paintings due to their enormous metabolic activity and ability to grow at low a_w values [32]. However, to the best of our knowledge, despite being one of the most numerous objects exhibited and stored in museums and warehouses worldwide, relatively fewer number of studies to date have been engaged in describing the fungal communities dwelling on canvas oil paintings compared to the other forms of art. Using culture-dependent methods, the most commonly documented fungi on painted surface, canvas and wooden frame are Ascomycota of genera *Alternaria*, *Aspergillus*, *Aureobasidium*, *Botrytis*, *Cladosporium*, *Drechslera*, *Epicoccum*, *Fusarium*, *Penicillium*, *Ulocladium*, *Scopulariopsis*, *Stemphylium*, *Trichoderma* and *Wardomyces*, with occasional observations of teleomorphs from *Chaetomium*, *Emericella* and *Eurotium* genera [32, 63, 69, 70, 72, 73]. Review of the above referenced literature data has highlighted fungi that were most frequently isolated from the infested oil paintings: *Alternaria alternata*, *Aspergillus flavus*, *A. niger*, *A. versicolor*, *Aureobasidium pullulans*, *Chaetomium globosum*, *Cladosporium cladosporioides*, *Eurotium chevalieri*, and *Penicillium chrysogenum*. Aside from the Ascomycota, the only other documented species are Zygomycota of *Cunninghamella*, *Mucor* and

Rhizopus genera, but they are in most cases regarded as transients only, i.e. part of the surface dust deposits, and not partaking in the complex process of biodeterioration [32, 74]. Air mycobiota of the depots and exhibition rooms, where paintings are stored and exhibited, respectively, was shown to correlate well with fungal communities observed on the surface of the paintings indicating airborne origin of the infestations. Furthermore, higher density and diversity of the fungal communities on the obverse of the painting is a direct result of the higher amount of airborne fungal propagules deposited on the painted surface due to gravitational settling of propagules from the air [22]. It has to be noted, however, that there are discrepancies between results of the studies utilizing culture-dependent and -independent methods. Results of the limited number of studies on non-culturable fraction of the oil paintings fungal communities showed discrepancies in numbers, with unculturable or not viable part of the community being more dense and prevalent, as well as in composition: Ascomycota of genera *Alternaria*, *Ascoidea*, *Aspergillus*, *Blastomyces*, *Candida*, *Chaetomium*, *Coccidioides*, *Diplodia*, *Eutypa*, *Exophiala*, *Fusarium*, *Gaeumannomyces*, *Histoplasma*, *Marssonina*, *Microsporum*, *Neofusicoccum*, *Paracoccidioides*, *Parastagonospora*, *Penicillium*, *Penicilliopsis*, *Pestalotiopsis*, *Pichia*, *Rasamsonia*, *Lodderomyces*, *Neurospora*, *Sordaria*, *Talaromyces*, *Thermothelomyces*, *Thielavia*, *Trichoderma*, *Tuber*, and *Verticillium* were dominant, followed by Mucoromycota of genera *Phycomyces* and *Lobosporangium* and Basidiomycota of genus *Puccinia* [63, 75, 76].

Fungal induced deterioration of canvas oil paintings can occur on both the obverse and the reverse side. It usually starts on the reverse side as canvas components are more readily degraded than those found on the obverse side. In addition, support polymers and the glue sizing in the canvas act as supplementary substrates for fungal growth [22]. Canvas was shown to be one of the most susceptible painting materials (only surpassed by linseed oil), with the susceptibility depending on the percentage content of cellulose, lignin, and other organic components [74, 77]. The higher the percentage of cellulose and lignin, the more resistant it will be to fungal attack [78]. Due to their ability to produce cellulolytic enzymes responsible for cellulose fibers dissolution, fungi of the genus *Alternaria*, *Aspergillus*, *Cladosporium* and *Ulocladium* are considered to be the main agents of canvas deterioration [63, 69, 76].

On the other hand, the degree of deterioration on the obverse side depends on the oil paints and their mode of application. Varnish, added to provide protection against environmental attacks, is the least susceptible painting material to fungal attack, while many pigments are known to possess antifungal properties [74]. Fungal communities are found to be less dense and diverse in pictorial layers containing pigments with heavy metals (e.g. Pb, Cu and Hg), compared to those found in pictorial layers without such compounds [75]. On the obverse side, hydrolytic activities that the fungi undertake to sustain growth results in the detachment of the paint layer from the support, with further increase in the loss of material happening due to excretion of destructive metabolites, i.e. organic or inorganic acids and the additional production of extracellular enzymes: lipases, esterases, endo-N-acetyl-glucosaminidases and proteases [22]. Such activities lead to formation of structural impairments usually manifested as exfoliation of paint layers, cracking, peeling, formation of paint blisters, detachment of the paint layer from the support, deformations and loss of strength of support. Strongly linked to these damages are the esthetic impairments (preceding the structural damages or forming as a resulting consequence) manifesting as the change of the original coloration due to pigment alterations, biofilm formation on the painted surface or staining as the result of pigment excretion by fungi [75]. Fungi of *Aspergillus* and *Penicillium* genera degrade glues and oil binders, and dissolve paints, contributing to chromatic alterations of the painted surfaces and detachment of the support, while *Aureobasidium pullulans* is considered by many to be the main biological agent of paint deterioration [1, 45, 69].

6. Wood and paper

Wood as a material has been widely used as a structural element for many types of constructions or for ritual, religious and decorative purpose. Historically, human usage of wood is embedded in wood cultural heritage reflecting past and present human life, culture, ideals, symbols and values. Wood cultural heritage objects can be classified as: moveable (musical instruments, frames, furniture, sculptures, iconic altar etc.), immovable (temples, churches, chapels, royal palaces, pagodas, wooden bridges etc.) and underwater (shipwrecks, foundation piles, wooden cargo or contents which were partially or totally underwater, periodically or continuously for at least 100 years), according to UNESCO [79]. Lignocellulose is the major component of wood biomass and consists of three types of polymers, cellulose (40–55%), hemicelluloses (24–40%) and lignin (18–35%) that are strongly intermeshed and chemically bonded by non-covalent bonds and by covalent crosslinkages [19]. Organic nature and optimal water content make the wooden substrate suitable for microbial attack [80]. However, microbial deterioration of these materials occurs only under poor conservation conditions: high humidity level, soil contact, poor ventilation, and rare maintenance [81]. Even though deterioration of wood cultural heritage is a process conducted by all groups of microorganisms, fungi have the most significant potential to affect this type of historic artworks [82]. Biodegradation and biodeterioration of wood materials is predominantly dependent on its moisture content, requiring a minimum of 20% of water. Despite dry wooden objects are considered to be resistant to fungal degradation due to low moisture content, occasional wetting, leaks and flooding can increase humidity, enabling conditions for fungal growth. The mechanism of biodeterioration implies the development of fungi on the surface (**Figure 1G** and **H**) or between internal structures, the production of extracellular enzymes, the structural change of basic biopolymers, which ultimately results in visible changes of the object [81]. Generally, fungi that attack wooden material can be distinguished as white-rot, brown-rot and soft rot fungi. White rot fungi are the only organisms that can completely depolymerize and degrade all lignin components as well as cellulose and hemicellulose. The largest number, of about 1500 species, belongs to the Basidiomycota and a smaller number belongs to Ascomycota. Most commonly found species are from genera *Bjerkandera*, *Donkioporia*, *Fomes*, *Irpex*, *Phanerochaete*, *Pholiota*, *Pleurotus* and *Trametes*. Brown rot fungi decompose cellulose and hemicellulose while lignin degradation is limited to the process of demethylation of methoxyl groups, partial oxidation and depolymerization in a non-enzymatic catalytic cycle of the Fenton type where the free radical reaction is initiated by hydroxyl radicals (OH•). Only 6% of the total number of species that have been confirmed to be able to decompose wood mass belong to this group, and almost all representatives inhabit coniferous wood (species from the genera *Antrodia*, *Aspergillus*, *Coniophora*, *Coriolellus*, *Fusarium*, *Gloeophyllum*, *Merulis*, *Paxillus*, *Poria*, *Postia*, *Serpula*, etc.). Soft rot fungi decompose cellulose and hemicellulose, while the process of lignin modification is limited to demethylation. It is typical for this group of fungi to attack wood mass with high levels of humidity and low lignin content, forming a microscopic cavities inside the wood, sometimes leading to discoloration and occurrence of cracking pattern similar to brown rot (species from the genera *Alternaria, Chaetomium, Daldinia*, *Humicola*, *Stemphylium*, *Xylaria,* etc.) [79, 82, 83]. Soft-rot decay has been described frequently from construction timbers, ancient Egyptian wooden coffins, wooden structures of Buddhist temples, waterlogged archeological wooden material, which can be related to more tolerant growth conditions. In comparison to soft-rot species, other two groups of wood decay fungi have a relatively narrow spectrum of growth conditions (preferring moisture content between 35% and 50%) but were documented in wooden churches, historic timbers and various historic buildings [79]. Objects of wood cultural heritage in the outdoors have been more disposed to deterioration process than those in the indoor as they are exposed to the relative humidity of 70% or higher, which is stimulatory for fungal growth. Additionally, the physical contact with moisture absorbing surface can also provoke fungal decay [79]. On the other hand, fungal degradation of waterlogged and buried wood is much slower than for that found in dry environment, but once excavated decay can occur rapidly [82]. Finally, it should also be emphasized that in some cases fungal decay can be observed in extreme conditions such as in wood with 17.4% of moisture content [79].

Additionally, paper, which is mostly produced by mechanical and chemical processing of cellulose fibers, originating from wood, is the most important material on which cultural achievements in the whole world are recorded and preserved. Since it is created as a product of the wood industry and consists of 90–99% of cellulose fibers, in the ecological sense, paper is considered to be a cellulose substrate. Books, documents, writings, old maps, photographs, etc. are objects made of paper that are most often kept in libraries, archives and museums. Apart from paper, cotton and linen are fabrics which main components are cellulose fibers. Also, paraments, defined as hangings or ornaments used for decorations of Christian churches' interiors are often tailored of cotton and linen. In that sense, it should be emphasized that, art objects made of cellulose fibers can be colonized by cellulolytic fungi. These fungi can degrade cellulose fibers via process off cellulolysis, defined as an enzymatic hydrolysis of cellulose polymer into glucose units. In that sense, the fungi capable for production of cellulolytic enzymes are frequently isolated from paper, especially from old books or documents kept in libraries, archives and museum depots. Among the frequently encountered species on the paper substrates are the members of genera *Chaetomium*, *Penicillium*, *Aspergillus*, *Eurotium*, and *Trichoderma* [84, 85]. Some authors reported the presence of *Fusarium* sp., *Humicola* sp., *Paecilomyces variotti* and *Trichoderma viride* on deteriorated art photographs which were part of the collection of the Museum of Contemporary Art (Belgrade, Serbia) [86]. Due to their ability to degrade cellulose fibers, these fungi are referred to soft rot fungi [87]. Ascospores and conidia of different cellulolytic microfungi ubiquitously present in the environment worldwide could easily be deposited on papers and other cellulolytic materials (books), and when optimal conditions are met, they could germinate, elongate and proliferate and consequently lead to fast book decay. Also, fungi can deteriorate the paper-based materials mechanically via hyphal penetration or through production and excretion of pigments and organic acids [88]. A specific and irreversible phenomenon in the form of brown to red spots on paper material has been described in the literature [89]. Since the color of these spots resembles the color of fox fur, the phenomenon is called "foxing". The origin of this phenomenon on paper documents is explained by two theories - abiotic and biotic. According to the abiotic theory, "foxing" is a consequence of natural chemical processes, most often oxidation, which takes place on paper material, as well as a consequence of the deposition of certain compounds on the paper surface [90]. According to the biotic theory, "foxing" is caused by microorganisms, especially fungi that produce organic acids which deteriorate the paper, permanently damaging it [89]. Isolation of a large number of fungal species from the parts of the paper on which the symptoms of "foxing" are observed, speaks in favor of biotic theory. The famous piece of art affected with foxing symptoms is self-portrait of Leonardo da Vinci's drawn in red chalk on paper and deposited in Royal Library in Turin. As a main culprit responsible for "foxing" spots on this famous piece of art, certain authors reported the fungus *Eurotium halophilicum*, which spores are documented near the foxing spots via SEM analyses, along with oxalates of fungal origin [91]. Additionally, they pointed out that tonophilic fungi can germinate on paper materials and also can metabolize organic acids, oligosaccharides and proteins, which react chemically with the material at a low water activity, forming brown products and, via oxidative reactions, leading to foxing spots.

7. Textile

Textile is defined as elastic material produced by spinning of natural or synthetic raw fibers and which are in final form composed of interlocked network of threads or yarns. Apart from synthetic fibers, materials used for textile production could be of plant or animal origin. Cotton, linen, hemp and jute are widely used fabrics of plant origin and hence they are composed of cellulose fibers. Many old-fashioned and vintage attires and garments, worn by our ancestors were woven from these fabrics and deposited in museum depots and exhibition rooms, or still worn during traditional festivities. In that sense, those fibers could be easily attacked by cellulolytic fungi, and mechanisms of biodeterioration are similar to those of the fungi capable of degrading paper-based materials. On the other hand, animal fibers include wool and silk [92]. Wool is a textile fiber obtained from various hairy mammals, but mostly from sheep, and main constituent of wool is a protein keratin. When compared with textile fibers of plant origin, wool is more resistant to fungal attack due to its specific cross-linked structure with disulphide bonds [93]. However, fungi capable for keratinolysis can attack wool fibers and cause wool degradation. Pioneer research by some authors demonstrated that fungi are the main "culprits" responsible for wool degradation, in much higher degree then bacteria, and members of genera *Aspergillus*, *Chaetomium*, *Fusarium*, *Microsporum*, *Penicillium*, *Rhizopus* and *Trichophyton* are among the most frequent wool colonizers [94]. In an *in vitro* study, other investigators showed that fungi do not grow directly on the wool fibers but rather between the fibers, and reported that proteolytic fungi *Cladosporium cladosporioides* and *Penicillium corylophilum* have the most intensive growth when inoculated on wool and also the highest impact on degradation and aging of wool fibers [93]. Although wool is very resilient to microbial attack, the silk is considered to be a natural fiber most resistant to the biodeterioration [92]. Silks are defined as fibrous proteins spun into fibers through activity of spiders and insects. The main producer of commercial silk is a domestic silkworm *Bombyx mori*. Chemically, raw silks are composed of highly crystalline polypeptide fibers, fibroin, linked to one another by a gum-like protein, sericin [95]. The amino acid composition of silk polypeptides results in a very stable β-pleated crystal structure, making fibroin totally insoluble in aqueous solvents and hence very resistant to enzymatic hydrolysis [96]. In that sense, scientific reports regarding the microbial deterioration of silk material are scarce. Still, some authors reported for the first time the mechanical deterioration of Japanese silk from Serbian museum collections caused by proteolytic fungus *Chaetomium globosum*. Scanning electron microscopy analysis of the analyzed scroll indicated that *C. globosum* hyphae are capable of the mechanical deterioration of silk, causing cracks and gaps in fibroin fibers and consequently lead to visible impairment of silk artifact itself (**Figure 1A** and **B**) [97].

8. Conclusion

Specific morphology and physiology of fungi enables them to colonize multifarious substrates, including cultural heritage artifacts. Due to their pronounced metabolic capacity, fungal deteriogens are able to significantly influence both aesthetical appearance and integrity of monuments, sculptures, murals, paintings, textile and documentary heritage. Nowadays, conversance of fungal biology is becoming crucial in proper assessment of contamination and colonization of artworks but also in their adequate storage and protection. Since mycology as a science gains more and more application in the conservation and restauration procedures, the investigations in this scientific filed become essential in cultural heritage safeguard.

Author details

Željko Savković*, Miloš Stupar, Nikola Unković, Aleksandar Knežević,
Jelena Vukojević and Milica Ljaljević Grbić
Faculty of Biology, Institute of Botany and Botanical Garden "Jevremovac",
University of Belgrade, Belgrade, Serbia

*Address all correspondence to: zsavkovic@bio.bg.ac.rs

References

[1] Sterflinger K, Piñar G. Microbial deterioration of cultural heritage and works of art - tilting at windmills? Applied Microbiology and Biotechnology. 2013;97:9637-9646. DOI: 10.1007/s00253-013-5283-1

[2] di Carlo, E., Barresi, G., Palla, F. Biodeterioration. In: Palla, F., Barresi, G., editors. Biotechnology and conservation of cultural heritage. Springer, Berlin; 2017. p. 49-66.

[3] Savković Ž, Stupar M, Unković N, Ivanović Ž, Blagojević J, Vukojević J. *In vitro* biodegradation potential of airborne *Aspergilli* and *Penicillia*. The Science of Nature. 2019;106(3-4):8. DOI:10.1007/s00114-019-1603-3

[4] Kasprzyk, I. Aeromycology - Main research fields of interest during the last 25 years. Annals of Agricultural and Environmental Medicine. 2008; 15:1-7.

[5] Savković Ž, Stupar M, Unković N, Ivanović Ž, Blagojević J, Popović S, Vukojević J., Ljaljević Grbić M. Diversity and seasonal dynamics of culturable airborne fungi in a cultural heritage conservation facility. International Biodeterioration & Biodegradation. 2021;157:105163. DOI: 10.1016/j.ibiod.2020.105163

[6] Florian MLE. Fungal facts: solving fungal problems in heritage collections. London: Archetype; 2002. 146 p.

[7] Garg KL, Kamal KJ, Mishra AK. Role of fungi in the deterioration of wall paintings. The Science of the Total Environment. 1995;167:255-271. DOI: 10.1016/0048-9697(95)04587-Q

[8] Pinna D, Salvadori O. Processes of biodeterioration: general mechanisms. In: Caneva G, Nugari MP, Nugari MP, Salvadori O, editors Plant biology for cultural heritage: biodeterioration and conservation. Los Angeles: Getty Conservation Institute; 2008. p. 15-34.

[9] Diakumaku E, Gorbushina AA, Krumbein WE, Panina L, Soukharjevski S. Black fungi in marble and limestones — an aesthetical, chemical and physical problem for the conservation of monuments. Science of The Total Environment. 1995;167 (1-3):295-304. DOI: 10.1016/0048-9697(95)04590-W

[10] Unković N, Ljaljević Grbić M, Stupar M, Savković Ž, Jelikić A, Stanojević D. Vukojević J. Fungal-induced deterioration of mural paintings: *in situ* and mock-model microscopy analyses. Microscopy and Microanalysis. 2016;22(2):410-421. DOI: 10.1017/S1431927616000544

[11] Caneva G, Maggi O, Nugari MP, Pietrini AM, Piervittori V, Ricci S, Roccardi A. The biological aerosol as a factor of biodeterioration. In: Mandrioli P, Caneva G, Sabbioni C, editors. Cultural heritage and aerobiology. Methods and Measurement Techniques for Biodeterioration Monitoring. Dordrecht: Springer Science+Business Media; 2003. p. 3-29.

[12] Warscheid Th, Braams J. Biodeterioration of stone: a review. International Biodeterioration & Biodegradation. 2000;46(4):343-368. DOI: 10.1016/S0964-8305(00)00109-8

[13] Gomez-Alarcon G, de la Torre MA. The effect of filamentous fungi on stone monuments - the Spanish experience. In: Singh J, Singh J, editors, Building mycology, management of decay and health in buildings. London: E & FN Spon; 1994. p. 295-309.

[14] Gadd GM. Fungal production of citric and oxalic acid: importance in metal speciation, physiology and

biogeochemical processes. Advances in microbial physiology. 1999;41:47-92. DOI: 10.1016/S0065-2911(08)60165-4

[15] Knežević A, Stajić M, Jovanović VM, Kovačević V, Ćilerdžić J, Milovanović I, Vukojević J. Induction of wheat straw delignification by *Trametes* species. Scientific Reports. 2016;6:26529. DOI: 10.1038/srep26529

[16] Sabatini L, Sisti M, Campana R. Evaluation of fungal community involved in the bioderioration process of wooden artworks and canvases in Montefeltro area (Marche, Italy). Microbiological Research. 2018;207:203-210. DOI: 10.1016/j.micres.2017.12.003

[17] Ćilerdžić J, Stajić M, Duletić-Laušević S, Vukojević J, Knežević A. Potential of *Trametes hirsuta* to produce ligninolytic enzymes during degradation of agricultural residues. Bioresources. 2011;6(3):2885-2895. DOI: 10.15376/biores.6.3.2885-2895

[18] Ferreira JA, Mahboubi A, Lennartsson PR, Taherzadeh MJ. Waste biorefineries using filamentous Ascomycetes fungi: Present status and future prospects. Bioresource Technology. 2016;215:334-345. DOI: 10.1016/j.biortech.2016.03.018

[19] Sánchez C. Lignocellulosic residues: Biodegradation and bioconversion by fungi. Biotechnology Advances. 2009;27:185-194. DOI: 10.1016/j.biotechadv.2008.11.001

[20] Souza SCMM, Melo IS, Oliveira PR. Ligninolytic enzyme production by Ganoderma spp. Enzyme and Microbial Technology. 2005;37:324-329. DOI: 10.1016/j.enzmictec.2004.12.007

[21] Knežević A, Stajić M, Milovanović I, Vukojević J. Degradation of beech wood and wheat straw by *Trametes gibbosa*. Wood Science and Technology. 2017;51:1227-1247. DOI: 10.1007/s00226-017-0921-x

[22] López-Miras M, Piñar G, Romero-Noguera J, Bolivar-Galiano FC, Ettenauer J, Sterflinger K, Martin-Sanchez I. Microbial communities adhering to the obverse and reverse sides of an oil painting on canvas: identification and evaluation of their biodegradative potential. Aerobiologia. 2013;29:301-314. DOI: 10.1007/s10453-012-9281-z1

[23] Savković ŽD, Stupar MČ, Ljaljević Grbić MV, Vukojević JB. Comparison of anti-*Aspergillus* activity of *Origanum vulgare* L. essential oil and commercial biocide based on silver ions and hydrogen peroxide. Acta Botanica Croatica. 2016;75(1):121-128. DOI: 10.1515/botcro-2016-0011

[24] Jacobson ES. Pathogenic roles for fungal melanins. Clinical Microbiology Reviews. 2000;13(4):708-717. DOI: 10.1128/CMR.13.4.708

[25] Butler MJ, Day AW. Fungal melanins: a review. Canadian Journal of Microbiology. 1998;44(12):1115-1136. DOI: 10.1139/w98-119

[26] Savković Ž, Unković N, Stupar M, Franković M, Jovanović M, Erić S, Šarić K, Stanković S, Dimkić I, Vukojević J, Ljaljević Grbić M. Diversity and biodeteriorative potential of fungal dwellers on ancient stone stela. International Biodeterioration & Biodegradation. 2016;115:212-223. DOI: 10.1016/j.ibiod.2016.08.027

[27] Pinna D, Salvadori O. Stone and related materials. In: Caneva G, Nugari MP, Nugari MP, Salvadori O, editors. Plant biology for cultural heritage: biodeterioration and conservation. Los Angeles: Getty Conservation Institute; 2008. p. 128-144.

[28] Jenkins KA, Smith BJ. Daytime rock surface temperature variability and its implications for mechanical rock weathering: Tenerife, Canary Islands. CATENA. 1990;17(4-5):449-459. DOI: 10.1016/0341-8162(90)90045-F

[29] Gorbushina AA, Krumbein WE. The poikilotrophic micro-organism and its environment. In: Seckbach J, editor. Enigmatic microorganisms and life in extreme environments. Cellular origin and life in extreme habitats. Vol. 1. Dordrecht: Springer; 1999. p. 175-185.

[30] Ramage G, Rajendran R, Sherry L, Williams C. Fungal biofilm resistance. International Journal of Microbiology. 2012;2012:1-14. DOI: 10.1155/2012/528521

[31] Cooke RC, Rayner ADM. Ecology of saprotrophic fungi. London; New York: Longman; 1984. 415 p.

[32] Sterflinger K. Fungi: Their role in deterioration of cultural heritage. Fungal Biology Reviews. 2010;24:47-55. DOI: 10.1016/j.fbr.2010.03.003

[33] Trovão J, Gil F, Catarino L, Soares F, Tiago I, Portugal A. Analysis of fungal deterioration phenomena in the first Portuguese King tomb using a multi-analytical approach. International Biodeterioration & Biodegradation. 2020;149:104933. DOI: 10.1016/j.ibiod.2020.104933

[34] Gorbushina AA, Krumbein WE, Hamman CH, Panina L, Soukharjevski S, Wollenzien U. Role of black fungi in color change and biodeterioration of antique marbles. Geomicrobiology Journal. 1993;11(3-4):205-221. DOI: 10.1080/01490459309377952

[35] Sterflinger K, Krumbein WE. Dematiaceous fungi as a major agent for biopitting on Mediterranean marbles and limestones. Geomicrobiology Journal. 1997;14(3):219-230. DOI:10.1080/01490459709378045

[36] Rikkinen J. What's behind the pretty colours? a study on the photobiology of lichens. Helsinki: Finnish Bryological Society; 1995. 239 p.

[37] Piervittori R, Nimis P, Tretiach M. Lichens. In: Caneva G, Nugari MP, Nugari MP, Salvadori O, editors. Plant biology for cultural heritage: biodeterioration and conservation. Los Angeles: Getty Conservation Institute; 2008. p. 77-81.

[38] Salvadori O. Characterisation of endolithic communities of stone monuments and natural outcrops. In: Ciferri O, Tiano P, Mastromei G, editors. Of Microbes and Art. Boston, MA: Springer; 2000 p. 89-101.

[39] Chen J, Blume H-P, Beyer L. Weathering of rocks induced by lichen colonization — a review. CATENA. 2000;39(2):121-146. DOI: 10.1016/S0341-8162(99)00085-5

[40] Adamo P, Marchetiello A, Violante P. The weathering of mafic rocks by lichens. The Lichenologist. 1993;25(03):285. DOI: 10.1017/S0024282993000349

[41] Adamo P. Weathering of rocks and neogenesis of minerals associated with lichen activity. Applied Clay Science. 2000;16(5-6):229-256. DOI: 10.1016/S0169-1317(99)00056-3

[42] Del Monte M, Sabbioni C, Zappia G. The origin of calcium oxalates on historical buildings, monuments and natural outcrops. Science of The Total Environment. 1987;67(1):17-39. DOI: 10.1016/0048-9697(87)90063-5

[43] Unković N, Erić S, Šarić K, Stupar M, Savković Ž, Stanković S, Stanojević O, Dimkić I, Vukojević J, Ljaljević Grbić M. Biogenesis of secondary mycogenic minerals related

to wall paintings deterioration process. Micron. 2017;100:1-9. DOI: 10.1016/j.micron.2017.04.004

[44] Stanojević D. Kreč kao istorijski materijal. In: Zbornik radova Seminara i radionice "Kreč kao istorijski materijal"; 25-26 August 2014; Sopoćani. Srbija: Republički zavod za zaštitu spomenika kulture – Beograd; 2014. p. 3-13.

[45] Ciferri O. Microbial Degradation of Paintings. Applied and Environmental Microbiology. 1999; 65: 879-885. DOI: 10.1128/AEM.65.3.879-885.1999

[46] Milanesi C, Baldi F, Borin S, Vignani R, Ciampolini F, Faleri C, Cresti M. Biodeterioration of a fresco by biofilm forming bacteria. International Biodeterioration and Biodegradation. 2006;57:168-173. DOI: 10.1016/j.ibiod.2006.02.005

[47] Heyn C, Petersen K, Krumbein WE. Investigation of microbial degradation of synthetic polymers used in the conservation and restoration of art objects. In: Bousher A, Chandra M, Edyvean R, editors. Biodeterioration and Biodegradation. UK: IchemE; 1995. p. 73-79.

[48] Warscheid T. The evaluation of biodeterioration processes on cultural objects and approaches for their effective control. In: Koestler RJ, Koestler VH, Charola AE, Nieto-Fernandez FE, editors. Art, Biology, and Conservation: Biodeterioration of Works of Art. USA: The Metropolitan Museum of Art; 2003. p. 14-27.

[49] Miličić Lj. Aditivi krečnih maltera. In: Zbornik radova Seminara i radionice "Kreč kao istorijski materijal"; 25-26 August 2014; Sopoćani. Srbija: Republički zavod za zaštitu spomenika kulture – Beograd; 2014. p. 89-92.

[50] Stanojlović M. Kreč kao polazni materijal za slikanje al fresco. In: Zbornik radova Seminara i radionice "Kreč kao istorijski materijal"; 25-26 August 2014; Sopoćani. Srbija: Republički zavod za zaštitu spomenika kulture – Beograd; 2014. p. 3-13

[51] Giustetto R, Gonella D, Bianciotto V, Lumini E, Voyron S, Costa E, Diana E. Transfiguring biodegradation of frescoes in the Beata Vergine del Pilone Sanctuary (Italy): Microbial analysis and minero-chemical aspects. International Biodeterioration & Biodegradation. 2015;98:6-18. DOI: 10.1016/j.ibiod.2014.10.020

[52] Gorbushina AA, Petersen K. Distribution of microorganisms on ancient wall paintings as related to associated faunal elements. International Biodeterioration & Biodegradation. 2000;46:277-284. DOI: 10.1016/S0964-8305(00)00103-7

[53] Gorbushina AA, Heyrman J, Dornieden T, Gonzalez-Delvalle M, Krumbein WE, Laiz L, Petersen K, Sáiz-Jiménez C, Swings J. Bacterial and fungal diversity and biodeterioration problems in mural painting environments of St. Martins church (Greene-Kreiensen, Germany). International Biodeterioration & Biodegradation. 2004;53:13-24. DOI: 10.1016/j.ibiod.2003.07.003

[54] Saarela M, Alakomi HL, Siuhko ML, Maunuksela L, Raaska L, Mattila-Sandholm T. Heterotrophic microorganisms in air and biofilm samples from roman catacomb, with special emphasis on actinobacteria and fungi. International Biodeterioration & Biodegradation. 2004;54:27-37. DOI: 10.1016/j.ibiod.2003.12.003

[55] Ripka K. Identification of microorganisms on stone and mural paintings using molecular methods [thesis]. Austria: Faculty of Natural Sciences; 2005.

[56] Pepe O, Palomba S, Sannino L, Blaiotta G, Ventorino V, Moschetti G, Villani F. Characterization in the archaeological excavation site of heterotrophic bacteria and fungi of deteriorated wall painting of Herculaneum in Italy. Journal of Environmental Biology. 2011; 32:241-250.

[57] Ruga L, Orlandi F, Romano B, Fornaciari M. The assessment of fungal bioaerosols in the crypt of St. Peter in Perugia (Italy). International Biodeterioration & Biodegradation. 2015;98:121-130. DOI: 10.1016/j.ibiod.2014.12.010

[58] Nevalainen A, Täubel M, Hyvärinen A. Indoor fungi: companions and contaminants. Indoor Air. 2015;25:125-156. DOI: 10.1111/ina.12182

[59] Karbowska-Berent J. Microbiodeterioration of mural paintings: a review. In: Koestler RJ, Koestler VH, Charola AE, Nieto-Fernandez FE, editors. Art, Biology, and Conservation: Biodeterioration of Works of Art. USA: The Metropolitan Museum of Art. 2003; p. 266-301.

[60] Pangallo D, Bučková M, Kraková L, Puškárová A, Šaková N, Grivalský T, Chovanová K, Zemánková M. Biodeterioration of epoxy resin: a microbial survey through culture-independent and culture-dependent approaches. Environmental Microbiology. 2014;17:462-479. DOI: 10.1111/1462-2920.12523

[61] Dornieden T, Gorbushina AA, Krumbein WE. Biodecay of cultural heritage a space/time-related ecological situation - an evaluation of a series of studies. International Biodeterioration & Biodegradation. 2000;46:261-270. DOI: 10.1016/S0964-8305(00)00107-4

[62] Sayer JA, Gadd GM. Binding of cobalt and zinc by organic acids and culture filtrates of *Aspergillus niger* grown in the absence or presence of insolubile cobalt or zinc phosphate. Mycological Research. 2001;105:1261-1267. DOI: 10.1016/S0953-7562(08)61998-X

[63] del Mar López-Miras M, Martín-Sánchez I, Yebra-Rodríguez Á, Romero-Noguera J, Bolívar-Galiano F, Ettenauer J, Sterflinger K, Piñar G. Contribution of the microbial communities detected on an oil painting on canvas to its biodeterioration. PloS ONE. 2013;8:e80198. DOI: 10.1371/journal.pone.0080198

[64] Rosado T, Candeias A, Caldeira AT, Mirão J, Gil M. Evaluation of mural paintings biodeterioration by oxalate formation. In: Rogerio-Candelera MA, Lazzari M, Cano E, editors. Science and Technology for the Conservation of Cultural Heritage. UK: Taylor & Francis Group; 2013. p. 147-150. DOI: 10.1201/b15577-35

[65] Sterflinger K. Fungi as geologic agents. Geomicrobiology Journal. 2000;17:97-124. DOI: https://doi.org/10.1080/01490450050023791

[66] Piñar G, Ripka K, Weber J, Sterflinger K. The micro-biota of a sub-surface monument the medieval chapel of St. Virgil (Vienna, Austria). International Biodeterioration & Biodegradation. 2009;63:851-859. DOI: 10.1016/j.ibiod.2009.02.004

[67] Roldán M, Clavero E, Hernãndez-Mariné M. Aerophytic biofilms in dim habitats. In: Sáiz-Jiménez C, editor. Molecular Biology and Cultural Heritage. The Netherlands: Swets & Zeitlinger; 2003. p. 163-169. DOI: 10.1201/9780203746578-21

[68] Urzì C, Realini M. Color changes of Noto's calcareous sandstone as related to

its colonization by microorganisms. International Biodeterioration & Biodegradation. 1998;42:45-54. DOI: 10.1016/S0964-8305(98)00045-6

[69] Salvador C, Bordalo R, Silva M, Rosado T, Candeias A, Caldeira AT. On the conservation of easel paintings: evaluation of microbial contamination and artists materials. Applied Physics A. 2017;123:80. DOI: 10.1007/s00339-016-0704-5

[70] Caselli E, Pancaldi S, Baldisserotto C, Petrucci F, Impallaria A, Volpe L, D'Accolti M, Soffritti I, Coccagna M., Sassu G, Bevilacqua F, Volta A, Bisi M, Lanzoni L, Mazzacane S. Characterization of biodegradation in a 17th century easel painting and potential for a biological approach. PLoS ONE. 2018;13:e0207630. DOI: 10.1371/journal.pone.0207630

[71] Capodicasa S, Fedi S, Porcelli AM, Zannoni D. The microbial community dwelling on a biodeteriorated 16th century painting. 2010;64:727-733. DOI: 10.1016/j.ibiod.2010.08.006

[72] Inoue M, Koyano M. Fungal contamination of oil paintings in Japan. International Biodeterioration. 1991;28:23-35. DOI: 10.1016/0265-3036(91)90031-L

[73] Vukojević J, Ljaljević Grbić M. Moulds on paintings in Serbian fine art museums. African Journal of Microbiology Research. 2010;4:1453-1456. DOI: 10.5897/AJMR.9000517

[74] Paner CM. Chemical control of fungi infesting easel oil paintings at the University of Santo Tomas, Museum of Arts and Sciences. Manila: University of Santo Tomas Graduate School; 2009.

[75] Santos A, Cerrada A, García S, San Andrés M, Abrusci C, Marquina D. Application of Molecular Techniques to the Elucidation of the Microbial Community Structure of Antique Paintings. Microbial Ecology. 2009;58:692-702. DOI: 10.1007/s00248-009-9564-2

[76] Piñar G, Poyntner C, Lopandic K, Tafer H, Sterflinger K. Rapid diagnosis of biological colonization in cultural artefacts using the MinION nanopore sequencing technology. International Biodeterioration & Biodegradation. 2020;148:104908. DOI: 10.1016/j.ibiod.2020.104908

[77] Rivera LEC, Ramos AP, Sánachez JIC, Serrano MED. Origin and Control Strategies of Biofilms in the Cultural Heritage. In: Kırmusaoğlu S, editor. Antimicrobials, Antibiotic Resistance, Antibiofilm Strategies and Activity Methods. London: IntechOpen; 2018. p. 51. DOI: 10.5772/intechopen.79617.ch4

[78] Poyatos F, Morales F, Nicholson AW, Giordano A. Physiology of biodeterioration on canvas paintings. Journal of Cellular Physiology. 2018;233:2741-2751. DOI: 10.1002/jcp.26088

[79] Kim YS, Singh AP. Wood as cultural heritage material and its deterioration by biotic and abiotic agents, In: Kim YS, Funada R, Singh AP, editors. Secondary Xylem Biology: Origins, Functions, and Applications. London, UK: Academic Press; 2016;233-257. Chapter 12. DOI: 10.1016/B978-0-12-802185-9.00012-7

[80] Goodell B, Winandy JE, Morrell JJ. Fungal degradation of wood: Emerging data, new insights and changing perceptions. Coatings. 2020;10(12):1210. DOI:10.3390/coatings10121210

[81] Tiano P. Biodegradation of cultural heritage: Decay mechanisms and control methods, 9th ARIADNE Workshop "Historic Material and their Diagnostic", ARCCHIP, Prague, 22 to 28 April 2002.

Available at: http://www.arcchip.cz/w09/w09_tiano.pdf

[82] Pyzik A, Ciuchcinski K, Dziurzynski M, Dziewit L. The bad and the good - Microorganisms in cultural heritage environments - An update on biodeterioration and biotreatment approaches. Materials. 2021;14:177. DOI:10.3390/ma14010177

[83] Knežević A, Milovanović I, Stajić M, Lončar N, Brčeski I, Vukojević J, Ćilerdžić J. Lignin degradation by selected fungal species. Bioresource Technology. 2013;138:117-123. DOI: https:10.1016/j.biortech.2013.03.182

[84] Szczepanowska H, Cavaliere AR. Fungal deterioration of 18th and 19th century documents: a case study of the tilghman family collection, wye house, easton. Maryland. International Biodeterioration & Biodegradation. 2000;46(3):245-249. DOI: 10.1016/S0964-8305(00)00061-5

[85] Corte AM, Ferroni A, Salvo VS. Isolation of fungal species from test samples and maps damaged by foxing, and correlation between these species and the environment. International Biodeterioration & Biodegradation. 2003;51(3):167-173. DOI: 10.1016/S0964-8305(02)00137-3

[86] Ljaljević Grbić M, Stupar M, Vukojević J, Maricić I, Bungur N. Molds in museum environments: Biodeterioration of art photographs and wooden sculptures. Archives of Biological Sciences. 2013;65(3):955-962. DOI: 10.2298/ABS1303955G

[87] Erickson KEL, Blanchette RA, & Ander P. Microbial and enzymatic degradation of wood and wood components. Berlin: Springer – Verlag. 1990: p. 407.

[88] Micheluz A, Manente S, Tigini V, Prigione V, Pinzari F, Ravagnan G, et al. The extreme environment of a library: Xerophilic fungi inhabiting indoor niches. International Biodeterioration & Biodegradation. 2015;99:1-7. DOI: 10.1016/j.ibiod.2014.12.012

[89] Arai H. Foxing caused by Fungi: twenty-five years of study. International Biodeterioration & Biodegradation. 2000;46(3):181-188. DOI: 10.1016/S0964-8305(00)00063-9

[90] Cain E, Miller BA. Photographic, spectral and chromatographic searchesi nto the nature of foxing. Milwaukee: Preprints American Institute for Conservation, 10th Annual Meeting. 1982: pp. 54-52.

[91] Piñar G, Tafer H, Sterflinger K, Pinzari F. Amid the possible causes of a very famous foxing: molecular and microscopic insight into L eonardo da V inci's self-portrait. Environmental Microbiology Reports. 2015;7(6):849-859. DOI: 10.1111/1758-2229.12313

[92] Szostak-Kotowa J. Biodeterioration of textiles. International Biodeterioration & Biodegradation. 2004;53(3):165-170. DOI: 10.1016/S0964-8305(03)00090-8

[93] Kavkler K, Demšar A. Impact of fungi on contemporary and accelerated aged wool fibres. Polymer Degradation and Stability. 2012;97(5):786-792. DOI: 10.1016/j.polymdegradstab.2012.02.002

[94] Agarwal PN, Puvathingal JM. Microbiological deterioration of woollen materials. Textile Research Journal 1969;39-38. DOI: 10.1177/0040517569 03900107

[95] Seves A, Romanò M, Maifreni T, Sora S, Ciferri O. The microbial degradation of silk: a laboratory investigation. International Biodeterioration & Biodegradation. 1998;42(4):203-211. DOI: 10.1016/S0964-8305(98)00050-X

[96] Otterburn MS. The chemistry and reactivity of silk. In: Asquith RS, editor.

The Chemistry of Natural Fibres. New York, London: Plenum Press; 1997. p. 53-79.

[97] Ljaljević Grbić M, Unković N, Stupar M, Vukojević J, Nedeljković T. Implementation of ATP biolumenescence method in the study of the fungal deterioration of textile artefacts. Fibres & Textiles in Eastern Europe. 2014;22(6):132-136.

Chapter 16

Bioremediation Techniques for Soil Pollution: An Introduction

Anita Verma

Abstract

Environmental pollution has been on the rise in the past few decades owing to increased human activities on energy reservoirs, unsafe agricultural practices and rapid industrialization. Soil pollution is one of the major worry among all because soil contamination can harm the humans by consumption of food grown in polluted soil or it can cause infertility to soil and lower the productivity, Among the pollutants that are of environmental and public health concerns due to their toxicities are: heavy metals, nuclear wastes, pesticides, greenhouse gases, and hydrocarbons. So this chapter will include; Sources of soil pollution and remediation of polluted sites using biological means has proven effective and reliable due to its eco-friendly features. Bio-remediation can either be carried out ex situ or in situ, depending on several factors, which include site characteristics, type and concentration of pollutants. It also seen as a solution for emerging contaminant problems.

Keywords: soil pollution, bio-remediation, ex situ bio-remediation, in situ bio-remediation

1. Introduction

Soil is an essential a neighborhood of the common habitat. It's pretty much as significant as plants, creatures, rocks, landforms, loch and waterways. It is a living space for a genuine scope of living beings. It goes about as stream control for water and synthetic substances between the environment and along these lines the world, and furthermore both as a source and store for gases (like oxygen and carbon dioxide) inside the climate. Soils do not simply influence characteristic cycles yet additionally record human exercises both at this and inside the past.

Soil is dynamic organically and a permeable medium that has created inside the highest layer of Earth's covering. Soil is one of the corpus foundations of life on Earth, which might be a supply of water and supplements, as a mechanism for the filtration and breakdown of squanders, and as a functioning member inside the cycling of carbon and different components through the environment accessible universally. It's gotten from enduring cycles driven by natural, climatic, geologic, and geographical impacts.

Soil is the linkage between the different ecosystems like biosphere, atmosphere, and hydrosphere. So, the soils are fundamental in the preservation of environmental quality at local, regional, and worldwide level. For example, its buffering capacity contributes to water quality, since the ability to act as a sink for contaminants can have an important role in controlling the negative impacts of pollution on other environment. Researchers are trying to develop and model different bioremediation techniques; however, there is no single bioremediation technique that treats all types of contamination and to restore polluted environments. Bioremediation is a natural process, which relies on bacteria, fungi, and plants to remove, reduce, degrade, or immobilize environmental pollutants from soil and water, thus restoring contaminated sites to a relatively clean nontoxic environment [1].

It is now recognized that, soil is considered a vital resource, and due to its slow formation, it can be considered nonrenewable. Moreover, it has impacts on environmental, economic, and cultural activities. These techniques are environmentally friendly and cost effective features are the major advantages of bioremediation compared to both chemical and physical methods of remediation. Thus far, several good definitions have been given to bioremediation, with particular emphasis on one of the processes.

2. Soil pollution

Soil contamination is that the decrease inside the efficiency of soil in light of the presence of soil toxins. Soil toxins adversely affect the actual substance and organic properties of the dirt and decrease its profitability. Pesticides, composts, natural excrement, synthetic substances, radioactive squanders, disposed of food, garments, cowhide merchandise, plastics, paper, bottles, tins-jars and cadavers all contribute towards causing soil contamination. Synthetic substances like iron lead mercury, copper, zinc, cadmium, aluminum, cyanides, acids and soluble bases and so on are available in modern squanders and arrive at the dirt either straightforwardly with water or in a roundabout way through air. (for example through corrosive downpour).

Soil contamination can cause contamination if harmful synthetics drain into groundwater, or whenever sullied spillover arrives at streams, lakes, or seas. Soil additionally normally adds to contamination by delivering unstable mixtures into the environment. Nitrogen escapes through alkali volatilization and denitrification. The disintegration of natural materials in soil can deliver sulfur dioxide and other sulfur compounds, causing corrosive downpour. Substantial metals and other possibly poisonous components are the principal genuine soil toxins in sewage. Sewage slop contains substantial metals and, whenever applied over and again or in huge sums, the treated soil may gather weighty metals and thus it become incapable to try and support blossoms.

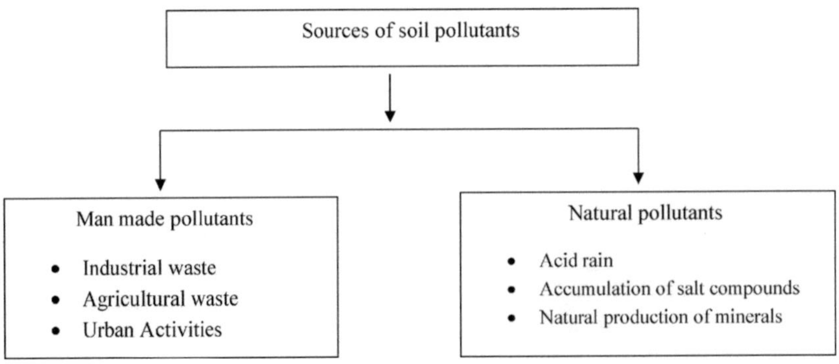

3. Man made pollutants

3.1 Agricultural pollution

Agricultural processes contribute to soil pollution. For increasing increase crop yield fertilizers are used which also cause pollution that impacts soil quality. Use of pesticides also harms plants and animals by contaminating the soil, these chemicals get deep inside the soil and poison the ground water system and runoff of these chemicals by rain and irrigation also contaminate the local water system and causes eutrophication of fresh water body. Phosphate is the main contributor to eutrophication its high concentration promotes Cyanobacteria and Algae growth which ultimately reduces dissolved oxygen in water [2].

3.2 Industrial waste

Most of the pollution is caused by industrial waste products and improper disposal of waste contaminates the soil with harmful chemicals. These pollutants affect plant and animal species and local water supplies and drinking water. On the other hand toxic fumes from the regulated landfills contain chemicals that can fall back to the earth in the form of acid rain and can damage the soil profile. Industrial activities like leads to acidification of soil and contamination due to the disposal of industrial waste, heavy metals, toxic chemicals, dumping oil and fuel, etc.

3.3 Urban activities

Human activities can lead to soil pollution directly and indirectly. For example improper drainage and increase run-off contaminates the nearby land areas or streams. Unorganized disposal of trash breaks down into the soil and it deposits in a number of chemical and pollutants into the soil. These may again seep into groundwater or wash away in local water system and excess waste deposition increases the presence of bacteria in the soil which leads to the generation of methane gas from decomposition activities by bacteria contributing to global warming and poor air quality. It also creates foul odors and can impact quality of life [3].

3.4 Acid rain

Acid rain primarily caused by Sulfur dioxide (SO2), oxides of nitrogen and ozone to some extent. Acid rain is caused when pollutants present in the air mixes up with the rain and fall back on the ground. Sulfuric and nitric acid solutions cause acidity in rainwater. Acid rain decreases the pH of the soil, causing its acidity to increase, which decreases the level of important nutrients found in the soil [4]. Soils low in cation exchange capacity and base saturation are the most sensitive to acid precipitation [5].

4. Natural source of soil pollution

Some of natural event also can be the cause of soil pollution like earthquakes, landslides, hurricanes, and flood. These natural disasters cause transpose to the composition of soil which leads to the contamination. For example weathering of naturally occurring sulphide-bearing rock make mineralized zones of arsenopyrite (gossans), Most of these minerals present a high spatial variability and many of them can be found in higher concentrations in deeper layers. However, As is slightly bioaccessible if getting from natural sources [6]. Soils and rocks are also natural sources of the radioactive gas Radon (Rn). High natural radioactivity is common in acidic igneous rocks, mainly in feldspar-rich rocks and illite-rich rocks.

However, There are other numerous of ways of soil contamination, for example, • Seepage from a landfill • Discharge of mechanical waste into the dirt • Percolation of defiled water into the dirt • Rupture of underground stockpiling tanks • Excess utilization of pesticides, herbicides or compost • Solid waste drainage • The most well-known synthetics associated with causing soil contamination are: • Petroleum hydrocarbons • Heavy metals • Solvents Soil contamination happens when these synthetic substances hold fast to the dirt, either from being straightforwardly spilled onto the dirt or through contact with soil that has effectively been tainted.

4.1 Effects of soil pollution

Impacts of soil pollution are not confined to soil and its biota but are carried over to every aspect of the environment and affect every organism from the earthworm to humans. Some of adverse effects are as follows:

a. **Human health**

Since we are dependent on the land for our food, pollution from the soil is transferred to us in this manner. Bioaccumulation of toxins occurs in our bodies, causing chronic poisoning, and leading to various diseases. Reproductive health, birth and developmental defects, neurologic effects, malnutrition, and mutations in the cells of the body leading to cancers; all these are on the increase today [7]. Considering direct impact of soil on human health because inhalation of polluted soil which have vaporized and contamination of it. Crops and plants grown on polluted soil absorb much of the pollution and then pass these on to us [8]. This could explain the sudden surge in small and terminal illnesses. Long term exposure to such soil can affect the genetic make-up of the body, causing congenital illnesses and can be carcinogenic, due to this congenital disorder or other chronic health problem created that cannot be cured easily. For example leukemia disease which is associated with higher concentration of benzene and its exposure is chronic to human health. Due to high concentration of mercury and cyclodienes, induce sufficient concentration of PCBs and cyclodienes can damage Kidney and liver toxicity. Carbamates and organophosphates can cause Neurological disorders. Arsenic, asbestos or dioxins, cause cancer and lower IQ caused by lead or arsenic, bone diseases through lead, fluoride or cadmium In fact, it can sicken the livestock to a considerable extent and cause food poisoning over a long period of time. The soil pollution can even lead to widespread famines if the plants are unable to grow in it.

b. **Growth of plants**

Contamination of soil can affect the ecological balance. Plants are mostly unable to adapt to the abrupt changes in the chemistry of the soil and this affects the microorganisms which are found in soil. This Substantial change causes soil disintegration. Enormous plots of land become infertile; unfit to help any life on it. Indeed, even the plants that do develop on these terrains will retain the poisons and move to the natural way of life. The natural equilibrium of any framework gets influenced because of the inescapable tainting of the soil. Most plants cannot adjust when the science of the soil changes so fundamentally in a brief timeframe. Growths and microbes found in the dirt that dilemma it together start to decrease, which makes an extra issue of soil disintegration. The fruitfulness gradually reduces, making land inadmissible for horticulture and any neighborhood vegetation to endure. The soil contamination makes enormous plots of land become dangerous to wellbeing. In contrast to deserts, which are appropriate for its local vegetation, such land cannot uphold most types of life.

c. **Air pollution**

Poisonous residue ascends from landfills alongside foul scent, contaminates the air and makes unfriendly impacts individuals who live close to them.

d. **Diminished Soil Fertility:**

The poisonous synthetics present in the dirt can diminish soil fertility and subsequently decline in the dirt yield. The defiled soil is then used to deliver leafy foods which needs quality supplements and may contain some harmful substance to cause genuine medical conditions in individuals burning-through them.

e. **Impact on scene and Odor contamination:**

Huge heaps of decline and trash being open unloaded and littered over a space ruins the serenity of the scene. The emanation of harmful and foul gases from landfills dirties the climate and causes genuine consequences for wellbeing of certain individuals. The horrendous smell makes burden others.

f. **Changes in Soil Structure:**

The passing of many soil living beings (for example night crawlers, creepy crawlies and microorganisms) in the dirt can prompt modification in soil structure. Aside from that, it could likewise compel their hunters to move to different spots looking for food.

g. **Impact on Ecosystem and Biodiversity:**

Soil contamination can prompt the absence of biodiversity in an environment. The existence of bird, creepy crawly, well evolved creature and reptile species that live in the dirt can get influenced by contamination. The dirt is a significant environment.

h. **Tainting of Water Sources:**

When it downpours, surface run-off conveys debased soil into water sources causing water contamination. Toxins can likewise penetrate down to debase ground water. The defiled water is subsequently unsuitable for both creature and human utilization. It will likewise influence amphibian daily routine since the living beings that experience in these water bodies will discover their living spaces inhabitable.

5. Bioremediation

Remediation means to get rid of an issue and if it is associated with taking care of an ecological issue like soil and groundwater contamination is called bio-remediation. Bioremediation is a mechanism which utilizes the living microorganisms to reduce natural contaminations or to anticipate contamination [9]. It is an evolution towards elimination of toxins from the climate in this way reestablishing the first characteristic environmental factors and forestalling further contamination. Bioremediation also can be a permanent in situ solution for contamination instead of simply translocating the problem. Remediation of heavy metals, metalloids, or other inorganic pollutants from soil or water can be done by this technique [10]. It is a cost-effective, efficient, novel, eco-friendly, and solar-driven technology with good public acceptance as compared with other engineering techniques.

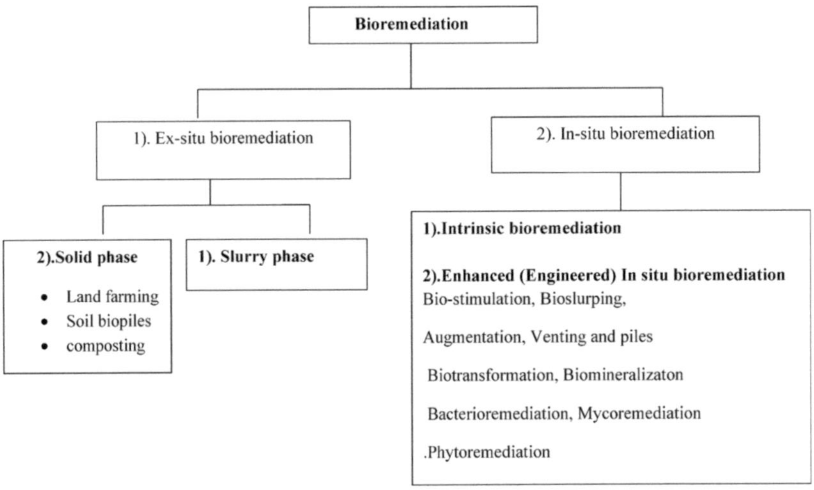

A. Based on applied strategies: bioremediation techniques applied on the basis of strategies can be classified in two categories.
Ex-situ bioremediation.
In-situ bioremediation.

5.1 Ex-situ bioremediation

Ex-situ as name suggests its mean to remove contamination mat to a remote treatment location. This classification is not much popular because it involves the big task of excavating polluted soil and transports it to offsite. The basic principal of ex situ remediation is to introducing the correct soil oxygen, moisture and nutrient conditions on offsite [11]. However, Ex situ bioremediation process poses a hazard to spreading contamination or risking an accidental spill during transport [12]. There are two technique classes can be applied explained bellow.

5.1.1 Slurry phase

This technique involves the process of combining contaminated soil with water and other additives in a large bio-reactor and mixed to keep the indigenous microorganisms in contact with the contaminants. Essential nutrients, oxygen are added and the conditions in the bio-reactor are ensured at optimum environment for the micro-organisms to degrade the contaminants. After completion of the treatment, the water is removed from the solids -wastewater is disposed and further treated if still contaminated. Slurry-phase is a relatively rapid process compared to other biological treatment processes specifically for contaminated clays [13].

5.1.2 Solid phase

Solid phase treatment use to treats soils in above-ground treatment area. This area equipped with collection systems to check the contaminants from escaping the treatment. The parameters like moisture, heat, nutrients, and oxygen are controlled to enhance rate of degradation. Solid-phase systems are simple to process and maintain in spite of, it require a large amount of space and more time of treatment than slurry-phase processes. This treatment can be achieved by following techniques [14].

5.1.2.1 Land farming

This technique basically stimulates biodegradation through indigenous microorganisms and facilitate aerobic degradation of contaminates. It is done by a simple methodology technique in which contaminated soil is excavated and spread over a prepared bed and regularly until pollutants are degraded. For promoting the growth of the indigenous species some nutrients and minerals are also added.

5.1.2.2 Soil biopiles

This biodegradation technique used for the remediation of excavated soil contaminated with petroleum contents. Soil biopiles also known as biocells. This technology involves the accumulation of contaminated soil into piles and the stimulation of microbial activity either aerobically or by adding nutrients, minerals or moisture [13]. A typical height of biopiles can be three and ten feet. This technology also uses oxygen as a method to stimulate bacterial growth. Biopiles are aerated by forcing air to move by injection through perforated piping placed throughout the pile [14].

5.1.2.3 Composting

Composting involves mixing the contaminated soil with a biomass such as straw, hay, or corncobs which make it suitable to deliver the optimum levels of air and water to the microorganisms. Composting involves the locating of the contaminated soil in treatment vessels and it is mixed there for aeration. Window composting a type of composting process in which the soil is placed in long piles named as windows and mixed by tractors regularly. A ratio of 75% contaminated soil to 25% compost use for composting. This ratio is depending on the variability of soil type, contaminants level and characteristics. Compost remediation is known as a faster remediation because it can remediate in weeks [15].

5.2 In-situ bioremediation

Bioremediation process is done at the contamination site defines the in-situ method. In situ is the preferred bioremediation method, as it requires less mechanical efforts to eliminates spreading contaminants and prevent the spread of pollutant through transportation or pumping away to other treatment locations.

In situ bioremediation are biological processes which include microorganisms metabolize organic contaminants to inorganic material, such as carbon dioxide, methane, water and inorganic salts. This process can be achieved either in natural or engineered conditions [16].

5.2.1 Types of In situ bioremediation

5.2.1.1 Intrinsic bioremediation

Intrinsic bioremediation is a process for converting environmental pollutants degrades to non-toxic forms through the immanent abilities of naturally occurring microbial population at the site. This process is usually employed in underground places as such underground petroleum tanks. Intrinsic bioremediation manages the innate capabilities of naturally occurring microbial communities to degrade environmental pollutants without modified or taking any engineering steps to accelerate the process [11]. This technique deals with stimulation of indigenous microbial population by feeding them nutrients and oxygen to increase their metabolic activity.

5.2.1.2 Enhanced (engineered) In situ bioremediation

As the name suggested this technique involves the introduction of specific microorganism to the contaminated site. Engineered in situ bioremediation accelerates the degradation process by enhancing the physicochemical conditions to increase the growth of microorganism.

A. **Bio-venting**

Bio-venting is an in situ remediation technique that uses microorganisms to degrade organic constituents adsorbed on soils [17]. This technique involves regulated stimulation of airflow for increasing oxygen to unsaturated zone for enhances the bioremediation, by increasing activities of indigenous microbes. In the process of bio-venting, amendments are done by adding nutrients and moisture to increase bioremediation to achieve microbial transformation of pollutants to a nontoxic state. This technique has gained popularity among other in situ bioremediation techniques especially in restoring sites polluted with light spilled petroleum products. Bioventing primarily use for the degradation of adsorbed fuel residuals, and also can use in the degradation of volatile organic compounds (VOCs) through biologically active soil.

B. **Bioslurping**

Bioslurping technique is the combination of bioventing and vacuum-enhanced pumping of soil and groundwater remediation by indirect provision of oxygen and stimulation of contaminant biodegradation [18]. This technique uses a "slurp" that extends into the free product layer, which draws up liquids (free products and soil gas) from this layer in a manner similar to that of how a straw draws liquid from any vessel. The bioslurping system is constituted by a well connected to an adjustable length called "slurp tube" is installed, and this slurp tube, connected to a vacuum pump, which is lowered into the light non-aqueous phase liquids (LNAPL) layer, and pumping begins to remove free product along with some groundwater. The vacuum-induced negative pressure zone in the well promotes LNAPL flow towards the well and also draws LNAPL trapped in small pore spaces above the water table. This technique used to remediate soils contaminated with volatile and semi-volatile organic compounds

C. **Biosparging**

Biosparging is basically a biological approach which removes the aromatic compounds contamination like benzene, toluene, ethylbenzene, xylene and neptthalene from an area. This process involves the loading of specific aerobic bacteria to break down the mineral oil and aromatic compounds into simpler and useful form. This technique is similar to bioventing where air is incorporate into soil subsurface to stimulate microbial activities to enhance pollutant removal from polluted sites. In biosparging air is injected at the saturated zone, which can cause upward movement of volatile organic compounds to the unsaturated zone to promote biodegradation [19]. There are two major factors which affect the biosparging process namely:

- Soil permeability (which determines pollutant bioavailability to microorganisms)
- Pollutant biodegradability

D. **Bioaugmentation**

Bioaugmentation is arrangement to enrich the existing microorganism population and make it more effective in reducing the level of contamination. This technique refers to the addition of organic culture to the contaminated soil and make environment of the site similar to a bioreactor. There are two common options can be used one is addition of a pre-adapted pure bacterial strain and second is addition of a pre-adapted consortium to the contaminated site. Bioaugmentation is mainly used in oil contaminated site for bioremediation. Bioaugmentation is a low-cost method in comparision of other methods of treating wastewater and soil contamination [20].

E. **Phytoremediation**

The direct use of green plants and their associated microorganisms to stabilize or reduce contamination in soils, sludges, sediments, surface water, or ground water is defined as Phytoremediation. This technique depends on the use of plant interactions (physical, biochemical, biological, chemical and microbiological) to contaminated sites to mitigate the toxic effects of pollutants. It is an alternative technology that can be used along with or in place of mechanical conventional clean-up technologies that often require high capital inputs and are energy intensive. Area with low concentrations of contaminants over large cleanup areas and at shallow depths presents especially favorable conditions for phytoremediation. Depending on pollutant type (elemental or organic), there are several mechanisms (accumulation or extraction, degradation, filtration, stabilization and volatilization) involved in phytoremediation [21]. Elemental pollutants (toxic heavy metals and radionuclides) are mostly removed by extraction, transformation and sequestration.

i. **Phytostabilization** - using plants to reduce heavy metal bioavailability in soil.

ii. **Phytoextraction** — using plants to extract and remove heavy metals from soil.

iii. **Phytovolatilization** — using plants to absorb heavy metal from soil and release into the atmosphere as volatile compounds.

iv. **Phytofiltration** — using hydroponically cultured plants to absorb or adsorb heavy metal ions from groundwater and aqueous waste.

1. **Phytostabilization**

   The phytostabilization process involves plants which established and function primarily to accumulate metals into tissues of root or aid in their precipitation in the root zone. This technique is based on the chemical stabilization of heavy metals using various non-organic and/or organic soil additives in connection with adequately chosen plant species [22]. Species which will be resistant to specific conditions present in the soil, such as low pH and high concentrations of heavy metals, ought to be selecting. Phytostabilization reduces the mobility of contaminants, and help to minimize the risk, of inorganic contaminants within the site. This technology does not generate contaminated secondary waste that needs further treatment. This technique basically limits the bioavailability of heavy metals and to restore adequate soil quality.

2. **Phytoextraction**

   Phytoextraction is a phytoremediation technique that uses plants to uptake and removes metals and other contaminants from soil or water [22]. This technology can be used to reduce both organic and inorganic pollutants from the soil, water and the air as well. This technology seems to be similar as solar driven pumps which can extract and concentrate certain elements from their environment. This should be achieved at a lower cost than any alternate technology as it only requires the identification and planting of such plant which possess the ability of hyperaccumulation. The ability to accumulate heavy metals varies significantly between species and between cultivars within a species [23].

3. **Phytovolatilization**

   Phytovolatilization, employs the plant-mediated uptake of contaminants and transforms them into volatile compounds, and subsequently releases these compounds in the atmosphere. In this technique plant absorbs organic pollutants an water while growing it travels from root to other parts of the plants as same or in an altered form due to its metabolic and transpiration pull.

4. **Phytofiltration**

   Phytofiltration technique is manly use to treat contaminated water. This technique involves, high metal-accumulating plants which function as biofilters, and it can be also effective in sequestering metals from polluted waters [24]. In this technique the polluteded water is either collected from a waste site or brought to the plants, or the plants are planted in the contaminated area, where the roots take up the waste water and the dissolved contaminants [25]. Many plant species naturally uptake heavy metals and other contaminant due to this it is a cost effective procedure for remediation.

5. **Phytodegradation**

   Phytodegradation technique refers to the degradation of organic contaminants through the enzymatic activates of plants. The plant releases enzymes from roots, or through metabolic activities within plant tissues. In phytodegradation organic contaminants are taken up by roots and metabolized in plant tissues to less toxic substances [26]. Phytodegradation process can degrade hydrophobic organic contaminants more efficiently.

6. **Mycoremediation**

   Mycoremediation is a technique of using fungus as a bioremediator. This biotechniques uses particular fungi that release enzymes which can degrade several pollutants and found to be promising strategies in the removal of contaminant with in a site. Mycoremediation is an efficient and economical technique as well [27].

6. Conclusion

Bioremediation is an effective technique available to clean up contaminated sites. The idea of bioremediation has a long history. However, other applications are relatively new and many other applications are emerging or being developed. This process can be aerobic or anaerobic depending on the microorganisms and the electron acceptors available. This process may be natural (intrinsic bioremediation) or it may be enhanced by man (engineered bioremediation). Several remediation approaches, particularly physical systems, involve the treatment of aqueous phase pollutants and, here, the distinction between soil and groundwater is of limited practical significance. Remediation approaches aimed primarily at treating or containing groundwater within 'geological' materials will be mentioned only briefly, whereas those commonly used for dual purposes will be considered in more detail. These technologies offer an efficient and cost effective way to treat contaminated ground water and soil.

There are other common methods of preventing soil pollution include reforestation and recycling of waste materials. De forestation often leads to erosion of the soil, which leads to soil pollution due to the loss of fertility of the soil. Thus, reforestation is an effective method of preventing soil pollution. In addition, reducing the volume of refuse or waste in landfills by recycling materials such as plastics, papers and various other materials is another effective and common method of preventing the phenomenon of soil pollution.

Overall study suggested that Pollution is a threat to our health and damages the environment and damage to soils which affects the ability to grow crops. Bioremediation can help to reduce and remove the pollution and to provide clean water, air and healthy soils for future generations. The bioremediation process is completely natural process with very less harmful side effects. It carried out in situ for most applications which do not require dangerous transport. It creates relatively few harmful byproducts. Bioremediation is way cheaper than most remediation methods because it does not require substantial equipment or labor.

Author details

Anita Verma
Department of Environmental Sciences, Babasaheb Bhimrao Ambedkar University, Lucknow, Uttar Pradesh, India

*Address all correspondence to: verma.02anita@gmail.com

References

[1] Sara P.B. Kamaludeen, Mallavarapu Megharaj, Albert L. Juhasz, Nabrattil Sethunathan, and Ravi Naidu *et. Al.,* from "Chromium–Microorganism Interactions in Soils: Remediation Implications(2003)",The University of Adelaide, Department of Soil and Water, Waite Campus, Glen Osmond, SA 5064, Australia and Tamil Nadu Agricultural University, Trichy Campus, Trichy, Tamil Nadu, India.

[2] Chaudhry F.N and Malik et.al.,from "Factors Affecting Water Pollution: A Review"(2017)Chaudhry FN, Department of Zoology, University of Gujrat, Hafiz Hayat Campus, Gujrat, Pakistan.

[3] Dr. Rajesh Kumar Mishra, Dr. Naseer Mohammad and Dr. N. Roychoudhury "Soil pollution: Causes, effects and control"Tropical Forest Research Institute P.O. RFRC, Mandla Road, Jabalpur (M.P.)– 482 021. India

[4] Mueller, E. et.al, The Effect of Acid Rain on Soil Nutrient Levels and Plant Growth.

[5] Tabatabai. M. Ali, Ames, Iowa et al, "Effect of acid rain on soils" Department of Agronomy Iowa State University.

[6] Albert L. Juhasz a,*, Euan Smith a , John Weber a , Matthew Rees b , Allan Rofe b , Tim Kuchel b , Lloyd Sansom c , Ravi Naidu et.al.,from "In vitro assessment of arsenic bioaccessibility in contaminated (anthropogenic and geogenic) soils"(2007).

[7] B.E. Johansen, The Dirty Dozen: Toxic Chemicals and the Earth's Future, Praeger Publishers, Westport, CT, 2003.

[8] Zaware Sandeep, et al, Environmental Impact Assessment on Soil Pollution Issue about Human, Health Department of Chemistry, Pacific Academy of Higher Education and Research University, Udaipur, Raj., INDIA

[9] Antonio Cristaldi a,b, *, Gea Oliveri Conti a , Eun Hea Jho c , Pietro Zuccarello a,b , Alfina Grasso a , Chiara Copat a , Margherita Ferrante et.al., from "Phytoremediation of contaminated soils by heavy metals and PAHs. A brief review".

[10] Hazrat Ali a,⇑ , Ezzat Khan and Muhammad Anwar Sajad et.al., from "Phytoremediation of heavy metals—Concepts and applications".

[11] Jim C. Philp and Ronald M. Atlas et. al., from "BIOREMEDIATION OF CONTAMINATED SOILS AND AQUIFERS".

[12] Gundula Prokop and Martin Schamann et.al.,from "Management of contaminated sites in Western Europe".

[13] EPA, (2003), Underground Storage Tanks. www.epa.gov/swerust1/ustsystm/erpdoc.pdf.

[14] M. Hyman, R. R. Dupont, Groundwater and Soil Remediation. Process Design and Cost Estimating of Proven Technologies, ASCE Press, 2001.

[15] C. J. Cunningham, J. C. Philip, Comparison of Bioaugmentation and Biostimulation in ex situ Treatment of Diesel Contaminated Soil, Land Contamination and Reclamation, University of Edinburgh, Scotland. 2000.

[16] Arpita Kulshreshtha, Ranu Agrawal, Manika Barar, Shilpi Saxena, "A Review on Bioremediation of Heavy Metals in Contaminated Water" Department of Chemistry, Jiwaji University, Gwalior (M.P.), India, Department of Chemistry, C.C.S. University, Meerut (U.P.), India

[17] Patrick Höhener and Violaine Ponsin et.al., from "In situ vadose zone bioremediation".

[18] E. Gidarakos and M. Aivalioti et.al., from "Large scale and long term application of bioslurping: The case of a Greek petroleum refinery site".

[19] Jila Baharlouei Yancheshmeh 1 *, Kazem khavazi 2 , Ebrahim Pazira 3 and Mahmood Solhi et.al.,from "Evaluation of inoculation of plant growth-promoting rhizobacteria on cadmium and lead uptake by canola and barley".

[20] MARGESIN*, R., WALDER, G., SCHINNER, F.et.al.,from "Bioremediation Assessment of a BTEX-Contaminated Soil".

[21] Irene Kuiper, Ellen L. Lagendijk, Guido V. Bloemberg, and Ben J. J. Lugtenberg et.al., from "Rhizoremediation: A Beneficial Plant-Microbe Interaction".

[22] Nadeem Sarwar , Muhammad Imran, Muhammad Rashid Shaheen, Wajid Ishaq, Asif Kamran et.al., from "Phytoremediation strategies for soils contaminated with heavy metals: Modifications and future perspectives".

[23] Abdul R. Memon &Peter Schröder et.al., from "Implications of metal accumulation mechanismsto phytoremediation'(2009).

[24] S. Dhanam et.al from "Strategies of Bioremediation of Heavy Metal Pollutants Toward Sustainable Agriculture" (2017).

[25] N. P. Singh , Jitendra Kumar Sharma , and Anita Rani Santal et.al.,from "Biotechnological Approaches to Remediate Soil and Water Using Plant–Microbe Interactions"(2015).

[26] P. Bulak, A. Walkiewicz, and M. Brzezinska et.al., from "Plant growth regulators-assisted phytoextraction" (2014).

[27] I.O..Fasidia, S.G. Jonathan, et.al. From "Biodegradation of Nigerian wood wastes by Pleurotus tuber-regium (Fries) Singer" (2008).

Chapter 17

Phytoremediatiation of Metal and Metalloid Pollutants from Farmland: An *In-Situ* Soil Conservation

Selvanayaki Krishnasamy, Ramkumar Lakshmanan and Mythili Ravichandran

Abstract

Phytoremediation is an effective technology for *in-situ* remediation of high level polluted soils. Phytoremediation is a plant-mediated approach, which involves the use of plants to absorb and remove elemental pollutants or lower their concentration or bioavailability to soil. Plants have efficacy to absorb compounds in the soil even at low concentration through their root system. Plant root system has geotropism which helps them to extend into the soil matrix and hyper accumulate heavy metals to increase their bioavailability considerably and thereby the polluted soil is domesticated and the soil fertility is enhanced. The heavy-metal-resistant endophytes give the promising effect on plant growth, by decreasing metal phytotoxicity and affecting metal translocation and accumulation in plants. It is an eye opening for researches to implement the phytoremediation of organic contaminants through endophytes that produce various enzymes to metabolize organic contaminants and reduce both the phytotoxicity and evapotranspiration of volatile contaminants. Here, we focus on the most widely used phytoremediation strategies, phytostabilization, phytoextraction, phytovolatilization, and phytofiltration in the remediation of heavy metal-polluted soil.

Keywords: phytoremediation, endophytes, phytostabilization, phytoextraction, phytovolatilization, phytofiltration

1. Introduction

Urbanization and industrialization lead to pollution that make water, air, and soil contaminated with high levels of heavy metals, organic and inorganic materials. These cause bioaccumulation and biomagnification in the ecosystem which in turn reflect in many health issues like colon cancer, heart diseases, liver, and kidney malfunction. These pollutions are solved by various methods such as removal, isolation, incineration, solidification –stabilization, vitrification, thermal treatment, solvent extraction, chemical oxidation, etc. To implement these methods several sophisticated techniques with skilled manpower are needed. They involve the transport of contaminated materials to treatment sites thus, adding ricks of secondary contamination. *In situ* techniques that are eco-friendly and more economical can be used to minimize the problem. In this scenario, biotechnology offers a technique for phytoremediation [1].

Toxic substances that are released from various industrial effluents are loaded in the water bodies. When they enter into the surrounding agricultural fields during irrigation with heavy metals (Pb, Zn, Cd, Cu, Ni, Hg), metalloids (As, Sb), inorganic compounds, radioactive chemical elements (U, Cs, Sr), petroleum hydrocarbons (BTEX), pesticides and herbicides (atrazine, bentazone, chlorinated and nitroaromatic compounds), explosives (TNT, DNT), chlorinated solvents (TCE, PCE) and industrial organic wastes (PCPs, PAHs) pollute the land [2]. Phytoremediation can be considered as one of the effective phenomenons in regenerating soil fertility.

Phytoremediation is an efficient phenomenon in which the plant (trees, shrubs, grasses, and aquatic plants) and their associated microorganisms undergo metabolic pathway to remove, degrade or isolate toxic substances from the environment using effective enzymes including both intra and extracellular enzymes [2, 3]. The word "phytoremediation" is coined from the Greek word 'phyton', meaning 'plant', and Latin 'remedium', which means 'to remedy' or 'to correct'. As the meaning indicates heavy metals and the unusual compounds that are transported to cultivated land by the polluted water bodies are converted into nontoxic through phytoremediation. When they are bio-accumulated they are metabolized by the heavy-metal-resistant endophytes. Endophytes play a key role in the reduction and in the decrease of metal phytotoxicity and affect metal translocation which is accumulated in plants. The plant role in phytoremediation and the removal of accumulated toxicity in soil is as follows: modifying the physical and chemical properties of contaminated soils, releasing root exudates and thereby increasing organic carbon, improving aeration by releasing oxygen directly to the root zone, as well as increasing the porosity of the upper soil zones, intercepting and retarding the movement of chemicals, effecting co-metabolic microbial and plant enzymatic transformations of recalcitrant chemicals and decreasing vertical and lateral migration of pollutants to groundwater by extracting available water and reversing the hydraulic gradient [4, 5]. Strategies of photoremediaton and the efficacy of endophytes will enhance the understanding level paving way for further study.

2. Phytoremediation-based strategies

2.1 Phytodegradation (phytotransformation)

A number of plant and microbial enzymes play a major role in degradating (metabolized) or mineralizing the contaminants which are hyper accumulated inside the plant cells. Phytoremediation mostly mediated by the group of enzymes are well documented. It is understood from Nitroreductases degradation of nitroaromatic compounds and glycosyltransferase that bioactivity of plant hormones are altered by glycosylation. This has been reviewed for plant hormones such as auxins, cytokinins, gibberellins and abscisic acid [6] and glutathione transferases (GSTs) that controls the internal cell pressure due to chemical-induced toxicity. It protects cell and provides tolerance by catalyzing S-conjugation between the thiol group of GSH and electrophilic moiety in the hydrophobic and toxic substrate [2].

Oxidases (Metal-modifying enzymes) which is involved in the assimilation of heavy metals into organic molecules (e.g., selenate is metabolized to dimethyl selenide), or in changing the oxidation state of metals e.g., toxic Cr (VI) is reduced to nontoxic Cr (III) [3]. Phosphatases, nitrilases and dehalogenases play a vital role in the transformation and conjugation of explosives and dehalogenases degradation. These enzymes are involved in the transformation of toxic xenobiotic compounds such as explosives, pesticides, nerve gases, and halogenated organic compounds. Nitro reductases are involved in the degradation of nitroaromatic compounds, chlorinated solvents and pesticides. Many diverse organophosphates detoxify other contaminants by reducing either halogen groups or organically bound phosphate [7].

Many endophytes are resistant to heavy metals and are capable of degrading organic contaminants. The endophyte-assisted phytoremediation has been documented in formulating biofertilizers which are providing promising result for *in situ* remediation of contaminated soils accompanied by phosphate solubilizing, biosurfactant activity in degradation of oil-contaminated soil, siderophore production, and antimicrobial activity. In addition, plants and many microorganisms contain abundance of oxidases such as laccases (degradation of anilines) and peroxidases. These enzymes are involved in forming a defense layer in many plant processes. *Populus* species and *Myriophyllium spicatum* are examples of plants that have these enzymatic systems [8]. Phytoremediation essentially comprises of six different strategies, though more than one may be used by the plant simultaneously. They are as shown in **Figure 1** and **Table 1**.

2.2 Phytostabilization

Metals are precipitated as insoluble forms by the direct action of roots which secrete phenolic and low molecular weight organic exudates subsequently trapped in the soil matrix as contaminants. Later when get accumulated organic or inorganic

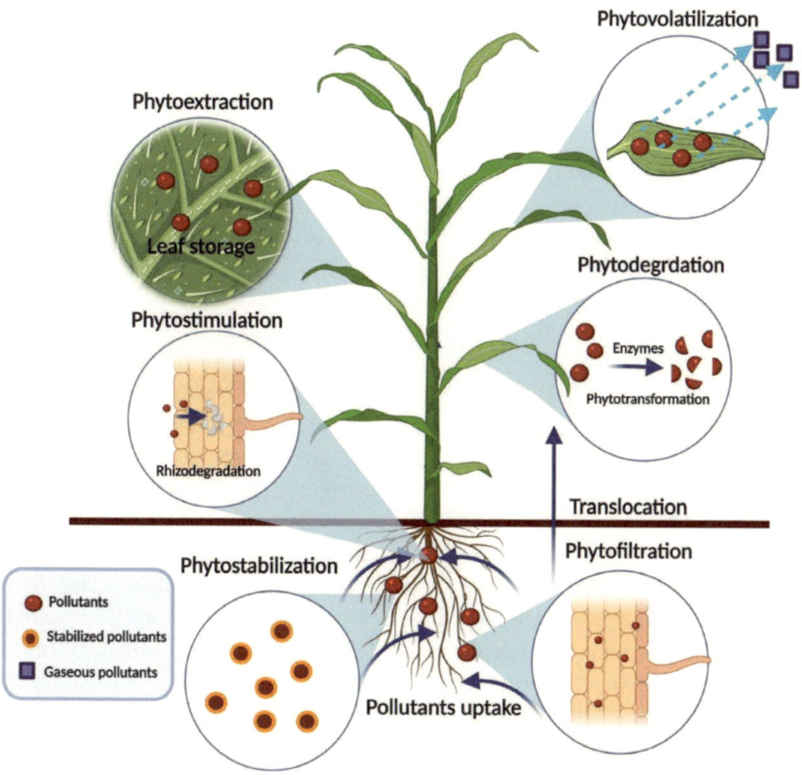

Figure 1.
Schematic representation of phytoremediation.

S.No	Type of contaminants	Medium/mode of remediation	Plant source	References
1.	1,2,4,-trichlorobenzene, Aniline, Benzene ethyl benzene, *m*-xylene, Nitrobenzene, Pentachlorophenol, Phenol, Trichloroethylene (TCE), Toluene, Methyl Tertiary Butyl Ether (MTBE), Perchloroethylene	Atmosphere/ Phytovolatilization through leaves, trunk, soil	Poplar, Russian olive, Eucalpytus, Pine and Willow trees	[9–12]
2.	Herbicides, Trichloroethylene and Methyl tert-butyl ether	Plant/ Phytodegradation through root enzymes	Cannas/ Microorganisms	[13]
3.	DDT, Polybrominated diphenyl ethers (PBDEs)	Soil/Rhizofiltration	Sunflower, Tobacco, Spinach, Rye and Indian Mustard	[14]
4.	Dichlorodiphenyl trichloroethane (DDT), Polybrominated diphenyl ethers (PBDEs) and Dichlorodiphenyl dichloroethylene	Soil/ Rhizodegradation	Rhizospheric bacterial population associated with plants	[15]
5.	Pesticides, Hydrocarbons and Animal manure	Soil/ Phytostabilization	Birch, Black locust, Oak, Scots pine and Douglas fir	[16]

Table 1.
Concise view of various phytoremediation strategies.

pollutants are incorporated into the lignin of the cell wall of cells or in humus. The main intention is to cultivate plants like *Haumaniastrum*, *Eragrostis*, *Ascolepis*, and *Gladiolus* in polluted agricultural fields to limit the mobilization and diffusion of contaminants in the soil [16–18].

The plants are involved in absorbing many toxic elements from rock, soil, and polluted water by the root system. Plant exudates aggregate metals in the soil. Soil microbes which are symbionts can decrease the toxic effects of contaminants in the soil. For example, exudate peptides from the bacterium *Pseudomonas putida* and Arbuscular Mycorrhizal Fungi (AMF) have great potential in phytostabilization and in removing metal contaminants in the soil can decrease Cd toxicity in plants. Plants can also convert contaminants into less toxic forms as well as decrease their bioavailability [19].

Siderophores, organic acids, and phenolics secreted by the microbes associated with the roots of certain plants are natural chelating compounds that form complexes with metals in the rhizosphere. In addition, plants, and their associated soil microbes play a major role in releasing chemicals that act as biosurfactants in the soil that increase the uptake of hydrocarbon toxic pollutants. These contaminants are stabilized in natural and constructed wetlands through a process called phytofiltration. It includes rhizofiltration where metals are precipitated within the rhizosphere zone and in the root membrane. Metal uptake by plants that is generally active diffusion takes place by specific protein transporters (channel proteins) or $H+$ coupled carrier proteins located along the cell membrane of the root. For example, the Fe regulated transporter (IRT1) allows the uptake of Fe. Uptake of other metals also occurs via IRT1 transporters, especially even in very low concentrations of Fe exist in the soil. By expelling the proton gradient, more ions are concentrated near the root zone. Inadvertent uptake of non-essential metals also takes place via other cell membrane transporters.

2.3 Phytovolatilization

Plants can absorb a high level organic, inorganic and heavy metals through their root system which later is metabolized and converted to nontoxic and also as volatile compounds that are released to the atmosphere by evapotranspiration. Removal of the water-soluble compounds like aldehydes takes up easily than ketones [20]. Distinctively Hg, Se, and As taken up by the roots are converted into non-toxic forms, and then released into the atmosphere during transpiration. The plant species like *Astragalus bisulcatus* and *Stanleya pinnata* for Se or transgenic plants (with bacterial genes) of *Arabidopsis thaliana*, *Nicotiana tabacum, Liriodendron tulipifera,* or *Brassica napus* for Hg can be mentioned as examples [17, 21].

2.4 Phytoextraction associated with endophytes

Phytoextraction is either a continuous process by cultivating metal hyper-accumulating plants as well as fast-growing plants or an induced process by using chemicals to increase the bioavailability of metals in the contaminated soil. This phenomenon uses the ability of plants to accumulate contaminants in the above ground. It is applied to heavy metals contaminants like Cd, Ni, Cu, Zn, Pb, Se, and As from industrial effluent mainly from the leather industry, paper and textile industries and organic compounds. Phytosequestration and phytoaccumulation are the techniques that preferentially use hyper-accumulator plants. They can store high concentrations of specific metals in their aerial parts at the rate of 0.01–1% dry weight depending on the metal. Plants such as *Elsholtzia splendens*, *Alyssum bertolonii*, *Thlaspi caerulescens,* and *Pteris vittata* are preferred for hyperaccumulator for Cu, Ni, Zn/Cd [22–24]. This process involves systematic harvesting and renewal of the biomass to lower the concentration of contaminants in the soil. Phytoextraction is a process that takes place in certain plants which undergo the accumulation of contaminants gradually (mainly metals) into their biomass. Certain plants can hyper accumulate metals without any toxic effects. These plants are adapted to naturally occurring metalliferous soils. More than 400 plant species can hyper accumulate various metals. However, most plants have the capability to hyper accumulate at least one specific metal [19].

Physiological, biochemical and molecular approaches are employed to identify the underlying mechanisms such as heavy metal accumulation and tolerance and adaptive mechanisms to cope up with heavy metal stress. Some adaptive mechanisms evolved by tolerant plants with the association of endophytes are the reason behind their gene encoded proteins and enzymes that involve in phytoremediation. This is organized by various factors including immobilization, plasma membrane exclusion, restriction of uptake and transport, synthesis of specific heavy metal transporters, chelation and sequestration of heavy metals by particular ligands, induction of mechanisms contrasting the effects of ROS and MG (such as upregulation of antioxidant and glyoxalase system), induction of stress proteins, the biosynthesis of polyamines and signaling molecules such as salicylic acid and nitric oxide [25–28].

Endophytes are ubiquitous and have been residing in all species of plants. In general, bacterial endophytes colonize the internal tissues of the plant that are nonpathogenic for their host [29]. Endophytes could produce different plant hormones like IAA, Cytokinin and gibbrellic acid to enhance the growth of the host plants. Endophytes have better adaptations against intrinsic and extrinsic stress factors, which lead to enhanced plant growth [30]. Many endophytes are the common rhizospheric bacteria which include *Pseudomonas, Bacillus, Acinetobacter, Actinobacteria, Sphingomonas* etc. that are found to be more predominant. They produce various secondary metabolites, volatile compounds and antibiotics to counteract the detrimental effect of pathogens through mechanisms parallel to that of PGPR [31]. Endophytic bacteria are developed as biocontrol agent against the fungal and bacterial phytopathogens [32]. For the phytoremediation of organic contaminants, endophytes have different enzymology to metabolize various organic contaminants and they reduce both the phytotoxicity and evapotranspiration of volatile contaminants.

Although heavy metals are toxic to plants, it has been proved that many plants are metal tolerant and some of them are metal hyperaccumulators [33]. The hyperaccumulator-associated endophytes are metal resistant, due to long-term adaptation to the high concentration of metals accumulated in the plants [34]. Hyperaccumulator associated endophytes and many metal-resistant endophytes were isolated from hyperaccumulating plants, such as *Alyssum bertolonii, Alnus firma, Brassica napus, Nicotiana tabacum, Thlaspi caerulescens, T. goesingense, and Solanum nigrum*. The reported metal-resistant endophytes belong to a wide range of taxa; in bacteria, these include *Arthrobacter, Bacillus, Clostridium, Curtobacterium, Enterobacter, Leifsonia, Microbacterium, Paenibacillus, Pseudomonas, Staphylococcus, Stenotrophomonas* and *Sanguibacte* and in fungi Microsphaeropsis, *Mucor, Phoma, Alternaria, Peyronellaea, Steganosporium* and *Aspergillus* [35].

Case studies emphasize the role of endophytic microbes that involve in the production of IAA. It helps in the plant growth promotion and the production of siderophore which means 'ron carrier' in Greek. They are small, high-affinity iron-chelating compounds that are secreted by microorganisms such as bacteria and fungi and serve primarily to transport iron across cell membranes. Biosurfactant activity of endophytic microbes enhances the emulsification of hydrocarbons and thus they have the potential to solubilize hydrocarbon contaminants and increase their availability for microbial degradation activity in oil-contaminated soil. Antimicrobial activity of endophytes gives promising effect against a broad spectrum of phytopathogen. Inoculation of plant growth-promoting bacteria (PGPR) and AMF can increase plant biomass. The AMF-plant symbionts usually reduce the accumulation of metals in the above ground tissue biomass of plants.

The role of AMF in regulating metal uptake by plants appears to vary depending on numerous factors like AMF population, plant species, nutrient availability and metal content in the soil. Even the application of specific soil fungicides, the AMF activity has resulted in increased metal accumulation in plants. Endophytes excel in the metabolism of unusual compound degradation including *Achromobacter violaceum, Pseudomonas, Bacillus, Acinetobacter, etc.* They induce phytoextraction which in turn promotes the use of fast-growing crops and chemical manipulation of the soil. The bioavailability of metals in less concentration in the soil is a limiting factor in phytoextraction. The bioavailability of metals can be increased by the use of natural chelators of low molecular weight organic acids or synthetic chelates like ethylenediaminetetracetic acid (EDTA) or acidifying chemicals like NH_4SO_4 as well as by the microbial activity like phosphate solubilizers, nitrogen fixers and complex organic contaminants degradation microbes.

The use of chelators increases the absorption of metals by the roots and helps in the translocation of metals from the roots to the foliage. The timing of chelate and its efficacy are directly proportional to the biomass production. To chelate Pb from contaminated soil, using EDTA is found to be a promising option and it can be applied to growing corn (*Zea mays*) in Pb-contaminated soil treated with

10 mmol kg^{-1} EDTA. This in turn will result in the accumulation of Pb at higher rate and facilitate the translocation of Pb from the roots to the parts of the plants. One of the limitations of using synthetic chelates enhances solubility of the metals within the soil and this increases the risk of metal migration into the soil profile and enters the groundwater. This can be avoided by treating the contaminated soil *ex-situ* in a confined site with an impermeable surface. The periodic application of low doses of synthetic chelates reduces the risk of metal migration [19].

2.5 Phytofiltration

Plants absorb soil ionic compounds by their root system through capillary action even at low concentrations. Phytofiltration is a phenomenon where plant roots (rhizofiltration), shoots (caulofiltration), or seedlings (blastofiltration) absorb water along with minerals and pollutants from contaminated or wastewaters [36].

Plants broaden their root system in search of water which deepens in the soil profile. It can establish a network in the root ecosystem. As it mounts up contaminants it aids in regaining the polluted soil and stabilizing soil fertility through the plant exudates. Root exudates often involve in altering the pH of rhizosphere, precipitating ions of heavy metals on plant root and minimizing the movement of heavy metals to underground water. When the roots become saturated, they are harvested and disposed of which minimize the soil contamination but can be dumped from one form to another form. Ideally, plants used for rhizofiltration should have a dense root system, high biomass production, and be tolerant to heavy metal.

In general the terrestrial and aquatic plants can be used for rhizofiltration. Plants cultivated in specific condition achieve biomass with effective root system while other potential submerged organs concentrate on the contaminants, especially heavy metals, radioactive elements and organic pollutants from contaminated water bodies. The plants kept in a hydroponic system absorb the concentrated contaminants when the effluents are passed and 'filtered' by the roots (Rhizofiltration) [17, 37]. Plants with high root biomass or high absorption surface with more accumulation capacity (aquatic hyperaccumulators) and tolerance to contaminants will achieve the best results. Promising examples include *Helianthus annus*, *Brassica juncea*, *Phragmites australis*, *Fontinalis antipyretica* and several species of *Salix*, *Populus*, *Lemna,* and *Callitriche* [38, 39].

Some terrestrial plants such as Indian mustard (*Brassica juncea*) and sunflower (*Helianthus annuus*) have longer, deeper and hairy root systems with good capacities to accumulate heavy metals during rhizofiltration [40, 41]. Unpredictably woody species make easy accumulation of heavy metals in their shoot system above the soil level. Their deep root system, participate effectively in preventing the soil erosion as well as the distribution of the contaminated soil to the surrounding [42]. In this concern biofiltration, bioaccumulation and biomagnification are threats related to the food chain and food web. Because of their in-edible nature this approach prevents the availability and probability of the heavy metals entering into the food chain through trees [43].

2.6 Rhizodegradation

Microbes that harbor inside plant parts like endophytic bacteria and fungal population that grows in roots are tend to promote the growth of plants and involve in degrading rhizosphere pollutants. They utilize exudates and metabolites of plants as a source of carbon and energy. In addition, plants provide biodegrading enzymes. The application of phytostimulation is targeted to organic contaminants [37]. The microbial community in the rhizosphere is diverged genetically and physiologically. This varies according to the spatial distribution of nutrients irrespective of factors [17].

There are other strategies, which are of rhizodegradation. These include:

2.6.1 Hydraulic barriers

Some large trees like *Populus* sp. have deep roots which have a major role in transpiration of groundwater in large quantities. Plant enzymes play a key role in eliminating contaminants after metabolized and vaporized together with water in plant tissues.

2.6.2 Vegetation covers

Herbs including grasses, shrubs, or trees planted on landfills, pits, trenches, or tailings minimize the infiltration of rainwater and the spread of pollutants. The roots facilitate soil aeration and in turn enhance the biodegradation, evaporation, and transpiration [44, 45]. Organic soil composed of sawdust, plant remains and NPK-fertilizers promote plant growth which helps in phytoremediation. Many field trials are emphasized at the end of a single biological cycle with 76 different plant species including cereals, shrubs, fruit trees, and even large trees like oaks and pines.

2.6.3 Constructed wetlands

The components of ecosystems comprise of organic soils, microorganisms, algae and vascular aquatic plants. All are involved in the effluent treatment through evaporation, filtration, ion exchange, adsorption and precipitation [46]. Here all the components are interlinked to phytoremediation and the entire system is given a promising effect [47]. The advantages are good cleaning efficacy, less cost of designing along with easy operation and maintenance. It is widely focused in the treatment of domestic, agricultural, and industrial wastewater, and also for treating acid mine drainages [48, 49]. Herbs (grasses, shrubs) or trees planted on landfills or tailings are used to reduce the infiltration of rainwater which is loaded with pollutants from various areas. Since there are difficulties in establishing rooting in tailings some other techniques must be evaluated for future prospective. For example, plants like *Hungarian agronomists* (*Biological Reclamation Process, BRP*) are propagated to utilize residues of organic soil that is composed of sawdust, plant remains and NPK-fertilizers [50].

2.6.4 Phytodesalination

The cultivation of halophytes on salt-rich soil is to improve the productivity of the soil and to remove the excess salt from saline soil [51, 52]. The potential of *Suaeda maritima* and *Sesuvium portulacastrum* is used in removal and accumulation of NaCl from highly saline soil. The plants in saline soil accumulate sodium in shoots and aerial parts which is based on the soil nature and the climatic conditions. The upper horizon of the soil layer is leached by halophytes [53].

3. Recent advancements in phytoremediation

To enhance the rate of phytoremediation, to improve the adaptation to various environmental conditions and to minimize their limitations such as slow-growth several strategies are developed through recombinant DNA technology to create transgenic plants or plant hybridization with fast-growing hyperaccumulators and microbe-associated phytoremediation. The hyperaccumulators can accumulate high levels of contaminants including heavy metals and other pollutants. Electrofusion is used for the fusion of protoplasts between two plants namely *T. caerulescens* which has a high-level Zn accumulator and *Brassica napus* which has a biomass production capability. The resulted somatic hybrids have the properties like hyperaccumulation capability, tolerance derived from *T. caerulescens* and higher biomass production derived from *B. napus* [54]. Moreover in comparison with rhizosphere microorganisms, endophytes have close interaction with their host plants. Genetically modified endophytes can be used as bio-fertilizers that could more efficiently improve phytoremediation [28].

3.1 Genetic engineering

Genetic engineering is a tool for improving strains in industries and clinics. It also enhances the phytoremediation abilities of plants in removing heavy metals. To generate genetically modified plants, a foreign source of the gene of interest which can be obtained from an organism, such as a plant species or even bacteria or animals, is transferred and inserted into the genome of a target plant through a proper vector system. After DNA recombination, the foreign gene gets integrated and inherited that confers specific traits to the plants. Moreover, genetic engineering has tools to transfer desirable characters from hyperaccumulator source plant to sexually incompatible plant species, which is impossible through traditional methods including vegetative propagation [55].

Therefore, creating transgenic plants with the desired gene expression in traits has attracted the researchers in the field of phytoremediation. Genetically modifying traits are fast-growing and high-biomass species with high tolerance against heavy metals. Their accumulation ability is more desirable than hyperaccumulators because sometimes hyperaccumulation may be harmful for biomass. Therefore, the selection of genes for genetic engineering should be based on heavy metal tolerance, construction of metabolic pathways in detoxifying heavy metals and accumulation mechanisms in plants. As heavy metals accumulation may create oxidative stress due to excessive Reactive Oxygen Species (ROS), a defense system provide heavy metal tolerance. To increase heavy metal accumulation through genetic engineering, genes are put under the control of strong promoter. The signal sequences facilitate in the uptake, translocation, and sequestration of heavy metals in elevated levels [56].

As metal chelators act as metal-binding ligands to improve heavy metal bioavailability, they promote heavy metal uptake and root-to-shoot translocation, as well as mediate intracellular sequestration of heavy metal ions in organelles. By over expression of genes encoding natural chelators, heavy metal uptake and translocation can be improved [57]. For example the supply of histidine is a nickel chelating agent and when it is supplied to plants which are originally non-accumulating species for metals greatly increases both its nickel tolerance and nickel transport to the shoot. It indicates the role of histidine in the hyper accumulation of nickel in *Alussum* plants [58].

Although the genetic engineering approach is a promising one, a few setbacks are there when the concentration is toxic to cells. On other hand, construction of all desired genes (that involved in mechanisms of detoxification and accumulation of heavy metals) is a time as well as effort consuming process and hence it is not providing a promising effect in the present scenario. Moreover it is difficult to get approval for the cultivation of genetically modified plants in test fields due to its toxicity, allergic levels and risk factor to ecosystem. Therefore, the researchers focus on alternate approaches to improve plants' role in phytoextraction.

3.2 Role of endophytes in phytoremediation

The role of plant-associated microorganisms (rhizospheric microorganisms) can be considered as an alternative approach to improve plant performance for phytoremediation as expressed in **Figure 2** and **Table 2**. The microbial communities

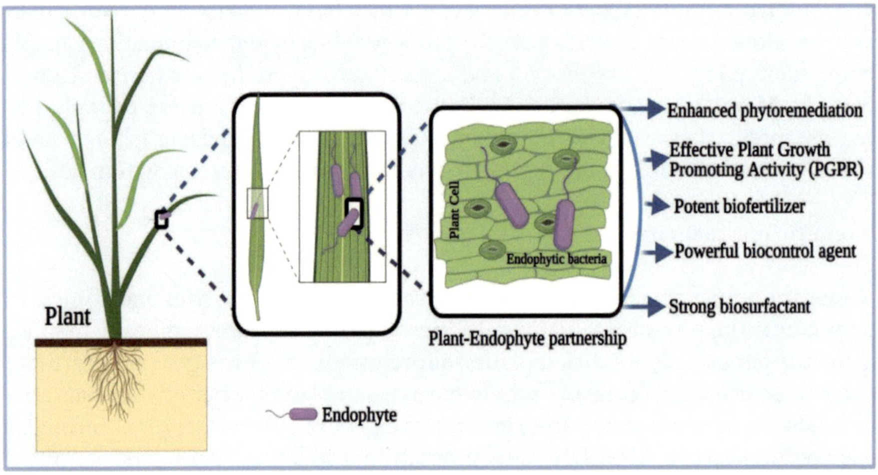

Figure 2.
Role of endophytes in sustainable ecological perspective.

S.No.	Endophytes	Plant Source	Tissue parts used	Resistant metal	References
1.	Bacteria	*Prosopis laevigata, Sphaeralcea angustifolia*	Root	Pb-Zn	[59]
2.	*K. ascorbata*	*Solanum lycopersicum, Brassica napus*	Seed tissue	Ni	[60]
3.	Rhizobacteria	*Thalaspi caerulescens, Alyssum bertolonii*	Root	Zn and Ni or Ni	[61, 62]
4.	*Microbacterium, Bacillus, Arthrobacter, Flavobacterium*	*Solanum nigrum*	Roots	Cd	[63]
5.	Bacteria	*Populus alba*	Leaves	Cd, Co, Pb	[64]
6.	*P. fluorescens, Microbacterium* sp. *Enterobacter* sp., *Xanthomonadaceae Pseudomonas* sp.,	*Brassica napus*	Roots	Pb	[65]
7.	*Pseudomonas fulva, Stenotrophomonas* sp., *Clostridium aminovalericum, Sanguibacter* sp.	*Nicotiana tabacum*	Seed	Cd	[66]
8.	Arbuscular mycorrhizal fungi (AMF), *Pseudomonas putida*	*Agrostis tenuis, Festuca rubra*	Roots	Cd. Cr(lll)	[19]
9.	*Kluyvera ascorbata*, *Pseudomonas tolaasii*, *P. fluorescens*, *Variovorax paradoxus*, *Rhodococcus* sp. and *Flavobacterium* sp.	*Graminaceae*, and *Brassica juncea*	Leaves and Roots	Cd, Zn, Cu, Ni, Co, Cr, Pb	[67, 68]
10	Plant growth promoting bacteria and Arbuscular Mycorhizal Fungi (AMF)	*Liriodendron tulipifera*	Roots	methyl-Hg	[19]

Table 2.
Documented records of role endophytes- based phytoremediation.

have symbiotic association with rhizosphere stimulating root proliferation and thus, promoting plant growth. They have increased heavy metal tolerance and plant fitness among the flora in local biosphere [69]. Plant growth-promoting rhizobacteria (PGPR) have a key role in phytoremediation. PGPR can promote plant growth via IAA production, antimicrobial activity, increase plant tolerance against heavy metals and improved nutrient uptake through diffusion as well as uptake of heavy metal from contaminated soil, translocation etc., [22]. This is achieved by producing various compounds such as organic acids for organic pollutant degradation, iron chelating siderophores, antibiotics, various enzymes involved in phytoremediation and growth promoting phytohormones [22]. PGPR can degrade the ethylene precursor ACC by synthesizing the 1-aminocyclopropane-1-carboxylate (ACC) deaminase. PGPR minimizes ethylene production and thus in turn promotes plant growth [70, 71].

Plants inoculated with PGPR containing extensive root and shoot densities result in enhanced uptake of heavy metals by the influence of ACC deaminase which promote phytoremediation efficiency [70, 72]. PGPR induces the formation of lateral root and root hair development, thus promoting plant growth and improving phytoremediation with bacterial indole acetic acid (IAA) [73]. Arbuscular mycorrhizal fungi (AMF) are the vast group of fungi, an important microbial community are predominant in soil profile that support plants for phytoremediation. AMF in rhizospheres increases the surface area for root absorption with an extensive hyphal network. They improve the uptake of water, nutrients and heavy metal bioavailability [74]. AMF can also produce phytohormones to promote plant growth and biosurfactant aids in phytoremediation [75].

A plant employs various strategies to enhance heavy metal bioavailability for better absorption. Root exudates promote desorption of heavy metals by making insoluble complexes of contaminants to free ions, by decreasing soil pH, which thus facilitate the accumulation of heavy metals in the soil for easy absorption near the roots [76]. Plants secrete metal-mobilizing compounds such as phytosiderophores, carboxylates, and organic acids in rhizosphere. According to the bioavailability heavy metals/metalloids in the soil are classified as high, moderate and low bioavailable heavy metals/metalloids. The high bioavailable are Cd, Ni, Zn, As, Se, Cu, moderately bioavailable heavy metals are Co, Mn, Fe, and least bioavailable are Pb, Cr [77].

4. Conclusion

Heavy metal pollution is a major issue which invade even in the breast milk of mother. Their toxic effect leads to multi organ failure and several cancers. They readily enter into the agricultural products and food. Health deteriorates due to the toxic effects and rapid accumulation in the environment through irrigation of contaminated water bodies. To prevent or mitigate heavy metal contamination and renovate the contaminated soil, a variety of techniques have been developed including phytoremediation. It has been proved to be a promising technique to remediate heavy metal-polluted soil.

Hyper-accumulation is the most straightforward approach for phytoremediation, and hundreds of hyper-accumulator plants have been identified so far. Phytoremediation has a few limitations including time-consuming process of plants in clearing the contaminants due to their slow growth in altered soil. But the genetic engineering is a powerful tool to modify the plants with resistant traits like fast growth even in polluted soil, high biomass production, heavy metal tolerance by designing their metabolic pathways and good adaption for surviving in various climatic and geological conditions.

Hence, a better understanding of the plant mechanisms for phytoremediation is more essential which comprises of absorption, translocation, and detoxification of pollutant in plants. These are mediated by different biomolecules and metabolic pathways. Their limitations can be overcome by genetic engineering of plants and endophytes to promote more effective way to create sustainable ecosystem. These engineered microbial consortiums can be used to improve soil health and further promote plant growth and fitness. Practically, a single approach will never be effective to the revival of heavy metal-polluted soil. So the combination of different new approaches such as genetic engineering, microbe-assisted bio fertilizer for plant growth promotion as well as detoxification of pollutants, and chelate-assisted approaches to concentrate the pollution near to rhizosphere are vital for highly effective and extensive phytoremediation in future.

Acknowledgements

Author thank Biorender for making drawing in this chapter.

Author details

Selvanayaki Krishnasamy*, Ramkumar Lakshmanan and Mythili Ravichandran
Department of Microbiology, K.S. Rangasamy College of Arts and Science, Tiruchengode, Tamil Nadu, India

*Address all correspondence to: selbiofilms@gmail.com

References

[1] Gomez Orea D. Recovery of degraded spaces. Madrid, Barcelona, Mexico: Ediciones Mundi-Prensa. 2004; Pp- 583.

[2] Deavall DG, Martin EA, Horner JM, Roberts R. Drug-induced oxidative stress and toxicity. Journal of Toxicology. 2012; 645-460. doi: 10.1155/2012/645460.

[3] De Souza MP, Pilon-Smits EAH, Terry N. "The physiology and biochemistry of selenium volatilization by plants," in Phytoremediation of Toxic Metal-Using Plants toClean up the Environment, I. Raskin and B. D. Ensley, Eds., Wiley, New York, NY, USA. 2000; pp. 171-190,

[4] Chang Y, Corapcioglu MY. Plant-enhanced subsurface bioremediation of nonvolatile hydrocarbons. Journal of Environmental Engineering. 1998; 112: 162-169.

[5] Evanko CR, Dzombak DA. 1997. Remediation of metals-contaminated soils and groundwater. Technology Evaluation Report. Pittsburgh: GWRTAC-Ground-Water Remediation Technologies Analysis Center.

[6] Kleczkowski K, Schell F. Phytohormone conjugates: nature and function. Critical Reviews in Plant Sciences. 1995; 14: 283-298.

[7] Lytle CM, Lytle PW, Yang N. "Reduction of Cr(VI) to Cr(III) by wetland plants: potential for in situ heavy metal detoxification," Environmental Science & Technology. 1998; 32(20): 3087-3093.

[8] Rylott EL, Bruce NC. Plants disarm soil: engineering plants for the phytoremediation of explosives. Trends in Biotechnology. 2008; 27(2): 73-81.

[9] Arnold CW, Parfitt, DG, Kaltreider, M: Phytovolatilization of oxygenated gasoline-impacted groundwater at an underground storage tank site via conifers. International Journal of Phytoremediation. 2007; 9 (1): 53-69.

[10] Wang X, Dossett MP, Gordon MP, Strand SE. Fate of carbon tetrachloride during phytoremediation with poplar under controlled field conditions. Environmental Science & Technology. 2004; 38(21): 5744-5749.

[11] James CA, Xin G, Doty SL, Muiznieks I, Newman L, Strand SE. A mass balance study of the phytoremediation of perchloroethylene-contaminated groundwater. Environmental Pollution. 2009; 157 (89): 2564-2569.

[12] Marr LC, Booth EC, Andersen RG, Widdowson MA, Novak JT. Direct volatilization of naphthalene to the atmosphere at a phytoremediation site. Environmental Science & Technology. 2006; 40 (17): 5560-5566.

[13] Sandermann HJ. Higher plant metabolism of xenobiotics: The 'green liver' concept. Pharmacogenetics, 1994; 4 (5): 225-241.

[14] Risky Ayu Kristanti, Wei Jie Ngu, Adhi Yuniarto, Tony Hadibarata. Rhizofiltration for Removal of Inorganic and Organic Pollutants in Groundwater: a Rev. Biointerface Research in Applied Chemistry. 2021; 11(4): 12326-12347.

[15] Sivaram AK, Logeshwaran P, Lockington R, Naidu R, Megharaj M. Low molecular weight organic acids enhance the high molecular weight polycyclic aromatic hydrocarbons degradation by bacteria. Chemosphere. 2019; 222: 132-140.

[16] Berti WR, Cunningham SD. Phytostabilization of metals. In: Raskin I, Ensley BD. (ed.) Phytoremediation of toxic metals.

Using plants to clean up the environment.New York: John Wiley & Sons, Inc. 2000.; 71-88.

[17] Ali H, Khan E, Sajad MA: Phytoremediation of heavy metals – Concepts and applications. *Chemosphere*. 2013; 91: 869-881.

[18] Domínguez MT, Madrid F, Marañón T, Murillo JM. Cadmium availability in soil and retention in oak roots: potential for phytostabilization. Chemosphere. 2009; 76: 480-486.

[19] Greipsson S. Phytoremediation. National Educational Knowledge. 2011; 3(10): 7.

[20] Tani A, Hewitt CN. Uptake of aldehydes and ketones at typical indoor concentrations by houseplants. Environmental Science & Technology. 2009; 43: 8338-8343.

[21] Ruiz ON, Daniell H. Genetic engineering to enhance mercury phytoremediation. Current Opinion in Biotechnology. 2009; 20: 213-219.

[22] Ma Y, Prasad M, Rajkumar M, Freitas H. Plant growth promoting rhizobacteria and endophytes accelerate phytoremediation of metalliferous soils. Biotechnology Advances. 2011; 29: 248-258. doi: 10.1016/j.biotechadv.2010.12.001

[23] Xie QE, Yan XL, Liao XY, Li X. The arsenic hyperaccumulator fern *Pteris vittata* L. Environmental Science & Technology. 2009; 43(22): 8488-8495.

[24] Van der Ent A, Baker AJM, Reeves RD, Pollard AJ, Schat H. Hyperaccumulators of metal and metalloid trace elements: Facts and fiction. Plant and Soil. 2013; 362: 319-334.

[25] Dalcorso G, Farinati S, Furini A. "Regulatory networks of cadmium stress in plants," Plant Signaling and Behavior, 2010; 5: (6):1-5.

[26] Hossain MA, da Silv JAT, Fujita M. "Glyoxalase system and reactive oxygen species detoxification system in plant abiotic stress response and tolerance: an intimate relationship," in *Abiotic Stress in Plants-Mechanisms and Adaptations*, A. K. Shanker and B. Venkateswarlu, Eds., 2011;pp. 235-266.

[27] Hossain MA, Hossain MD, Rohman MM, da Silva JAT, Fujita M. "Onion major compounds (flavonoids, organosulfurs) and highly expressed glutathione-related enzymes: possible physiological interaction, gene cloning and abiotic stress response," in *Onion Consumption and Health*, C. B. Aguirre and L. M. Jaramillo, Eds., Nova Science Publishers, 2012; New York, NY, USA.

[28] Zhang Y, He L, Chen Z, Zhang W, Wang Q, Qian M, Sheng X. Characterization of lead-resistant and ACC deaminase-producing endophytic bacteria and their potential in promoting lead accumulation of rape. Journal of Hazard Materials. 2011; 186:1720-1725.

[29] Schulz BJE, Boyle CJC, Sieber TN. 2006. Microbial root endophytes. 1994; pp. 1-13.

[30] Pillay VK, Nowak J. Inoculum density, temperature, and genotype effects on *in vitro* growth promotion and epiphytic and endophytic colonization of tomato (*Lycopersicon esculentum* L.) seedlings inoculated with a pseudomonad bacterium. Canadian Journal of Microbiology. 1997; 43: 354-361.

[31] Lodewyckx C, Vangronsveld J, Porteous F, Moore E R, Taghavi S, Mezgeay M, der Lelie DV. Endophytic bacteria and their potential applications. Critical Reviews in Plant Sciences. 2002; 21:583-606.

[32] Berg G, Krechel A, Ditz M, Faupel A, Sikora RA, Ulrich A, Hallmann J. Endophytic and ectophytic potatoassociated bacterial communities differ in structure and antagonistic function against plant pathogenic fungi. FEMS Microbiology Ecology. 2005; 51: 215-229.

[33] Rosa G, Peralta-Videa JR, Montes M, Parsons JG, Cano-Aguilera I, Gardea-Torresdey JL. Cadmium uptake and translocation in tumbleweed (*Salsola kali*), a potential Cd-hyperaccumulator desert plant species: ICP/OES and XAS studies. Chemosphere. 2004; 55: 1159-1168.

[34] Idris R, Trifonova R, Puschenreiter M, Welzel WW, Seissitsch A. Bacterial communities associated with flowering plants of the Ni hyperaccumulator Thlaspi goesingense. Applied and Environmental Microbiology. 2004; 70:2667-2677.

[35] Li HY, Li DW, He CM, Zhou ZP, Mei T, Xu HM. Diversity and heavy metal tolerance of endophytic fungi from six dominant plant species in a Pb-Zn mine wasteland in China. *Fungal Ecol.* 2011; doi:10.1016/j.funeco.2011.06.002.

[36] Mesjasz-Przybyłowicz J, Nakonieczny M, Migula P, Augustyniak M,Tarnawska M, Reimold U. Uptake of cadmium, lead nickel and zinc from soil and water solutions by the nickel hyperaccumulator Berkheya coddii. *Acta Biologica Cracoviensia Botanica*. 2004; 46: 75-85.

[37] Frers C. El uso de plantas acuáticas en el tratamiento de aguas residuales. Carmen de Areco, Argentina: El Planeta Azul., 2009; 11: 301-305

[38] Favas PJC, Pratas J, Prasad MNV. Accumulation of arsenic by aquatic plants in large scale field conditions: Opportunities for phytoremediation and bioindication. Science of the Total Environment. 2012; 433: 390-397.

[39] Pratas J, Favas PJC, Paulo C, Rodrigues N, Prasad MNV. Uranium accumulation by aquatic plants from uranium-contaminated water in Central Portugal. International Journal of Phytoremediation. 2012; 14: 221-234.

[40] Dhanwal P, Kumar A, Dudeja S, Chhokar V, Beniwal V. "Recent advances in phytoremediation technology," in *Adv. Environ.Biotechnol.,* eds (Singapore: Springer), 2017; 227-241. doi: 10.1007/978-981-10-4041-2_14.

[41] Rezania S, Taib SM, Md Din MF, Dahalan FA, Kamyab H. Comprehensive review on phytotechnology: heavy metals removal by diverse aquatic plants species from wastewater. Journal of Hazardous Materials. 2016; 318: 587-599. doi: 10.1016/j.jhazmat.2016.07.053.

[42] Suman J, Uhlik O, Viktorova J, Macek T. Phytoextraction of heavy metals: a promising tool for clean-up of polluted environment? Frontiers in Plant Science. 2018; 9:1476. doi: 10.3389/fpls.2018.01476.

[43] Burges A, Alkorta I, Epelde L, Garbisu C. From phytoremediation of soil contaminants to phytomanagement of ecosystem services in metal contaminated sites. International Journal of Phytoremediation. 2018; 20: 384-397.

[44] Brooks RR, Chiarucci A, Jaffre T. Revegetation and stabilization of mine dumps and other degraded terrain. In: Brooks RR. (ed.) Plants those hyperaccumulate heavy metals: their role in phytoremediation, microbiology, archaeology, mineral exploration and phytomining. New York: CAB International. 1998; 227-247.

[45] Jorba M, Vallejo R. La restauracion ecologica de canteras: un caso con aplicación de enmiendas orgánicas y riegos. Ecosistema. 2008; 17(3): 119-132.

[46] Fonder N, Headley T. The taxonomy of treatment wetlands: A proposed classification and nomenclature system. Ecological Engineering. 2013; 51: 203-211.

[47] Horne AJ. Phytoremediation by constructed wetlands. In: Terry N, Bañuelos G. (eds.) Phytoremediation of contaminated soil and water. New York: Lewis Publishers. 2000; Pp13.

[48] Adams A, Raman A, Hodgkins D: How do the plants used in phytoremediation in constructed wetlands, a sustainable remediation strategy, perform in heavy-metal-contaminated mine sites?. Water and Environment Journal. 2013; 27(3): 373-386.

[49] Lopez Pamo E, Aduvire O, Barettino D.Tratamientos pasivos de drenajes acidos de mina: estado actual y perspectivas de futuro. Boletín Geologicoy Minero. 2002; 113(1): 3-21.

[50] Gonzalez V. A indústria extractiva e o ambiente. Boletim de Minas. 1990; 27(3): 311-323.

[51] Khan HE, Sajad MA. Phytoremediation of heavy metals – Concepts and applications. Chemosphere. 2013; 91: 869-881.

[52] Zhu XF, Zheng C, Hu YT. "Cadmium-induced oxalate secretion from root apex is associated with cadmium exclusion and resistance in *Lycopersicon esulentum*," Plant, Cell & Environment. 2011; 34(7):1055-1064.

[53] Ravindran KC, Venkatesan K, Balakrishnan V, Chellappan KP, Balasubramanian T. 2007.Restoration of saline land by halophytes for Indian soils. Soil Biology and Biochemistry. 39: 2661-2664.

[54] Brewer EP, Saunders JA, Angle JS, Chaney RL, Mcintosh MS. Somatic hybridization between the zinc accumulator *Thlaspi caerulescens* and *Brassica napus*. Theoretical and Applied Genetics. 1999; 99: 761-771. doi: 10.1007/s001220051295.

[55] Berken A, Mulholland MM, Leduc DL, Terry N. Genetic engineering of plants to enhance selenium phytoremediation. Critical Reviews in Plant Sciences. 2002; 21: 567-582. doi: 10.1080/0735-260291044368.

[56] Das N, Bhattacharya S, Maiti MK. Enhanced cadmium accumulation and tolerance in transgenic tobacco over expressing rice metaltolerance protein gene OsMTP1 is promising for phytoremediation. *Plant Physiology and Biochemistry*. 2016; 105: 297-309. doi: 10.1016/j.plaphy.2016.04.049.

[57] Wu G, Kang H, Zhang X, Shao H, Chu L, Ruan C. A critical review on the bio-removal of hazardous heavy metals from contaminated soils: issues, progress, eco- environmental concerns and opportunities. Journal of Hazardous Materials. 2010; 174: 1-8. doi: 10.1016/j.jhazmat.2009.09.113.

[58] Ute Kramer, Janet D, Cotter-Howells, Andrew C.Smith. Free histidine as a metal chelator in plants that accumulate nickel. Nature. 1996; 379: 635-638.

[59] Brenda Roman-Ponce, Juan Ramos-Garza, María Soledad Vasquez-Murrieta, Flor Nohemí Rivera-Orduna, Wen Feng Chen , Jun Yan, Paulina Estrada-de Los Santos and En Tao Wang. Cultivable endophytic bacteria from heavy metal (loid)-tolerant plants. 2016; 198(10): 941-956. DOI: 10.1007/s00203-016-1252-2.

[60] Burd GI, Dixon DG Glick BR. A plant growth-promoting bacterium that decreases nickel toxicity in seedlings. Applied and Environmental Microbiology. 1998; 64(3): 3663-3668.

[61] Delorme TA, Gagliardi JV, Angle JS, Chaney R. Influence of the zinc hyperaccumulator *Thalaspi caerulescens* J. and C. Presl and the nonmetal accumulator *Trifolium pratense* L. on soil microbial populations. Canadian Journal of Microbiology. 2001; 47(8): 773-776.

[62] Mengoni A, Barzanti R, Gonnelli C, Gabbrielli R, Bazzicalupo M. Characterization of nickel-resistant bacteria isolated from serpentine soil. Environmental Microbiology. 2001; 3(11): 691-698.

[63] Luo S, Chen L, Chen J, Xiao X, Xu T,Wan Y, Rao C, Liu C, Liu Y, Lai C, Zeng G. Analysis and characterization of cultivable heavy metal-resistant bacterial endophytes isolated from Cd hyperaccumulator *Solanum nigrum* L. and their potential use for phytoremediation. Chemosphere 2011; 85: 1130-1138.

[64] Balestrazzi A, Bonadei M, Quattrini E, Carbonera D. Occurrence of multiple metal-resistance in bacterial isolates associated with transgenic white poplars (*Populus alba* L.). Annals of Microbiology. 2009; 59:17-23.

[65] Sheng X, Xia J, Jiang C, He L, Qian M. Characterization of heavy metal-resistant endophytic bacteria from rape (*Brassica napus*) roots and their potential in promoting the growth and lead accumulation of rape. Environmental Pollution. 2008b; 156:1164-1170.

[66] Mastretta C, Taghavi S, van der Lelie D,Mengoni A, Galardi F, Gonnelli C, Barac T, Boulet J,Weyens N, Vangronsveld J. Endophytic bacteria from seeds of *Nicotiana tabacum* can reduce cadmium phytotoxicity. International Journal of Phytoremediation. 2009;11: 251-267.

[67] Dell'Amico E, Cavalca L, Andreoni V. Analysis of rhizobacterial communities in perennial *Graminaceae* from polluted water meadow soil, and screening of metal-resistant, potentially plant growth-promoting bacteria. FEMS Microbiology Ecology. 2005; 52(2): 153-162. [doi:10.1016/j.femsec.2004.11.005]

[68] Belimov AA, Hontzeas N, Safronova VI, Demchinskaya, SV, Piluzza G, Bullitta S and Glick BR. Cadmium- tolerant plant growth-promoting bacteria associ ated with the roots of Indian mustard (*Brassica juncea* L. Czern.). Soil Biology and Biochemistry. 2005; 37 (2): 241-250.

[69] Fasani E, Manara A, Martini F, Furini A, DalCorso G. The potential of genetic engineering of plants for the remediation of soils contaminated with heavy metals. Plant, Cell & Environment. 2018; 41: 1201-1232. doi: 10.1111/pce.12963.

[70] Arshad M, Saleem M and Hussain S. Perspectives of bacterial ACC deaminase in phytoremediation. Trends in Biotechnology. 2007; 25: 356-362.

[71] Glick BR. Bacteria with ACC deaminase can promote plant growth and help to feed the world. Microbiological Research. 2014; 169: 30-39. doi: 10.1016/j.micres.2013.09.009.

[72] Huang XD, El-Alawi Y, Penrose DM, Glick BR, Greenberg BM. Responses of three grass species to creosote during phytoremediation. Environmental Pollution. 2004; 130: 453-463. doi: 10.1016/j.envpol.2003.12.018

[73] Glick BR. Using soil bacteria to facilitate phytoremediation. Biotechnology Advances. 2010; 28: 367-374. doi: 10.1016/j.biotechadv.2010.02.001

[74] Gohre V, Paszkowski U. Contribution of the arbuscular mycorrhizal symbiosis to heavy metal

phytoremediation. Planta. 2006; 223: 1115-1122. doi: 10.1007/s00425-006-0225-0

[75] Vamerali T, Bandiera M, Mosca G. Field crops for phytoremediation of metal-contaminated land. Environmental Chemistry Letters. 2010; 8: 1-17. doi: 10. 1007/s10311-009-0268-0

[76] Thangavel P, Subbhuraam C. Phytoextraction: role of hyperaccumulators in metal contaminated soils. Proceedings of the Indian National Science Academy. 2004; 70: 109-130.

[77] Prasad MNV. Phytoremediation of metal-polluted ecosystems: hyper for commercialization. Russian Journal of Plant Physiology. 2003; 50: 686-701. doi: 10.1023/A: 1025604627496.

Chapter 18

Phytochelatin Synthase in Heavy Metal Detoxification and Xenobiotic Metabolism

Ju-Chen Chia

Abstract

Phytochelatin synthase (PCS) is well-known for its role in heavy metal detoxification in plants, yeasts and non-vertebrate animals. It is a protease-like enzyme that catalyzes glutathione (GSH) to form phytochelatins (PCs), a group of Cys-rich and non-translational polypeptides with a high affinity to heavy metals. In addition, PCS also functions in xenobiotic metabolism by processing GS-conjugates in the cytosol. Because PCS is involved in GSH metabolism and the degradation of GS-conjugates, it is one of the important components in GSH homeostasis and GSH-mediated biodegradation. This chapter reviews the biochemical mechanism of PCS, how the enzyme activity is regulated, and its roles in heavy metal detoxification as well as GS-*S*-conjugate metabolism. This chapter also highlights the potential applications of PCS in the improvement of plant performance under combined stresses.

Keywords: Phytochelatin synthase, heavy metal stress, GS-conjugate metabolism, glutathione, combined pollution

1. Introduction

Phytochelatins (PCs, $(\gamma\text{Glu-Cys})_n\text{-Gly}$, n = 2–11) are cysteine-rich polypeptides that are synthesized non-translationally from the tripeptide glutathione (GSH, γGlu-Cys-Gly); this process is catalyzed by phytochelatin synthase (PCS, EC 2.3.2.15) [1–4]. PCs play essential roles in heavy metal detoxification because of their high affinities to a broad range of metal ions, e.g. cadmium (Cd), mercury (Hg), arsenic (As), zinc (Zn), lead (Pb), silver (Ag), nickel (Ni) and copper (Cu) [1–3]. Upon exposure to heavy metals, PCs are synthesized in the cytosol to chelate free metal ion and to prevent the generation of hydroxyl radicals [4–6] (**Figure 1**). These PC-metal complexes eventually are transferred into the vacuole through specific tonoplast ABCC-type transporters for sequestration [22–26] (**Figure 1**). In plants, PCS is constitutively expressed in the cytosol and can be activated by multiple types of metal ions [1, 3, 6]. For example, AtPCS1 isolated from *Arabidopsis thaliana* can be activated by the metal ions mentioned above [6]. In addition, some PCS homologs, such as the model legume *Lotus japonicus* LjPCS1 and LjPCS3, can be activated by iron (Fe) and aluminum (Al) [27].

PCS can be found in plants, yeasts and non-vertebrate animals and plays a critical role in responding to heavy metal stress in these organisms [28–32]. It was first partially purified from the suspension cells of bladder campion (*Silene cucubalus*) for its ability to synthesize PCs from GSH in the presence of Cd^{2+} [4]. Soon after the isolation of the enzyme, the genes coding PCS were cloned from plant and yeast

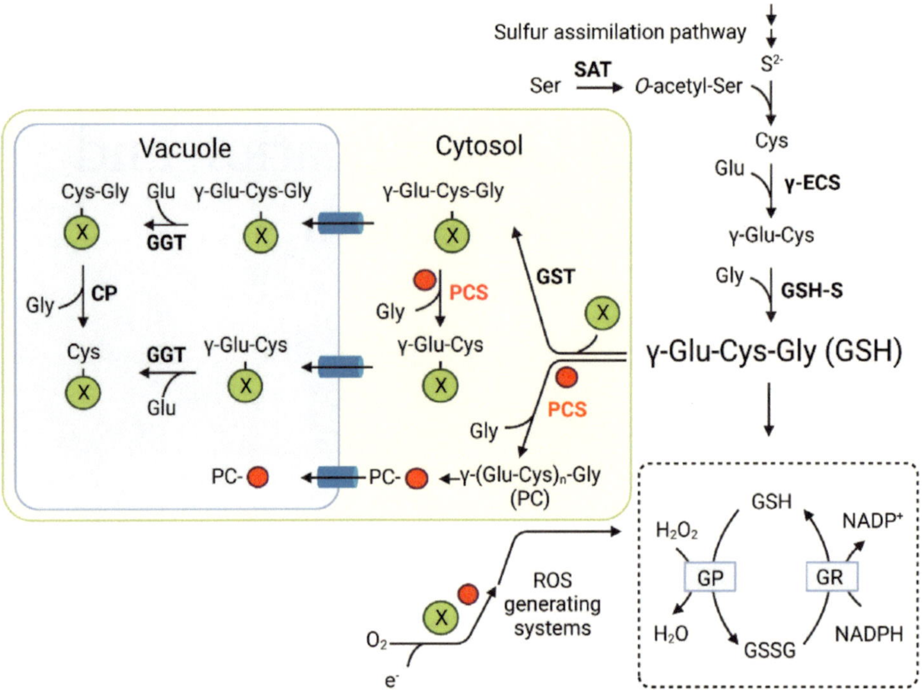

Figure 1.
The involvement of phytochelatin synthase in glutathione metabolism, heavy metal detoxification, and glutathione-S-conjugate degradation.
An overview of the roles of phytochelatin synthase (PCS) in the metabolic pathways of glutathione (GSH, γ-Glu-Cys-Gly). The brief pathway of GSH biosynthesis and the major route of glutathione-S-conjugate (GS-conjugate) degradation are also shown in the figure [7–9]. The presence of xenobiotic compounds (X, marked as green circles) and free heavy metal ions (red circles) induces ROS generation and causes oxidative stress. The cytosolic xenobiotic compound is transferred to GSH by glutathione S-transferase (GST) to initiate the detoxification, and then the GS-conjugates enter vacuoles for further degradation [10–13]. In the vacuoles, GS-conjugates are first catalyzed to Cys-Gly-conjugates by γ-glutamyl transpeptidase (GGT) before the final deglycination catalyzed by carboxypeptidase (CP) [14–16]. In the presence of heavy metals, PCS uses GSH as substrates to synthesize phytochelatins (PCs), which chelate free metal ions in the cytosol [4]. Heavy metals also activate PCS to initiate the cleavage of GS-conjugates [17–20]. The cytosolic γ-Glu-Cys-conjugates then enter vacuoles and serve as the substrates of GGT for the second step of degradation [21]. The blue cylinders represent tonoplast ABCC transporters that facilitate the import of GS-metabolites and PC-metal complexes [12, 13, 22–24]. In the brief biosynthetic pathway of GSH, the rate-limiting enzymes are indicated in bold. Note that the cellular compartmentation of GSH synthesis or redox reactions is not included in this figure. SAT, serine acetyltransferase; γ-ECS, γ-glutamylcysteine synthetase; GSH-S, GSH synthetase. GP, GSH peroxidase; GR, GSH reductase. The figure was created with BioRender.com.

sources, including AtPCS1, TaPCS1 from wheat (*Triticum aestivum*), and SpPCS from *Schizosaccharomyces pombe* by three independent research groups [5, 33, 34]. Since then, PCS sequences from various model organisms have been largely characterized, such as *Caenorhabditis elegans* (CePCS1) [31], the Cd hyperaccumulator *Thlaspi caerulescens* (TcPCS1) [35], and *Oryza sativa* (OsPCS1, OsPCS2, OsPCS5, OsPCS15) [36–38]. Besides eukaryote PCS sequences, a gene encoding a PCS-like protein, NsPCS, was identified from the genome of cyanobacterium *Nostoc* sp. [39–42].

PCS is a key component for the heavy metal tolerance in plants. Its importance was first confirmed in the Arabidopsis mutants locking AtPCS1 activity, as these mutants show severe growth defects when challenged by heavy metals such as Cd^{2+}, Hg^{2+}, Zn^{2+}, Pb^{2+}, and As^{3+} [28, 43–45]. The synthesis of PCs is crucial to the local response to heavy metal stress and is also involved in the roots-to-shoots translocation of heavy metals [26, 37, 43, 46]. The first evidence of the long-distance transfer of PC-metal complexes is that the PCs synthesized in the roots can be translocated to the shoots via phloem loading and *vice versa* [26, 43, 46]. Additionally, plants defective in PC synthesis show altered patterns of heavy metal accumulation at the whole-plant level while being sensitive to heavy metal stress. For example, the Arabidopsis AtPCS1-deficient mutant, *cad1–3*, accumulated significantly less Cd in the shoots than the wild type or the transgenic line heterologously overexpressing TaPCS1 [43], and the rice *OsPCS2* RNAi plants failed to transfer As^{3+} from the roots to the shoots [37]. Overall, the phenotypes of these PCS-deficient mutants suggest heavy metal ions absorbed through the roots can be loaded into the shoots in the form of PC-chelates.

PCS is a well-known multitasker involved in different biological processes [21, 47]. Besides its significant role in synthesizing PCs from GSH, PCS can catalyze the deglycination of GSH-*S*-conjugates (GS-conjugates), and thus, it is involved in the GS-conjugate catabolism [17–21, 48]. In addition, PCS is also associated with indole glucosinolates metabolism and immune responses [49–51]. Intriguingly, the catalytic-site mutants of PCS are still functional in this pathway, which suggests that the role of PCS in the indole glucosinolate metabolism is independent of PC synthesis and GS-metabolism [51]. Among these PCS-involving biological processes, this chapter focuses on the catalytic mechanism of PCS and its functions in both heavy metal stress and GSH metabolism. The potential applications of PCS in combating multiple stresses are also discussed.

2. The biochemical mechanism of phytochelatin synthase

2.1 The domain organization of phytochelatin synthase

The eukaryotic PCS has two domains with distinct functions: a conserved N-terminal domain that shows γ-glutamylcysteine dipeptidyl transpeptidase activity and a variable C-terminal domain involved in metal sensing [52–54]. Using AtPCS1 as a model, the molecular functions of the N- and C-domains as well as the catalytic mechanism of eukaryotic PCS were revealed [6, 53–55]. The N-terminal half AtPCS1 is sufficient for deglycination of GSH and elongating PC molecules, indicating that the N-terminal domain carries out the core catalysis [53, 56]. However, the truncated AtPCS1 without the C-terminal domain is less thermostable and has lower PC synthetic activity than the full-length enzyme [53, 55, 56]. Notably, the deletion of the C-terminal domain completely impairs the PC synthesis activity of the enzyme in the presence of Zn^{2+} and partially inactivates PC synthesis in the Cd- or Hg-containing reactions [53, 55]. These findings suggest that the C-terminal domain is essential for stabilizing the protein and functions as a metal sensor [53, 55]. More evidence has shown that the C-terminal end of AtPCS1 is required for the augmentation of PC synthetic activity. One example is that the residues from Asp373 to the C-terminal end of AtPCS1 contain multiple regions involved in Zn-dependent and As-dependent activation of PC synthesis [45, 57, 58]. (Also see Section 2.4: The activation of phytochelatin synthase through the chelation of heavy metal ions).

PCS-like sequences also exist in prokaryotes with moderate sequence homology to the N-domain of eukaryotic PCS [39–41]. However, prokaryotic PCS likely has unique functions apart from PC synthesis. For example, the PCS homolog found in cyanobacterium *Nostoc* sp. PCC 7120 (NsPCS) shows distinct characters from its eukaryotic counterparts that efficiently catalyze PC synthesis. NsPCS is a “half-PCS molecule” that does not have a C-terminal domain [39, 40, 52] and catalyzes the hydrolysis of GSH at a high rate and the synthesis of PCs at a relatively low rate [39, 40, 42]. Besides, the enzyme activity of NsPCS seems insensitive to the absence or presence of Cd^{2+}, which suggest that the prokaryotic PCS is involved in GSH metabolism in the cells rather than the responses to heavy metal stress [39, 42].

2.2 The core catalytic mechanism

Vatamaniuk et al. [6] first confirmed that the synthesis of PCs occurs through a ping-pong mechanism and involves two substrates: one GSH as the low-affinity substrate for the first step of PC synthesis and one metal-GSH conjugate as the high-affinity substrate for the second step [6]. In the standard PC synthesis reactions *in vitro*, which resemble the concentrations of the GSH and metal ions in the cytosol, GSH exists at a considerably higher level (millimolar) than heavy metal ions (micromolar) [6, 7, 59]. Presumably, more than 98% of total metal ions in this condition are associated with GSH as bis(glutathionato)metal ions (metal·GS_2), and the free Cd concentration can be as low as 10^{-6} μM [6]. Under these circumstances, GSH and Cd·GS_2 are two separate compounds for PC synthesis.

Right after GSH enters the catalytic site of PCS, a Gly residue is removed to form the γGlu-Cys acyl-enzyme intermediate, and then a metal-GS_2 accepts the γGlu-Cys unit to generate a PC_2 ((γGlu-Cys)$_2$-Gly) [6, 21, 54, 55]. Following the synthesis of PC_2, the elongation of PCs occurs using previously synthesized PCs as acceptors to receive γGlu-Cys [6, 21, 54, 55, 60]. The whole process can be described as two equations:

$$PCS + \gamma\mathrm{Glu} - Cys - Gly \rightarrow PCS - \gamma\mathrm{Glu} - Cys + Gly \quad (1)$$

$$\begin{aligned} &PCS - \gamma\mathrm{Glu} - Cys + \mathrm{metal} \cdot (\gamma\mathrm{Glu} - Cys - Gly)_2 \rightarrow PCS \\ &+ \mathrm{metal} \cdot (\gamma\mathrm{Glu} - Cys)_2 - Gly + \gamma\mathrm{Glu} - Cys - Gly \end{aligned} \quad (2)$$

Overall, PCS catalyzes the peptide chain elongation from C-to-N terminus [4, 6, 21, 54]:

$$\begin{aligned} &(\gamma\mathrm{Glu} - Cys) - Gly + (\gamma\mathrm{Glu} - Cys)_n - Gly \\ &\rightarrow (\gamma\mathrm{Glu} - Cys) \cdot (\gamma\mathrm{Glu} - Cys)_n - Gly + Gly \end{aligned} \quad (3)$$

PCS and Cys proteases share similar core catalytic mechanisms to hydrolyze a GSH molecule and form a γ-Glu-Cys-acyl–enzyme intermediate [41, 52]. The Cys protease-like catalytic triad of PCS was confirmed based on the mutagenic studies of AtPCS1 and the crystal structures of NsPCS [41, 54, 55]. Vatamaniuk et al. and Romanyuk et al. reported that Cys56, His162, and Asp180 of AtPCS1 are the three residues of the catalytic triad among divergent PCS sequences [54, 55]. The molecular interaction between these residues and GSH was further revealed by Vivares et al. with the crystal structures of native NsPCS and the γ-Glu-Cys-acyl–enzyme intermediate at a 2.0-Å resolution [41]. Although NsPCS only shares 36% identity at the amino acid level with AtPCS1, it contains the conserved catalytic triad and can catalyze the deglycination of GSH [39–42]. These crystal structures provide details about the hydrolysis of the peptide bond that involves Cys56 and the 3D structure of the Cys-His-Asp catalytic triad [41]. It is worth mentioning that NsPCS formed homodimers in the crystallization experiments [26]. This is in agreement with the dimerization of the partially purified PCS from *Silene cucubalus*, which was confirmed by determining the native molecular weight of the protein [4].

2.3 Critical amino acids contributing to the enzyme activity

Based on the high-resolution crystal structure of NsPCS, multiple research groups have simulated putative 3D structures of eukaryotic PCS using various programs, and these structure models provide valuable information that uncovers the conserved molecular mechanism of PCS [56, 61–65]. For example, the molecular models of AtPCS1 reveal the key amino acids that contribute to the mechanism of the second substrate recognition and the enzyme activation through Thr phosphorylation [56, 61]. Chia et al. first reported how AtPCS1 might attract and stabilize the second substrate, metal-GS_2, after the γ-Glu-Cys-acyl–enzyme intermediate is formed [61]. In this study, the modeled AtPCS1 structure revealed a pocket in proximity to the first substrate-binding site, consisting of three loops containing several conserved amino acids, including Arg152, Lys185, and Tyr55. Mutations on Arg152 or Lys185 (Arg-to-Lys or Lys-to-Arg substitutions) resulted in the complete abrogation of enzyme activity, indicating that the arrangement of these positive charges is crucial for the binding of the second substrate. Mutations at Tyr55 did not completely impair the enzyme activity, but the Tyr55 mutant AtPCS1 showed lower catalytic activities than the wild-type enzyme due to a reduced affinity to metal-GS_2. In addition, the mutation at Tyr55 reduced Cd^{2+} binding ability of the AtPCS1 protein. It was therefore suggested that Tyr55 binds to the Cd ion of metal-GS_2 through cation-π interaction and thus contributes to the recognition of the second substrate. Besides these three amino acids, other conserved residues on the loops constituting the second substrate-binding pocket, including Gln50, Glu52, Glen157, Phe184, and Tyr186, are also important for the PC synthesis activity [61, 62].

Wang et al. identified that Thr49 is the phosphorylation site related to the activation of AtPCS1 [56]. The mutant AtPCS1 with Thr49-to-Ala49 substitution could not be phosphorylated, and its PC synthesis activity was significantly lower than that of the wild-type enzyme. According to the proposed 3D model of AtPCS1, Thr49 is within proximity to Arg183, which is also crucial for the catalytic activity of AtPCS1, and both residues are next to the catalytic site and substrate binding pockets. It was proposed that the phosphorylated Thr49 interacts with Arg183, and that this interaction serves as a "molecular clip" to give the active site a conformation appropriate for catalysis. Because Thr49 and Arg183 are both highly conserved among PCS sequences, the activity of eukaryotic PCS may as well be regulated by similar phosphorylation modifications [56, 66].

2.4 The activation of phytochelatin synthase through the chelation of heavy metal ions

As a key component of early response to heavy metal stress, PCS protein is constitutively expressed in the cytosol for rapid activation stimulated by heavy metals [2, 3, 6]. The heavy metal ions entering cytosol are essential for forming the second substrate [6, 60]. They can also bind to PCS, resulting in augmentative activation [6, 55, 67, 68]. Moreover, heavy metals could be a critical factor that triggers PCS phosphorylation [56, 69]. For example, AtPCS1 phosphorylation only occurred in the presence of Cd^{2+} in the *in vitro* experiments [56].

PCS, confirmed to be a metalloenzyme *in vitro*, is also likely to be one *in vivo* [6, 17]. Equilibrium analyses show that one AtPCS1 molecule binds seven Cd^{2+} in solutions containing 10 μM $CdCl_2$ [6, 61]. Apart from Tyr55, which is proposed to bind the Cd^{2+} on the metal-GS_2, the Cd binding capability of PCS presumably comes from conserved Cys pairs and CysXXCys motifs also found in metallothionein [30, 61]. Peptide screening of SpPCS and TaPCS showed that the core sequences containing consensus Cys-rich motifs could bind Cd^{2+} *in vitro* [67]. The subsequent site-direct mutagenesis analysis indicated that conserved Cys pairs at the N-terminal domain were critical for PCS activity, while the Cys-rich motifs at the C-terminal domain only slightly affected the PC synthesis rate [67]. It is not yet clear how these

Cys-rich motifs enhance the PC synthesis rate. It is possible that they bind metal-GS_2 complexes or free metal ions to stabilize the protein structure [6, 30, 55, 67]. More investigations are still needed to explain the molecular functions of these Cys-rich motifs and how they participate in the metal activation of PCS.

3. Phytochelatin synthase-targeting genetic engineering approaches in phytoremediation of heavy metals

3.1 The effects of phytochelatin synthase overexpression on the accumulation of heavy metals in plants

PC synthesis plays a critical role in heavy metal tolerance and accumulation. It is therefore no surprise that the breeding and engineering approaches for phytoremediation requiring heavy metal hyperaccumulators have focused on strategies to enhance PC biosynthetic capacity [48, 70–76]. Studies have shown that the transgenic plants expressing functional PCS usually have a higher tolerance to heavy metal stress. For example, the overexpression of AtPCS1 in Arabidopsis, tobacco or Indian mustard (*Brassica juncea*) enhanced Cd, Zn, and As tolerance and accumulation [71, 77–80]. Other PCS homologs e.g., CePCS, TaPCS1, NtPCS1, CdPCS from an aquatic As-accumulator plant (*Ceratophyllum demersum*), MaPCS1/MaPCS2 from mulberry (*Morus alba*), and VsPCS1 from legume *Vicia sativa* were also used to develop transgenic plants that accumulate higher concentrations of heavy metals than their natural variants [43, 46, 81–86]. These reports on improving heavy metal accumulation and tolerance of the plants indicate the potential applications of PCS on phytoremediation approaches.

Although PCS can be a molecular tool for phytoremediation of heavy metal-contaminated soils and waters, the overexpression of PCS promotes the catabolism of GSH, which also plays essential roles in redox reactions and heavy metal stress [87, 88]. If the metabolic pathways supplying GSH cannot maintain specific levels in the presence of highly expressed PCS, the consumption of GSH usually leads to changes in the GSH/GSSG ratio that exacerbate oxidative stress [62, 72]. The increased GSH demand driven by PC synthesis may also affect other metabolic pathways requiring GSH [89]. In these regards, the use of functional PCS with diminished catalytic activity could reduce the depletion of GSH, maintain redox homeostasis and supporting PC synthesis during exposure to heavy metals at the same time [62]. Indeed, the Arabidopsis and *Brassica juncea* transgenic lines expressing a partially deactivated AtPCS1 mutant, AtPCS1-Y186C, showed enhanced Cd^{2+} tolerance and higher GSH/GSSG ratio than the transgenic lines expressing wild-type AtPCS1 [62]. These results suggest that PC synthesis and redox homeostasis are both important for successful heavy metal resistance.

Besides the imbalance of cellular redox state, PCS overexpression could result in an unknown disruption in cellular metal homeostasis under heavy metal stress because PCS itself is a metalloenzyme and can bind a wide range of metal ions [90, 91]. Expressing synthetic genes encoding peptide analogs of PCs with a general structure of Met(Glu-Cys)$_n$Gly (n = 16–20) could be an alternative way to enhance the accumulation of heavy metals in the plants without the overexpression of PCS [92]. The Arabidopsis transgenic plants transformed with the artificial genes encoding these PC-like polypeptides resulted in hyperaccumulation of Cd^{2+} and $As^{3+/5+}$ in the plants [92]. However, the impact of accumulating synthetic PC-like polypeptides on the overall metal homeostasis is yet to be determined.

3.2 Phytochelatin synthase-involving pathway engineering for enhancing heavy metal accumulation

Pathway engineering involved in the co-expression of both GSH synthesis and PC synthesis pathways is another strategy to preserve the balance of GSH metabolism in the cells with constitutive PC synthesis [72, 74]. A kinetic model of GSH and phytochelatin synthesis in plants suggests that at least two enzymes, γ-glutamylcysteine synthetase (γ-ECS) and PCS, should be increased to enhance PC synthesis without depleting the GSH pool [89]. In fact, the effects of modified GSH/PC synthesis pathways have been tested in *Escherichia coli* and tobacco plants, respectively [93, 94]. In these experiments, the activities of SpPCS and two enzymes catalyzing the rate-limiting steps of GSH biosynthesis, including serine acetyltransferase (SAT) and γ-ECS, were enhanced (**Figure 1**) [7, 8, 93, 94]. The *E. coli* cells co-overexpressing these enzymes accumulated significantly higher concentrations of PCs and Cd^{2+}, while the single-gene expression in the PC synthesis pathway had limited effects [93]. These findings support the "gene stacking" approaches to enhancing heavy metal metabolism. Although the same strategy co-overexpressing these three genes in tobacco increased some classes of non-protein thiol, the Cd^{2+} accumulation in the transgenic plants did not change compared to the wild type [94]. These findings suggest that other mechanisms, in addition to the availability of precursors for PC synthesis, limit Cd accumulation in plants [94].

Overall, the genetic engineering approaches involved in manipulating PC synthesis have shown promising prospects for improving the performance of plants in the phytoremediation of heavy metals. However, there are also setbacks pointing at the complexity of the stress response induced by heavy metals [75]. Thus, while enhanced PC synthesis can contribute to the heavy metal chelating, other factors, such as the subsequent vacuolar sequestration or the delicate balance of GSH metabolic pathways under heavy metal stress, should be considered in order to achieve heavy metal tolerance.

4. The role of phytochelatin synthase in glutathione-*S*-conjugate metabolism

4.1 The involvement of phytochelatin synthase in the catabolism of glutathione derivates

PCS has a broad substrate selectivity and can use GS-derivates as substrates. For example, PCS isolated from plant species can accept *S*-alkylated GSH such as *S*-methyl-GS and *S*-hexyl-GS [6, 95] and large side residues like xenobiotic GS-conjugates (abbreviated as GS-conjugates) [17, 19, 20]. The bulky *S*-residues of GSH that can be converted to γ-Glu-Cys-conjugates by PCS include benzyl-, nitrophenyl-, phenylbenzyl-, uracil-, bimane-, and acetamido-fluorescein-groups [17, 19]. However, when PCS uses GS-derivates with these bulky *S*-linked side residues, it tends to transfer the γ-Glu-Cys-conjugate intermediate to a hydrogen group [17, 19, 20]. As a result, PCS processes the hydrolysis of GS-conjugates instead of their polymerization.

Besides its significant role in heavy metal detoxification, PCS also participates in the biodegradation of xenobiotic compounds because of its capability to process GS-conjugates [17–20]. Glutathione conjugation is a major pathway to inactivate xenobiotic compounds in plant cells [7]. Glutathione transferase (GST) detoxifies xenobiotics in the cytosol by transferring these compounds to GSH [10, 11, 96, 97]. These GS-conjugates enter vacuoles rapidly for sequestration and further degradation [12, 13]. In Arabidopsis, the transport of GS-conjugates for vacuolar sequestration is facilitated by AtABCC1/AtMRP1 and AtABCC2/AtMRP2, which also transfer PC-metal complexes into vacuoles [12, 13, 22, 24]. Because of the high efficiency of this sequestration mechanism, the subsequent catabolism of GS-conjugates is presumably processed in the vacuoles [18]. First, vacuolar γ-glutamyl-transpeptidase (GGT) initializes the degradation of GS-conjugates by removing the γ-Glu-residue from the GS-conjugates to form Cys-Gly-conjugates [14, 15], and then, carboxypeptidase cleaves the Gly residue and results in the accumulation of the Cys-conjugates [16]. Alternatively, the GS-conjugate degradation can be initiated by PCS when vacuolar sequestration is not an available route [17–20]. The pathways of GS-conjugate metabolism are summarized in **Figure 1**.

4.2 Phytochelatin synthase may participate in initiating the first step of glutathione-*S*-conjugates degradation in the cytosol

Monochlorobimane (MCB) is widely used as a model xenobiotic for Arabidopsis to study the catabolism of GS conjugates [14, 15, 17–20, 98]. The bimane-labeled thiols can be analyzed by high performance liquid chromatography [99]. In addition, the fluorescent GS-bimane can be directly monitored *in situ*, which indicates the compartmentation and the turnover of GS-conjugates [15, 18, 20]. Data have shown that AtPCS1 initiates the first step of GS-bimane degradation in cytosol by removing the Gly residue and providing substrates for the vacuolar GGT [17, 19–21] (**Figure 1**). This detour could be a functionally alternative route to detoxify xenobiotics when the major pathway is blocked [17–20].

The direct evidence showing the involvement of PCS in GS-conjugate metabolism is the defects in the turnover of GS-bimane shown in the Arabidopsis AtPCS1-deficient mutants [21, 47]. The AtPCS1-deficient mutant, *ΔPCS1*, and the AtPCS1/AtPCS2 double-deficient mutant, *ΔPCS*, are impaired in the degradation of GS-bimane to γ-Glu-Cys-bimane [19, 20]. Blum et al. (2007) report that in the absence of Cd^{2+}, the abundance of the γ-Glu-Cys-bimane in both *ΔPCS1* and *ΔPCS* mutants was significantly reduced compared to the wild type after the plants were challenged by the xenobiotic bimane [19]. Moreover, the induction of γ-Glu-Cys-bimane was not observed in AtPCS1-deficient lines in the plants treated with Cd^{2+}, which resulted in a > 10-fold lower γ-Glu-Cys-bimane accumulation compared with the wild type grown in the same conditions [19]. The GS-baimane concentration could be rescued by transfecting AtPCS1 cDNA into PCS-deficient protoplasts, suggesting that this process is indeed PCS-dependent [19]. The inhibited γ-Glu-Cys-bimane accumulation in the mutant lines indicates that AtPCS1 efficiently catalyzes GS-conjugates in the presence of Cd^{2+} [19, 20].

Although the GS-bimane conversion is altered in the AtPCS1-deficient mutants, the GS-bimane in these mutants still can be degraded through the major detoxification pathway in the vacuoles [18–20]. Besides, the overall turnover of GS-bimane in the mutants is only slightly affected without blocking the vacuolar transport pathway [18–20]. These findings underline that the vacuolar GGT-initiated GS-conjugates degradation is the major pathway among two compensatory routes responsible for the turnover of the xenobiotics [18].

In plant cells, both the cytosolic PCS and the vacuolar carboxypeptidase can catalyze the formation of γ-Glu-Cys-bimane [16–19]. However, the vacuolar carboxypeptidase tends to catalyze the Cys-Gly-conjugates following the cleavage of γ-Glu-residue initiated by GGT [15]. In this regard, PCS is supposed to be the primary component responsible for the γ-Glu-Cys-bimane formation observed in the process of GS-conjugate conversion. Another example showing the importance of PCS in the initiation of the cytosolic xenobiotic compound is the metabolism of the herbicide safener fenclorim [100]. Fenclorim enhances GST activity in Arabidopsis and is subsequently degraded via the GS-conjugation pathway [97, 100]. In the Arabidopsis suspension cells, GS-fenclorim was sequentially processed to γ-Glu-Cys-fenclorim and Cys-fenclorim, suggesting that deglycination is the initial step to the catabolism of fenclorim [21, 100]. However, more evidence is needed to confirm the direct involvement of PCS in this process.

4.3 The glutathione-*S*-conjugate conversion via phytochelatin synthase is metal-dependent

The presence of metal ions is a critical requirement for PCS-dependent catalysis of GS-bimane [17–20]. Intriguingly, the efficiency of GS-bimane hydrolysis activated by different metal ions is separate from that of metal-stimulated PC synthesis [17]. For example, the PC formation of AtPCS1 activated by Cd^{2+} is usually 2–5 times more efficient than the PC synthesis rate measured in the presence of $Cu^{+/2+}$ [4, 6, 53]. On the other hand, AtPCS1 could catalyze the deglycination of GS-bimane 60% more efficiently in $Cu^{+/2+}$ solutions than in the presence of Cd^{2+} [17]. It was suggested that *in vivo* AtPCS1 is a Cu-containing metalloenzyme in unstressed conditions, and consequently, the Cu-bound PCS favors the catalysis of GS-conjugate over PC synthesis in the normal growth conditions [17]. Evidence supporting this hypothesis is that in Arabidopsis, the deglycination of GS-bimane was PCS-dependent in the absence of heavy metals [19]. However, considering AtPCS1 binds Cu^{2+} only at a low affinity [6, 21], and the majority of cytosolic Cu is usually associated with Cu chaperons [101], it is possible that the concentration of free cytosolic Cu ions is not sufficient to fully activate PCS for the catalysis of GS-conjugates.

5. Can phytochelatin synthase play a part in the phytoremediation of combined pollutions?

With global industrialization and the development of modern cropping systems, massive amounts of toxic substances such as pesticides, heavy metals and inorganic fertilizers have been released into the environment and caused massive pollutions [97, 102]. Inevitably, the combined contaminations have damaged the ecosystems and become global issues [103, 104].

In the case of soil pollution, the primary sources of heavy metals include pesticides, fertilizers, mining, industrial processing and wastewater [102, 105]. One example of combined pollutants is glyphosate-based herbicides, which are highly toxic to the environment yet are the most-used pesticides in the world [106]. Heavy metals such as As, Ni, and Pb, which activate the catalytic activity of PCS, can be found as contaminants in many commercial glyphosate-based herbicides [3, 6, 106]. Interestingly, the *in vivo* chronic regulatory experiments showed that the toxicity of these herbicides might come from the heavy metals included in formulants instead of the active ingredients [106]. These findings suggest that heavy metal toxicity may occur in the biological materials used in the phytoremediation of xenobiotic compounds. Thus, heavy metal detoxification mechanisms in phytoremediation plants also need to be considered to improve their performance in co-contaminated soils and groundwaters.

GSH and its derivates are widely involved in plant development and stress response, and GSH itself serves as a hub for the mechanisms of heavy metal detoxification, xenobiotics biodegradation and oxidative stress response [7, 9]. Because PCS is a key enzyme in GSH metabolism, it is not surprising that PCS should be involved in both heavy metal stress and the turnover of xenobiotics. Other critical enzymes in the GSH metabolic pathway such as GST have been used in combating multiple stresses, including heavy metal and xenobiotics degradation, and biotic stress [107]. However, the role of PCS is primarily emphasized in heavy metal stress response despite its contribution to the degradation of GS-metabolites and innate immunity. Based on the knowledge about the diverse functions of PCS, it is worth exploring whether PCS can be a useful tool in enhancing the tolerance and performance of the plants challenged by combined stresses.

6. Conclusion

Both heavy metals and pesticides significantly arrest plant growth and development. The co-contamination of soils by both heavy metals and pesticides has raised concerns regarding crop safety and productivity, and is therefore crucial to remediate. Phytoremediation presents the advantages of high efficiency, low cost, and sustainability. Thus, it has been one of the most common strategies for the remediation of polluted soils. This chapter summarizes the critical role of PCS in heavy metal detoxification and the involvement of PCS in GS-conjugate degradation. In the presence of heavy metals, PCS catalyzes the synthesis of PCs and the initiation of GS-conjugate metabolism. Despite a large body of literature illustrating the function of PCS in heavy metal resistance, there has been less emphasis on the participation of PCS in the detoxification of xenobiotic compounds and its potential application in biodegradation. Given that PCS has diverse functions in different types of stress, this chapter discusses the potential inclusion of PCS to achieve phytoremediation for combined pollutions.

The key question related to PCS overexpression in plant materials for phytoremediation is how GSH homeostasis can be balanced. Although pathway engineering enhancing GSH metabolism and PCS activity seems a promising approach, the consequences of manipulating these pathways may not directly lead to improving the performance of plants exposed to stress, due to the complexity of the cellular GSH network. Thus, the challenge for the future is not only to characterize the involvement of PCS in stress responses but also to broaden our knowledge in PCS as a factor that regulates GSH status and cellular redox homeostasis.

Acknowledgements

I sincerely thank Dr. Rong-Huay Juang (National Taiwan University of Science and Technology) for stimulating discussion and critical reading of the manuscript.

Author details

Ju-Chen Chia
Soil and Crop Sciences Section, School of Integrative Plant Science,
Cornell University, Ithaca, New York, USA

*Address all correspondence to: jc2969@cornell.edu

References

[1] Grill E, Winnacker EL, Zenk MH. Phytochelatins: the principal heavy-metal complexing peptides of higher plants. Science. 1985;230(4726):674-6.

[2] Grill E, Winnacker EL, Zenk MH. Synthesis of seven different homologous phytochelatins in metal-exposed *Schizosaccharomyces pombe* cells. FEBS Lett. 1986;197(1-2):115-20.

[3] Grill E, Winnacker EL, Zenk MH. Phytochelatins, a class of heavy-metal-binding peptides from plants, are functionally analogous to metallothioneins. Proc Natl Acad Sci USA. 1987;84(2):439-43.

[4] Grill E, Löffler S, Winnacker EL, Zenk MH. Phytochelatins, the heavy-metal-binding peptides of plants, are synthesized from glutathione by a specific γ-glutamylcysteine dipeptidyl transpeptidase (phytochelatin synthase). Proc Natl Acad Sci USA. 1989;86(18):6838-42.

[5] Clemens S, Kim EJ, Neumann D, Schroeder JI. Tolerance to toxic metals by a gene family of phytochelatin synthases from plants and yeast. EMBO J. 1999;18(12):3325-33.

[6] Vatamaniuk OK, Mari S, Lu YP, Rea PA. Mechanism of heavy metal ion activation of phytochelatin (PC) synthase. J Biol Chem. 2000;275:31451-9.

[7] Noctor G, Mhamdi A, Chaouch S, Han Y, Neukermans J, Marquez-Garcia B, et al. Glutathione in plants: an integrated overview. Plant Cell Environ. 2012;35(2):454-84.

[8] Mendoza-Cózatl D, Loza-Tavera H, Hernández-Navarro A, Moreno-Sánchez R. Sulfur assimilation and glutathione metabolism under cadmium stress in yeast, protists and plants. FEMS Microbiol Rev. 2005;29(4):653-71.

[9] Noctor G, Queval G, Mhamdi A, Chaouch S, Foyer CH. Glutathione. Arabidopsis Book. 2011;9:e0142.

[10] Hatton PJ, Dixon D, Cole DJ, Edwards R. Glutathione transferase activities and herbicide selectivity in maize and associated weed species. Pestic Sci. 1996;46(3):267-75.

[11] Andrews CJ, Skipsey M, Townson JK, Morris C, Jepson I, Edwards R. Glutathione transferase activities toward herbicides used selectively in soybean. Pestic Sci. 1997;51(2):213-22.

[12] Lu Y-P, Li Z-S, Rea PA. *AtMRP1* gene of *Arabidopsis* encodes a glutathione *S*-conjugate pump: Isolation and functional definition of a plant ATP-binding cassette transportergene. Proc Natl Acad Sci USA. 1997;94(15):8243-8.

[13] Lu Y-P, Li Z-S, Drozdowicz YM, Hörtensteiner S, Martinoia E, Rea PA. AtMRP2, an Arabidopsis ATP binding cassette transporter able to transport glutathione *S*-conjugates and chlorophyll catabolites: Functional comparisons with AtMRP1. Plant Cell. 1998;10(2):267-82.

[14] Ohkama-Ohtsu N, Zhao P, Xiang C, Oliver DJ. Glutathione conjugates in the vacuole are degraded by γ-glutamyl transpeptidase GGT3 in Arabidopsis. Plant J. 2007;49(5):878-88.

[15] Grzam A, Martin MN, Hell R, Meyer AJ. γ-Glutamyl transpeptidase GGT4 initiates vacuolar degradation of glutathione S-conjugates in Arabidopsis. 2007;581(17):3131-8.

[16] Wolf AE, Dietz KJ, Schröder P. Degradation of glutathione *S*-conjugates by a carboxypeptidase in the plant vacuole. FEBS Lett. 1996;384(1):31-4.

[17] Beck A, Lendzian K, Oven M, Christmann A, Grill E. Phytochelatin synthase catalyzes key step in turnover

of glutathione conjugates. Phytochemistry. 2003;62(3):423-31.

[18] Grzam A, Tennstedt P, Clemens S, Hell R, Meyer AJ. Vacuolar sequestration of glutathione *S*-conjugates outcompetes a possible degradation of the glutathione moiety by phytochelatin synthase. FEBS Lett. 2006;580(27):6384-90.

[19] Blum R, Beck A, Korte A, Stengel A, Letzel T, Lendzian K, et al. Function of phytochelatin synthase in catabolism of glutathione-conjugates. Plant J. 2007; 49(4):740-9.

[20] Blum R, Meyer KC, Wünschmann J, Lendzian KJ, Grill E. Cytosolic action of phytochelatin synthase. Plant Physiol. 2010;153(1):159-69.

[21] Rea PA. Phytochelatin Synthase. eLS2020. p. 1-15.

[22] Song WY, Park J, Mendoza-Cózatl DG, Suter-Grotemeyer M, Shim D, Hörtensteiner S, et al. Arsenic tolerance in *Arabidopsis* is mediated by two ABCC-type phytochelatin transporters. Proc Natl Acad Sci USA. 2010;107:6.

[23] Mendoza-Cózatl DG, Zhai Z, Jobe TO, Akmakjian GZ, Song W-Y, Limbo O, et al. Tonoplast-localized Abc2 transporter mediates phytochelatin accumulation in vacuoles and confers cadmium tolerance. J Biol Chem. 2010;285(52):40416-26.

[24] Park J, Song W-Y, Ko D, Eom Y, Hansen TH, Schiller M, et al. The phytochelatin transporters AtABCC1 and AtABCC2 mediate tolerance to cadmium and mercury. Plant J. 2012;69(2):278-88.

[25] Vögeli-Lange R, Wagner GJ. Subcellular localization of cadmium and cadmium-binding peptides in tobacco leaves: implication of a transport function for cadmium-binding peptides. Plant Physiol. 1990;92(4):1086-93.

[26] Mendoza-Cózatl DG, Jobe TO, Hauser F, Schroeder JI. Long-distance transport, vacuolar sequestration, tolerance, and transcriptional responses induced by cadmium and arsenic. Curr Opin Plant Biol. 2011;14(5):554-62.

[27] Ramos J, Naya L, Gay M, Abián J, Becana M. Functional characterization of an unusual phytochelatin synthase, LjPCS3, of *Lotus japonicus*. Plant Physiol. 2008;148(1):536-45.

[28] Howden R, Goldsbrough PB, Andersen CR, Cobbett CS. Cadmium-sensitive, *cad1* mutants of *Arabidopsis thaliana* are phytochelatin deficient. Plant Physiol. 1995;107(4):1059-66.

[29] Cobbett CS. Phytochelatins and their roles in heavy metal detoxification. Plant Physiol. 2000;123(3):825-32.

[30] Cobbett C, Goldsbrough P. Phytochelatins and metallothioneins: Roles in heavy metal detoxification and homeostasis. Ann Rev Plant Biol. 2002;53(1):159-82.

[31] Vatamaniuk OK, Bucher EA, Ward JT, Rea PA. A new pathway for heavy metal detoxification in animals. J Biol Chem. 2001;276(24):20817-20.

[32] Bundy JG, Kille P, Liebeke M, Spurgeon DJ. Metallothioneins may not be enough—The role of phytochelatins in invertebrate metal detoxification. Environ Sci Technol. 2014;48(2):885-6.

[33] Vatamaniuk OK, Mari S, Lu YP, Rea PA. AtPCS1, a phytochelatin synthase from *Arabidopsis*: Isolation and *in vitro* reconstitution. Proc Natl Acad Sci USA. 1999;96(12):7110-5.

[34] Ha SB, Smith AP, Howden R, Dietrich WM, Bugg S, O'Connell MJ, et al. Phytochelatin synthase genes from Arabidopsis and the yeast *Schizosaccharomyces pombe*. Plant Cell. 1999; 11(6):1153-64.

[35] Küpper H, Parameswaran A, Leitenmaier B, Trtílek M, Šetlík I.

Cadmium-induced inhibition of photosynthesis and long-term acclimation to cadmium stress in the hyperaccumulator *Thlaspi caerulescens*. New Phytol. 2007;175(4):655-74.

[36] Das N, Bhattacharya S, Bhattacharyya S, Maiti MK. Identification of alternatively spliced transcripts of rice phytochelatin synthase 2 gene *OsPCS2* involved in mitigation of cadmium and arsenic stresses. Plant Mol Biol. 2017;94(1):167-83.

[37] Yamazaki S, Ueda Y, Mukai A, Ochiai K, Matoh T. Rice phytochelatin synthases OsPCS1 and OsPCS2 make different contributions to cadmium and arsenic tolerance. 2018;2(1):e00034.

[38] Park HC, Hwang JE, Jiang Y, Kim YJ, Kim SH, Nguyen XC, et al. Functional characterisation of two phytochelatin synthases in rice (*Oryza sativa* cv. Milyang 117) that respond to cadmium stress. Plant Biol (Stuttg). 2019;21(5):854-61.

[39] Harada E, von Roepenack-Lahaye E, Clemens S. A cyanobacterial protein with similarity to phytochelatin synthases catalyzes the conversion of glutathione to γ-glutamylcysteine and lacks phytochelatin synthase activity. Phytochemistry. 2004;65(24):3179-85.

[40] Tsuji N, Nishikori S, Iwabe O, Shiraki K, Miyasaka H, Takagi M, et al. Characterization of phytochelatin synthase-like protein encoded by alr0975 from a prokaryote, *Nostoc* sp. PCC 7120. Biochem Biophys Res Commun. 2004;315(3):751-5.

[41] Vivares D, Arnoux P, Pignol D. A papain-like enzyme at work: Native and acyl–enzyme intermediate structures in phytochelatin synthesis. Proc Natl Acad Sci USA. 2005;102(52):18848-53.

[42] Tsuji N, Nishikori S, Iwabe O, Matsumoto S, Shiraki K, Miyasaka H, et al. Comparative analysis of the two-step reaction catalyzed by prokaryotic and eukaryotic phytochelatin synthase by an ion-pair liquid chromatography assay. Planta. 2005;222(1):181-91.

[43] Gong JM, Lee DA, Schroeder JI. Long-distance root-to-shoot transport of phytochelatins and cadmium in *Arabidopsis*. Proc Natl Acad Sci USA. 2003;100(17):10118-23.

[44] Estrella-Gómez N, Mendoza-Cózatl D, Moreno-Sánchez R, González-Mendoza D, Zapata-Pérez O, Martínez-Hernández A, et al. The Pb-hyperaccumulator aquatic fern *Salvinia minima* Baker, responds to Pb^{2+} by increasing phytochelatins via changes in *SmPCS* expression and in phytochelatin synthase activity. Aquat Toxicol. 2009;91(4):320-8.

[45] Tennstedt P, Peisker D, Böttcher C, Trampczynska A, Clemens S. Phytochelatin synthesis is essential for the detoxification of excess zinc and contributes significantly to the accumulation of zinc. Plant Physiol. 2009;149(2):938-48.

[46] Chen A, Komives EA, Schroeder JI. An Improved grafting technique for mature Arabidopsis plants demonstrates long-distance shoot-to-root transport of phytochelatins in Arabidopsis. Plant Physiol. 2006;141(1):108-20.

[47] Clemens S, Peršoh D. Multi-tasking phytochelatin synthases. Plant Sci. 2009;177(4):266-71.

[48] Stephan C. Evolution and function of phytochelatin synthases. J Plant Physiol. 2006;163(3):319-32.

[49] Clay NK, Adio AM, Denoux C, Jander G, Ausubel FM. Glucosinolate metabolites required for an *Arabidopsis* innate immune response. Science. 2009;323(5910):95-101.

[50] De Benedictis M, Brunetti C, Brauer EK, Andreucci A, Popescu SC, Commisso M, et al. The *Arabidopsis*

thaliana knockout mutant for *Phytochelatin Synthase1* (*cad1-3*) is defective in callose deposition, bacterial pathogen defense and auxin content, but shows an increased stem lignification. Front Plant Sci. 2018;9(19).

[51] Hématy K, Lim M, Cherk C, Piślewska-Bednarek M, Sanchez-Rodriguez C, Stein M, et al. Moonlighting function of Phytochelatin Synthase1 in extracellular defense against fungal pathogens. Plant Physiol. 2020;182(4):1920.

[52] Rea PA, Vatamaniuk OK, Rigden DJ. Weeds, worms, and more. Papain's long-lost cousin, phytochelatin synthase. Plant Physiol. 2004;136(1):2463-74.

[53] Ruotolo R, Peracchi A, Bolchi A, Infusini G, Amoresano A, Ottonello S. Domain organization of phytochelatin synthase. J Biol Chem. 2004;279(15): 14686-93.

[54] Vatamaniuk OK, Mari S, Lang A, Chalasani S, Demkiv LO, Rea PA. Phytochelatin synthase, a dipeptidyltransferase that undergoes multisite acylation with γ-glutamylcysteine during catalysis. J Biol Chem. 2004; 279(21):22449-60.

[55] Romanyuk ND, Rigden DJ, Vatamaniuk OK, Lang A, Cahoon RE, Jez JM, et al. Mutagenic definition of a papain-like catalytic triad, sufficiency of the N-terminal domain for single-site core catalytic enzyme acylation, and C-terminal domain for augmentative metal activation of a eukaryotic phytochelatin synthase. Plant Physiol. 2006;141(3):858-69.

[56] Wang HC, Wu JS, Chia JC, Yang CC, Wu YJ, Juang RH. Phytochelatin synthase is regulated by protein phosphorylation at a threonine residue near its catalytic site. J Agric Food Chem. 2009;57(16):7348-55.

[57] Kühnlenz T, Hofmann C, Uraguchi S, Schmidt H, Schempp S, Weber M, et al. Phytochelatin synthesis promotes leaf Zn cccumulation of *Arabidopsis thaliana* plants grown in soil with adequate Zn supply and is essential for survival on Zn-contaminated soil. Plant Cell Physiol. 2016;57(11):2342-52.

[58] Uraguchi S, Sone Y, Ohta Y, Ohkama-Ohtsu N, Hofmann C, Hess N, et al. Identification of C-terminal regions in *Arabidopsis thaliana* Phytochelatin Synthase 1 specifically involved in activation by arsenite. Plant and Cell Physiol. 2018;59(3):500-9.

[59] Pettersson O. Heavy-metal ion uptake by plants from nutrient solutions with metal ion, plant species and growth period variations. Plant and Soil. 1976;45(2):445-59.

[60] Ogawa S, Yoshidomi T, Yoshimura E. Cadmium(II)-stimulated enzyme activation of *Arabidopsis thaliana* phytochelatin synthase 1. J Inorg Biochem. 2011;105(1):111-7.

[61] Chia JC, Yang CC, Sui YT, Lin SY, Juang RH. Tentative identification of the second substrate binding site in Arabidopsis phytochelatin synthase. PLoS One. 2013;8(12):e82675.

[62] Cahoon RE, Lutke WK, Cameron JC, Chen S, Lee SG, Rivard RS, et al. Adaptive engineering of phytochelatin-based heavy metal tolerance. J Biol Chem. 2015;290(28):17321-30.

[63] Zayneb C, Imen RH, Walid K, Grubb CD, Bassem K, Franck V, et al. The phytochelatin synthase gene in date palm (*Phoenix dactylifera* L.): Phylogeny, evolution and expression. Ecotoxicol Environ Saf. 2017;140:7-17.

[64] Kolahi M, Yazdi M, Goldson-Barnaby A, Tabandeh MR. *In silico* prediction, phylogenetic and bioinformatic analysis of *SoPCS* gene, survey of its protein characterization and gene expression in response to cadmium in *Saccharum officinarum*. Ecotoxicol Environ Saf. 2018;163:7-18.

[65] Filiz E, Saracoglu IA, Ozyigit II, Yalcin B. Comparative analyses of phytochelatin synthase (PCS) genes in higher plants. Biotechnol Biotechnol Equip. 2019;33(1):178-94.

[66] Moudouma C, Gloaguen V, Riou C, Forestier L, Saladin G. High concentration of cadmium induces *AtPCS2* gene expression in *Arabidopsis thaliana* (L.) Heynh ecotype Wassilewskija seedlings. Acta Physiol Plant. 2012;34(3):1083-91.

[67] Maier T, Yu C, Küllertz G, Clemens S. Localization and functional characterization of metal-binding sites in phytochelatin synthases. Planta. 2003;218(2):300-8.

[68] Vestergaard M, Matsumoto S, Nishikori S, Shiraki K, Hirata K, Takagi M. Chelation of cadmium ions by phytochelatin synthase: role of the Cystein-rich C-terminal. Anal Sci. 2008;24(2):277-81.

[69] Kulik A, Anielska-Mazur A, Bucholc M, Koen E, Szymańska K, Żmieńko A, et al. SNF1-related protein kinases type 2 are involved in plant responses to cadmium stress. Plant Physiol. 2012;160(2):868-83.

[70] Peterson AG, Oliver DJ. Leaf-targeted phytochelatin synthase in *Arabidopsis thaliana*. Plant Physiol Biochem. 2006;44(11-12):885-92.

[71] Gasic K, Korban S. Transgenic Indian mustard (*Brassica juncea*) plants expressing an *Arabidopsis* phytochelatin synthase (*AtPCS1*) exhibit enhanced As and Cd tolerance. Plant Mol Biol. 2007; 64(4):361-9.

[72] Guo J, Dai X, Xu W, Ma M. Overexpressing *GSH1* and *AsPCS1* simultaneously increases the tolerance and accumulation of cadmium and arsenic in *Arabidopsis thaliana*. Chemosphere. 2008;72(7):1020-6.

[73] Koźmińska A, Wiszniewska A, Hanus-Fajerska E, Muszyńska E. Recent strategies of increasing metal tolerance and phytoremediation potential using genetic transformation of plants. Plant Biotechnology Reports. 2018;12(1):1-14.

[74] Suman J, Uhlik O, Viktorova J, Macek T. Phytoextraction of heavy metals: A promising tool for clean-up of polluted environment? Front Plant Sci. 2018;9(1476).

[75] Yan A, Wang Y, Tan SN, Mohd Yusof ML, Ghosh S, Chen Z. Phytoremediation: A promising approach for revegetation of heavy metal-polluted land. Front Plant Sci. 2020;11(359).

[76] Török A, Gulyás Z, Szalai G, Kocsy G, Majdik C. Phytoremediation capacity of aquatic plants is associated with the degree of phytochelatin polymerization. J Hazard Mater. 2015;299:371-8.

[77] Pomponi M, Censi V, Girolamo V, Paolis A, Toppi L, Aromolo R, et al. Overexpression of Arabidopsis phytochelatin synthase in tobacco plants enhances Cd^{2+} tolerance and accumulation but not translocation to the shoot. Planta. 2006;223(2):180-90.

[78] Gasic K, Korban S. Expression of Arabidopsis phytochelatin synthase in Indian mustard (*Brassica juncea*) plants enhances tolerance for Cd and Zn. Planta. 2007;225(5):1277-85.

[79] Brunetti P, Zanella L, Proia A, De Paolis A, Falasca G, Altamura MM, et al. Cadmium tolerance and phytochelatin content of *Arabidopsis* seedlings over-expressing the phytochelatin synthase gene *AtPCS1*. J Exp Bot. 2011;62(15): 5509-19.

[80] Zanella L, Fattorini L, Brunetti P, Roccotiello E, Cornara L, D'Angeli S, et al. Overexpression of *AtPCS1* in tobacco increases arsenic and arsenic plus cadmium accumulation and detoxification. Planta. 2016;243(3): 605-22.

[81] Wojas S, Clemens S, Hennig J, Skłodowska A, Kopera E, Schat H, et al. Overexpression of phytochelatin synthase in tobacco: distinctive effects of *AtPCS1* and *CePCS* genes on plant response to cadmium. J Exp Bot. 2008;59(8):2205-19.

[82] Wojas S, Clemens S, Skłodowska A, Antosiewicz DM. Arsenic response of *AtPCS1*- and *CePCS*-expressing plants – effects of external As(V) concentration on As-accumulation pattern and NPT metabolism. J Plant Physiol. 2010; 167(3):169-75.

[83] Lee BD, Hwang S. Tobacco phytochelatin synthase (NtPCS1) plays important roles in cadmium and arsenic tolerance and in early plant development in tobacco. Plant Biotechnol Rep. 2015; 9(3):107-14.

[84] Zhang X, Rui H, Zhang F, Hu Z, Xia Y, Shen Z. Overexpression of a functional *Vicia sativa* PCS1 homolog increases cadmium tolerance and phytochelatins synthesis in Arabidopsis. Front Plant Sci. 2018;9(107).

[85] Fan W, Guo Q, Liu C, Liu X, Zhang M, Long D, et al. Two mulberry phytochelatin synthase genes confer zinc/cadmium tolerance and accumulation in transgenic *Arabidopsis* and tobacco. Gene. 2018;645:95-104.

[86] Shri M, Dave R, Diwedi S, Shukla D, Kesari R, Tripathi RD, et al. Heterologous expression of *Ceratophyllum demersum* phytochelatin synthase, *CdPCS1*, in rice leads to lower arsenic accumulation in grain. Sci Rep. 2014;4(1):5784.

[87] Schützendübel A, Polle A. Plant responses to abiotic stresses: heavy metal-induced oxidative stress and protection by mycorrhization. J Exp Bot. 2002;53(372):1351-65.

[88] Singh S, Parihar P, Singh R, Singh VP, Prasad SM. Heavy metal tolerance in plants: Role of transcriptomics, proteomics, metabolomics, and ionomics. Front Plant Sci. 2016;6(1143).

[89] Mendoza-Cózatl DG, Moreno-Sánchez R. Control of glutathione and phytochelatin synthesis under cadmium stress. Pathway modeling for plants. J Theor Biol. 2006;238(4):919-36.

[90] Lee S, Moon JS, Ko TS, Petros D, Goldsbrough PB, Korban SS. Overexpression of Arabidopsis phytochelatin synthase paradoxically leads to hypersensitivity to cadmium stress. Plant Physiol. 2003;131(2):656-63.

[91] Kim JH, Lee S. Overexpression of *Arabidopsis* phytochelatin synthase (*AtPCS1*) does not change the maximum capacity for non-protein thiol production induced by Cadmium. J Plant Biol. 2007;50(2):4.

[92] Shukla D, Tiwari M, Tripathi RD, Nath P, Trivedi PK. Synthetic phytochelatins complement a phytochelatin-deficient *Arabidopsis* mutant and enhance the accumulation of heavy metal(loid)s. Biochem Biophys Res Commun. 2013;434(3):664-9.

[93] Wawrzyńska A, Wawrzyński A, Gaganidze D, Kopera E, Piatek K, Bal W, et al. Overexpression of genes involved in phytochelatin biosynthesis in *Escherichia coli*: effects on growth, cadmium accumulation and thiol level. Acta biochimica Polonica. 2005;52(1):109-16.

[94] Wawrzyński A, Kopera E, Wawrzyńska A, Kamińska J, Bal W, Sirko A. Effects of simultaneous expression of heterologous genes involved in phytochelatin biosynthesis on thiol content and cadmium accumulation in tobacco plants. J Exp Bot. 2006;57(10):2173-82.

[95] Oven M, Page JE, Zenk MH, Kutchan TM. Molecular characterization of the homo-phytochelatin synthase of soybean *Glycine max*. J Biol Chem. 2002;277(7):4747-54.

[96] Dixon DP, Cummins I, Cole DJ, Edwards R. Glutathione-mediated

detoxification systems in plants. Curr Opin Plant Biol. 1998;1(3):258-66.

[97] Del Buono D, Terzano R, Panfili I, Bartucca ML. Phytoremediation and detoxification of xenobiotics in plants: herbicide-safeners as a tool to improve plant efficiency in the remediation of polluted environments. A mini-review. Int J Phytoremediation. 2020;22(8): 789-803.

[98] Fricker MD, May M, Meyer AJ, Sheard N, White NS. Measurement of glutathione levels in intact roots of Arabidopsis. J Microsc. 2000;198(3): 162-73.

[99] Newton GL, Fahey RC. Determination of biothiols by bromobimane labeling and high-performance liquid chromatography. Method Enzymol. 1995;251:148-66.

[100] Brazier-Hicks M, Evans KM, Cunningham OD, Hodgson DRW, Steel PG, Edwards R. Catabolism of glutathione conjugates in *Arabidopsis thaliana*: Role in metabolic reactivation of the herbicide safener fenclorim. J Biol Chem. 2008;283(30):21102-12.

[101] Printz B, Lutts S, Hausman J-F, Sergeant K. Copper trafficking in plants and its implication on cell wall dynamics. Front Plant Sci. 2016;7(601).

[102] Alengebawy A, Abdelkhalek ST, Qureshi SR, Wang M-Q. Heavy metals and pesticides toxicity in agricultural soil and plants: Ecological risks and human health implications. Toxics. 2021;9(3):42.

[103] Uwizeyimana H, Wang M, Chen W, Khan K. The eco-toxic effects of pesticide and heavy metal mixtures towards earthworms in soil. Environ Toxicol Pharmacol. 2017;55:20-9.

[104] Zhang H, Yuan X, Xiong T, Wang H, Jiang L. Bioremediation of co-contaminated soil with heavy metals and pesticides: Influence factors, mechanisms and evaluation methods. Chem Eng J. 2020;398:125657.

[105] Srivastava V, Sarkar A, Singh S, Singh P, de Araujo ASF, Singh RP. Agroecological responses of heavy metal pollution with special emphasis on soil health and plant performances. Front Environ Sci. 2017;5(64).

[106] Defarge N, Spiroux de Vendômois J, Séralini GE. Toxicity of formulants and heavy metals in glyphosate-based herbicides and other pesticides. Toxicol Rep. 2018;5:156-63.

[107] Kumar S, Trivedi PK. Glutathione S-Transferases: Role in combating abiotic stresses including arsenic detoxification in plants. Front Plant Sci. 2018;9(751).

Chapter 19

Sustainable Textile Processing by Enzyme Applications

Shekh Md. Mamun Kabir and Joonseok Koh

Abstract

Enzymatic treatments have gained popularity in the textile industry because of environmental friendly and energy conserving alternatives. Advancement in biotechnology and modification of enzymes has been focused based on various textile process applications. All the manufacturing steps of textile chemical processing, enzymes are using for implementations of the green technology to meet up the challenge of fourth industrial revolution. In this category, amylases, peroxidase used for desizing and bleaching, cellulase activates for bio polishing and denim finishing. This chapter summarizes the current developments of enzyme technology and highlights the environment-friendly and sustainable enzymatic textile processing in the textile industry.

Keywords: enzyme, microorganisms, textile fibre, bio-processing, finishing

1. Introduction

Enzymes are biocatalysts obtained from living cells through biochemical reactions specifically metabolic process of the cells [1]. Enzymes obtained from the natural source since ancient times in the production of food products, such as cheese, sourdough, beer, wine, vinegar and indigo formation [2]. The development of fermentation processes has grown during the last century, specifically for the production of purified enzymes in a large scale [3]. The use of recombinant gene technology has improved enzyme-manufacturing processes. Most industrial enzymes occurred hydrolysis for degrading the natural substances [4]. Enzymes are used in not only food production but also pharmaceuticals, textiles, leather processing [5–7]. There are prominent enzyme manufacturer for textile processing are listed in **Table 1**.

2. Enzyme structure and its mechanism

Enzymes are amino acid based globular proteins that range in size from less than 100 to more than 2000 amino acid residues. One or more polypeptide chains can be arranged and folded to form a specific three-dimensional structure, called active site incorporate with substrate. The active site may involve a small number (less than 10) of the constituent amino acids [9] (**Figure 1**).

The hypothesis of an enzyme-substrate complex was first proposed by the German chemist Emil Fischer in 1894. The lock and key theory explained, as a key is

Industrial enzymes	Manufacturer	Established	Applications
Protease, xylanase, glucoamylase	Novozymes, Denmark	1921	Household care, textiles, food and beverages, oil and fats
Amylase, protease, phytase, xylanase, β-mannanase	Genencor, Denmark	1982	Food and Beverages, Textiles, Detergents, Biofuels
Amylases, Proteases, Cellulases, Xylanase, Pectinase	AB Enzymes, Germany	1907	Feed additives, food, textile, detergent, pulp and paper, biofuels
Protease, Amylase, Laccase, Catalase, Cellulase, Lipases	Dyadic, USA	1979	Food, brewing and animal feed enzymes, pulp and paper, textile enzymes

Table 1.
Industrial enzyme for textile applications [8].

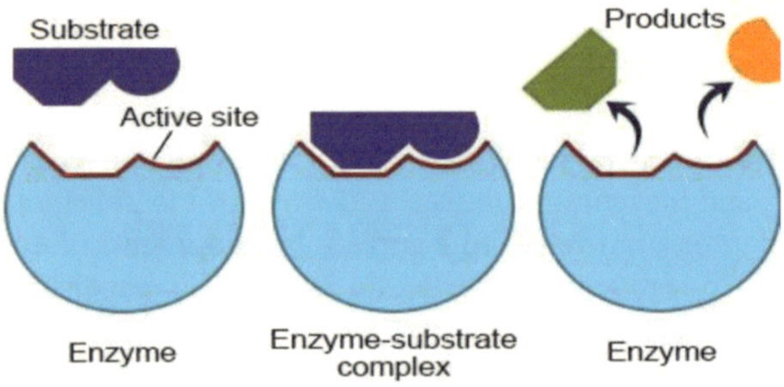

Figure 1.
Mechanism of enzyme-substrate complex [9].

the substrate and lock is an enzyme. Enzyme are not shown rigid structures in a crystallographic x-ray but quite flexible in shape. In 1958, Daniel Koshland presented the 'induced-fit model' of substrate and enzyme binding, which is also known as 'hand-in-glove model' [10].

3. Enzyme classification for textile processing

Enzymes are biocatalysts, which can speed up the chemical processes [11]. Enzymes activates like other inorganic catalysts such as acids, bases, metals, and metal oxides. The molecule that an enzyme acts on is known as its substrate, which is converted into a product. The original attempt to classify enzymes was done according to different function. The International Commission on Enzymes (EC) was established in 1956 by the International Union of Biochemistry (IUB) in consultation with the International Union of Pure and Applied Chemistry (IUPAC) recommended hundreds of enzymes that had been discovered. The EC classification system is divided into six categories [12]:

- EC1 Oxidoreductases: catalyze oxidation/reduction reactions.
- EC 2 Transferases: transfer a functional group.
- EC 3 Hydrolases: catalyze the hydrolysis of various bonds.

- EC 4 Lyases: cleave various bonds by means other than hydrolysis and oxidation.

- EC 5 Isomerases: catalyze isomerization changes within a single molecule.

- EC 6 Ligases: Join two molecules with covalent bonds.

In textile industry mainly hydrolases and Oxidoreductases are engage for various enzymatic applications. Most of the enzyme applications in textiles are confined to cotton processing: removal of impurities (desizing, scouring, bleaching); bio-finishing to improve appearance; bio-stoning or stone washing of denims to produce the fashionable aged look; bleaching cleanup to remove residual H_2O_2 before dyeing [13–17]. In addition to that there are efforts to substitute conventional processes of anti-shrinking, anti-pilling wool & degumming of silk with protease enzyme, retting of bast fibers with pectinase or hemicelluloses the several studies have been reported on modification of synthetics using hydrolases class of enzymes to impart hydrophilicity and antistatic properties [18, 19]. Moreover, detergent with the mixture of enzymes to remove varieties of stains in garments laundering [20]. Textile chemical processing is highly chemical intensive, and a variety of complex chemicals & auxiliaries are regularly used. So, mixed color to water causes toxicity for different form of life. It is essential to treat the effluent specially the residual colorants before discharging in environment. Hence, different enzymes employed for the textile effluent treatment containing synthetic dyes as colorants has been used. International Union of Biochemistry and Molecular Biology set up the Enzyme Commission for solving the complexity of and inconsistency in the naming of enzymes. They proposed almost 7000 enzymes; however, 75 are commonly used in the textile industry [21] According to the use of enzymes in textile industries, enzyme can be classified as shown in **Table 2**.

3.1 Hydrolases

The group of hydrolases enzyme includes amylases, cellulases, pectinases, proteases and lipases. In addition, hydrolases enzymes mainly act as hydrolysis. For isolating microbial strains that produce the desired enzyme and optimizing the conditions for growth, commercial quantities can be obtained. This technique, well known for more than 3,000 years is called fermentation [8].

EC digit	Enzyme groups	Enzyme class	Reaction type
a	Hydrolases	Amylases	Hydrolysis
		Cellulase	
		Proteases	
		Pectinase	
		Lipases/Esterase's	
b	Oxidoreductases	Catalases	Oxidation/reduction
		Peroxidase	
		Laccases	

Table 2.
Classification of enzyme based on textile applications.

3.1.1 Amylases

Amylases hydrolyze starch molecules to give dextrins and maltose, which composed of glucose units [22]. The various starch-splitting enzymes are known as α-amylases and β-amylases [23] (**Figure 2**).

α-amylases are produced from different fungi, yeasts and bacteria. The different microorganisms from α-amylases are listed in **Table 3**.

α-amylases are quite stable over a wide range of pH from 4 to 11. Optimum temperature for the activity of α-amylases is usually applied for modified microorganism. Addition of Ca^{2+} cans enhances thermo stability [24].

Commercial desizing compound under name Rhozyme DX, Rhozyme GC, Diastafor LCD (Bacterial α-Amylases) is more suitable for desizing compared to β-amylase from crude Barley & amyloglucosidase from (*Rhiszopus genus mold*) [25]. Amylase obtained from cheap waste animal pancreas are efficient in desizing for exhaust and pad batch application [26]. Microbial α- Amylase obtained from *Bacillus amyloliquefaciens* performed 100% desizing efficiency with pH 6.5 & 60° C for 1 hour [27]. α- Amylase obtained from *Aspergillus niger* sp. MK 07concentration of 300 U/ml performed at 75° C, pH 6.5 with 0.3 M $CaCl_2$ [28]. Glucoamylase (Multifect GA 10 L) and α- Amylase (Optisize Next) enzymes mixed with chelating agent in citric acid perform simultaneous acid-demineralization and desizing [29]. Ultrasound assisted α-amylase save half desizing process time [30] and α-amylase in winch machine performs highest desizing quality [31]. Ca^{2+} ion independent α-Amylase (*Bacillus sp. KR8104*) activate at moderate temperature (30-70°) desizing with acid-demineralization possible under acidic condition, presence of salts could

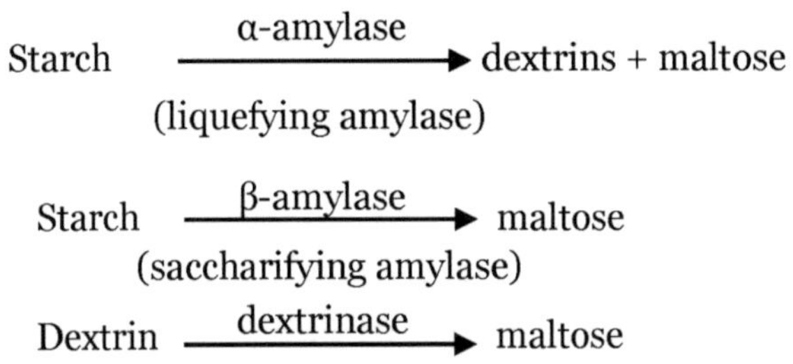

Figure 2.
Degradation of starch by α-amylases and β-amylases.

Microorganisms	Enzyme
1. Bacteria	
Bacillus subtilis	Amylase
B. coagulans	α-amylase
B. licheniformis	α-amylase
2. Fungi	
A. niger	Amylase
A. oryzae	Amylase
Ascomycetes	α-amylase
Basidomycetes	α-amylase

Table 3.
α-amylase from different microorganisms.

decrease enzyme dosage and process time [32]. Amylase obtained from *Aspergillus niger & Aspergillus flavus* shows higher desizing efficiency (*A. niger – 96%, A. flavus – 90%*) with significant improvement in absorbency & extractable impurities [33]. α- Amylase (*Bacillus sp. KR 8104*) shows simultaneous enzyme production and desizing optimized using inexpensive raw materials for α- Amylase [34]. Amylase assisted with other enzyme like *Aquazym* amylase & Lipase improve starch removal and shorten desizing time [35]. α-Amylase (*Aquazym*) incorporate of H_2O_2 or neutral cellulase improves desizing with whiteness and dye-ability presence of wetting agent governed extent of desizing [36]. Alkaline amylase (commercial enzyme Novozymes) performs both desizing & bio-scouring in single bath [37]. Polygalacturonase (*Trichoderma harzianum*) optimize combined pre-treatment in terms of weight loss, residual starch %, absorbency, strength loss and copper number [38]. Thermophile α-Amylase obtained from *Bacilluslicheniformis* optimizes cotton desizing using 3 g/l acidic with 6 pH at 85° C for 40 min [39]. α-Amylase obtained from *mesophilic* shows thermal stability in various additives for high temperature desizing, chitosan saves 2/3 enzyme dosage & improve desizing effect [40]. Amylase from *Aspergillus niger* immobilize with alkylamine glass beads are effective for removal of starch stains along with various detergents [41]. α-Amylase from soyabean seeds entrapped in agarose and agar matrices with 75% & 77% activity reused up to 5 cycles in starch stain removal [42].

3.1.2 Cellulases

Cellulases are hydrolytic enzymes that breakdown the cellulose to form oligosaccharides and finally glucose. Cellulase combining at least three types of enzyme for working synergistically on cotton (**Figure 3**).

The length of cellulose chains cleaves endogluconases or endocellulases in the middle of the amorphous region. However, exo-cellulases start their action from the crystalline ends of cellulose chains and convert to glucose by β-4-glucosidase [43]. These enzymes are commonly produced by soil-dwelling fungi and bacteria (**Table 4**).

A temperature range from 30 to 60° C is an active condition for cellulase. According to pH sensitivity, cellulase enzymes are classified in different categories

$$\text{Cellulose} \xrightarrow{\text{Cellulase}} \text{Oligosaccharide} \xrightarrow{\text{Cellulase}} \text{Glucose}$$

Figure 3.
Degradation of cellulose by cellulase enzyme

Microorganisms	**Enzyme**
1. Bacteria	
Pseudomonas fluorescents	Cellulase
E. coli	Cellulase
2. Fungi	
Trichodermareesei	Cellulase
Aspergillus niger	Cellulase

Table 4.
Cellulase enzyme from different microorganisms.

such as acid stable (pH 4.5–5.5), neutral (pH 6.6–7) or alkali stable (pH 9–10) [44]. Cellulase obtained from *Denimax L*& Pectinase (Pectinex USP) Novo Nordisk, V1-4 Xylanase (*Bacillus sp.*) optimized bio-polishing of jute-cotton blended fabrics for fuzz removal [45]. Cellulase, Endo-enriched cellulase, Exo-endo mixed cellulase are best suited for cotton and lyocell fabric especially for cotton, knits, linen and rayon [46]. Ecostone L883042 and Denimax L used for desorption the cellulase from cotton using ultra filtration for recovering and recycling [47]. Commercial cellulase like Gempil 4 L used for ring spun colored knitted fabrics for pill and fuzz removal [48]. G-ZYME VGB ST trade mark from Rossari are very useful for the action of cellulases on reactive dyed cotton and also show very good effect of bio-polishing on various spun yarn knitted fabrics [49, 50]. On the other hand, Acid cellulase from Genencor, USA performed better enzymatic hydrolysis on viscose, lyocell, modal &cotton fabrics were examined in-terms of degradation rate and weight loss [51]. Acid cellulase obtained from *Talaromyces emersonii* are thermostable cellulase obtained & applied for jute-based fabrics finishing to exhibit improved lusture, handle and durable softness [52]. Commercial cellulase Biopolish EC used for combined scouring-bleaching by cellulase treatment for knit fabrics [53]. Cellulase from *Trichoderma Vride G* optimizes cotton fabric for bio-polishing to improve its smoothness with minimum weight loss [54]. Indiage44L (Gencor) complex with mixed cellulase act as bio-scouring followed by bleaching either peroxide or peracetic acid is efficient for towels and endoglucanase is effective instead of cellulase as additive for terry towel washing [55]. Cellulase from *Trichoderma reesei* performed cotton bio-polishing & its effect on the morphologies [56]. Cellulase from *Chaetomium globosum* performed cotton bio-polishing in-terms of breaking strength, weight loss, thickness, drape and abrasion resistance [57]. Cellulase from *Aspergillus niger* immobilized on maleic anhydride modified PVA coated chitosan beads improves stability of acidic cellulase in neutral pH range [58]. Endoglucanase II (*Trichoderma reesei*) is effective for removing color from denim, producing a good stonewashing effect with lowest hydrolysis level [59].

Alkali Cellulase (*Alkalothermophilic Thermomonospora sp.*) are first time alkali stable endoglucanase used for bio-polishing denims, provide abrasive effect and softness with lower backstaning& negligible weight loss [60]. Cellulase from *Hypocrea jecorina* with nonionic surfactant and dispersing agents provide double benefits of reduction in backstaning and increased cellulase activity [61].

Suhong 89 from Acid cellulase efficiently removes indigo from denim surface with minimum hydrolysis & possibility of cellulase reuse [62].

3.1.3 Pectinases

Pectinase are complex enzymatic group that degrade the pectic substances. They are produced from saprophytes and plant pathogens which can degrade the plant cell walls. There are three major classes of pectin degrading enzymes are pectin esterases, polygalacturonases and polygalacturonate lyases [63].

Pectin Esterases: Pectin esterases liberate pectin and methanol by de-esterifying the methyl ester. Their activity is highest on 65–75% methylated pectin, is to act on methoxy group adjacent to free carboxyl group. Its action has very little effect on the molecular weight of the pectin (**Figure 4**).

Pectin esterases active in the pH range of 4–8 and optimal temperature range for maximum activity is 40-50° C.

Polygalacturonases: Polygalacturonases reduce the molecular weight of the pectins. They catalyze the hydrolytic cleavage with the introduction of water across the oxygen. They are classified further as endo-galacturonases and exo-galacturonases (**Figure 5**).

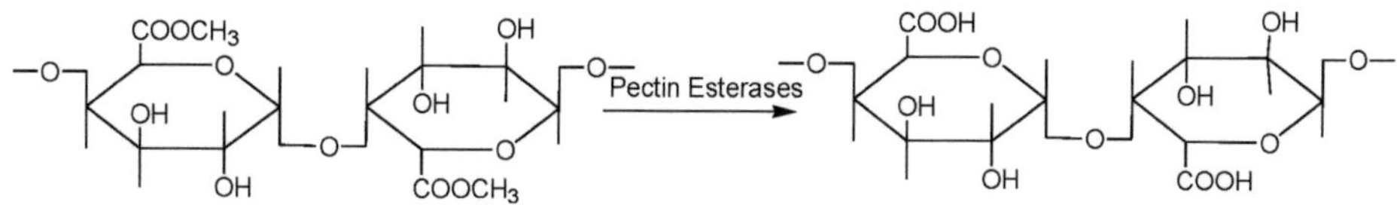

Figure 4.
Degradation of pectin by pectin esterases.

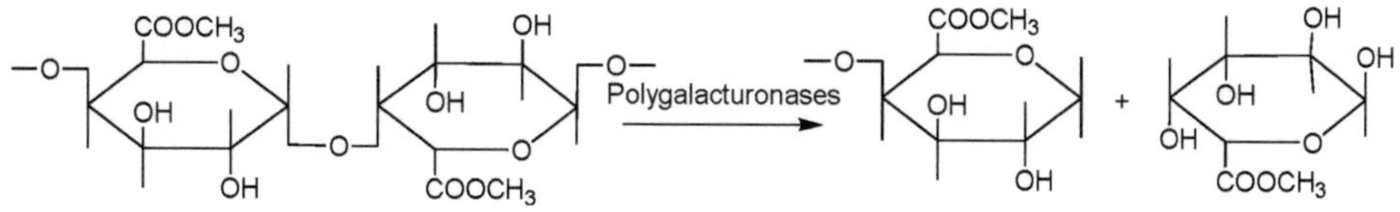

Figure 5.
Degradation of pectin by polygalacturonases.

Polygalacturonases obtained from different natural sources with respect to physiochemical and biological properties as well as their mode of actions.

Pectin Lyases: Pectin lyases depolymerise the pectin. These catalyse the trans-eliminative cleavage of the galacturonic acid polymer. It can break down the glycosidic linkages at C-4 and eliminate H from C-5 position (**Figure 6**).

Pectin esterases, polygalacturonases and pectin lyases are mainly produced in plants such as banana, citrus fruits and tomato, but also by bacteria and fungi (**Table 5**).

Bioprep 3000 L, Novozyme acts as alkaline pectinase are efficiently remove impurities formed uniform dyeing consistency & equivalent color depth with different direct dyes [64]. Pectinase from *Aspergillus niger* named as Bioprep 3000 L agitate improve efficiency and optimize enzymatic scouring provide less damage & superior fabric quality [65, 66]. Bioprep 3000 L Pect062L, Biocatalyst (Acid Pectinase) performs both acid and alkali pectinase are equally efficient but acid pectinase works with lower concentration [67]. Bioprep 3000 L scoured cotton knit fabric at 80° C for removing wax and higher dye uptake optimize bio-scouring for

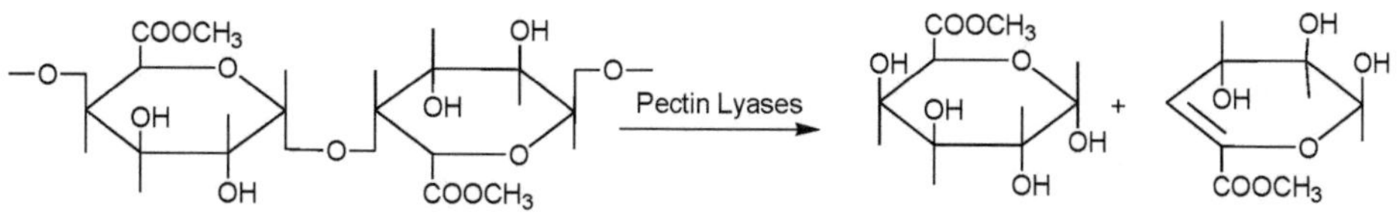

Figure 6.
Degradation of pectin-by-pectin lyases.

Microorganisms	Enzyme
1. Bacteria	
Erwinia	Pectin esterases
Bacteroides	Pectin lyase
Pseudoalteromonas	Polygalacturonase
2. Fungi	
Aspergillus niger	Polygalacturonases

Table 5.
Pectinase enzyme from different microorganisms.

physical, chemical & low stress mechanical properties [68]. Pectinase obtained from *B.macerans* strain V-2692 contains cellulose & hemi-cellulase remove pectin but improve fabric capillary much greater which can substitute cotton boil-off [69]. Multifect cellulose GC obtained from *Trichoderma longibrachiatum* and Multifect pectinase PL from *Aspergillus niger* synergism of cellulose efficiency removes pectin and protein, mechanical agitation and compatible surfactants also play important role [70]. Viscozyme 120 L (Pectinase + hemicellulose) treatment prior to alkaline scouring along with chelating agent at acidic pH, efficiently lightens the seed-coat fragments to improve whiteness and gives better results of wet ability, pectin removal and dyeing with 60 min treatment time [71]. Pectinase, lipase and cellulase enzyme combinedly perform successful scouring, dye and water absorbency with some fiber damage [72]. Alkaline Pectinase (*Bacilus*), Acidic pectinase (Microorganism) along with neutral cellulose (*Apergillus aculeatus*) performs best in wax removal & high absorbency [73]. Alkali Pectinase & Cellulase combinedly scoured knitted fabrics in two step one bath process [74]. Bioprep 3000 L Alkali Pectinase (*Bacillus sp.*) & Lipolase 100 L (*T. lanuginosus*) combined lipase in one-step reduce time required & fabrics with superior properties and excellent dyeing performance obtained [75]. Xylanase from *Bacillus pumilus* are thermostable enzyme provide simultaneous desizing and scouring, addition of chelating & wetting agent increases hydrolysis and allowed reduction of H_2O_2 consumption in consecutive bleaching [76]. Bioprep 3000 L and Forylase KP. Cognis (Acidic Pectinase) simultaneous scouring & bleaching using Pectinase & PAA sufficiently remove pectin and wax to achieve excellent absorbance with medium degree whiteness without damaging fiber and good dye-ability with less energy & water use in enzymatic and/or PAA treatments because of 60° C & pH 6–8 [77–79].

3.1.3.1 Proteases

Proteolytic enzymes produced by microorganisms are mixtures of endopeptidases and exopeptidases. The simplified form the action of the proteases are (**Figure** 7).

Microbial proteases obtained from the plant source such as papain, ficin, and the animal proteases obtained from pepsin and trypsin. Microbial proteolytic enzymes obtained from different fungi and bacteria. Most fungal proteases activate in a pH range (about 4 to 8), and bacterial proteases generally work best over a range of about pH 7 to 8 [80] (**Table 6**).

Proteases with cellulase (Commercial Enzyme) mixture perform for bioscouring and optimized using ANN technique to achieve desired absorbency and pectin removal [81]. Proteases (*Bacillus*) and cellulase provide successful scouring, dye, and water absorbency with some fiber damage in presence of cellulase [82].

3.1.4 Lipases/Esterases

Esterases represent a group of hydrolases that catalyse the cleavage of fats, oils and formed ester bonds. They are widely obtained from animals, plants and microorganisms. These enzymes make attractive biocatalysts for the production of

Proteins —endopeptidases→ proteoses, peptones, polypeptides —exopeptidases→ Amino acids

Figure 7.
Degradation of protein by proteases.

Microorganisms	Enzyme
1. Bacteria	
Bacillus clausii	Proteases
Pseudoalteromonas sp.	Proteases
2. Fungi	
Aspergillus flavus-oryzae	Proteases
Aspergillus tamarii	Proteases

Table 6.
Protease enzyme from different microorganisms.

optically pure compounds in fine-chemicals synthesis. Lipases have three-dimensional structure with the characteristic α/β-hydrolase fold [83]. The phenomenon of interfacial activation can be distinguished by lipases and esterases. The interfacial activation is due to hydrophobic domain covering the lipase active site and the presence of a substrate concentration will lid open, making the active site accessible. It can be used for elimination of natural triglycerides in scouring and tallow compounds in desizing process [84]. The phytopahthogenic fungus is the best examples of lipases are shown in **Table 7**.

Arylesterase obtained from Bio-bleach system HUNTSMAN and H_2O_2 in situ generate peracetic acid for mild temperature at 65° C, neutral bleaching of cotton [85]. Lipase enzymes perform both bio-scouring and bio-bleaching, which provide high degree of whiteness [86].

3.2 Oxidoreductases

The enzymes catalyze oxido reduction reactions, transfer electrons through substrates like cellulose. In the majority of cases, the substrate that is oxidized is regarded as a hydrogen donor. The systematic name of oxidoreductases is based on donor acceptor groups. The enzymes named oxidase, which contains molecular oxygen (O_2) is the acceptor.

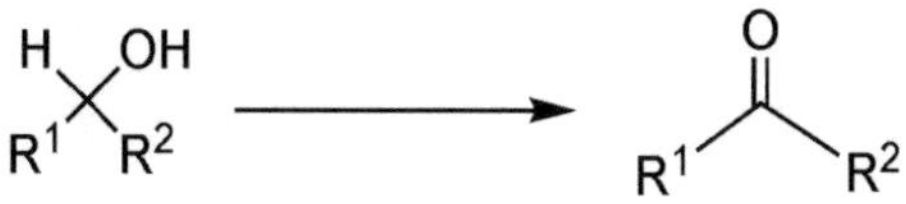

R^1 = hydrogen, organic residue

R^2 = hydrogen, organic residue, alcoxy residue

Microorganisms	Enzyme
1. Bacteria	
Pseudomonas	Lipases
Burkholderia	Lipases
2. Fungi	
Fusarium solani pisi	Lipases
Cunninghamella verticillata	Lipases

Table 7.
Lipase enzyme from different microorganisms.

Oxidoreductases enzymes categorize two nonhydrolytic enzymes such as peroxidase and catalase.

3.2.1 Peroxidase/glucose oxidase

Glucose oxidase or peroxidase acts in the presence of oxygen to convert glucose to gluconic acid and hydrogen peroxide. It can oxidize only β-D-glucose (**Figure 8**).

The galactose oxidase (GO) from *Dactylium deudroides* the oxidation of D-galactose at the C-6 position in the presence of oxygen to give D-galactohexodialdose and hydrogen peroxide. The enzyme contains one atom of Cu^{2+} per molecule as co-factor. Recent investigations indicate that the enzyme catalyses the stereo specific oxidation of glycerol, 3-halogenopropane-1-2-diols and polyols to the corresponding aldehydes (**Figure 9**).

Peroxidase is synthesized in several species of fungi and bacteria are illustrated in **Table 8**.

Glucose-Oxidase desize cotton fabric and enzymatically produce peroxide for bleaching at elevated temperature with high pH [87]. Combined glucose, glucose-oxidase & peroxidase for cotton bleaching [88]. Glucose-oxidase (Commercial Novo Nordisk- Denmark) optimized bio-bleaching of cotton, linen and their blends with 25 U/ml GOE, 10 g/l D-glucose at 85° C & pH 10 for 90 min [89]. Glucose-

$$\underset{\text{Glucose}}{C_6H_{12}O_6} + O_2 + H_2O \xrightarrow{\text{glucose oxidase}} \underset{\text{Gluconic acid}}{C_6H_{12}O_7} + H_2O_2$$

Figure 8.
Degradation of glucose-by-glucose oxidase.

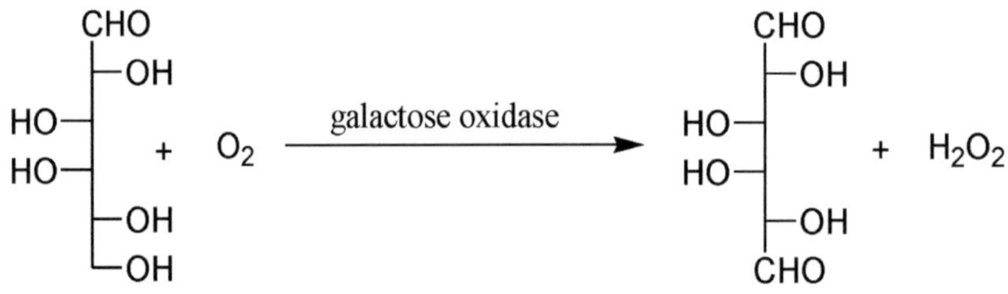

D-galactose D-galactohexodialdose

Figure 9.
Degradation of D-galactose by galactose oxidase.

Microorganisms	Enzyme
1. Bacteria	
Penicillium notatum	Peroxidase
Staphylococcus aureus	Peroxidase
2. Fungi	
Botrytis cinerea	Peroxidase
Aspergillus oryzae	Peroxidase

Table 8.
Peroxidase enzyme from different microorganisms.

Oxidase (*Aspergillus niger*) Biozyme with pullanase mixture used for sufficient (800 mg/l) H_2O_2 for bleaching and maximum whiteness obtained in alkaline pH compared to neutral and acidic pH [90]. Glucose-oxidase from (*Aspergillus niger*) with external oxygen supply and mechanical agitation essential for H_2O_2 generation cotton bleached at room temperature & acidic pH with high enzyme concentration [91]. 6% increase in whiteness index with comparable mechanical properties using peroxide produced by glucose oxidase [92]. One bath low temperature of cotton pretreatment by glucose oxidase where liberated H_2O_2 converted to peracetic acid using TAED as activator [93]. Glucose-Oxidase GC 199 combine desizing, scouring with enzymes followed by bleaching with in-situ generated peracetic acid using different activators [94]. Assistance of ultrasound with glucose-oxidase improves whiteness due to increase enzyme reaction at 90° C with pH 11 [95]. Multifect GO 5000 L, Genecor performed desizing, bleaching and reactive dyeing for cotton towel [96]. Glucose-oxidase from *Aspergillus niger* immobilized on porous carriers-glass & alumina, low enzyme concentration provides sufficient H_2O_2 release which further activated for textile bleaching [97].

3.2.2 Catalases

Catalases (CATs) also known as hydroperoxidases, catalyse the degradation of H_2O_2 to H_2O and O_2. Catalase, which is also found in commercial fungal glucose oxidase preparations [98] (**Figure 10**).

Catalases are ubiquitous oxidoreductases enzymes present in archaea, bacteria, fungi, plants and most have optimum temperatures (20-50° C) and neutral pH. Catalases obtained from animal sources (bovine liver) are generally cheap; therefore, the production of microbial catalase will be economically advantageous when recombinant technology is used. Catalases have special properties such as thermosatbility and operate both in alkaline or acidic pH. The chloroperoxidase from *Caldariomyces fumago* also catalyses the oxidation of halide ions except fluoride (**Table 9**).

Glucose-oxidase (multifect GO 5000 L, Genecor), Catalase (Terminox Ultra 10 L) integrated desizing, bleaching and reactive dyeing of cotton towel was performed [96].

$$2\,H_2O_2 \xrightarrow{\text{Catalase}} 2H_2O + O_2$$

Figure 10.
Degradation of hydrogen peroxide by catalase.

Microorganisms	Enzyme
1. Bacteria	
Pseudomonas putida	Catalase
Neurospora crassa	Catalase
2. Fungi	
Aspergillus terreus	Catalase
Aspergillus niger	Catalase

Table 9.
Catalases enzyme from different microorganisms.

Microorganisms	Enzyme
1. Bacteria	
S.lavendulae	Laccases
Theiophora terrestris	Laccases
2. Fungi	
Trametes villosa	Laccases
Botrytis cinerea	Laccases

Table 10.
Laccases enzyme from different microorganisms.

3.2.3 Laccases

Laccase originated from blue-multicopper oxidase family. It oxidizes a variety of aromatic and non-aromatic phenolic compound also depolymerizes the substrate by a radical-catalyze reaction mechanism.

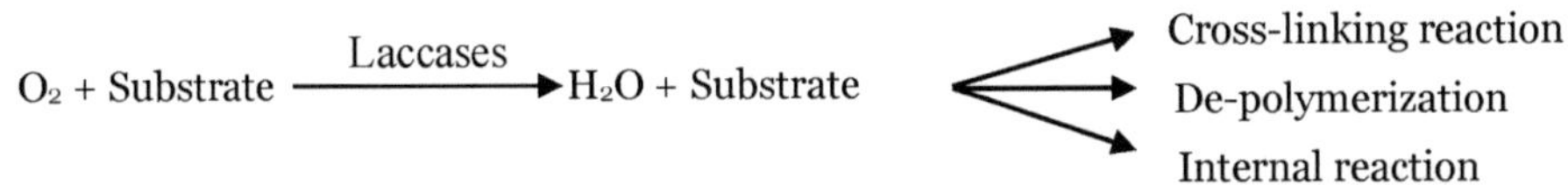

Laccases have been found in plants, fungi, insect and bacteria. However, more than 60 fungal strains have found laccase activity. Fungal laccase is a protein approximately 60–70 KDa, which activate in the acidic pH range and optimal temperature between 50 and 70° C. Few laccases enzymes activate with optimum temperature below 35° C (**Table 10**).

Combined laccase/peroxide bleaching applied in batch & pad dry method [99]. Laccase from *Trametes hirsute* with mediator improve whiteness of cotton due to oxidation of flavonoids [100]. Complex enzyme Laccase & Peroxidase (*Ph Chrysoporium & Trichosporon cutaneum R57)* efficiently degrade & remove lignin from flax fiber to provide whiteness [101]. Laccase, Novozyme obtained from *Trametes villosa* assistance with ultrasound and PVA addition stabilize laccase and improve bleaching [102]. Ecolite II (Commercial Laccase, Jeans are company) and H_2O_2 performed bleaching of linen allows better dye uptake for both reactive and cationic dyes [103].

4. Textile applications

The enzymatic textile processing has started in the middle of nineteenth century. The enzymes were introduced in de-sizing purposes for the first time in 1857; however, enzymatic de-sizing process was successfully introduced in 1912 [104]. In addition, cellulases were introduced in 1980s for de-pilling and de-fuzzing of cellulose-based fabrics [105]. In the early 1990s, catalases were entered into the bleaching and pectin-degrading enzymes to replace traditional alkaline scouring [106]. In biotechnological research is underway, around the globe, introduce environmental friendlier strategies for textile processing is extensively to the modern industry. The potential of enzymatic textile processing is illustrated in **Figure 11**.

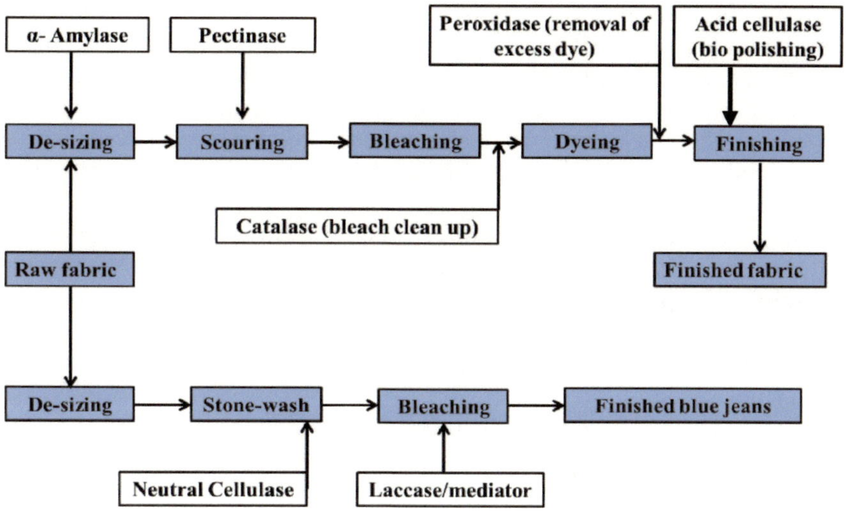

Figure 11.
Enzymes used in various operations in textile wet processing [107].

5. Enzyme used in textile wet processing

5.1 De-sizing

The cotton warp yarns are sized to improve the yarn strength. Besides help in the interweaving during the procedure, it protects yarn against abrasion and snagging. Mostly starch-based products are used to apply sizes, synthetic and semi-synthetic sizes are polyvinyl alcohol (PVA) and carboxymethylcellulose (CMC) used [104–106]. The purpose of size is to protect the yarn from the abrasive action of weaving loom.

Desizing is the first step for wet processing in textile finishing technology employed to remove sizing material from the fabric. The size must be removed before bleaching and dyeing, for uniformity of wet processing. Chemically, starch is poly-α-glucopyranose in which amylase and amylopectin are present. However, they are insoluble in water. They can be solubilized by hydrolyzing them to shorter chain compounds. The object of desizing is to convert starch to soluble dextrin. The stages of hydrolyzing are mentioned below:

Starch (insoluble) → dextrin (insoluble) → dextrin (soluble) → maltose (soluble) → α-glucose (soluble)

Types of desizing methods

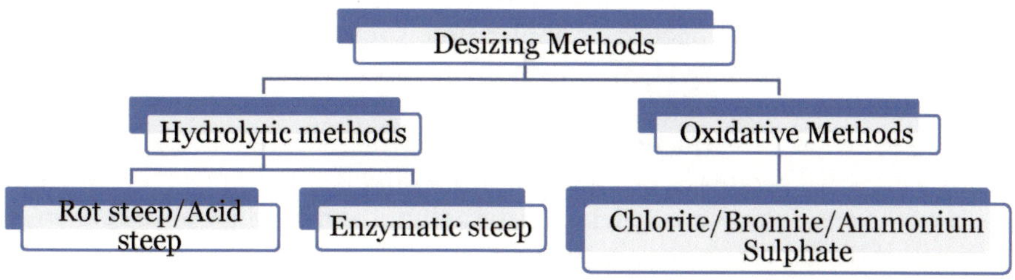

Enzyme desizing is the most widely practiced method of desizing starch. *Amylase* can catalyses the breakdown of starch form sugars, dextrin and maltose. The advantage of these enzymes is specific for starch, removing it without damaging to the support fabric. An enzymatic desizing process at low-temperature (30-60°C) and optimum pH is 5.5–6.5 is required for amylase [108]. Rising the temperature desizing facilities starch removal as well reduces process duration therefore the thermophile amylases have gained wider acceptance. Thermophile *Bacillus licheniformis* α-amylase can provide high temperature efficient starch removal efficiency with improved absorbency [109]. The enzymatic desizing process can be divided into three steps:

Impregnation: Enzyme solution is absorbed by the fabric. This stage involves through wetting of fabric with enzyme solution at a temperature of 70° C or higher with a liquid pick up of 1 liter per kg fabric. During this stage, gelatinization of the size is to the highest possible extent.

Incubation: The size is broken down by the enzyme. Long incubation time allows a low enzyme concentration.

After-wash: The breakdown products from the size are removed from the fabric. The desizing process is not finished until the size breakdown products have been removed from the fabric. This is best obtained by a subsequent detergent wash at the highest possible temperature.

5.2 Scouring

In textile terminology "scouring" applies to the impurities removal process. Raw cotton contains about 90% of cellulose and various non-fibrous impurities such as dirt, oils, waxes, gums and seed fragments. Pectins are complex polysaccharides comprised of α-(1, 4) linked D-galacturonic acid backbone. Pectin is non-cellulosic substance in cotton acts as cementing/adhesive material; therefore, removing pectin, will enhance to remove other non-cellulosic substances. In scouring process, the target fabric is usually boiled in the presence of alkali solution using large iron-made vessels called kiers. Classic alkaline scouring using sodium hydroxide removes most of such contaminations, essential to achieve satisfactory wet-ability. In textile processing, the process based on extensive alkali consumption requires heavy rinsing which ultimately leads several by-products in the wastewater effluent and poses severe damage to the cellulose contents of the fabric [110, 111].

The process of bio scouring is based on the concept of decomposition of pectin using enzyme. Bio scouring is ecofriendly, energy conserving alternative based on the idea of specially targeting the non-cellulosic impurities with appropriate enzymes without adversely affecting the substrate. Natural properties of the cotton fiber are preserved; the fabric is softer to the touch than after classic scouring. Pectinases enzyme activate in two medium acidic and alkaline. Acidic pectinases that function in a slightly acidic medium (pH between 4 and 6), as well as alkaline pectinases that function in a slightly alkaline medium (pH between 7 and 9) [112]. Optimum enzyme concentration varies from pectinase to pectinase but in general, pectinases are effective in low concentrations 0.005–2% range. In addition, the optimum temperature of pectinases application is form 40-60° C beyond which enzyme reduces its activity [113]. A high temperature rinsing after bio scouring is required for the removal of waxes (**Figure 12**).

Mixed enzymatic treatments of unscoured cotton fabric conducted by German scientists involved pectinase, cellulase, protease, and lipase. Beside the temperature, the pH of the environment is crucial for the activity and stability of the enzyme. An assistance of lipases removes natural fats & lubricants for better absorbency and levelness in dyeing. Bio-scouring for mutations containing lipase are more effective in attaining good hydrophilicity for cellulosic textiles [114]. In recent years, pectinases have been immobilized by ion exchange resins, aminated silica gel and macroporuos polyacrylamide for cotton scouring.

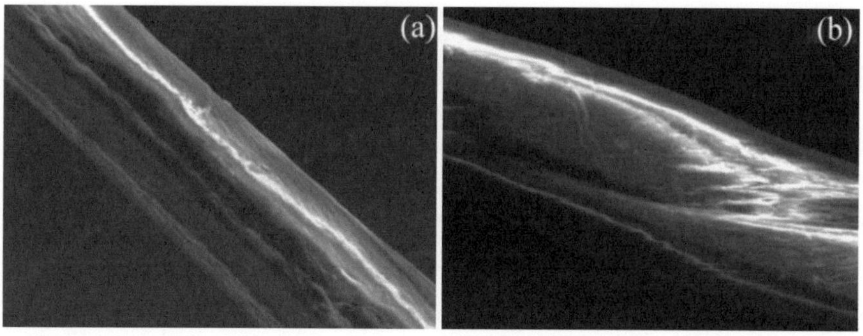

Figure 12.
SEM images of a Bio-scoured cotton fabric (a) Bio-scoured with pectinase enzyme, (b) Bio-scoured with cellulase and pectinase mixed enzyme.

$$H_2O_2 \rightleftharpoons H^+ + HO_2^-$$

Figure 13.
Dissociation of hydrogen peroxide.

5.3 Bleaching

Bleaching is a process for improving the whiteness of textile materials with or without removing the natural coloring matter or extraneous substances. Bleaching produces permanent and basic white effect on fabric, which is required for level dyeing and sharp printing. Among the different oxidizing and reducing bleaching agent, H_2O_2 is mostly used as a universal bleaching agent from last two decades. The dissociation of hydrogen peroxide increased with rising temperature and form perhydroxyl anion shown in **Figure 2**. Perhydroxyl ions (HO_2^-) demobilize the mobile electrons of conjugated double bonds in chromophores and caused decolorization. However, hydrogen peroxide bleaching process required high temperature and long processing time, which leads to higher energy consumption and increased fiber damage, which would cause problems in dyeing [115] (**Figure 13**).

Many researchers explored the alternative eco-friendly bleaching method for cotton processing, such as laccase/mediator or glucose-oxidase/peroxidase and bleaching with enzymatically in situ generated per acids. Lacasses with copper containing oxidoreductases enzymes used for bio bleaching to bleach textiles, modify fabric surfaces and coloration of cotton [116]. Another important bio-bleaching method for producing H_2O_2 is glucose oxidase. Generation of peroxidase with glucose oxidase requires slightly acidic to neutral conditions at low temperatures, however these conditions is insignificant. In addition, at the temperature 80-90° C and alkaline pH 11, glucose oxidase provides efficient results for improving the whiteness of the cotton fabric [117]. On the other hand, in addition of bleach activators such as Tetraacetylethylenediamine (TAED), nanoyloxybenzene sulphonate (NOBS), N-[4-(triethyl ammoniomethyl) benzoyl] caprolactum chloride (TBCC) enhances the bleaching performance. Combined laccase and glucose oxidase can perform better bleaching effect on linen fabric (**Table 11**).

5.4 Bleach clean up

In textile industry, bleaching is carried out by H_2O_2 after scouring and before dyeing. However, 10–15% of H_2O_2 retains on fabric, which can degrade the cellulose and formed pinhole on the fabric surface, it can reduce the strength of fiber.

Condition of Cotton Fabric		Whiteness (stensby degree)
Grey Fabric		52 ± 0.5
H_2O_2 bleached		80.0 ± 0.5
Enzymatic bleach		71.0 ± 1.2

Table 11.
Whiteness of different bleaching process.

Different reducing agent is used to destroy the hydrogen peroxide, or water to rinse out the hydrogen peroxide bleach. However, catalase enzyme can now be used to decompose excess H_2O_2 [118]. This eliminates the use of strong reducing agent and minimizes the water consumption. The cost of enzyme for degradation of hydrogen peroxide in bleaching effluents could be reduced by the immobilized catalase enzymes [119]. The process of bleach clean-up is very straightforward. A summary of the methodology is- 1) Drain the bleach liquor after bleaching 2) Fresh cold water filled 3) Maintain the pH is in the range 6.5–7 and the temperature 45° C 4) Add catalase (*Terminox Ultra*) enzyme 5) After 10–20 minutes checking the H_2O_2 removed by using Merck peroxide test strips 6) Start the dyeing process. Dyeing without and with bleach clean-up has shown in **Table 12.**

5.5 Removal of excess dye

The dye removal process from over dyed fabric is called as "back stripping" or "destructive stripping". Bio-enzymes used for decolorization is lignin peroxidase, manganese peroxidase and laccase. Previous studies have been investigated for color stripping from dyed textile fabric by microbial strains and their non-specific enzymatic system. The catalase enzymes from *G. lucidum* showed the ability for color stripping from the reactive black B dyed cotton fabric (**Table 13**) [120, 121].

The enzymes used for textile ETP are laccase, manganese peroxidase, lignin peroxidase and tyrosinase. These enzymes can catalyze the chlorinated phenolic compounds and halogenated organic compounds [122].

Dyeing without bleach clean-up	Dyeing with bleach clean-up

Table 12.
Bleach clean-up process on cotton fabric and dyeing effect.

Dyed Fabric	Bio-stripping

Table 13.
Bio-stripping process on dyed cotton fabric.

Before cellulase treatment	After cellulase treatment
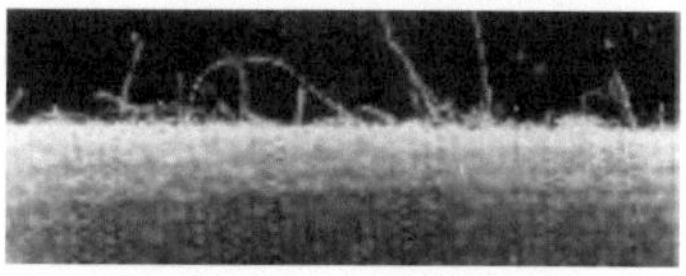	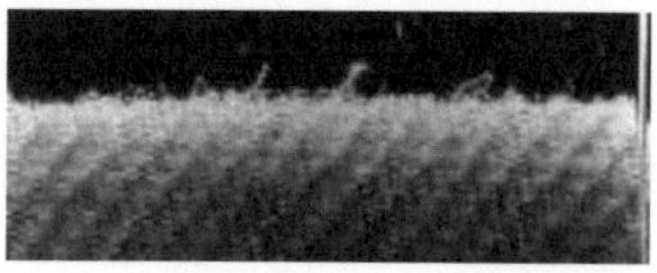

Table 14.
Bio-polishing process on cotton fabric.

5.6 Bio-polishing

The process of removal of micro, fuzzy fibrils from the fabric surfaces through the action of cellulase enzyme is called bio polishing. It enhances the color brightness, hand feel, water absorbance property of fibers; strongly reduce the tendency for pill formation [123]. Cellulase enzymes are widely used for bio polishing.

Cellulases enzymes hydrolyze the cellulose structure by degrading β-(1-4) glycosidic linkages. Endocellulases cleave bonds along the length of cellulose chains in the middle of the amorphous region, exoglucanases act from the crystalline ends of cellulose chains and convert soluble oligosaccharides to glucose. Commercially available cellulases enzyme for bio polishing are a mixture of endogluconases, exoglucanases & cellobioses (**Table 14**).

Bio polishing is done before or after dyeing to the cotton, fabrics influence dyeability, besides improving the appearance and handle properties. Cellulase enzyme treatment enhances the post dyeing effect and resin finishing with increases softness. Endoglucanases with acid cellulase are suited for bio polishing of cellulosic fabrics [124]. The enzymatic hydrolysis of cotton also enhanced by mechanical action with the addition of surfactant. For bio polishing acid and neutral cellulases bath is maintained at 4.5–5.5 and 7 respectively, the process is heated at 55° C. Finally, the process is terminated at the temperature 85° C. The immobilization of cellulase can restrict its action to fiber surface. Various immobilization of cellulases enzyme methods improve thermal stability and reusability [125].

6. Denim manufacturing

Due to special fading effect and aging process, denims became the most popular fashion trends in recent times. For increasing softness of denim garments usually pumice stones washing are using. However, the use of natural pumice stones has many disadvantages; it can cause severe physical damage to garments, machine and stone dust can clog the machine drainage lines and large amount of back staining on the fabric. In addition, the complete rid of pumice stones; several wash is required for denim fabrics. It causes high water consumptions [126, 127].

6.1 Denim wash using neutral cellulase

Cellulase enzyme application on denim finishing has started in the late 1980s. Cellulase enzyme can remove trapped indigo dye from the denim fabric, which causes non uniform shade fading & worn looking. Bio washing with cellulase enzyme is an ecofriendly and superior quality for denim fabric. Application of neutral cellulases enzyme is active in a wide temperature range from 30° to 60° C and based upon their application pH 6.6–7 cellulase [128, 129].

6.2 Denim bleaching using laccase/mediator

Eco friendly denim bleaching has been originated in the late 1980s due to adverse effect of the conventional chemical bleaching process. By oxidizing the flavonoids of denim fabric, laccase enzyme can enhance the whiteness of the fabric. Enzymatic bleaching system leads not only damage of the denim fabric but also save water consumption. There are some successful industrial laccase enzymes for denim bleaching are used as DenLite®from Novozyme (Novo Nordisk, Denmark) and Zylite from the company Zytex (Zytex Pvt. Ltd., Mumbai, India) [130].

7. Modification of synthetic fiber by enzyme

Even though synthetic fibers are mostly hydrophobic, however improving the hydrophilicity along with some properties such as weaving comfort, anti- pilling, dyeability, antistatic charge generation by Enzymes are used. Different classes of enzymes such as cutinases, lipases and esterases are suitable for polyester modification. Less potential for surface hydrolysis of polyester modification were esterases. Esterases from *Thermobifida halotolearans* have performed for both PET and PLA surface hydrolysis [131]. Hydrolyzing the polyester has been carried out by cutinases obtained from *Aspergillus oryzae, Penicillium citrinum, Fusarium solani, Thermobifida fusca. Thermobifida fusca, Thermobifida celluloysitica* [132, 133]. Lipases obtained from *Humicola sp., Candida Antarctica, Thermomyces lanuginosus, Triticum aestivum. And Rhizopus delemar* considered as a suitable for polyester hydrolysis [132]. Digital printed polyester fabric was treated by lipases for improving the color fastness [133].

Laccases with a mediator have been performed to increase the hydrophilicity of nylon 66 fabrics [134]. For improving dye bath exhaustion with reactive and acid dyes on Nylon 66 fabrics were treated by proteases from *Beauveria sp.,* an amidase from *Nocardia sp.* and a cutinase from *F. solani pisi* [135]. It was confirmed that acid and disperse dyes showed higher exhaustion on the protease & lipase treated Nylon 6 [136]. Proteases from a novel *Bacillus* isolate improve hydrophilicity and cationic dye affinity of nylon fabric without affecting the mechanical properties [137]. The vinyl acetate moieties in PAN can be hydrolyzed by cutinases and lipases [138].

8. Enzymatic treatments of ETP in textile industry

Textile effluents are usually highly colored, presenting different chemical substances when discharged into waters after finishing processes. Textile effluents can be discolored through physical, chemical and biological technologies. Several researches have been carried out for the removal of dyes from industrial effluents by chemical, physical and biological techniques. The decoloration dyestuffs and recalcitrant compounds using some biological techniques such as anaerobic, aerobic and combined process [139]. WRF are the principal organisms which have been investigated for dye degradation and decolorization purposes. The major lignin mineralizing enzymes of WRF are LiP, MnP and laccase that are involved in dye degradation [140]. Other than these enzymes many oxidase including versatile peroxidase, glyoxal oxidase, aryl alcohol oxidase and oxalate decarboxylase can perform dye degradation [141]. Toxic organic compounds can be detoxified through oxidative coupling is mediated with oxidoreductases. The detoxification of toxic organic compounds through oxidative coupling is mediated with oxidoreductases [142]. Enzymes like laccase, manganese peroxidase and lignin peroxidase catalyze the removal of chlorinated phenolic compounds [143]. Microbial oxygenases, such as monooxygenases and dioxygenases are active against a wide range of compounds [144].

9. Conclusion

Enzymes have tremendous progress in textile chemical processing to meet up the green and sustainable demand in 21st century. There are several commercially successful enzymes are amylases, cellulases, pectinases and catalase for textile wet processing. Enzyme immobilization is another important technique for highly efficient textile processes. This chapter highlights the integration of enzyme based biotreatments in textile processing. In this context, different enzymatic processes have already been developed or in the process of development for textile processing. In this regard, this chapter summarizes current developments and highlights the environment-friendly enzymatic applications. So, extensive research is required for the implementation of enzyme-based processes for both synthetic and natural fibers. Due to wide variations in the properties of individual enzymes and their reaction mechanism, there are still considerable and reliable tools for potential applications in different textile processing.

Conflict of interest

The authors have declared no conflicts of interest.

Author details

Shekh Md. Mamun Kabir[1*] and Joonseok Koh[2]

1 Department of Wet Process Engineering, Bangladesh University of Textiles, Dhaka, Bangladesh

2 Division of Chemical Engineering, College of Engineering, Konkuk University, Seoul, South Korea

*Address all correspondence to: mamunkabir.butex@gmail.com

References

[1] Robinson P. K., Enzymes: principles and biotechnological applications. Essays Biochem. 2015; 59:1-41. DOI: 10.1042/bse059001

[2] Godfrey T, West SI: Introduction to industrial enzymology. In Industrial Enzymology, edn 2. Edited by Godfrey T, West S. London: Macmillan Press; 1996:1-8

[3] Voigt CA, Kauffman S, Wang ZG: Rational evolutionary design: the theory of in vitro protein evolution. Adv Protein Chem 2000, 55:79-160.

[4] Vogel A., May Oliver: Industrial Enzyme Applications. Wiley; 2019. 25-45 p. DOI: 10.1002/9783527813780

[5] McCoy M: Novozymes emerges. Chem Eng. News 2000, 19:23-25

[6] Choudhury RB, Jana AK, Jha MK: Enzyme technology applications in leather processing. Ind J Chem Technol 2004, 11:659-671.

[7] Araujo R, Casal M, Cavaco-Paulo A: Application of enzymes for textiles fibers processing. Biocatal Biotechnol 26:332-349

[8] Singh R., Kumar M., Mittal A. & Mehta P. K.: Microbial enzymes: industrial progress in 21st century. Biotech 2016, 6:174.

[9] Changeux J. P., 50 years of allosteric interactions: the twists and turns of the models. Nat, Rev, Mol. Cell Biol. 2013; 14: 819-829.

[10] Kamata K., Mitsuya M., Nishimura T., Eiki J., Nagag Y. Structural basis for allosteric regulation of the monomeric allosteric enzyme human glucokinase. Structure. 2004; 12: 429-438.

[11] Cavaco-Paulo A. and Gübitz G. M., Textile processing with enzyme, Woodhead Publishing; 2003. ISBN 18557366101

[12] Mojsov K., Enzyme Applications in Textile Preparatory Process: A Review. 2012; 2:272-295p.

[13] Achwal WB., Enzymetic removal of cotton pectin, Colourage; 1992;39:35p

[14] Lenting HBM, Warmoeskerken MMCG., A fast, continuousenzyme-based pretreatment process concept for cotton containing textiles, Biotransformation; 2004; 22(5-6): 361-368.

[15] Saravanan P., Sivasaravanan S., Sudharshan P. M., Vasanthi N. S., Senthil R. K., Das A., Ramachandran T., One-step process for desizing and bleaching of cotton fabrics using the combination of amylase and glucose oxidase enzymes, J. Appl. Polym Sci, 2012; 123(4): 2445-2450.

[16] Tzanov T., Calafell M., Guebitz G. M., Cavaco-Paulo A., Bio-preparation of cotton fabrics, Enzym Microb Technol, 2001; 29(6):357-362.

[17] Shao-Wei D., Da-Nian L., Kinetics of the thermal inactivation of Bacillus subtilis α-amylase and its application on the desizing of cotton fabrics, J. Appl. Polym. Sci, 2008; 109(6): 3733-3738.

[18] Shen J., Rushforth M., Cavaco-Paulo A., Guebitz G., Lenting H., Development and industrialization of enzymatic shrink-resist process based on modified proteases for wool machine washability, Enzyme Microb. Technol, 2007; 34, 1-6.

[19] Silva C. J. S. M., Gubitz G., Cavaco-Paulo A., Immobilization of proteases with a water soluble-insoluble reversible

polymer for treatment of wool, Enzyme Microb. Technol., 2006b; 39: 634-640.

[20] Galante, Y. M., Cristina, F., Enzyme applications in detergency and in manufacturing industries, Curr. Org. Chem., 2003; 7(13): 139-1422.

[21] Hoh C., Villela Filho M., The Enzyme Classification. Industrial Biotransformation's, 2nd Edition, Wiley; 2006; 37-39p.

[22] Windish, W. W., and Mhatre, N. S., Microbial amylase, Adv. Appl. Microbiol. 1965; 7: 273-304.

[23] Pandey A, Nigam P, Soccol CR, Soccol VT, Singh D, Mohan R., Advances in microbial amylases, Biotechnol Appl Biochem; 2000; 31: 135-152.

[24] Heinen W., Lauwers A. M., Amylase Activity and Stability at High and Low Temperature Depending on Calcium and Other Divalent Cations, Enzymes and Proteins from Thermophilic Microorganisms Structure and Function; 1976 77-89p.

[25] Hanson, M. A., Gilbert, R. D., A new look at desizing with enzymes, Text. Chem. Colorist Am. Dyest. Report; 1974; 6(12): 28-31.

[26] Khan, A. F., Arif, S., Development and applications of animal amylases for enzymatic desizing of woven fabric, Pak. J. Sci. Ind. Res., 2006; 49(2), 103-105.

[27] Haq, I., Ali, S., Javed, M. M., Hameed. U., Saleem, A., Adnan, F., Qadeer, M. A., Production of alpha amylase from a randomly induced mutant strain of *Bacillus amyloliquefaciens* and its application as a desizer in textile industry, Pak. J. Bot., 2010; 42(1), 473-484.

[28] Chimata, M. K., Chetty, C. S., Suresh, C., Fermentative production and thermostability characterization of α-amylase from Aspergillus species and its application potential evaluation in desizing of cotton cloth, Biotechnol. Res. Int., 2011; 2011:323891.

[29] Dehabadi, V. A., Opwis, K., Gutmann, J., Combination of acid-demineralization and enzymatic desizing of cotton fabrics by using industrial acid stable glucoamylases and α-amylases, Starch/Starke., 2011; 63 (12), 760-764.

[30] Wang W., Yu, B., Zhong, C., Use of ultrasonic energy in the enzymatic desizing of cotton fabric, J. Clean. Prod., 2012; 33, 179-182.

[31] Ul-Haq, Nasir, H., Cleaner production technologies in desizing of cotton fabric, J. Text. Inst., 2012; 103 (3), 304-310.

[32] Chand, N., Natri, A. S., Sajedi, R. H., Mahadavi, A., Rassa, M., Enzymatic desizing of cotton fabric using a Ca^{2+} independent Amylase with acidic pH profile, J. Mol. Catal. B Enzym., 2012; 49 (11), 1885-1888.

[33] Sreelaakshmi, S. N., Paul, A., Vasanthi, N. S., Sarvanan, D., Low temperature acidic amylases from *Aspergillus* for desizing of cotton fabrics, J. Text. Inst., 2014, 105(1), 59-66.

[34] Chand, N., Sajedi, R. H., Nateri, A. S., Khajeh, K., Mehdi, Rassa., Fermentative desizing of cotton fabric using an α-amylase producing Bacillus strain: Optimisation of simultaneous enzyme production and desizing, Process Biochem., 2014; 49(11): 1885-1888.

[35] Lange, N. K., Lipase-assisted desizing of woven cotton fabrics, Text. Chem. Colorist., 1997; 29(6): 23-26.

[36] Ibrahim, N. A., El-Hossamy, M. Morsy, M. S., Eid, B. M., Development of new eco-friendly options for cotton

wet processing, J. Appl. Polym. Sci., 2004; 93(4), 1825-1836.

[37] Kuilderd, H., Wu. G., Simultaneous Desizing and Scouring with Enzymes, AATCC Review., 2008; June, 33-35.

[38] Aly, A. S., Sayed, M., Zahran, M. K., One-step process for enzymatic desizing & bio-scouring of cotton fabrics, J. Nat. Fibers; 2010, 7(2), 71-92.

[39] Saravanan, D., Prakash, A. A., Jagadeeshwaran, D., Nalankilli, G., Ramachandran, T., Prabakaran, C., Optimization of thermophile *Bacillus lichneformis*α-amyulase desizing of cotton fabrics, IJFTR., 2011, 36(3), 253-258.

[40] Fu, K., Wang, D., Li, Y., Lu, D., Effect of additives on mesophilic α-amylase and its application in the desizing of cotton fabrics, J. Text. Inst., 2015, 106(12), 1322-1327.

[41] Dhingra, S., Khanna, M., Pundir, C. S., Immobilization of α-amylase onto alkylamine glass beads affixed inside a plastic beaker: kinetic properties and application, Indian J. Chem. Technol., 2006, 13, 119-121.

[42] Prakash, O., Jaiswal, N., Immobilization of a thermostable amylase on agarose and agar matrices and its application in starch stain removal, World Appl. Sci. J., 2011, 13 (3), 572-577.

[43] Teeri T. T., Crystalline cellulose degradation: new insight into the function of cellobiohydrolases, Trends in Biotechnology; 1997;5:160-167.

[44] Araújo R., Casal M. & Cavaco-Paulo A., Applications of enzymes from textile fibres processing, Biocatalysts and Biotransformation; 2008; 5: 332-349.

[45] Sreenath, H. K., Shah, A. B., Yang, V. W., Gharia, M. M., Jeffries, T. W., Enzymatic polishing of jute/cotton blended fabrics, J. Ferment. Bioeng, 1996, 81(1): 18-20.

[46] Kumar, A., Yoon, M., Charles, P., Optimizing the use of cellulase enzymes in finishing cellulosic fabrics, Text. Chem. Colorist Am. Dyest. Report., 1997, 29(4):37-42.

[47] Azevedo, H., Bishop, D., Cavaco-Paulo, A., Possibilities for recycling cellulases after use in cotton processing, Appl. Biochem. Biotechnol. 2002, 101, 61-75.

[48] Ozdil, N., Ozdogan, E., Oktem, T., Effects of enzymatic treatment on various spun yarn fabrics, FIBRES Text. East. Eur., 2003, 114(43), 58-61.

[49] Yamada, M., Amano, Y., Horikawa, E., Nozaki, K., Kanda, T., Mode of action of cellulase on dyed cotton with a reactive dye, Biosci. Biotechnol. Biochem. 2005, 69(1), 45-50.

[50] Chinta, S. K., Landage, S. M., Verma, K., Effect of biopolishing treatment on various spun yarn knitted fabrics, Gob. J. Bio-Science Biotechnol. 2012, 1(2), 287-295.

[51] Schimper, C. B., Ibanescu, C., Bechtold, T., Technical aspects in enzymatic hydrolysis of cellulosics, Lenzing, Berichte, 2006, 85, 107-112.

[52] Gomes, I., Sarkar, P.K., Rahman, S. R., Rahim, M. A., Gomes, D. J. Production of cellulase from *Talaromyces emersonii* and evaluation of its application in eco-friendly functional finishing of jute-based fabrics, Bangladesh J. Microbiol, 2007, 24(2), 109-114.

[53] Uddin, M. G., Effect of biopolishing on dye ability of cotton fabric-a review, Trends Green Chem, 2016, 2(1): 1-5.

[54] Bai, G., Fu, K., Jin, N., Zhu, L., Chai, H., Lu, D., Bio-polishing of cotton fabrics with cellulase, Adv. Material Res., 2012, 468-471, 46-49.

[55] Tavcer, P. F., Effects of cellulase enzyme treatment on the properties of cotton terry fabrics, FIBRES Text. East. Eur., 2013, 21, 6(102): 100-106.

[56] Saravanan, D., Laxmi, S. N. S., Vasanthi, N. S., Raja, K. S., Biopolishing of cotton fabric with fungal cellulase and its effect on the morphology of cotton fibres, Indian J. Fibre & Text. Res., 2013, 38(2), 156-160.

[57] Chinnamma, S. K., Antony, V. A. R., Production and application of cellulase enzyme for biopolishing of cotton, Int. J. Sci. Technol. Manag., 2015, 4(1), 1606-1612.

[58] Dincer, A., Telefoncu, A., Improving the stability of cellulose by immobilization on modified polyvinyl alcohol-coated chitosan beads, J. Mol. Catal. B Enzym, 2007, 45, 10-14.

[59] Heikinheimo, L., Buchert, J., Miettinen-Oinonen, A., Suominen, P., Treating denim fabrics with Trichoderma reesei cellulases, Text. Res. J., 2000, 70(11), 969973

[60] Anish, R., Rahman, M. S., Rao, M., Application of cellulases from an Alkalothermophilic Thermomonospora sp. in biopolishing of denims, Biotechnol, Bioeng, 2007, 6(1), 48-56.

[61] Zilz, L., Rao. M., Budag, N., Scharf, M., Cavaco-Paulo, A., Andreaus, J., Nonionic surfactants and dispersants for cellulase treatment of cotton textiles, Color technol, 2012, 129, 49-54.

[62] Yu, Y., Yuan, J., Wang, Q., Fan, X., Ni, X., Wang, P., Cui, L., Cellulase immobilization onto the reversibly soluble methacrylate copolymer for denim washing, Carbohydr. Polym. 2013, 95(2), 675-680.

[63] Kertesz Z. I., The Pectic Substances, Interscience Publisher Inc., New York, 1951.

[64] Etters, J.N., Cotton preparation with alkaline pectinase: an environment advance. Text. Chem. Colorist Am. Dyest. Report, 1999, 10(3), 33-36.

[65] Kim. J., Kim, S. Y., Choe, E. K., The beneficial influence of enzymatic scouring on cotton properties, J. Nat. Fibers, 2005, 2(4), 39-52.

[66] Kim, J., Choe, E. K., Kim, S. Y., Nam, S.W., Optimization of the enzymatic scouring, J. Nat. Fibers, 2006, 3(2/3), 155-168.

[67] Tzanov, T., Calafell, M., Guebitz, G. M., Cavaco-Paulo, A., Bio-preparation of cotton fabrics, Enzyme Microbol. Technol., 2001, 29, 357-362.

[68] Kalantzi, S., Mamma, D., Christakopoulos, P., Kekos, D., Effect of pectate lyase bio scouring on physical, chemical and low stress mechanical properties of cotton fabrics, Bioresour. Technol., 2008, 99(17), 8185-8192.

[69] Morozova, V. V., Semenova, M.V., Salanovich, T. N., Okunev, O.N., Koshelev, A. V., Bubnova, T. V., Krichevskii, G. E., Timatkov, A. G. Barysheva, N. V., Sinitsyn, A. P., Application of neutral-alkaline pectate lyase to cotton fabric boil off, Appl. Biochem. Microbiol., 2006, 42(6), 603-608.

[70] Li, Y., Hardin, I. R., Enzymatic scouring of cotton-surfactants, agitation & selection of enzymes, Text. Chem. Colorist, 1998, 30(9), 23-30.

[71] Csiszar, E., Losonczi, A., Szakacs, G., Rusznak, I., Bezur, L., Reicher, J., Enzymes and chelating agent in cotton pretreatment, J. Biotechnol, 2001, 89, 271-279.

[72] Sangwatanaroj, U., Choonukulpong, K., Ueda, M., Cotton Scouring with Pectinase and Lipase/ Protease/ Cellulase, AATCC Review, 2003, May, 17-20.

[73] Ismail, O. E., Influence of Wax and Pectin Removal on Cotton Absorbency, AATCC Review, 2008, June, 37-42.

[74] Anis, P., Eren, H. K., Comparison of alkaline scouring of cotton vs alkaline pectinase preparation, AATCC review, 2002, 22-26.

[75] Kalantzi, S., Mamma, D., Christakopoulos, P., Kekos, D., Effect of pectate lyase bioscouring on physical chemical and Low stress mechanical properties of cotton fabrics, Bioresour. Technol., 2008, 99(17), 8185-8192.

[76] Battan, B., Dhiman, S. S., Ahlawat, S., Mahajan, R., Sharma, J., Application of thermostable xylanase of *Bacillus pumilus* in textile processing, Indian J. Microbiol, 2012, 52(2), 222-229.

[77] Persa, P., Tavcer, P. F., Low water and energy saving process for cotton pretreatment, Text. Res. J., 2009, 79(1), 76-88.

[78] Shafie, A. E., Fauda, M. M. G., Hashem, M., One step process for bioscouring and peracetic acid bleaching of cotton fabric, Carbohydrate Polym., 2009, 78(2), 302-308.

[79] Tavcer, P. F., Dyeing of environmentally friendly pretreated cotton fabric, In: Hauser, Petter (Ed.), Textile Dyeing, ISBN: 978-953-307-565-5.

[80] Underkofler L. A., Barton R. R. & Rennert S. S., Production of Microbial Enzymes and Their Applications, Microbiological Process Report, Applied Microbiology., 1958; 6(3):212-221.

[81] Vigeneswaran, C., Anbumani, N., Ananthasubramanian, M., Kandhavadivu, P., Ecofriendly approach to improve pectinolytic reaction and process optimization of bioscouring of organic cotton textiles. J. Eng. Fibers Fabr, 2013, 8(2), 121-123.

[82] Traore, M. K., Buschle-Diller, G., Environmentally friendly scouring process, Text. Chem. Colorist Am. Dyes. Report., 2000, 32(12), 40-43.

[83] Ollis D. L., Cheah E., Cygler M., Dijkstra B., Frolow F., Franken S. M., Harel M., Remington S. J., Silman I., Schrag J., Sussman J. L.Verschuren K. H. G., Goldman A., The α/β hydrolase fold, Protein Engineering, 1992; 5:197-211.

[84] Schmid R. D., Verger R., Lipases: interfacial Enzymes with Attractive Applications, Angew Chem Int Ed Engl. 1998; 37:1608-1633.

[85] Spicka, N., Tavcer, P. F., New combined bio-scouring and bio-bleaching process of cotton fabrics, Mater. Technol., 2013, 47(4), 409-412.

[86] Siddiquee, A. B., Bashar, M., Sarker, P., Tohfa, T. T., Hossan, M. A., Azad, M. I., Akhtar, N., Comparative study of conventional and enzymatic pretreatment (scouring and bleaching) of cotton knitted fabrics, Int. J. Eng. Technol., 2014, 3(1), 37-43.

[87] Tzanov, T., Calafell, M., Guebitz, G. M., Cavaco-Paulo, A., Bio-preparation of cotton fabrics, Enzyme Microbiol. Technol., 2001, 29, 357-362.

[88] Opwis, K., Knittel, D., Schollmeyer, E., Cordes , A., Simultaneous application of glucose oxidases and peroxidases in bleaching processes, Eng. Life Sci., 2008, 8(2), 175-178.

[89] Ramadan, A. R., Characterization of biobleaching of cotton/linen fabrics, JTATM., 2008, 6(1), 1-12.

[90] Anis, P., Davulcu, A., Eren, H. A., Enzymatic pre-treatment of cotton Part 2: peroxide generation in desizing liquor and bleaching, FIBRES Text. East. Eur., 2009, 17(2), 87-90.

[91] Saravanan, D., Vasanthi, N. S., Raja, K. S., Das, A., Ramachandran, T.,

Bleaching of cotton fabrics using peroxide produced by glucose oxidase, Indian J. Fibre & Text. Res., 2010, 35 (3), 281-283.

[92] Farooq, A., Ali, S., Abbas, N., Fatima, G. L., Ashraf, M. A., Comparative performance evaluation of conventional bleaching and enzymatic bleaching with glucose oxidase on cotton fabric, J. Clean. Prod., 2013, 42, 167-171.

[93] Tavcer, P. F., Low temperature bleaching of cotton induced by glucose oxidase enzymes and hydrogen peroxide activators, Biocatal. Biotransformation, 2012, 30(1), 20-26.

[94] Li, M., Hinks, D., An Environmentally Benign Approach to Cotton Preparation: One Bath Enzymatic Desizing, Scouring & Activated Bleaching, AATCC Review, 2012, Sept/Oct, 46-51.

[95] Davulcu, A., Eren, H. A., Avinc, O., Erismis, B., Ultrasound assisted biobleaching of cotton, Cellulose, 2014, 21, 2973-2981.

[96] Ali, S., Khatri, A., Tanwari, A., Integrated desizing-bleaching-reactive dyeing process of cotton towel using glucose oxidase enzyme, J. Clean. Prod., 2014, 66, 562-567.

[97] Tzanov, T., Costa, S. A., Gubitz, G. M., Cavaco-Paulo, A., Hydrogen peroxide generation with immobilized glucose oxidase for textile bleaching, J. Biotechnol., 2002, 93, 87-94.

[98] Lončar, N., Fraaije, M. W., Catalases as biocatalysts in technical applications: current state and perspective, Applied Microbiology and Biotechnology, 2015, 99(8), 3351-3357.

[99] Tzanov, T., Basto, C., Gubitz, G. M., Cavaco-Paulo, A., Laccases to improve the whiteness in a conventional bleaching of cotton, Macromol. Mater. Eng., 2003, 288(10), 807-810.

[100] Pereira, L., Bastos, C., Tzanov, T., Cavaco-paulo, A., Guebitz, G. M., Environmentally friendly bleaching of cotton using lacases, Environ. Chem. Lett., 2005, 3, 66-69.

[101] Beteheva, R., Georgieva, N., Yotova, L., Valchev, I., Chadjiska, C., Biobleaching of flax fibers by degradation of lignin with phanerochaete chrysosporium and Trichosporon cutaneum R57, J. Nat. Fibers, 2007, 4(4), 31-40.

[102] Basto, C., Tzanov, T., Cavaco-Paulo, A., Combined ultrasound-laccase assisted bleaching of cotton, Ultrason. Sonochemistry, 2007,14, 350-354.

[103] Abou-Okeil, A., El-Shafie, A., El Zawahry, M. M., Eco-friendly laccase-hydrogen peroxide/ultrasound-assisted bleaching of linen fabrics and its influence on dyeing efficiency, Ultrason. Sonochemistry, 2010, 17, 383-390.

[104] Aly A. S., Moustafa A. B., Hebeish A., Bio-technological treatment of cellulosic textiles, Journal of Cleaner Production, 2004; 12: 697-705

[105] Shahid Chatha S. A., Asgher M., Iqbal H. M. N., Enzyme-based solutions for textile processing and dye contaminant biodegradation- a review, Environ Sci Pollut Res, 2017, 24: 14005-14018.

[106] Cavaco-Paulo A., Gübitz G. M., Textile processing with enzymes, The Textile Institute, CRC Press, Woodhead Publishing Ltd,2003, England

[107] Krik O., Borchert T. V. and Fuglsang C. C., Industrial enzyme applications, Protein technologies and commercial enzymes, 2002; 13: 345-351

[108] Shukla S. R., Sharma U., Kulkarni K. S., Enzymes and their use in textile processes, Colourage, 2000; 47: 19-26.

[109] Saravanan D., Ramanathan V. A., Karthick P., Murugan S. V., Nalankilli G., Ramachandran T., Optimisation of multi-enzyme scouring process using Taguchi methods, IJFTR, 2010; 35: 164-171

[110] Sójka-Ledakowicz J., Lichawska J., Pyć R., Integrated Enzymatic Pre-Treatment of Cotton Fabrics, Journal of Natural Fibers, 2006,3:199-207

[111] Wang Q., Fan X., Hua Z., Gao W., Chen J., Degradation kinetics of pectins by an alkaline pectinase in bio scouring of cotton fabrics, Carbohydrate polymers, 2007, 67(4):572-575

[112] Persa P., Tavcer P. F., Bio-scouring and Bleaching of Cotton with pectinase enzyme and peracetic acid in one bath, Coloration Technology, 2008, 124(1): 36-42.

[113] Li Y., Hardin I. R., Enzymatic scouring of cotton-effects on structure & properties, Text. Chem. Colorist, 1997, 29(8), 71-76.

[114] Vigneswaran C., Ananthasubramanian M., Anbumani N., Ecofriendly Approach to Improve Pectinolytic Reaction and Process Optimization of Bioscouring of Organic Cotton Textiles, J. Eng. Fibers Fabr. 2013, 8(2), 121-133.

[115] Basto C., Tzanov T., Cavaco-Paulo A., Combined ultrasound-laccase assisted bleaching of cotton, Ultrason. Sonochemistry, 2007, 14: 350-354

[116] Tavcer P.F., Krizman P., Presa P., Combined bio scouring and bleaching of cotton fibres, J. Natural Fibers, 2006, 3: 83-97.

[117] Anis P., Davulcu A., Eren H. A., Enzymatic pre-treatment of cotton Part 2: peroxide generation in desizing liquor and bleaching, FIBRES Text. East. Eur., 2009, 17: 87-90.

[118] Fraser J., Peroxygens in environmental protection, Effluent Water Treatment Journal, 1986, 26:186-189.

[119] Costa S. A., Tzanov T., Paar A., Carneiro F., Gubitz G. M., Cavaco Paulo A., Recycling of textile bleaching effluents for dyeing using immobilized catalase, Biotechnol. Letter, 2002. 24: 173-176

[120] Chatha S.A.S., Asgher M, Ali S, Hussain A. I., Biological color stripping: a novel technology for removal of dye from cellulose fibers, Carbohydrate Polymer, 2012, 87:1476-1481

[121] Chatha S. A. S., Mallhi A., Hussain A., Asgher M., & Nigam S., A biological approach for color-stripping of cotton fabric dyed with CI reactive black 5 using fungal enzymes from solid state fermentation, Current Biotechnology, 2014, 3: 166-173.

[122] Le Roes-Hill M., Prins A., Biotechnological potential of oxidative enzymes from Actinobacteria, 2016, DOI: 10.5772/61321

[123] Madhu A., Chakraborty J. N., Developments in application of enzymes for textile processing, Journal of cleaner production, 2017, 145:114-133

[124] Bahtiyari M. I., Duran K., Usage of commercial cellulases in bio polishing of viscose fabrics, TEKSTIL ve KONFEKSIYON, 2010, 1:57-64.

[125] Kumar V. S., Meenakshisundaram S., Selvakumar N., Conservation of cellulase enzyme in bio polishing application of cotton fabrics, J. Text. Inst., 2008, 99: 339-346.

[126] Pazarlioglu N. K., Sariisik M., Telefoncu A., Treating denim fabrics with immobilized commercial cellulases, Process Biochem, 2005, 40: 767-771.

[127] Yu Y., Yuan J., Wang Q., Fan X., Ni X., Wang P., Cui L., Cellulase

immobilization onto the reversibly soluble methacrylate copolymer for denim washing, Carbohydrate Polymer, 2013, 95(2): 675-680

[128] Bhat M.K., Cellulases and related enzymes in biotechnology, Biotechnol Adv, 2000, 18:355-383.

[129] Sarkar A. K., Etters J. N., Kinetics of the enzymatic hydrolysis of cellulose, AATCC Rev, 2001, 48-52

[130] Pereira L., Bastos C., Tzanov T., Cavaco-Paulo A., Guebitz G. M., Environmentally friendly bleaching of cotton using laccases, 2005, Environ Chem Lett, 2005, 3:66-69.

[131] Ribitsch, D., Acero, E. H., Greimel, K., Dellacher, A., Zitzenbacher, S., Marold, A., Rodriguez, R. D., Steinkellner, G., Gruber, K., Schwab, H., Guebitz, G. M., A new esterase from Thermobifida halotolearans hydrolyses polyethylene terephthalate (PET) and polylactic acid (PLA), Polymers, 2012, 4, 617-629.

[132] Gubitz, G. M., Cavaco-Paulo, A., New substrates for reliable enzymes: enzymatic modification of polymers, Curr. Opin. Biotechnol. 2003, 14, 577-582.

[133] Kaneli, M., Vasilakos, S., Nikolaivits, E., Ladas, S., Christakopoulos, P., Topakas, E., Surafce modification of poly (ethylene terephthalate) (PET) fibers by a cutainase from Fusarium oxysporum, Process Biochem, 2015, 50(11), 1885-1892.

[134] Ibrahim, D. F., Abd El-Salam, S. H., Enzymatic treatment of polyester fabrics digitally printed, J. Text. Sci. Eng., 2012, 2(3), 1-4.

[135] Silva, C. J. S. M., Cavaco-Paulo, A., Biotransformations in synthetic fibres, Bio-catal. Biotransform., 2008, 26(5), 350-356.

[136] Parvinzadeh, M., Assfipour, R., Kiumarsi, A., Biohydrolysis of nylon 66 with different proteolytic enzyes, Polym. Degrad. Stab., 2009,94(8), 1197-1205.

[137] Begum, S., Wu, J., Takawira, C. M., Wang, J., Surface modification of polyamide 66 fabrics with an alkaline protease-subtilisin, J. Eng. Fibers Fabr., 2016, 11(1), 64-74.

[138] Gashti, M. P., Willoughby, J., Agrawal, P., In: Hauser, P. (Ed.), Surface and Bulk Modification of Synthetic Textiles to Improve Dyeability, Textile Dyeing, 2011, ISBN: 78-53-307-565-5.

[139] Pearce, C. I., Lloyd, J. R., Guthire, J. T., The removal of colour from textile wastewater using whole bacterial cells: a review, Dyes and Pigments, 2003, 58 (3), 179-196.

[140] Asgher, M., Bhatti, H. N., Ashraf, M., Legge, R. L., Recent developments in biodegradation of industrial pollutants by white rot fungi and their enzyme system, Biodegradation, 2008, 19(6), 771-783.

[141] Aguiar A., de Souza-Cruz, P.B., Ferraz, A., Oxalic acid, Fe^{3+} reduction activity and oxidative enzymes detected in culture extracts recovered from *Pinus taeda wood* chips bioreacted by Ceriporiopsis subvermispora, Enzym Microb. Technol., 2006, 38(7), 873-878.

[142] Karigar, C. S., Rao, S. S., Role of microbial enzymes in the bioremediation of pollutants: a review. Enzym Res., 2011, https://doi.org/10.4061/2011/805187

[143] Mai, C., Schormann, W., Milstein, O., Enhanced stability of laccase in the presence of phenolic compounds, Appl. Microbiol. Biotechnol. 2000, 54(4), 510-514.

[144] Fetzner, S., Lingens, F., Bacterial dehalogenases: biochemistry, genetics and biotechnological applications, Microbio Rev., 194, 58(4), 641-685.

Chapter 20

Recombinant Fungal Cellulases for the Saccharification of Sugarcane Bagasse

Raquel Guimarães Benevides, Sandra Aparecida de Assis, Alison Borges Vitor, Geise Camila Ribeiro, Cleidineia Souza de Santana and Larissa Emanuelle da Silva Almeida

Abstract

Cellulases are important enzymes in cellulose degradation that occurs in nature, this degradation involves a system of extracellular multienzymes and have wide application. The construction of a high-quality system for the production of these enzymes is important for its application in the process of saccharification of biomass involved in the biofuel production process. Several species of fungi are capable of synthesizing and secreting high amounts of cellulase, most studies with fungal species use linearized plasmid, since these are encompassed to chromosomal DNA, improving its stability and expression efficiency. Advances in the production of recombinant enzymes focus on the search for industrially viable microorganisms capable of producing enzymes under various conditions, expressing them in a highly efficient manner, aiming at the synthesis of several copies of genes and a strong promoter. To resay these restrictions, molecular biology combined with recombinant DNA technology is a viable tool in enzymatic production. In subsequent topics, the production of endoglucanases, exoglucanases and β-glucosidase of fungi cloned in *Escherichia coli*, *Pichia pastoris* and other different expression systems will be addressed.

Keywords: recombinants cellulases, fungal cellulases, lignocellulosic biomass, cellulose degradation, heterologous systems, CRISPR/Cas9

1. Introduction

Alternative renewable fuel as bioethanol in the form of biofuel derived from biomass can contribute sources to replace fossil fuel-based conventional energy sources [1].

Cellulases are important enzymes in cellulose degradation that occurs in nature, this degradation involves a system of extracellular multienzymes and have wide application [2, 3].

Cellulase enzymes play an important role in industrial processes, representing about 20% of the global enzyme market worldwide and presenting a wide range of application, from food, feed, textile, pulp and pulp industries. An application that has been growing in recent years is the conversion of biomass into fermentable sugars for the production of biofuels [4–6].

Cellulases act on cellulosic fiber, catalyzing the degradation of β-1,4-glycocydic bonds [7] and includes three different types that act synergistically, based on classification the mode of action and specificities of the substrate, these: endoglucanases (EC 3.2.1.4) that randomly hydrolyze β-1.4 bonds in the cellulose molecule; cellobiohydrolases or exoglucanases (EC 3.2.1.91) which release a cellobiose unit and act procedurally at the end of the chain; and β-glycosidases (EC 3.2.1.21) that hydrolysis cellobiose to glucose [2, 8].

The construction of a high-quality system for the production of these enzymes is important for its application in the process of saccharification of biomass involved in the biofuel production process [9]. Current efforts have focused on fungal cellulases to transform lignocellulosic biomass into fermentable sugars that can be converted into ethanol. This process will allow the production of renewable fuel from cellulosic biomass [10].

Advances in the production of recombinant enzymes focus on the search for industrially viable microorganisms capable of producing enzymes under various conditions, expressing them in a highly efficient manner, aiming at the synthesis of several copies of genes and a strong promoter. Several species of fungi are capable of synthesizing and secreting high amounts of cellulase; most studies with fungal species use linearized plasmid, since these are encompassed to chromosomal DNA, improving its stability and expression efficiency [11].

For genetic engineering, the main expression systems are: *E. coli*, a bacteria classified as belonging to the *Bacteria* Domain, *Proteobacteria philum; Gammaproteobacteria* class, *Enterobacteriales* order and *Enterobacteriaceae* family, which has a high rate of development, easy to manipulate, transform and capture of plasmids, and can grow with high cell density. *E. coli* is generally transformed with self-replicated plasmid that does not integrate with chromosomal DNA and continues to replicate independently of cell divisions [12, 13]; and *Pichia pastoris,* a yeast classified as belonging to the Fungi Kingdom, *Eucomycota* division*; Ascomycota* subdivision*;* class *Hemoascomycetes,* the order *Endomycetales,* family *Sacharomycetaceae* and subfamily *Sacharomycetoideae.* A remarkable physiological characteristic of this yeast is the fact that it is methyltrophic, that is capable of growing in culture medium containing methanol as the only source of carbon and the ability to secrete high amounts of extracellular proteins [14, 15].

Due to the advance in the techniques of recombinant expression, the production systems of recombinant enzymes are promising strategies for the efficient production of industrial cellulase that can increase productivity in several industrial applications, including biomass in the processing of biofuels and thus meet the increasing demands of this enzyme [16].

The cost of obtaining sugars from the biomass of sugarcane bagasse for fermentation is still high, mainly due to the low enzymatic yield of fungal production. Thus, it generates the need for cellulase supplementation to these enzymatic cocktails. To resay these restrictions, molecular biology combined with recombinant DNA technology is a viable tool in enzymatic production. In subsequent topics, the production of endoglucanases, exoglucanases and β-glucosidase of fungi cloned in *E. coli* and *Pichia pastoris* will be addressed.

2. Lignocellulosic biomass

Lignocellulosic biomass is characterized mainly by the presence of two carbohydrate polymers (cellulose and hemicellulose), as well as an aromatic polymer called lignin, in addition to other components found in smaller amounts, such as ash, pectin, proteins, non-structural carbohydrates (glucose, fructose and sucrose) and lipids. Most of the biomass of lignocellulosic materials is composed of cellulose (40–50%), hemicellulose (20–30%) and lignin (10–25%) and the specific composition of lignocellulosic biomass varies depending on different factors, mainly plant species, age, growth stage and environmental factors, genetic variability, and cultivation conditions of plant material [2, 17, 18].

Lignocellulosic biomass has a complex internal structure and several of its main components also have complex structures. Cellulose and hemicellulose are polysaccharides composed of simple sugars while lignin is a complex network of aromatic alcohols. In general, hemicelluloses and lignin provide an amorphous matrix in which crystalline cellulose microfibrils are dispersed [2, 18].

Corn straw, sugarcane bagasse, rice straw and wheat bran are promising and abundant lignocellulosic raw products from plant residues in the United States, South America, Asia and Europe [19].

3. Heterologous systems

A potential tool to develop better industrial production of cellulase are techniques of heterologous expression. This technology leads to enzyme yields at an economically viable level, since it allows the creation of microbial strains that express sets of adapted and synergistically active enzymes, within a single cell or combining different strains [20]. There are a variety of protein expression systems available, including bacterial and yeasts expression systems.

For the bacterial expression system, the most used is *Escherichia coli*, whose genetic characteristics are already well described. In addition, it has easy of manipulation, has an abundance of commercially available strains and vectors and has great ability to express recombinant genes with high yields [20–22].

As an alternative to the bacterial system, yeasts are often used, where *Pichia pastoris* yeast has become the most widely used host system for the expression of many heterologous proteins with relative ease of technique and at lower costs than those of most other eukaryotic systems [23–25].

Data from the last 15 years describing the recombinant fungal cellulases candidates for cellulose hydrolysis produced in the expression systems *E. coli*, *P. pastoris* or other different systems are presented below.

4. Recombinant fungal Cellulases produced by different expression systems

Numerous study techniques have been improved in recent years for cloning, heterologous expression and characterization of cellulases (**Table 1**). Several studies show efficient technologies to produce endoglucanases and β-glucosidases cloned in *E. coli* strains.

The Fungus *Trichoderma virens* ZY-01 expressed an endoglucanase cloned in a vector of expression pET-32-EG, being successfully elaborated, and expressed, in a heterologous way, in *Escherichia coli* and the target protein presented a weight of 39 kDa by electrophoresis SDS-PAGE [36].

A. fumigatus gene encoding endo-1,4-β-glucanase (Afu6g01800), studied by Bernardi et al. [46], was cloned in the vector pET-28a (+) and expressed in the strain of *E. coli* Rosetta ™ (DE3). The research results showed that the afegl7 enzyme belonged to the GH7 family, in which the Af-egl7 gene encodes the protein comprising 460 amino acids with a CBM1 domain in the 424–460 residues and molecular mass of 52 kDa.

Fungus	Vector	Enzyme	Gene	Molecular mass	Activity Enzyme	Substrate	Author
Lentil edodes	*E. coli*	celobiohydrolase	cel6B	46.4 kDa	0.12 U/min/mg	p-nitrophenyl-β-galactopyranoside	Taipakova et al. [26]
Aspergillus niger	*P. shepherds*	endoglucanase	EglA	~30 kDa	63.83 ± 4.68 U/mg	β-glucan CMC	Quay et al. [10]
Trichoderma harzianum	*Pichia Shepherds*	Endoglucanase III	rThEGIII	24.6 kDa	—	CMC	Generoso et al. [27]
Myceliophthora thermophila	*Pichia Shepherds*	endoglucanase	EG7A	46 kDa	468 U/ml	CMC	Karnaouri et al. [28]
Aspergillus niger	*Pichia Shepherds*	celobiohydrolase	CBH1	60 kDa		CMC	Li et al. [29]
Aspergillus niger	*Pichia Shepherds*	β-glycosidase	Bgl1	121 kD	45 U/mL	Celobiosis.	Zhao et al. [30]
Myceliophthora thermophila	*Pichia Shepherds*	β-glycosidase	Bgl3B	130.0 kDa	15 U ml – 1	CMC	Zhao et al. [31]
Penicillium funiculosum	*Pichia Shepherds*	β-glucosidase	Bgl4P	~130 kDa	1,354.3 U/mg	p-nitrophenyl-β-glucode and celobiosis	Ramani et al. [32]
Aspergillus niduluns	*E. coli TOP 10F*	endoglucanase	XegA	28 kDa	33.3, (U/mg) 75% / 2 h	Sugarcane Bagasse	Lima et al. [33]
Mr Harzianum IOC-3844	*E. coli (pET28a)*	β-glucosidase (GH1 and GH3)	RThBgl	54.72 kDa	relative (> 60%) between pH 5.0 and 7.0	pNPG (Sigma-Aldrich) as substrate	Santos et al. [34]
Aspergillus nidulans AN2227	*pPICZ*	β-glucosidase	—	100 kDa	0.52 µmole / ml / min	nitrophenyl-p-D-glucopiranoid (pNPG)	Auta et al. [35]
Trichoderma Virens ZY-01	*E. coli*	endo-1,4-β-d-glucanase (EG)	Eg	39 kDa 1069 bp	expression was Km = 13.71 mg / mL and Vmax = 0.51 µmol / min · Ml.	CMC	Zeng et al. [36]
Aspergillus niger F321	*E. coli vector pGEM-T*	β-glucosidase	ANRA 2.6 and ANRA12.9	1,190 bp and 1,950 bp,	—	—	Auta et al. [37]

Fungus	Vector	Enzyme	Gene	Molecular mass	Activity Enzyme	Substrate	Author
Myceliophthra thermophila	*Pichia Shepherds*	endoglucanase (EG)	MtEG5A	75 kDa	160 h – 53 U / ml	Wheat Straws, Sorb, Birch	Karnaouri et al. [38]
Athermophila Myceliopthor B1J	*pGAPZαA*	endoglucanase	Mt-Egl	47 kDa	70% activity after 3 h of pH exposure 5–12	wheat bran, corn cob, sunflower splints, rice straw and rice bran	Phadtare et al. [39]
Aspergillus fumigatus DBiNU-1	*E. coli DH5 (Kluyveromyces lactis)*	endoglucanase	Cel7	48.19 kDa	0.80 U/ml with a specific activity of 3.08 U/mg protein	*Li et al. 2016*	Rungrattanakasin et al. [40]
Trichoderma Reesei ZU-02	Dh5α *E. coli* (pUC18-PsT	β-glucosidase	Bgl (2.5 kb)	—	112.2 IU/mL after 84 h of fermentation and the FPA reached 89.76 FPU/mL after 96 h.	Corn straw	Xia et al. [41]
T. reesei Rut-C30 A. niger NL02	*pCAMBI*	glucosidase (BGA), endoglucanase	Pcbh1 CBH2	—	BGA = 1.93 ± 0.28 IU/mL and EG =716.11 ± 41.16 U/mL	Steam-blasted corn bagasse	Zhao et al. [42]
Thermophilum chaetomium	*PPIC9K Express Pichia Shepherds*	endoglucanase	ctendo7	48 kDa	3.05 IU/mg in optimal reaction condition of 55°C, pH 5.0	Pretreated Wheat Straw	Hua et al. [43]
Thermoascus aurantiacus	*P. pastoris X-33*	native endoglucanase	Reg	~ 33 kDa	142 IU / mg	CMC in sodium citrate buffer	Jain et al. [44]
Pinophilus talaromyces	*Saccharoys cerevisiae*	β-glucosidase	Bgl3B	92 kDa	0.56 nkat/mg pH 4.0 enzymatic activity: 1 to 60° C.	a wheat-rye hybrid	Trollope et al. [45]
Aspergillus fumigatus	*E. coli* (pET-28a) RosettaTM strain	endo-1,4-β-glucanase	AfEGL7	52 kDa	(51.98 ± 0.0069 U mg^{-1})	Sugarcane bagasse (SEB), Barley	Bernardi et al. [46]
Aspergillus glaucus CCHA	*P. Pastorals GS115*	endoglucanase	AgCM ase,	55.0 kDa	343.81 ± 2.77 μM /mg/min	Rice and corn straws	Li et al. [47]

Fungus	Vector	Enzyme	Gene	Molecular mass	Activity Enzyme	Substrate	Author
Aspergillus fumigatus	*Pichia pastoris X-33*	endo-1,4-β-glucanase	Af-EGL7	70 kDa	(40–45%) after 72 and 48 h	Sugarcane bagasse "in natura"	Bernardi et al. [48]
Penicillium verruculosum (PvBGL)	*P.canescens RN3–11–7).*	β-glucosidase	BGL1	90 kDa	85 and 124 units /mg of protein, respectively	pNPG) p-nitrophenyl-p-Dglucopiranodes hydrolysis of MCC	Volkov et al. [49]
Talaromyces leycettanus	*Pichia Shepherds*	TlCel5A and TlCel6A	TlCel5A and TlCel6A	45 kDa and 65 kDa	3,905.6 U/mg vs. 109.0 U/mg in liquenae and 840.3 and 0.09 U/mg in CMC.	steam-blasted corn straw (SECS), corn cob, soybean meal and wheat bran	Gu et al. [50]
Trichoderma Reesei	*P. pastoris SMD1168*	exoglucanase	Cel6A	50 kDa 750 bp	—	—	Anindyawati et al. [51]
Matsutake Tricholoma (TmEgl5A)	*Pichia pastoris KM71H*	endoglucanase	TmEgl	40 kDa	514 135 U /mg and 104 3 U/mg	grown in barley and vermicuite	Onuma et al. [52]
Hypocrea sp. W63	*Pichia Shepherds*	b-glucosidase (BGL)	EpB-BGL	76.5 kDa	194.25 IU / mg	p-nitrophenyl-b-D-glucoside (pNPG) hydrolysis in sugarcane bagasse	Liang et al. [53]
Trichoderma Reesei	*Pichia Shepherds*	endoglucanase	ReEG I	~ 45 kDa	34.3 U/mg	Pulp, carboxymethylated cellulose, oat xylan, birch xylan, corn straw	Tao et al. [54]
Trichoderma Harzianum	*E. coli* JM109	β glucosidase	cel7a	—	2.24 ± 0.05 IU / mL - 46.66 IU. L^{-1}. H	pretreated sugarcane bagasse	Delabona et al. [55]
Aspergillus fresenii.	*E. coli* (by *K. phaffii* X33)	β-1, 4-glucosidases	Bgl T2	35 kDa	bgl T2 under pH 5.0 per 1 h provides up to 60% activity.	pNPG, CMC-Na, pNPC and Avicel	Yang et al. [56]

Table 1.
Recombinant fungal Cellulases used in the Saccharification of sugarcane bagasse.

RNA-Seq of β-glucosidase and genomic data are instruments that can be used to express *T. harzianum* genes for the deterioration of lignocellulosic biomass. The target gene of the recombinant protein (rThBgl) cloned and expressed heterological in *Escherichia coli* Rosetta was purified with high yields. The results showed a significant increase in the activity of β-glicosidase and in the filter paper cellulase (FPA). The cellulase produced by the transformers reached higher hydrolysis yield, with less enzymatic load during the saccharification of pretreated corn straw [41].

Delabona et al. [55] reported that *Trichoderma harzianum* overexpressed the methyltransferase of the global regulator - LAE1, in order to improve the production of cellulases, considering that the evaluation of the impact of LAE1 to induce cellulases made use of soluble carbon sources and lignocellulose and low cost in an agitated bioreactor. Using sugarcane bagasse with sucrose, the overexpression of lae1 culminated in a significant increase in the expression of the gh61b (31x), cel7a (25x), bgl1 (20x) and xyn3 (20x) genes. As a result, reduction of sugar released from the pretreated sugarcane bagasse, hydrolyzed by the recombinant crude enzymatic cocktail, obtained 41% cloned improvement through plasmid in *Escherichia coli*.

Two new genes of the β-glucosidases of *Aspergillus niger* 321 were successfully cloned in the pGEM-T vector [37]. *Aspergillus fresenii* (JCM 01963) *Escherichia coli* TOP 10 and *K. phaffii* X-33 (Invitrogen, USA) were used as host strains. The bgl T2-opt gene was synthesized according to the *K. phaffii* codon trend and constructed in the (pPICZαA vector Invitrogen, USA) with the sites of the restriction enzymes EcoRI and XbaI. This article effectively discovered a new β-1,4-glucosidase bgl T2 and its *Aspergillus fresenii* ORF, under the help of high-yield sequencing of the mRNA technique. Such a method is more convenient than the traditional of obtaining a new enzyme and its genetic information, as deluded in the discussion section. The bgl T2 gene was expressed by *K. phaffii* X33. The properties of bgl T2 were tested, including pH and temperature optimums for catalysis, pH tolerance, thermostability, effects of unusual chemicals and kinetic properties against pNPG [56].

The gene of *M. thermophila* coding for endoglucanase (EG) was isolated from fungal genomic DNA and then cloned and amplified in *E. coli* strains and, finally, expressed heterologic in *P. pastoris* and two basic strategies were followed for the production of EG. These strategies include controlling proteolysis through low temperature and adding numerous amino acid supplements to the culture medium. The enzyme presented high thermostability and was able to hydrolyze several natural substrates, cellobiose as the main product, characteristics that reflect its potential use in different biotechnological applications [38].

The study by Bernardi et al. [48] used the vector *Pichia pastoris* X-33 on to improve the characterization of an endo-1,4-β-glucanase, thermostable GH7 of *Aspergillus fumigatus* (Af EGL7). The kinetic parameters Km and Vmax were estimated and evidenced a robust enzyme which provided an improved hydrolysis of sugarcane bagasse "in natura", exploded sugarcane bagasse, corn cob, rice straw and bean straw [48]. A recombinant thermoalkalin endoglucanase of *Myceliopthor thermophila* BJA (rMt-egl) was used in the application and enzymatic saccharine of agro residues. The gene of this codon-optimized endoglucanase (Mt-egl) was expressed, constitutively in *Pichia pastoris* under the regulation of the GAP promoter. It was confirmed that recombinant endoglucanase (rMt-egl), efficiently hydrolyzed industrial agro residues, which were tested, and wheat bran. The effort aims to improve the production of rMt-egl by various approaches to molecular biology and cultivation [39].

Zhao et al. [42] investigated the fungi *T. reesei* and *Aspergillus niger* with the intention of further improving cellulase production and performance in enzymatic hydrolysis. For this study, *Escherichia coli* DH5a was used for plasmid propagation. *Agrobacterium tumefaciens* AGL-1 was used for the transformation of recombinant *T. reesei*, constructed, and transformed into a pCAMBIA1300-PsCT vector. The vector *Pichia pastoris* was used for the production of recombinant CBH II by Zhao et al. [42] using a cloning vector of pUCm-T (Sangon, Shanghai, China) to obtain the *T. reesei* Rut-C30 and *A. niger* NL02. A vector pUC18-PsT containing fragments of 1.6 kb of Pcbh1-ss and 1.4 kb of Tcbh1 was used as vector structure to construct the set of DNA sequences with the expression information. As a result, a binary vector pCAMBIA1300-hph, in which it was added to this hygromycin gene, culminating in a final expression vector pCAMBIA1300-hph-PsCT. The results confirmed that the BG and CBHII genes provide a good performance in the hydrolysis of steam exploded corn pomace.

The recombinant endoglucanase (EG I) gene of *Trichoderma reesei* was successfully expressed in *Pichia pastoris*, with the objective of producing oligosaccharides from various biomass-derived substrates. Recombinant endoglucanase I (ReEG I) showed catalytic activity in relation to cellulose and xylan hydrolysis. Among several glucan and xylan substrates (paper pulp, carboxymethylated cellulose, oat xylan, birch xylan), birch xylan exhibited higher yield of xylooligosaccharides (XOS) [54].

An endoglucanase gene (ctendo7) of the fungus *Chaetomium thermophilum* was expressed in *Pichia pastoris*. The recombinant enzyme was purified by affinity chromatography with Ni^{2+} and subsequently characterized, through this analysis it was possible to conclude that the enzyme belongs to the family of glycosides hydrolase 7 and exhibited considerable activity against carboxytyl sodium cellulose (CMC-Na) and xylan of 1.91 IU / mg and 3.05 IU/mg in the ideal reaction condition of 55°C, pH 5.0, respectively, showed high hydrolytic efficiency in multiple lignocellulosic substrates at high temperatures [43].

The fungus *T. aurantiacus* RCKK was cloned in *P. pastoris* X-33 for overexpression. After the expression of recombinant endoglucanase (rEG), of molecular size of ~33 kDa confirmed by SDS-PAGE and western blotting, followed by determination of gel activity by zymogram analysis, the recombinant was successfully expressed in *P. pastoris* X-3 and bioreactor tests demonstrated that the enzyme is suitable for industrial applications [44].

An endoglucanase (TmEgl) was isolated from the solid-state culture of the ectomycorrhizal fungus *Tricholoma matsutake* (TmE-gl5A) cultivated in barley and vermiculite, which purified by fractionation of ammonium sulfate, ionic exchange, hydrophobic and gel filtration. TmEgl5A showed a molecular mass of approximately 40 kDa, as determined by SDS-PAGE. The gene encoding TmEgl was cloned and expressed in *Pichia pastoris* KM71H. These results suggested that *T. matsutake* produces a typical endoglucanase in solid state culture. *T. matsutake* presents itself as a strong candidate for the production of enzymes that degrade the cell wall of plants [52].

The unique candidate for GH5 cellulase of *A. glaucus* produced an endoglucanase called AgCMCase, which was cloned and expressed in the *Pichia pastoris* system [47]. The purified AgCMCase degraded the CMC-Na and was also able to hydrolyze the corn straw and rice to release sugar. The study showed that AgCMCase activity was retained by more than 95% after 4 h of incubation in the presence of NaCl 4 M, suggesting that it is a halotolerant enzyme. Thus, the interesting properties of AgCMCase can make it a potential candidate for industrial applications.

Recombinant β-glucosidase (EC 3.2.1.21) of *Aspergillus nidulans* AN2227 was expressed using buffered methanol complex medium (BMMY). Purification was performed using precipitation with ammonium sulfate and anionic exchange chromatography in the DEAE-Sephadex A-50 column. The enzyme was purified 2.58

times from the crude extract. The β-glucosidase was purified for electrophoretic homogeneity, containing a relative molecular weight of 100 kDa, as determined by electrophoresis in polyacrylamide gel, with sodium dodecile sulfate (SDS-PAGE). In the study, β-glucosidase was purified for electrophoretic homogeneity from the crude extract. The characteristics expressed from *P. pastoris* X33, with high-level expression, were described. The study suggests that the protein may be present in the monomeric form, with the enzyme having a good pH and temperature stability, making it an excellent candidate for cellulose hydrolysis [35].

A new bgl1 gene, which encodes a GH3 family of β-glucosidase of *Penicillium verruculosum* (PvBGL) was cloned and expressed heterological in the strain of *P. canescens* RN3–11–7 (niaD-) under the control of the xylAgene promoter. After the construction of the rPvBGL vector its properties were studied and compared with those of rAnBGL of *Aspergillus niger*, previously expressed in the same fungal host. It was observed that rPvBGL had an observed molecular mass of 90 kDa (SDS-PAGE data). It was possible to verify that rPvBGL converted polymeric substrates into glucose much faster than the recombinant BGL of *A. niger* (rAnBGL). Thus, this study showed the possibility of using rPvBGL for the construction of complex and balanced enzymatic preparations of cellulase based on the fungus *P. verruculosum* [49].

A β-glucosidase (BGL) of *Hypocrea sp.* W63 was cloned and expressed in *Pichia pastoris* and recombinant enzyme after purification presented a specific activity of 194.25 IU/mg. This study used *C. autoethanogenum* and *A. succinogenes* for the co-production of ethanol and succinic acid, using sugarcane bagasse as a source of fermentable sugars. The good conversion of epB-BGL suggests a great potential for the biorefining of cellulosic material [53].

Other study produced an exoglucanase (Cel6A) cloned in *Pichia pastoris*. The Cel6A gene was derived from *Trichoderma reesei* was produced synthetically, and the codon optimized for better expression in yeast *P. pastoris*. The gene was placed under the regulation of the GAP promoter, and the recombinant plasmid, called pLIPI-TrCel6A, inserted with the *T. reesei* Cel6A gene (TrCel6A), integrated into the genome of *P. pastoris* SMD1168H. The recombinant enzyme was successfully expressed by *P. pastoris*, with a main product that shows a molecular size of about 50 kDa. The recombinant Plasmid Cel6A selected was linearized with the enzyme BamHIO recombinant plasmid, pLIPI-TrCel6A, carrier of the *T. reesei* Cel6A gene (TrCel6A) integrated into the genome of *P. pastoris* SMD1168H [50].

4.1 Recombinant endoglucanases

Generoso et al. [27] using the expression system in *Pichia pastoris*, obtained a β-1,4-endoglucacase belonging to Glycosil hidrolases 12 (cel12a), cloned in the vector pPICZαA, isolated from the filamentous fungus *Trichoderma harzianum* IOCzianum-3844. The recombinant enzyme rThEGIII presented a molecular mass of 25 kDa, which is similar to the predicted mass, which is 24.6 kDa, as demonstrated by the authors. A large amount of rThEGIII was produced after 24 h of methanol induction, where in approximately 48 h, 300 mg of the purified enzyme was obtained from 1 L of medium. The optimum pH and temperature for rThEGIII activity were 5.5 and 48.2°C, respectively, similar to other EGIII already described. These characteristics indicate that rThEGIII is promising for simultaneous saccharification and fermentation, since the authors showed that the enzyme presented stability at temperatures close to ideal, lasting several days with acceptable activity.

Another recombinant endoglucanase was reported by Quay et al. [10], from the fungus *Aspergillus niger* ATCC 10574. The coding gene for the enzyme (EglA) was cloned in a pPICZαC vector and expressed in recombinant form in *P. pastoris* X-33.

After purification, the recombinant protein obtained presented a mass of ~30 kDa. Based on biochemical characterization, EglA had excellent activity at 50°C and ideal pH of 4.0, with a high stability at temperatures between 30 and 50°C and pH between 2.0 and 7.0. EglA showed greater affinity in the presence of β-glucan followed by carboxymethylcellulose (CMC) with a specific activity of 63.83 and 9.47 U/mg, respectively. Significant increase in activity was also observed with the presence of metal ions (Mn^2, Co^{2+}, Zn^{2+}, Mg^{2+}, Ba^{2+}, Fe^{2+}, Ca^{2+} and K^+). Based on these attributes, this enzyme can be signaled in order to be explored for enzymatic hydrolysis of agro-industrial residues.

A recombinant endoglucanase (MtEG7a), belonging to the family of glycosides hydrolase 7, was obtained by Karnaouri et al. [28], isolated from the fungus *Myceliophthora thermophila*; cloned in a pPICZαC vector and functionally expressed in the yeast *Pichia pastoris*. The purified recombinant enzyme (MtEG7a) was tested for its activity in relation to different substrates; where the enzyme showed high activity for β-glucan of barley (298 U/mg) and carboxymethylcelllulase (177 U/mg), also presenting activity for xylan-containing substrates, such as wheat arabinoxylan (5 U/mg). The highest activity levels were verified at pH 5.0 and the ideal activity temperature was 60°C, rapidly losing its activity at temperatures above 65°C. This study shows that the primary enzymatic activity of MtEG7 a hydrolysis the β-1.4 bonds of substrates because the activity of MtEG7a in β-1,3-glucan bonds was completely inhibited. In addition, the characteristics in terms of catalytic efficiency and thermostability of MtEG7a, makes it a good candidate for industrial applications, including the saccharification of lignocellulosic materials [28].

Rubini et al. [57], reported the isolation and cloning of the first cDNA of *P. echinulatum* (Pe-egl1) that encodes a supposed endoglucanase. This cDNA was expressed in a system of heterologous expression based on the methyl yeast trophic *Pichia pastoris*. *P. echinulatum* EGL1 secreted in the culture supernatant of a recombinant strain of *Picchia pastoris* revealed several characteristics of industrial interest, such as an optimal activity at 60°C and in a wide pH range. Recombinant *P. echinulatum* EGL1 is also interesting for its high thermostability.

Lahjouji et al. [58] described a cDNA of celobiohydrolase Tvcel7a de *Trametes versicolor* cloned and expressed in *Aspergillus niger*. The biochemical properties of purified TvCel7a obtained from both peaks were studied in detail. The optimum pH and temperature were 5.0 and 40°C, respectively. The enzyme is stable in a pH range extending from 3.0 to 9.0 and at temperatures below 50°C. Kinetic parameters with the p-nitrophenyl substrate β-D-cellobioside (pNPC) were 0.58 mM and 1.0 μmol/ min/mg of protein for purified TvCel7a found in peaks 1 and 2. TvCel7a catalyzes the hydrolysis of pNPC, filter paper, β-glucan and avicel in several degrees, but no detectable hydrolysis was observed when the substrates carboxymethylcellulose, laminarin and pNPG were used.

Nakazawa et al. [59] attempted to increase the specific activity of *T. reesei* EG III in *E. coli* by random gene mutagenesis using error-prone PCR followed by plate-assay activity screening. They reported that the yield in the active form of EG III was improved in transforming and the specific activity of their mutant (2R4) was increased. In addition, the stability in the pH and heat of these mutants increased unexpectedly.

Koseki et al. [60] produced an endoglucanase of the glycosyl hydrolase family 61 of *Aspergillus kawachii* (AkCel61) and a truncated enzyme only with the catalytic domain (rAkCel61ΔCBM) in *Pichia pastoris* and analyzed its biochemical properties. The proteins rAkCel61 and rAkCel61ΔCBM produced small amounts of oligosaccharides from soluble carboxymethylcellulose. They also exhibited a slight hydrolytic activity in relation to laminarin. However, they showed no detectable activity in relation to microcrystalline cellulose, arabinoxylan and pectin. Both recombinant enzymes also showed no detectable activity for p-nitrophenyl-β-D-glucosides, p-nitrophenyl-β-D-cellobiosides and p-nitrophenyl-β-D-celotriosides.

Igarashi et al. [61] report the identification of the gene encoding the endoglucanase (EG) of the family 45 (GH) of *Phanerochaete chrysosporium*, cloning the cDNA, determining its heterologous expression in the methylotrophic yeast *Pichia pastoris* and characterizing the recombinant protein. The recombinant protein showed hydrolytic activity in relation to amorphous cellulose, carboxymethylcellulose, liquena, barley-glucan and glucomannan, but not xylan. In addition, a synergistic effect was observed with cellobiohydrolase of the recombinant GH 6 family of the same fungus for amorphous cellulose as substrate, indicating that the enzyme can act together with other cellulolytic enzymes to hydrolyze cellulosic biomass in nature.

A new β-1,3-1,4-glucanase gene (designated as PtLic16A) of *Paecilomyces thermophila* was successfully cloned and expressed in *Pichia pastoris* as β-1,3-1,4-active extracellular glucanase. The purified enzyme had a molecular mass of 38.5 kDa in SDS-PAGE. It was optimally active at pH 7.0 and at a temperature of 70°C. In addition, the enzyme exhibited strict specificity for β-1,3-1,4-D-glucans. This was the first report on cloning and expression of a β-1,3-1,4-glucanase gene of *Paecilomyces sp* [62].

The gene encoding an endoglucanase of the glycosyl hydrolase (GH) family 45 (Cel45A) was cloned from *P. decumbens* and expressed in *Pichia pastoris* [63]. As far as we know, this is the first report of characterization of a protein of the GH 45 family in *Penicillium* species. The purified recombinant enzyme showed higher activity on glucomannan konjac (KGM) than on sodium carboxymethylcellulose (CMC-Na) or phosphoric acid cellulose (PASC). The highest hydrolytic activity was detected at pH 5.0 in KGM and pH 3.5 in CMC-Na, indicating that the mode of action in both substrates may be different for Cel45A. The optimum temperatures in both substrates were 60°C and about 90% of the relative activities were retained at 70° C. Products released from PASC and CMC-Na were mainly cellobiose, cellotriose. The protein with the highest glucomannanase activity can aid in the efficient degradation of lignocellulose by *P. decumbens* in the natural state.

4.2 Recombinant Exoglucanases

In a study conducted by Li et al. [29], a gene (cbh1) encoding a cellobiohydrolase (CBH) was isolated from the fungus *Aspergillus niger* NL-1. The cellobiohydrolase gene (cbh1) was successfully expressed in *Pichia pastoris* KM71H, presenting molecular mass of approximately 60 kDa. The amino acid sequence encoded by cbh1 shows high homology with the glycoside hydrolase sequence family 7. The recombinant cbh1 exhibited ideal activity at 60°C and pH 4.0 with Km and Vmax for CMC-Na of 13.81 mM and 0.269 μmol/min, respectively. When submitted to 2 h of incubation at 90°C, the enzyme retained more than 80% of its activity and was stable in the pH range 1.0 ± 10.0; due to moderate to high temperature stability and a wide pH range, the authors point out that this enzyme has potential in several industrial applications.

Taipakova et al. [26], obtained the cellobiohydrolase coding gene (Cel6B), belonging to the glycosyl hydrolase 6B family, lentinula edodes isolate cloned in vector pET11d and transformed into *E. coli* (Rosetta DE3). The recombinant protein obtained presented a mass of 46.4 kDa. However, there was the formation of an insoluble inclusion body, preventing enzymatic activity. Such a feature has been observed before, according to Chiang et al. [64], overexpressed proteins in *E. coli* can lead to the formation of the inclusion body. To obtain the recombinant protein in the active form, Taipakova et al. [26], denaturated with 6 M guanidine chloride.

After this stage, the enzyme showed activity of 0.12 U/min, being considered much lower when compared to other celobiohidolases, however, an optimization of this expression system in *E. coli* has a great possibility of obtaining that of active cellulases.

The genome of the basidiomycete *Phanerochaete chrysosporium* contains sequences encoding at least 166 putative hydrolase glycosides, many of which are predicted to β-1,3-glucanases [65]. Kawai et al. [66], cultivated *P. chrysosporium* with laminarin as the only carbon source and found that several β-1,3-glucanases were secreted in the medium. The cDNA encoding a new β-1,3-glucanase with molecular mass of 36 kDa was cloned and expressed in a heterologous way in the methylotrophic yeast *Pichia pastoris*. Based on the catalytic activity of the recombinant enzyme in relation to various substrates β-1, 3-glucan, the recognition pattern for the branched structure of β-1,3/16-glucan is discussed: Lam16A generates non-branched oligosaccharide from branched β-1,3/1,6-glucan.

Voutilainen et al. [67] characterized three new cellobiohydrolases originated from thermophilic ascomycetes fungi. The properties of these three cellobiohydrolases were compared to one of the best characterized celobiohydrolases, *T. reesei* Cel7A. *C. thermophilum* Cel7A showed the highest specific activity and optimum temperature in soluble substrates and these properties also correlate well with its high activity in polymeric substrates.

A gene(cel4) encoding for a cellobiohydrolase II (Ex-4) Ex-4 has been isolated from the basidiomycete of the white rot strain *Irpex lacteus* MC-2 and successfully expressed in yeast *Pichia pastoris*. The recombinant Ex-4 showed endo-processive degradation activity for cellulosic substrates and a synergistic effect on Avicel degradation was observed when the enzyme acted together with cellobiohydrolase I (Ex-1) or endoglucanase (En-1) produced by *I. lacteus* MC-2 [68].

4.3 β-Glycosidase recombinant

Ramani et al. [32] obtained a β-glucosidase (rBgl4) of *Penicillium funiculosum* successfully expressed in the expression system of *Pichia pastoris* KM71H. The recombinant protein rBgl4, after purified presented a weight of ~130 kDa. The rBgl4 activity test at different pH showed ideal activity at pH 5.0 and temperature of 60°C. The enzyme exhibited a high substrate conversion rate for p-nitrophenyl-β-glucosidase and cellobiose, being 3,332 and 2,083 μmol/min/mg, respectively. In addition, rBgl4 demonstrated glucose concentration tolerance of up to 400 mM.

A two-fold increase in glucose yield was observed when supplemented with crude cellulase of *Trichoderma reesei* Rut-C30 in cellulose hydrolysis, suggesting that the recombinant enzyme is a term β-glucosidase and glucose tolerant, and maybe a potential complement to commercial cellulases in cellulose hydrolysis, ensuring profitability in bioethanol production [32].

The gene of a β-glycosidase (bglI) of *Aspergillus niger* NL-1, expressed in *Pichia pastoris,* was obtained by Zhao et al. [30]. The recombinant enzyme showed high activity at pH 4.0 and temperature 60°C and was stable in a pH range of 3.0 to 7.0 and held more than 85% of activity after incubation at 60°C for 30 minutes. The β recombinant glucosidase presented molecular mass of 121 kDa. The authors determined glucose production from avicel compared to recombinant β-glycosidase, where, without the addition of recombinant β-glucosidase, glucose yield was only 49.3%, while with the addition of recombinant β-glucosidase, glucose yield was 63.4%, 70.5% and 78.6%, corresponding to 0.5, 0.75 and 1.0 U/mL, respectively. The results also indicate that BGLI was high glucose tolerant and organic solvent, presenting higher efficiency in the hydrolysis of cellobiose than β-glucosidases.

This study points to the use of β-glycosidase to improve the enzymatic conversion of cellulose to glucose through synergistic action.

Zhao et al. [31] expressed in *Pichia pastoris* a thermostable beta glycosidase of the thermophilic fungus *Myceliophthora thermophila*. The molecular mass of the enzyme after purification was 130.0 kDa the recombinant enzyme (MtBgl3b) MtBgl3b presented pH 5.0 as the ideal for activity at 60°C, and excellent thermostability at 60 or 65°C. The authors also determined the effects of some metal ions and chemical reagents on the activities of MtBgl3b, where Ca^{2+}, Pb^{2+}, K^{+}, Mn^{2+}, EDTA, β-ME and Triton X-100 improved the activity by 6.4–29.9%, while Fe^{3+} completely suppressed the enzymatic activity. In addition, the activities of MtBgl3b were determined in relation to different substrates, for which the enzyme had higher activity against pNPG (258.7 U mg – 1), followed by pNPC (164.5 U mg-1), celotetraosis (125.7 U mg-1), celotriosis (118.0 U mg-1), celobiosis (62.2 U mg-1) and gentilebiosis (63.9 mg U-1). These results indicate that the enzyme presented desirable industrial properties, in addition to thermostability, wide spectrum of substrates and the capacity resistant to ethanol, which makes this protein a great candidate for industrial applications [30].

A β-glucosidase from *A. niger* was successfully expressed in *P. pastoris* and recombinant produced gentileoligosaccharides from glucose. In addition, the main operating parameters of this enzymatic conversion were optimized. At 80% glucose, 60°C, pH 4.5, 1 mmol/ L K +, 60 U of beta-glucosidase per gram of substrate and reaction time of 48 h, the gentiooligosaccharides produced reached 50 g/L [69].

5. One-time expression with the new CRISPR/Cas9 system technology

In recent years, several genetic tools have been elaborated and applied in various fungi and widely shared in different sectors of the economy. However, there is still a certain limitation in the studies of functional genomics for the production of recombinant cellulases in fungi. For this logic, emerging tools stand out, including CRISPR-Cas9-based genome editing (Clustered Regularly Interspaced Short Palindromic Repeats), i.e., Grouped and Regularly Interspaced Palindromic Repeats, as an agile tool for genome-specific gene edits [70]. The CRISPR-Cas9 system contains two components: the effector protein, which is Cas9 endonuclease, and a single chimeric guide RNA (sgRNA). This tool was involved to allow rapid editing of the genome of several organisms, among them, some varieties of filamentous fungi [71–75].

However, such approaches are not as useful as those available for yeasts and bacteria, considering the complexity of fungi, such as multicellular morphology, cell differentiation, thick chitinous cell walls and lack of adequate plasmids [76]. Composing the need to establish a genome editing system that can be used to develop a hyper cell factory for preparations of lignocellulolytic enzymes and other heterologous proteins, as well as to characterize the mechanisms that regulate protein induction, synthesis, and secretion [77].

In studies with *Myceliophthora thermophile* Li et al. [78] used the CRISPR/Cas9 technique in five highly expressed genes encoding extracellular proteases, degrade extracellular proteins and reduce cellulase yield. The results attest that Mtalp1 is a gene that degrades protease that inhibits cellulase production. To perform this study, five genes were selected and constructed using the CRISPR-Cas9 technique, resulting in a Mutant DMtalp1 that demonstrated protease activity substantially lower than 58.4%, and may be a good initial strain for additional metabolic engineering in order to produce cellulases and other proteins.

Liu et al. [74] used the CRISPR/Cas9 system for effective multiplexed genome engineering, successfully developed in thermophilic species *M. thermophila* and *M. heterothallica*. CRISPR/Cas9 can efficiently modify the imported a mdS gene into the genome through non-homologous nhej end-mediated events. As evidence of principle, the genes of the cellulase production pathway including cre-1, res-1, gh1–1 and alp-1, were chosen as editing targets. Simultaneous multigenic fissures of up to four of these different loci were prepared with the integration of the neomycin selection marker by means of a single transformation, using the CRISPR/Cas9 system.

This genome engineering tool gave rise to several strains that exhibit marked production of hypercellulase, among which extracellular secret activities of protein and lignocellulase increased substantially (up to 5 and 13 times, respectively), in analogy with the parent lineage. In their research, Salazar-Cerezo et al. [79], used *Penicillium subrubescens*, which is an ascomycete fungus with a robust content of families of active enzymes specific to carbohydrates involved in the degradation of ligonocellulosoic biomass. First, a method was developed for the engendering and transformation of protoplasts, using hygromycin as a selection marker. Subsequently, the CRISPR/Cas9 system was established in *P. subrubescens* by successfully excluding the KU70 gene, which was directly involved in the non-homologous end of the DNA repair mechanism. According to Salazar-Cerezo et al. [79], it was possible to consider the implementation of the CRISPR/Cas9 system in the filamentous fungus *P. subrubescens* and the effective protocols for generating and transforming protoplasts were optimized. In this way the MUTATED KU70 gene showed no discrepant phenotypic differences with the wild-type strains reported in the study, enabling the use of these mutants as parental strains for subsequent transformation events.

Rantasalo et al. [80] made use of CRISPR/Cas9 multiplexed in combination with System (S), classified as synthetic expression producing large amounts of the highly pure calB gene; this combination allowed the production of strains in a shorter time. Rantasalo et al. [80], when using the SES tool, it was used by the calB gene indices in an inducing medium, with highly constitutive expression provided by the SES, being possible to produce approximately 4 grams of glucose per liter of calB, in a cellulase inducing medium.

Zheng et al. [81], was able to create a CRISPR/Cas9 system in *Aspergillus niger* for sgRNA expression based on an endogenous U6 promoter and two heterologous promoters of U6. The three u6 promoters tested made feasible the transcription of sgRNA and the interruption of the gene of polyketide synthase albA gene in *A. niger*. In addition, this system allowed the insertion of highly efficient genes in the target genomic locus in *A. niger*, using DNAs from donors with homologous arms of up to 40 bp.

One of the alternatives for bioprospecting enzymes potentially for industry and the use of CRISPR/Cas9 with non-model species, such as the fungus *Huntiella omanensis*, a filamentous fungus ascomycete belonging to the family Ceratocystidaceae, will allow cutting in metabolic and genetic pathways that have not yet been studied in model species [82].

However, filamentous fungi are considered the largest producers of cellulases so far, since for genome editing systems using CRISPR-Cas9, it was stipulated in more than 40 different species of filamentous fungi and oomycetes, being therefore an important strategy regarding the production of potential strains for applications in the industry [83]. Cellulases and heterologous of fungi, with great industrial potential in the manufacture of bioethanol, using one of the most efficient techniques of recent times for genetic engineering, CRISPR/Cas9. It can be concluded that this technique, combined with biotechnological advances, will result in the improvement of fungal cells capable of producing biofuels economically and on an industrial scale, resulting in higher yield and quality of products.

6. Conclusions

Cellulases are important enzymes in cellulose degradation that occurs in nature, this degradation involves a system of extracellular multienzymes and have wide application. Molecular biology combined with recombinant DNA technology is a viable tool in enzymatic production with high activity, what makes recombinant fungal cellulases good candidates for industrial applications, including the saccharification of lignocellulosic materials.

Acknowledgements

This work was supported by two doctoral scholarships (88882.447813/2019-2101 and 88882.447821/2019-2101) financed by CAPES (Coordination for the Improvement of Higher Education Personnel). We thanks the Biotechnology Graduate Program of State University of Feira de Santana (UEFS/FIOCRUZ), Fundação de Amparo à Pesquisa do Estado da Bahia (FAPESB) and CNPq.

Conflict of interest

The authors declare no conflict of interest.

Author details

Raquel Guimarães Benevides[1*], Sandra Aparecida de Assis[2], Alison Borges Vitor[1,2], Geise Camila Ribeiro[2], Cleidineia Souza de Santana[1] and Larissa Emanuelle da Silva Almeida[2]

1 LAPEM, Biology Department, State University of Feira de Santana, Feira de Santana, Bahia, Brazil

2 Enzymology and Fermentation Technology Laboratory, Health Department, State University of Feira de Santana, Feira de Santana, Bahia, Brazil

*Address all correspondence to: raquelgb@uefs.br

References

[1] Dey P, Pal P, Kevin JD, Das DB. Lignocellulosic bioethanol production: prospects of emerging membrane technologies to improve the process – a critical review. Rev Chem Eng 2020; 36(3): 333-367. DOI: 10.1515/revce-2018-0014.

[2] Shahzadi T, Mehmood S, Irshad, M, Anwar Z, Afroz A, Zeeshan N, Rashid U, Sughra K. Advances in lignocellulosic biotechnology: A brief review on lignocellulosic biomass and cellulases. Adv Biosci Biotechnol. 2014;5:246-251. DOI: 10.4236/abb.2014.53031.

[3] Bhattacharya AS, Bhattacharya A, Pletschke BI. Synergism of fungal and bacterial cellulases and hemicellulases: a novel perspective for enhanced bio-ethanol production. Biotechnol Lett. 2015;37:1117-1129. DOI: 10.1007/s10529-015-1779-3

[4] Mohapatra S, Padhy S, Das Mohapatra PK, Thatoi HN. Enhanced reducing sugar production by saccharification of lignocellulosic biomass, *Pennisetum* species through cellulase from a newly isolated *Aspergillus fumigatus*. Bioresour Technol. 2018;253:262-272. DOI: 10.1016/j.biortech.2018.01.023

[5] Wang H, Zhai L, Geng A. Enhanced cellulase and reducing sugar production by a new mutant strain Trichoderma harzianum USA20. J Biosci Bioeng. 2020;129:242-249. DOI: 10.1016/j.jbiosc.2019.08.016

[6] Zhang H, Hua S-F, Zhang L. Co-immobilization of cellulase and glucose oxidase on graphene oxide by covalent bonds: a biocatalytic system for one-pot conversion of gluconic acid from carboxymethyl cellulose. J Chem Technol Biotechnol. 2020;95:1116-1125. DOI: 10.1002/jctb.6296

[7] Ma Y, Han C, Chen J, Li H, He K, Liu A, Li D. Fungal cellulase is an elicitor but its enzymatic activity is not required for its elicitor activity. Mol Plant Pathol. 2015;16:14-26. DOI: 10.1111/mpp.12156

[8] Kuhad RC, Gupta R, Singh A. Microbial cellulases and their industrial applications. Enzyme Res. 2011;2011:1-10. DOI: 10.4061/2011/280696

[9] Kishishita S, Fujii T, Ishikawa K. Heterologous expression of hyperthermophilic cellulases of archaea *Pyrococcus* sp. by fungus *Talaromyces cellulolyticus*. J Ind Microbiol Biotechnol. 2015;42:137-141. DOI: 10.1007/s10295-014-1532-2

[10] Quay DHX, Bakar FDA, Rabu A, Said M, Illias RM, Mahadi NM, Hassan O, Murad AMA. Overexpression, purification and characterization of the *Aspergillus niger* endoglucanase, EglA, in *Pichia pastoris*. Afr J Biotechnol. 2011;10:2101-2111. DOI: 10.5897/AJB10.1046

[11] Kun RS, Gomes ACS, Hildén KS, Cerezo SS, Mäkelä MR, De Vries RP. Developments and opportunities in fungal strain engineering for the production of novel enzymes and enzyme cocktails for plant biomass degradation. Biotechnol Adv. 2019;37:1-19. DOI: 10.1016/j.biotechadv.2019.02.017

[12] Mohammad SF, Feng Y, Yang G. Optimization of cell culture and cell disruption processes to enhance the production of thermophilic cellulase FnCel5A in E.coli using response surface methodology. PLoS ONE. 2019;14:1-16. DOI: 10.1371/journal.pone.0210595

[13] Nawaz M, Zafar S, Shervani SK, Ray S, Buneen U, Rubab A, Kanwal M, Qandeel-e-Arsh, Khurram MF, Sajid SN. Optimization of conditions for the

production of recombinant cellulase by using *E. coli* BL21 Codon Plus in Fermenter. Biosci Biotech Res Asia. 2020;17:173-190. DOI: 10.13005/bbra/2822

[14] Valli M, Tatto NE, Peymann A, Gruber C, Landes N, Ekker H, Thallinger GG, Mattanovich D, Gasser B, Graf AB. Curation of the genome annotation of *Pichia pastoris* (*Komagataella phaffii*) CBS7435 from gene level to protein function. FEMS Yeast Res. 2016;16:1-12. DOI: 10.1093/femsyr/fow051

[15] Pekarsky A, Veiter L, Rajamanickam V, Herwig C, Grünwald-Gruber C, Altmann F, Spadiut O. Production of a recombinant peroxidase in different glyco-engineered *Pichia pastoris* strains: a morphological and physiological comparison. Microb Cell Fact. 2018;17:1-15. DOI: 10.1186/s12934-018-1032-6

[16] Garvey M, Klose H, Fischer R, Lambertz C, Commandeur U. Cellulases for biomass degradation: comparing recombinant cellulase expression platforms. Trends Biotechnol. 2013;31:581-593. DOI: 10.1016/j.tibtech.2013.06.006

[17] Koupaie EH, Dahadha S, Lakeh AAB, Azizi A, Elbeshbishy E. Enzymatic pretreatment of lignocellulosic biomass for enhanced biomethane production-A review. J Environ Manag. 2019;233:774-784. DOI: 10.1016/j.jenvman.2018.09.106

[18] Sajith S, Priji P, Sreedevi S, Benjamin S. An Overview on Fungal Cellulases with an Industrial Perspective. Int J Food Sci Nutr. 2016;6:1-13. DOI: 10.4172/2155-9600.1000461

[19] Laca A, Laca A, Díaz M. Hydrolysis: From cellulose and hemicellulose to simple sugars. In: Basile A, Dalena F, editors. Second and Third Generation of Feedstocks. 1 ed. Elsevier; 2019. p. 213-240. DOI: 10.1016/B978-0-12-815162-4.00008-2

[20] Lambertz C, Garvey M, Klinger J, Heesel D, Klose H, Fischer R, Commandeur U. Challenges and advances in the heterologous expression of cellulolytic enzymes: a review. Biotechnol biofuels. 2014;7:1-15. DOI: 10.1186/s13068-014-0135-5

[21] Juturu V, Wu JC. Microbial cellulases: engineering, production and applications. Renewable Sustainable Energy Rev. 2014;33:188-203. DOI: 10.1016/j.rser.2014.01.077

[22] Makino T, Skretas G, Georgiou G. Strain engineering for improved expression of recombinant proteins in bacteria. Microb cell fact. 2011;10:1-10. DOI: 10.1186/1475-2859-10-32

[23] Ahmad M, Hirz M, Pichler H, Schwab H. Protein expression in *Pichia pastoris*: recent achievements and perspectives for heterologous protein production. Appl Microbiol Biotechnol. 2014;98:5301-5317. DOI: 10.1007/s00253-014-5732-5

[24] Araújo JA, Ferreira TC, Rubini MR, Duran AGG, Marco JL, Moraes LMP, Torres FAG. Coexpression of cellulases in *Pichia pastoris* as a self-processing protein fusion. AMB Expr. 2015;5:1-10. DOI: 10.1186/s13568-015-0170-z

[25] Jiménez AV, Wang H, Siegfried BD. Expression and characterization of a recombinant endoglucanase from western corn rootworm, in *Pichia pastoris*. J Insect Sci. 2014;14:1-5. DOI: 10.1093/jisesa/ieu104

[26] Taipakova SM, Stanbekova G, Ischenko A, Saparbayev M, Bissenbaev AK. Cloning and expression of *Lentinula edodes* cellobiohydrolase CEL6B gene in *E. coli*. Int J Biol Chem. 2011;2011:19-26.

[27] Generoso WC, Malagó-Jr W, Pereira Jr N, Henrique-Silva F. Recombinant expression and characterization of an endoglucanase III (cel12a) from *Trichoderma harzianum* (Hypocreaceae) in the yeast *Pichia pastoris*. Genet Mol Res. 2012;11:1544-1557. DOI: 10.4238/2012.May.21.11

[28] Karnaouri AC, Topakas E, Christakopoulos P. Cloning, expression, and characterization of a thermostable GH7 endoglucanase from *Myceliophthora thermophila* capable of high-consistency enzymatic liquefaction. Appl Microbiol Biotechnol. 2014;98:231-242. DOI: 10.1007/s00253-013-4895-9

[29] Li GQ, Chai CS, Fan S, Zhao LG. Cloning of a cellobiohydrolase gene (cbh1) from *Aspergillus niger* and heterogenous expression in *Pichia pastoris*. Advanced Materials Research. 2012;347-353:2443-2447. DOI: 10.4028/www.scientific.net/AMR.347-353.2443

[30] Zhao L, Zhou T, Li X, Fan S, You L. Expression and characterization of GH3 β-Glucosidase from *Aspergillus niger* NL-1 with high specific activity, glucose inhibition and solvent tolerance. Microbiology. 2013;82:356-363. DOI: 10.1134/S0026261713030181

[31] Zhao J, Guo C, Tian C, Ma Y. Heterologous expression and characterization of a GH3 β-glucosidase from thermophilic fungi *Myceliophthora thermophila* in *Pichia pastoris*. Appl Biochem Biotechnol. 2015;177:511-527. DOI: 10.1007/s12010-015-1759-z

[32] Ramani G, Meera B, Vanitha C, Rajendhran J, Gunasekaran P. Molecular cloning and expression of thermostable glucose-tolerant β-glucosidase of *Penicillium funiculosum* NCL1 in *Pichia pastoris* and its characterization. J Ind Microbiol Biotechnol. 2015;42:553-565. DOI: 10.1007/s10295-014-1549-6

[33] Lima MS, Damasio ARL, Crnkovic PM, Pinto MR, Silva AM, Silva JCR, Segato F, Lucas RC, Jorge JA, Polizeli MLTM. Co-cultivation of *Aspergillus nidulans* recombinant strains produces an enzymatic cocktail as alternative to alkaline sugarcane berry pretreatment. Front Microbiol. 2016;7:1-9. DOI: 10.3389/fmicb.2016.00583

[34] Santos CA, Zanphorlin LM, Crucello A, Tonoli CCC, Ruller R, Horta MAC, Murakami MT, Souza AP. Crystal structure and biochemical characterization of the recombinant ThBgl, a GH1 β-glucosidase overexpressed in *Trichoderma harzianum* under biomass degradation conditions. Biotechnol Biofuels. 2016;9:2-11. DOI: 10.1186/s13068-016-0487-0

[35] Auta R, Wusu AD, Radecka I, Hooley P. Expression and characterization of recombinant β-glucosidases from *Aspergillus nidulans* AN2227. Science World Journal. 2016;11:7-15.

[36] Zeng R, Hu Q, Yin X-Y, Huang H, Yan J-B, Gong Z-W, Yang Z-H. Cloning a novel endo-1,4-β-d-glucanase gene from *Trichoderma virens* and heterologous expression in *E. coli*. AMB Expr. 2016;6:1-7. DOI: 10.1186/s13568-016-0282-0

[37] Auta R, Campbell A, Radecka I, Hooley P. Enzyme Assay, Cloning and Squencing of Novel β-glucosidase Gene From *Aspergillus niger* f321 (unidentified Nigerian strain). Science World Journal. 2016;11:44-52.

[38] Karnaouri A, Muraleedharan MN, Dimarogona M, Topakas E, Rova U, Sandgren M, Christakopoulos P. Recombinant expression of thermostable processive *Mt*EG5 endoglucanase and its synergism with *Mt*LPMO from *Myceliophthora thermophila* during the hydrolysis of lignocellulosic substrates. Biotechnol Biofuels. 2017;10:2-17. DOI: 10.1186/s13068-017-0813-1

[39] Phadtare P, Joshi S, Satyanarayana T. Recombinant thermo-alkali-stable endoglucanase of *Myceliopthora thermophila* BJA (rMt-egl): Biochemical characteristics and applicability in enzymatic saccharification of agro-residues. Int J Biol Macromol. 2017;104:107-116. DOI: 10.1016/j.ijbiomac.2017.05.167

[40] Rungrattanakasin B, Premjet S, Thanonkeo S, Klanrit P, Thanonkeo P. Cloning and expression of an endoglucanase gene from the thermotolerant fungus *Aspergillus fumigatus* DBiNU-1 in *Kluyveromyces lactis*. Braz J Microbiol. 2018;49:647-655. DOI: 10.1016/j.bjm.2017.10.001

[41] Xia Y, Yang L, Xia L. Combined strategy of transcription factor manipulation and β-glucosidase gene overexpression in *Trichoderma reesei* and its application in lignocellulose bioconversion. J Ind Microbiol Biotechnol. 2018;45:803-811. DOI: 10.1007/s10295-018-2041-5

[42] Zhao C, Deng L, Fanga H. Mixed culture of recombinant *Trichoderma reesei* and *Aspergillus niger* for cellulase production to increase the cellulose degrading capability. Biomass Bioenergy. 2018;112:93-98. DOI: 10.1016/j.biombioe.2018.03.001

[43] Hua C, Li W, Han W, Wang Q, Bi P, Han C, Zhu L. Characterization of a novel thermostable GH7 endoglucanase from Chaetomium thermophilum capable of xylan hydrolysis. Int J Biol Macromol. 2018;117:342-349. DOI: 10.1016/j.ijbiomac.2018.05.189

[44] Jain KK, Kumar S, Bhardwaj KN, Kuhad RC. Functional expression of a thermostable endoglucanase from *Thermoascus aurantiacus* RCKK in *Pichia pastoris* X-33 and its characterization. Mol Biotechnol. 2018;60:736-748. DOI: 10.1007/s12033-018-0106-3

[45] Trollope K, Nel DW, Volschenk H. The heterologous expression potential of an acid-tolerant *Talaromyces pinophilus* β-glucosidase in *Saccharomyces cerevisiae*. Folia Microbiol. 2018;63:725-734. DOI: 10.1007/s12223-018-0613-4

[46] Bernardi AV, Gouvêa PF, Gerolamo LD, Yonamine DK, Balico LLL, Uyemura SK, Dinamarco TM. Functional characterization of GH7 endo-1.4-β-glucanase from *Aspergillus fumigatus* and its potential industrial application. Protein Expression Purif. 2018;150:1-11. DOI: 10.1016/j.pep.2018.04.016

[47] Li Z, Pei X, Zhang Z, Wei Y, Song Y, Chen L, Liu S, Zhang S-H. The unique GH5 cellulase member in the extreme halotolerant fungus *Aspergillus glaucus* CCHA is an endoglucanase with multiple tolerance to salt, alkali and heat: prospects for straw degradation applications. Extremophiles. 2018;22:675-685. DOI: 10.1007/s00792-018-1028-5

[48] Bernardi AV, Yonamine DK, Uyemura SA, Dinamarco TM. A Thermostable *Aspergillus fumigatus* GH7 Endoglucanase Over-Expressed in *Pichia pastoris* Stimulates Lignocellulosic Biomass Hydrolysis. Int J Mol Sci. 2019;20:2261. DOI: 10.3390/ijms20092261

[49] Volkov PV, Rozhkova AM, Zorov IN, Sinitsyn AP. Cloning, Purification and Study of Recombinant GH3 Family β-glucosidase From *Penicillium verruculosum*. Biochimie. 2019;168:231-240. DOI: 10.1016/j.biochi.2019.11.009

[50] Gu Y, Zheng F, Wang Y, Su X, Bai Y, Yao B, Huang H, Luo H. Characterization of two thermophilic cellulases from *Talaromyces leycettanus* JCM12802 and their synergistic action on cellulose hydrolysis. PLoS ONE. 2019;14:1-15. DOI: 10.1371/journal.pone.0224803

[51] Anindyawati T, Putra R, Yuliawati, Dewi KS, Fuad AM, Sudiyani Y.

Heterologous Expression of *Trichoderma reesei* Exoglucanase (Cel6A) in *Pichia pastoris* Under the Control of GAP Promoter. AIP Conf Proc. 2019;2155: 020044. DOI: 10.1063/1.5125548

[52] Onuma H, Hara K, Sugita K, Kano A, Fukuta Y, Shirasaka N. Purification and characterization of a glycoside hydrolase family 5 endoglucanase from *Tricholoma matsutake* grown on barley based solid-state medium. J Biosci Bioeng. 2019;128:669-676. DOI: 10.1016/j.jbiosc.2019.05.012

[53] Liang CY, Xu JL, Xu HJ, Qi W, Zhang Y, Luo W, Chen XY, Wang ZM, Yuan ZH. Gene cloning and characterization of an organic solvent stimulated β-glucosidase and its application for the coproduction of ethanol and succinic acid. Cellulose. 2019;26:8237-8248. DOI: 10.1007/s10570-019-02477-y

[54] Tao Y, Yang L, Yin L, Lai C, Huang C, Li X, Yong Q. Novel approach to produce biomass-derived oligosaccharides simultaneously by recombinant endoglucanase from *Trichoderma reesei*. Enzyme Microb Technol. 2019;134:109481. DOI: 10.1016/j.enzmictec.2019.109481

[55] Delabona PS, Codima CA, Ramoni J, Zubieta MP, Araújo BM, Farinas CS, Pradella JGC, Seiboth B. The impact of putative methyltransferase overexpression on the *Trichoderma harzianum* cellulolytic system for biomass conversion. Bioresour Technol. 2020;313:123616. DOI: 10.1016/j.biortech.2020.123616

[56] Yang Y, Wang J, Guo H, Cao Y. The enzymatic characters of heterologous expressed novel β-1, 4-glucosidase originated from *Aspergillus fresenii*. Biotech. 2020;10:1-9. DOI: 10.1007/s13205-020-02229-x

[57] Rubini MR, Dillon AJP, Kyaw CM, Faria FP, Poças-Fonseca MJ, Silva-Pereira I. Cloning, characterization and heterologous expression of the first *Penicillium echinulatum* cellulase gene. J Appl Microbiol. 2010;108:1187-1198. DOI: 10.1111/j.1365-2672.2009.04528.x

[58] Lahjouji, K, Storms R, Xiao Z, Joung K-B, Zheng Y, Powlowski J, Tsang A, Varin L. Biochemical and molecular characterization of a cellobiohydrolase from Trametes versicolor. Appl Biotechnol Microbiol. 2007;75:337-346. DOI: 10.1007/s00253-006-0824-5

[59] Nakazawa H, Okada K, Onodera T, Ogasawara W, Okada H, Morikawa Y. Directed evolution of endoglucanase III (Cel12A) from *Trichoderma reesei*. Appl Microbiol Biotechnol. 2009;83:649-657. DOI: 10.1007/s00253-009-1901-3

[60] Koseki T, Mese Y, Fushinobu S, Masaki K, Fujii T, Ito K, Shiono Y, Murayama T, Iefuji H. Biochemical characterization of a glycoside hydrolase family 61 endoglucanase from *Aspergillus kawachii*. Appl Microbiol Biotechnol. 2008;77:1279-1285. DOI: 10.1007/s00253-007-1274-4

[61] Igarashi K, Ishida T, Hori C, Samejima M. Characterization of an Endoglucanase Belonging to a New Subfamily of Glycoside Hydrolase Family 45 of the Basidiomycete *Phanerochaete chrysosporium*. Appl Environ Microbiol. 2008;74:5628-5634. DOI: 10.1128/AEM.00812-08

[62] Hua C, Yan Q, Jiang Z, Li Y, Katrolia P. High-level expression of a specific β-1,3-1,4-glucanase from the thermophilic fungus Paecilomyces thermophila in *Pichia pastoris*. Appl Microbiol Biotechnol. 2010;88:509-518. DOI: 10.1007/s00253-010-2759-0

[63] Liu G, Wei X, Qin Y, Qu Y. Characterization of the endoglucanase and glucomannanase activities of a glycoside hydrolase family 45 protein from *Penicillium decumbens* 114-2. J Gen

Appl Microbiol. 2010;56:223-229. DOI: 10.2323/jgam.56.223

[64] Chiang, C-J, Chen PT, Yeh CY, Chao Y-P. Statistical optimization of one-step immobilization process for recombinant endoglucanase from *Clostridium thermocellum*. Process Biochem. 2013;48:1886-1892. DOI: 10.1016/j.procbio.2013.08.022

[65] Martinez D, Larrondo LF, Putnam N, Gelpke MDS, Huang K, Chapman J, Helfenbein KG, Ramaiya P, Detter JC, Larimer F, Coutinho PM, Henrissat B, Berka R, Cullen D, Rokhsar D. Genome sequence of the lignocellulose degrading fungus *Phanerochaete chrysosporium* strain RP78. Nat Biotechnol. 2004;22:695-700. DOI: 10.1038/nbt967

[66] Kawai R, Igarashi K, Yoshida M, Kitaoka M, Samejima M. Hydrolysis of β-1,3/1,6-glucan by glycoside hydrolase family 16 endo-1,3(4)-β-glucanase from the basidiomycete *Phanerochaete chrysosporium*. Appl Microbiol Biotechnol. 2006;71:898-906. DOI: 10.1007/s00253-005-0214-4

[67] Voutilainen SP, Puranen T, Siika-Aho M, Lappalainen A, Alapuranen M, Kallio J, Hooman S, Viikari L, Vehmaanperä J, Koivula A. Cloning, expression, and characterization of novel thermostable family 7 cellobiohydrolases. Biotechnol Bioeng. 2008;101:515-528. DOI: 10.1002/bit.21940

[68] Toda H, Nagahata N, Amano Y, Nozaki K, Kanda T, Okazaki M, Shimosaka M. Gene Cloning of Cellobiohydrolase II from the White Rot Fungus *Irpex lacteus* MC-2 and Its Expression in *Pichia pastoris*. Biosci Biotechnol Biochem. 2008;72:3142-3147. DOI: 10.1271/bbb.80316

[69] Liu L, Song Z, Zhu T, Zhang M, Wu J, Chen J. Production of gentiooligosaccharide by recombinant beta-glucosidase. Acta Microbiol Sin. 2009;49:597-602.

[70] Javed MR, Noman M, Shahid M, Ahmed T, Khurshid M, Rashid MH, Ismail M, Sadaf M, Khan F. Current situation of biofuel production and its enhancement by CRISPR/Cas9-mediated genome engineering of microbial cells. Microbiol Res. 2019;219:1-11. DOI: 10.1016/j.micres.2018.10.010

[71] Fuller KK, Chen S, Loros JJ, Dunlap JC. Development of the CRISPR/Cas9 system for targeted gene disruption in *Aspergillus fumigatus*. Eukaryotic Cell. 2015;14:1073-1080. DOI: 10.1128/EC.00107-15

[72] Liu R, Chen L, Jiang Y, Zhou Z, Zou G. Efficient genome editing in filamentous fungus *Trichoderma reesei* using the CRISPR/Cas9 system. Cell Discovery. 2015;1:1-11. DOI: 10.1038/celldisc.2015.7

[73] Matsu-ura T, Baek M, Kwon J, Hong C. Efficient gene editing in *Neurospora crassa* with CRISPR technology. Fungal Biol Biotechnol. 2015;2:1-7. DOI: 10.1186/s40694-015-0015-1

[74] Liu Q, Gao R, Li J, Lin L, Zhao J, Sun W, Tian C. Development of a genome-editing CRISPR/Cas9 system in thermophilic fungal *Myceliophthora* species and its application to hyper-cellulase production strain engineering. Biotechnol Biofuels. 2017;10:1-14. DOI: 10.1186/s13068-016-0693-9

[75] Shi T-Q, Liu G-N, Ji R-Y, Shi K, Song P, Ren L-J, Huang H, Ji X-J. CRISPR/Cas9-based genome editing of the filamentous fungi: the state of the art. Appl Microbiol Biotechnol. 2017;101:7435-7443. DOI: 10.1007/s00253-017-8497-9

[76] Graf R, Li X, Chu VT, Rajewsky K. sgRNA sequence motifs blocking efficient CRISPR/Cas9-mediated gene editing. Cell Reports. 2019;26:1098-1103. DOI: 10.1016/j.celrep.2019.01.024

[77] van Leeuwe TM, Arentshorst M, Ernst T, Alazi E, Punt PJ, Ram AFJ. Efficient marker free CRISPR/Cas9 genome editing for functional analysis of gene families in filamentous fungi. Fungal Biol Biotechnol. 2019;6:1-13. DOI: 10.1186/s40694-019-0076-7

[78] Li F, Liu Q, Li X, Zhang C, Li J, Sun W, Liu D, Xiao D, Tian C. Construction of a new thermophilic fungus *Myceliophthora thermophila* platform for enzyme production using a versatile 2A peptide strategy combined with efficient CRISPR-Cas9 system. Biotechnol Lett. 2020;42:1181-1191. DOI: 10.1007/s10529-020-02882-5

[79] Salazar-Cerezo S, Kun RS, Vries RP, Garrigues S. CRISPR/Cas9 technology enables the development of the filamentous ascomycete fungus *Penicillium subrubescens* as a new industrial enzyme producer. Enzyme Microb Technol. 2020;133:109463. DOI: 10.1016/j.enzmictec.2019.109463

[80] Rantasalo A, Vitikainen M, Paasikallio T, Jäntti J, Landowski CP, Mojzita D. Novel genetic tools that enable highly pure protein production in *Trichoderma reesei*. Sci Rep. 2019;9:1-12. DOI: 10.1038/s41598-019-41573-8

[81] Zheng X, Zheng P, Sun J, Kun Z, Ma Y. Heterologous and endogenous U6 snRNA promoters enable CRISPR/Cas9 mediated genome editing in *Aspergillus niger*. Fungal Biol Biotechnol. 2018;5:1-9. DOI: 10.1186/s40694-018-0047-4

[82] Wilson AM, Wingfield BD. CRISPR-Cas9-Mediated Genome Editing in the Filamentous Ascomycete *Huntiella omanensis*. J Visualized Ex. 2020;9:1-11. DOI: 10.3791/61367

[83] Schuster M, Kahmann R. CRISPR-Cas9 genome editing approaches in filamentous fungi and oomycetes. Fungal Genet Biol. 2019;130:43-53. DOI: 10.1016/j.fgb.2019.04.016

Chapter 21

Novel Acumens into Biodegradation: Impact of Nanomaterials and Their Contribution

Danushika C. Manatunga, Rohan S. Dassanayake and Renuka N. Liyanage

Abstract

Biodegradation is the most viable alternative for numerous health and environmental issues associated with non-biodegradable materials. In recent years, there has been considerable interest in biodegradable nanomaterials due to their relative abundance, environmental benignity, low cost, easy use, and tunable properties. This chapter covers an overview of biodegradation, factors and challenges associated with biodegradation processes, involvement of nanotechnology and nanomaterials in biodegradation, and biodegradable nanomaterials. Furthermore, current chapter extensively discusses the most recent applications of biodegradable nanomaterials that have recently been explored in the areas of food packaging, energy, environmental remediation, and nanomedicine. Overall, this chapter provides a synopsis of how the involvement of nanotechnology would benefit the process of biodegradation.

Keywords: Biodegradation, nanoparticles, food packaging, energy storage, environmental remediation, nanomedicine

1. Introduction to biodegradation

Sustainable development is a principle that is implemented to preserve the environment for the future generation while meeting the needs of the present generation. Environmental pollution is considered one of the significant barriers to sustainable development. Therefore, the drive for sustainable development must address environmental pollution by removing pollutants, restoring polluted areas, or using without affecting the unpolluted areas [1]. Biodegradation is identified as a key eco-friendly and economical way of sustainable development, which entails enzymatic degradation or a breakdown of complex organic matter into small molecules in the presence of microorganisms [2]. The microorganisms could also allow the biodegradation of organic matter in the presence of a growth substrate used as the primary source of energy and carbon source, a process called cometabolism [2]. The biodegradation process is an effective alternative for commonly applied waste disposal methods such as incineration and landfilling [3].

2. Key challenges associated with biodegradation process

Biodegradation may sometimes lead to incomplete mineralization of the total organic content, such as recalcitrant materials leaving unprocessed contaminants behind [4]. This could be due to the complex structure of the materials, higher molecular weight, crosslinking, shape, texture, surface area, and degradation rate [5]. For example, depending on the degree of crystallinity, orientation and packing of polymers, the degradation rate is severely affected. It has been observed that even under the same conditions, the degradation of amorphous regions of polycaprolactone (PCL) by filamentous fungi is much faster than the degradation of crystalline regions of PCL, where the amorphous regions may permit easy access to microbes during the degradation process [4]. Therefore, ensuring the complete or partial degradation of these complex substances to produce harmless products without secondary pollution is extremely important [5].

Additionally, the microbial biodegradation process also depends on various factors such as nutrient availability, microbe type, substrate properties, and environmental conditions such as pH, moisture content, and redox potential [6–8]. Generally, the redox potential relies on the presence of the electron acceptors at the active site, such as oxygen, nitrates, manganese oxides, iron oxides, sulfate, and triggering the aerobic or anaerobic biodegradation. Even though many of the microbes prefer physiological pH of 7.4 and a temperature of 37°C for their growth, certain microbes such as fungal species prefer an acidic environment. In contrast, some bacteria prefer relatively high temperatures for their optimal growth. Therefore, not exactly knowing the required growth conditions could be a significant factor contributing to the incomplete degradation of the substrate in some cases [7].

Furthermore, microbial metabolism in biodegradation is an energy transformation process that is solely governed by the functions of enzymes and the intermediates produced during the reactions [5]. Therefore, proper screening is required to identify the microbes with an inherent set of genes that are capable of degrading the contaminants, the factors and the conditions under which the population of these microbes could increase, and the synergic performance of these microbes with other technologies to establish an environmentally profitable biodegradation platform [9]. It has also been identified that more optimization procedures and scaling up are required for the biodegradation of large contaminated areas [10].

3. Role of nanotechnology in biodegradation

Biodegradable materials can be considered a preeminent group of materials that could also be called next-generation materials leading to zero global environmental pollution. Owing to this concern, the consumption of biodegradable materials, such as polymers, has increased two to three-fold in a broad spectrum of fields, including agriculture, automotive, packaging, energy and environment, and biomedical. But still, the contribution coming from the biodegradable polymers in said areas only accounts for 3–5% of total polymer consumption [11]. This lower contribution could be mainly due to the poorly addressed issues such as low durability, low performance, and high production cost [12].

In this context, combining the concept of biodegradation with nanotechnology could be identified as a more systemic and innovative approach to address the current issues with the biodegradation process [13]. Nanotechnology is an evolving branch of science that has diversified its application in many disciplines such as agriculture [14], healthcare [15], transportation [16], electronics [17], food [18], water purification [19–25], and security [26]. Nanotechnology involves manipulating matter in 1–100 nm nanoscale to create materials where at least one of the dimensions of the particles in the nano range [27]. The combined approach of nanotechnology-mediated biodegradation could address a wide range of potential applications in agriculture, food packaging, environmental remediation, and healthcare while accounting for reduced costs and no impact on environmental pollution [12]. Nanomaterials have proven effective as excellent adsorbents, sensors, and catalysts for biodegradation purposes due to their specific surface area and high reactivity [28].

Furthermore, the presence of nanoparticles and microbes that are actively engaged in the biodegradation process has paved the way in improving growth profiles of microbes by acting as biodegradation enhancers [29–32]. It has also been observed that the integration of nanotechnology with the enzymatic pathways in the biodegradation process could lead to profound activity and improved reusability of the enzymes [13]. Nanoparticles have also performed as effective sensor systems to detect the utilization of the raw materials and the production of specific products, which provided an inference on the progression of the biodegradation process [33]. Hence, this process of nano-biodegradation would ultimately involve the reduction of accumulating harmful non-biodegradable materials in the environment [34, 35].

3.1 Factors affecting the performance of nanomaterials during biodegradation

3.1.1 Properties of the nanomaterial

The chemical and physical interaction between the nanoparticles and the microbiota during the biodegradation is majorly influenced by the properties of nanomaterials such as size, shape, surface functionalization chemical structure, as they could influence the reactivity and stability of the nanomaterial [36, 37]. In addition, nanomaterials exhibit a quantum effect where less energy is required for associated chemical reactions [38]. Furthermore, the surface plasmon resonance exhibited by certain types of nanomaterials such as gold nanoparticles (Au NPs) [39] and silver nanoparticles (Ag NPs) [40] can also be used to detect and identify the contaminants. The smaller size of the nanomaterials also allows them to penetrate deeper into complex organic molecules [41].

3.1.2 Properties of the microorganisms and culture medium affecting the biodegradation

The performance of the nanomaterials during biodegradation also depends on the type of the organism, such as bacteria, fungi, protozoa and the type of enzymes used for the degradation of the contaminants [38]. Growth conditions such as pH, redox potential, temperature, ionic strength, solubility, presence or absence of electron acceptors in the culture medium also influence the activity and stability of the nanoparticles [13]. Therefore, proper control of the culture medium conditions to obtain a prolonged uninterrupted biodegradation procedure is necessary.

3.2 Types of nanomaterials used in biodegradation

Different types of nanomaterials have been utilized in the biodegradation process. **Table 1** summarizes the specific types of nanomaterials and their applications. Commonly used biodegradable nanomaterials include zero-valent metals, metal oxides, metal sulfides, nano clay, nanocomposites, carbon-based nanomaterials,

Type of the Nanomaterial	Type of degradation	Biodegradation process	Reference
Zero valent metals, oxides, sulfides			
I. Zero valent iron (nZVI)	Microorganism mediated- Organohalide-respiring bacteria (OHRB), sulfate reducing bacteria (SRB) and iron reducing bacteria (IRB)	nZVI provides suitable living conditions for the growth and activity of anaerobic bacteria to degrade organohalides, heavy metals	[42]
II. Zirconia (ZrO_2)	Microorganism mediated- *Pseudomonas aeruginosa*	Synthesis of ZrO_2 via *P. aeruginosa* for adsorption driven bioremediation of tetracycline	[43]
III. Silicon dioxide (SiO_2)	Microorganism mediated-Indigenous actinomycetes species isolated from the effluent contaminated site	Actinomycetes mediated synthesis of silica and use for adsorption and decolourisation of textile effluent	[44]
IV. Iron oxide (Fe_3O_4)	Microorganism mediated- *Microbacterium sp., Pseudomonas putida and Bacterium Te68R*	Enhance the consortium growth that involve in Low-Density Polyethylene (LDPE) degradation	[45]
V. Cadmium Zinc sulfide quantum dots (CdZnS QDs)	Microorganism mediated- *Escherichia coli*	Immobilization of nanoscale CdZnS QDs in to the extracellular matrix of bacterial biofilms which are later on used as catalysts for the degradation of nitro aromatic compounds	[46]
Nanoclay	Microorganism mediated- *Pseudomonas spp., Sphingomonas spp., Flavobacterium spp., Burkholderia spp., Rhodococcus spp., Mycobacterium spp., and Bacillus spp.*	Clay/modified clay minerals as effective adsorbents of PAHs/ volatile oxygen compounds (VOCs) to trigger the microbial mediated biodegradation	[47]
Nanocomposites			
I. Nanocellulose composites	Microorganism mediated- *Arthrobacter globiformis D47*	Bacteria decorated nanocellulose being used as a scaffold to grow the bacteria as well as to remove Diuron via biodegradation	[48]
II. Fe_3O_4/biochar composites	Microorganism mediated- *R. capsulatus*	Improve the adsorption capacity of photosynthetic bacteria as well as to improve the efficiency of bioremediation of wastewater	[49]
Carbon based nanomaterials			
I. Fullerene 60	Microorganism mediated- *Pseudomonas putida strain MK4 (DQ318885),* *Bacterium Te68R strain PN12 (DQ423487). P. aeruginosa strain* *PS1 (EU741797), P. putida strain PW1 (EU741798), and P. aeruginosa strain C1 (EU753182)*	Influence the growth cycle of LDPE, HDPE epoxy and epoxy silicon degrading bacteria and accelerate the polymer biodegradation process of bacterial consortia	[50]

Type of the Nanomaterial	Type of degradation	Biodegradation process	Reference
II. Carbon nanotubes (CNTs)	Microorganism mediated- *S. cerevisiae*, Actinomycetes	Immobilization of microbes for bioremediation of heavy metals	[51]
Biopolymer based nanomaterials			
I. Alginate beads	Microorganism mediated- *Acinetobacter sp., Bacillus circulans, Bacillus licheniformis, Brevibacillus brevis, Burkholderia cepacia, Leifsonia aquatica and Sphingomonas paucimobilis*	Improved the bacterial attachment required for oil bioremediation	[52]
II. Chitosan beads	Microorganism mediated- *Serratia sp. AC-11*	Remove polycyclic hydrocarbons by immobilizing the bacteria by improving the degradation rate	[53]
Nanofibrous materials			
I. Polyvinyl alcohol (PVA) and Polyethylene oxide (PEO) nanofibers	Microorganism mediated- *Pseudomonas aeruginosa ATCC 47085*	Provide suitable platforms for preservation of living bacterial cells and direct use for bioremediation of methylene blue	[54]
II. Cyclodextrin nanofibers	Microorganism mediated- *Lysinibacillus sp. NOSK*	Provide a matrix for the encapsulation of bacteria to perform bioremediation of heavy metals and reactive dyes	[55]
Biodegrading nanoparticles			
I. Polylactic acid (PLA) micelles	Physiological Enzymes mediated	Tumor targeting and efficient drug delivery	[56]
II. Polylactic glycolic acid (PLGA) micelles	Physiological Enzymes mediated	Use of thermosensitive and biodegradable triblock copolymer for temperature sensitive drug delivery for liver cancer	[57]
III. Polycarprolactone (PCL) nanoparticles	Physiological Enzymes mediated	Biodegradable nanocarriers for therapeutic compounds	[58]
IV. Chitosan nanoparticles	Physiological Enzymes mediated	Biodegradable nanocarriers for drug delivery diagnosis and other biological applications	[59]
V. Dendrimers	Physiological Enzymes mediated	Biocompatible, biodegradable delivery system against infections and cancer	[60]
VI. Liposomes	Physiological Enzymes mediated	Less toxic, biodegradable delivery systems for various diseases	[61]

Table 1.
Different types of nanomaterials used in biodegradation processes.

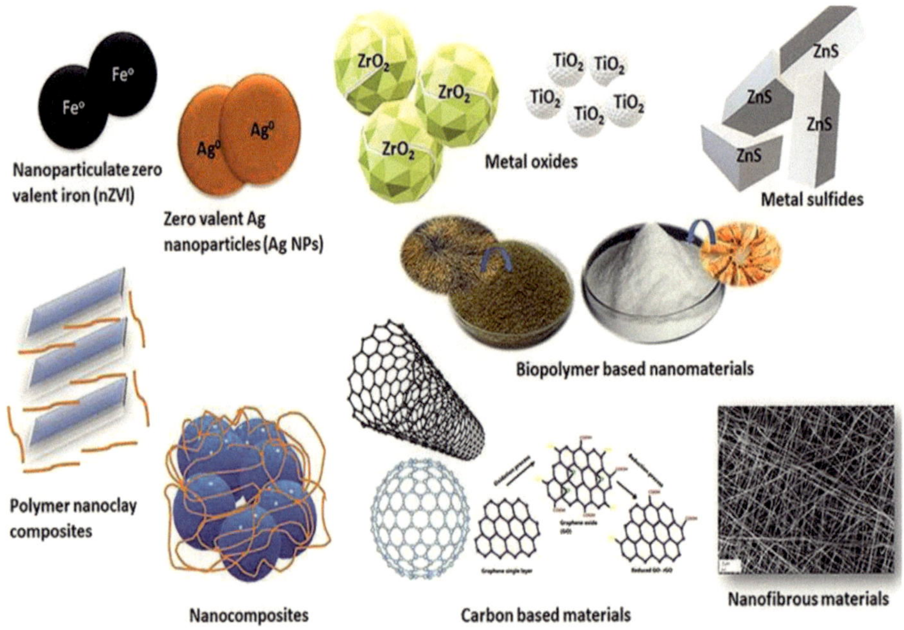

Figure 1.
Different types of nanomaterials widely used for biodegradation process.

biopolymer-based nanomaterials, and nanofibrous materials (see **Figure 1**). These nanomaterials can be synthesized using two different ways; one is the laboratory-mediated synthesis of nanoparticles (ex-situ) [29], and the other one is the in-situ synthesis of nanomaterials inside the microbes [62]. Besides, there could be another lineage of ex-situ synthesized nanomaterials, which are biodegradable in origin and mainly applied in the biomedical field as theragnostic agents [27]. After performing its' definite action including controlled drug delivery, imaging, implantation, tissue engineering) these nanomaterials undergo natural degradation upon the enzymatic attack inside the living cells [27].

However, the selection of the type of nanomaterial relies on the nature of the contaminants and the microorganism that mediates the biodegradation process [12].

4. Applications of biodegradable nanomaterials

Biodegradable nanomaterials or nanoparticles include two major types: nanomaterials directly synthesized from various biopolymers such as polypeptides, polysaccharides and polynucleotides; and metallic nanoparticles, which are colloidal particles encapsulated inside a polymer matrix. The selection of this biopolymer matrix is based on many factors, including the size of the nanoparticles, degree of biocompatibility and biodegradability, surface properties and functionality and the type of application [63]. These biodegradable nanoparticles are typically in the 10–500 nm size range. Widely used methods for the fabrication of biodegradable nanoparticles include emulsification, solvent evaporation, coprecipitation, desolvation, coacervation, electrospray and electrospinning [63]. Over the past few years, many studies have been conducted in various fields on the preparation and applications of biodegradable nanomaterial. However, the applications in food packaging, energy, environmental remediation, and nanomedicine are discussed in this section.

4.1 Food packaging

Packaging plays an imperative role in the food industry. The major function of packaging is protecting food from physical damage while handling, transporting and storage. Packaging materials also maintain the food quality by protecting against air, moisture, insects, light, and dust and prevent contamination from chemical and biological sources. Commonly used packaging materials include plastics, metals, paper and paper boards, glass, and other traditional materials. However, food packaging accounts for 50% of petroleum-based plastics [64]. Upon disposal, plastics remain in the environment taking many years to degrade. The fragments of plastics, also known as microplastics, enters the ecosystems via food chains causing growing environmental and health concerns. Therefore, there is a significant interest in the development of environmentally friendly food packing alternatives. Biodegradable nanoparticles have recently been employed for food packaging applications due to their simple synthesis route, non-toxicity, relative abundance, low cost, and eco-friendly nature. Following are recent food packing applications of biodegradable nanoparticles reported.

Pandey *et al.* prepared the biodegradable meat packaging material using fibrous composite nano-layers (PVA-CH-AgNPs-FCNLs) as an alternative for plastic packaging [65]. PVA-CH-AgNPs-FCNLs were synthesized by electrospinning of a blend of silver nanoparticles (AgNPs) incorporated chitosan (CH) and polyvinyl alcohol (PVA). PVA-CH-AgNPs-FCNLs showed bioactivity against *Escherichia coli* (gram-negative bacteria) and *Listeria monocytogens* (gram-positive bacteria) and extended the meat shelf-life by one week [65]. Ediyilyam and coworkers investigated biodegradable films prepared from silver nanoparticles (AgNPs) incorporated chitosan (CH) and gelatin (GE) polymer blend for food packaging applications [66]. They reported the improved physicochemical and biological functioning of the films upon incorporating the AgNPs. CH–GE–AgNPs films also displayed antimicrobial activity against bacteria and fungi and enhanced the shelf life of carrot pieces wrapped in them over ten days [66].

Kumar *et al.* developed low-cost biodegradable nanocomposite hybrid films containing chitosan, gelatin, and zinc oxide nanoparticles (ZnO NPs) [67]. ZnO NPs reinforced hybrid nanocomposites exhibited enhanced thermal stability, elongation-at-break (EAB), and compactness properties with antimicrobial activity against *Escherichia coli* (gram-negative) bacteria. The authors claimed that these hybrid nanocomposite films have the potential to be developed as biodegradable postharvest packaging of fresh fruits and vegetables [67]. Saral Sarojini and coworkers fabricated the biodegradable food packaging films from Mahua oil-based polyurethane (PU) and chitosan (CS), incorporated with zinc oxide nanoparticles [68]. They reported enhanced hydrophobicity of the film by about 63%, high UV-screening ability, high transparency, high degree of biodegradation of 86%, and antimicrobial resistance for the ZnO incorporated PU/CS films. ZnO-reinforced PU/CS films also extended their shelf life up to nine days upon wrapped with carrot pieces [68].

Starch-based (St) nanocomposite films prepared by incorporating silver (Ag), copper oxide (CuO) and zinc oxide (ZnO) nanoparticles (NPs) were tested for physicomechanical and antimicrobial properties by Peighambardoust *et al.* [69]. Ag/ZnO/CuO NPs incorporated starch-based films showed better antimicrobial and mechanical properties due to the synergistic effect. The authors reported the potential use of these starch-based nanocomposites as food packaging materials [69]. Colored biodegradable dye (methylene blue)-clay (montmorillonite)-nanopigment (DCNP)-polylactic acid (PLA) nanocomposite films were prepared and tested for various functional properties by Mahmoodi *et al.* [70]. The PLA-DCNP films exhibited high mechanical strength, barrier properties, blocking effect against destructive radiation, biodegradability properties, and potential food packaging applications [70].

4.2 Energy

Recent advancement in biodegradable nanomaterials has led to the development of energy-efficient devices including ignition engines, solar cells, supercapacitors, and rechargeable batteries. Current applications of biodegradable polymers in energy-efficient devices are discussed below.

Ettefaghi *et al*. investigated the biodegradable carbon-based quantum dots as alternatives for metal and metal oxide fuel additives [71]. The use of a combination of diesel-biodiesel-water-biodegradable carbon nanoparticles showed an increase in engine torque and power and a decrease in brake-specific fuel consumption. The bio-nano emulsion fuels also reduced the emission of nitrogen oxide and unburned hydrocarbons [71]. Abdalkarim and coworkers prepared biodegradable dipole responsive magnetic/solar-driven PCF composites reinforced with magnetic cellulose nanocrystals hybrids (MCNC) [72]. The PCF/MCNC composites showed enhanced latent heat phase change enthalpies, thermal stability, and increased magnetic/solar-driven thermal energy storage efficiencies. The authors also reported the potential of PCF/MCNC composites for drying and preservation of agriculture products, including fruits [72].

Shaheen *et al*. synthesized nanocomposites of molybdenum and zinc oxide [MoO_3@ZnO] via chemosynthetic and biomimetic routes and showed a direct bandgap of 4.5 and 3.5 eV, respectively [73]. They demonstrated the semi-conducting and capacitive properties of the biogenic nanocomposite using electrochemical studies included cyclic voltammetry (CV) and electrochemical impedance spectroscopy (EIS) suitable for applications in solar cells [73]. Aziz and coworkers fabricated a methylcellulose: dextran (MC: Dex) polymer blend-based electrolyte system with ammonium iodide (NH_4I) salt for electrical double-layer capacitor (EDLC) application [74]. The electrolyte system was ionic in nature and showed the maximum ionic conductivity as 1.12×10^{-3} S/cm with an electrochemical stability window of 1.27 V. The EDLC device offered an initial specific capacitance of 79 F/g, an energy density of 8.81 Wh/kg and power density of 1111.0 W/kg at a current density of 0.2 mA/cm^2 [74]. Youssef *et al*. prepared the conducting bionanocomposite hydrogels using chitosan (CS)/hydroxyl ethylcellulose (HEC)/polyaniline (PAni) loaded with graphene oxide (GO) doped by silver (Ag) nanoparticles as a semiconductor material for electrical storage devices [75]. CS/HEC/PAni/GO@Ag bionanocomposite hydrogels exhibited improved swelling percentage, capacitance, permittivity, antibacterial activities, and biodegradation properties. The bionanocomposite displayed the highest dc-conductivity of 8.53×10^{-2} S/cm [75].

4.3 Environmental remediation

The rapid industrialization and urbanization across the globe have significantly impacted the terrestrial and aquatic environments by releasing harmful industrial effluents, including colored organic dyes, heavy metals, polycyclic aromatic hydrocarbons (PHAs), chlorinated organics and perfluorosurfactants [76]. The release of these toxic substances imposes serious health concerns on all living beings. Biodegradable nanomaterials have recently been considered highly efficient agents for environmental remediation due to their high chemical reactivity, surface properties, catalytic activity, easy synthesis and fabrication, and environmental benignity. This section covers the applications of biodegradable nanoparticles in environmental remediation.

Rajeswari *et al.* reported the synthesis of biodegradable mixed matrix membranes (MMMs) using aluminum oxide (Al_2O_3) and nano zerovalent iron (nZVI) nanoparticles blended cellulose acetate-polysulfone (CA-PSF) for the removal of methylene blue (MB) dye and Cu (II) metal ions [77]. The authors reported the rejection values 91 and 94% for MB dye and for Cu (II) the rejection values of 84 and 88% using CA-PSF/Al_2O_3 and CA-PSF/nZVI membranes [77]. Pandey and coworkers fabricated slow-release microencapsulated zerovalent iron nanoparticles (ZVINPs) in polylactic acid (PLA)-based microparticles for in-situ groundwater remediation of hydrophilic (methyl orange dye) and hydrophobic (trichloroethylene) water contaminants by electrospraying technique [78]. The authors reported that approximately 8 wt% ZVINPs were slowly released from the biodegradable microparticles after 60 h and 32 h incubation to fully remediate methyl orange (25 mg/L) and trichloroethylene (0.2 vol%) from water, respectively [78]. The photocatalytic properties of Mg-doped ZnO nano-semiconductors for the decontamination of non-treated laundry wastewater were investigated by Oliveira *et al.* [79]. The authors showed the degrading of approximately 53% of pollutants after 240 min of UV–vis irradiation, reducing 31% in total organic carbon (TOC). The treated laundry wastewater promoted the growth of cucumber seeds and tomato roots [79].

Electrospun and thermally cross-linked poly(vinyl alcohol) (PVA) and konjac glucomannan (KGM)-based biodegradable nanofiber membranes loaded with zinc oxide (ZnO) nanoparticles were prepared by Lv *et al.* [80]. ZnO@PVA/KGM membranes exhibited photocatalytic decolorization of methyl orange dye (20 mg L − 1) with a removal efficiency of over 98% under 120 min of solar irradiation. They also investigated efficient air-filtration and antibacterial performances for the ZnO@PVA/KGM membranes [80]. **Figure 2(A)–(D)** shows the schematic presentation of the preparation of the ZnO@PVA/KGM membranes by electrospinning, air filtration process, Photocatalytic degradation, and (D) antibacterial activity of the membranes [80]. Barbosa and coworkers prepared the biodegradable poly(butylene adipate-co-terephthalate) membranes functionalized with cellulose nanoparticles

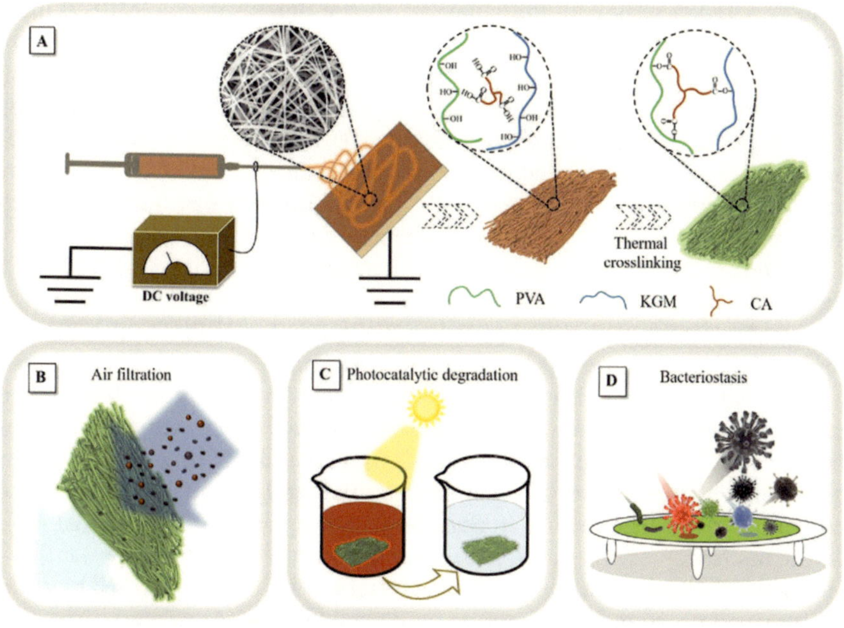

Figure 2.
Schematic representation of the (A) preparation, (B) air filtration process, (C) photocatalytic degradation, and (D) antibacterial activity of the biodegradable ZnO@PVA/KGM nanofiber membranes [80].

(CNS) via phase inversion technique for the removal of chromium (Cr) ions from contaminated drinking water [81]. The CNS functionalized membranes that were subjected to phosphorylation (CNS-P) displayed the removal of 93% and 88% of Cr(VI) and Cr(III), respectively, showing their application in domestic houses and water treatment stations [81].

4.4 Nanomedicine

Biodegradable nanomaterials have been recently investigated in nanomedicine due to their controlled drug release and targeted drug delivery, giving enhanced therapeutic effects and reduced side effects. Biodegradable nanomaterials impose less cytotoxicity on cells. Due to modifying and functionalizing ability, the biodegradable nanoparticles can also improve drug stability and solubility. The vital applications of biodegradable nanoparticles in nanomedicine include drug delivery, cancer therapy, imaging, and antimicrobial activity.

Far *et al.* synthesized biodegradable poly(lactic-co-glycolic acid) (PLGA) nanoparticles (NPs) loaded with mometasone furoate (MF) using the nanoprecipitation method [82]. They reported the controlled release of MF using PLGA NPs over 7 days in vitro with an initial burst release, demonstrating therapeutic potential in nasal delivery applications [82]. Gai and coworkers developed a drug delivery system (DDS) for rheumatoid arthritis (RA) therapy using benzoylaconitine (BAC) encapsulated methoxy-poly (ethylene glycol)-poly(lactide-co-glycolide) (mPEG-PLGA) nanoparticles (NPs) via hydrophobic interaction [83]. The mPEG-PLGA NPs (NP/BAC) system exhibited low cytotoxicity and good biocompatibility for lipopolysaccharide (LPS)-activated macrophages and efficient in vivo anti-inflammatory effect with the high ear (69.8%) and paw (87.1%) swelling suppressing rate. The authors mentioned the possible application of biodegradable NP/BAC system in anti-inflammation and RA therapy as an effective DDS [83].

Qin *et al.* reported the synthesis of tumor-sensitive biodegradable nanoparticles using fluorescent zeolitic imidazolate framework-8 nanoparticles loaded with doxorubicin (FZIF-8/DOX) as the core and a molecularly imprinted polymer (MIP) as the shell (FZIF-8/DOX-MIPs) [84]. FZIF-8/DOX-MIPs showed an inhibitory effect on the growth of MCF-7 tumors and served as a diagnostic agent giving stronger red fluorescence at the tumor sites [84]. A pH-sensitive biodegradable garcinol (GAR)-loaded poly (lactic–co–glycolic acid) (PLGA) coated with Eudragit® S100 (ES100) (GAR-PLGA-ES100 nanoparticles (NPs)) was designed for reducing inflammation caused by pro-inflammatory cytokines in the gastrointestinal tract [85], see **Figure 3**. The authors reported the site-directed release of the drug specifically from NPs at the colonic pH of 7.4, reducing the activation of inflammation that leads to inflammatory bowel disease (IBD) [85].

Han *et al.* developed hypericin encapsulated methoxy poly(ethylene glycol)-b-poly(ε-caprolactone) (PEG-PCL) biodegradable nanoparticles (Hyp-NP) with necrosis affinity and fluorescence imaging in vitro and in vivo [86]. The authors showed the cellular internalization with intracellular cytoplasmic localization and preserved fluorescence and necrosis affinity for Hyp-NPs, suggesting their potential applications in tumor imaging and therapy [86]. Fernández-Gutiérrez and coworkers reported the fabrication of a biocomposite polymeric system for the antibacterial coating of polypropylene mesh materials for hernia repair [87]. **Figure 4(a)–(d)** shows the microscopic and scanning electron microscopic (SEM) images of the meshes with different coatings. The antibacterial coating was performed by a film of chitosan containing poly(D,L-lactide-co-glycolide) (PLGA) nanoparticles loaded with antibiotic (rifampicin) or an antiseptic (chlorhexidine). Both biocomposite coatings exhibited antibacterial activity and cell compatibility, offering a potential strategy to protect meshes from bacterial adhesion following implantation [87].

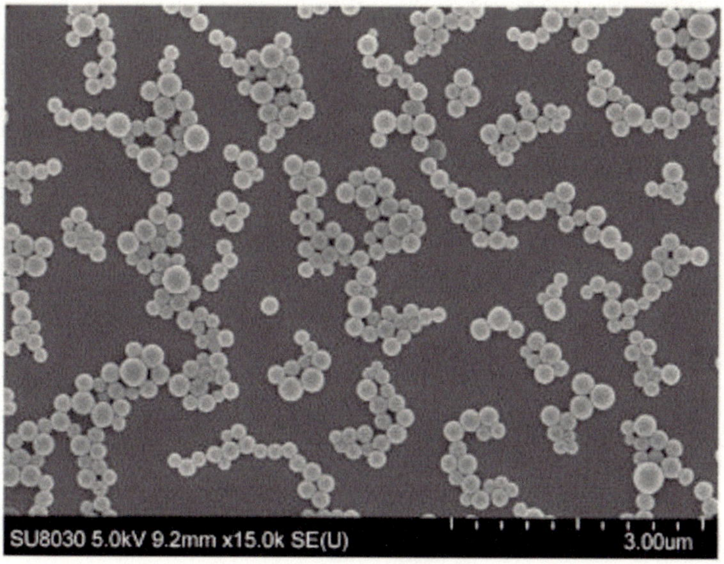

Figure 3.
SEM image of the biodegradable GAR-PLGA-ES100 NPs (At scale 3.00 μm) [85].

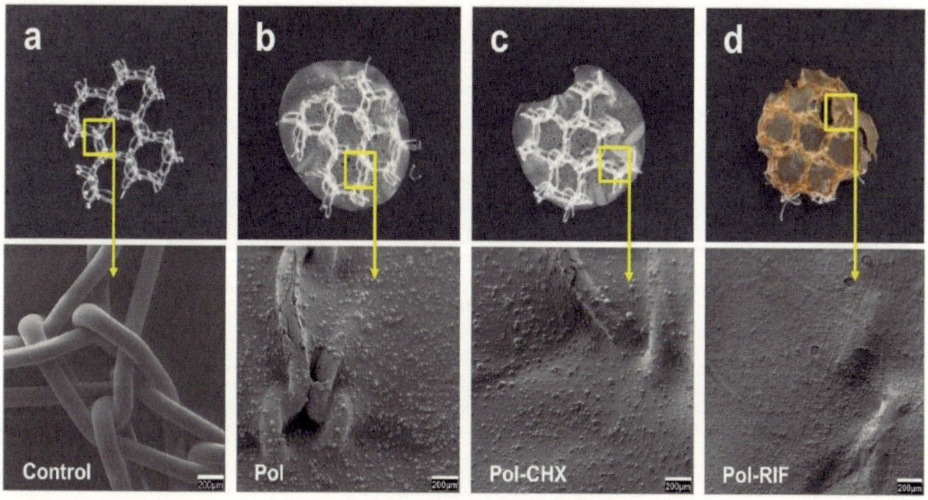

Figure 4.
Macroscopic pictures and SEM micrographs of different meshes (a) nude control (chitosan only), (b) coated with the unloaded biocomposite (chitosan-PLGA), (c) coated with chlorhexidine (CHX)-loaded biocomposite (chitosan-PLGA-CHX), and (d) coated with the rifampicin (RIF)-loaded biocomposite (chitosan-PLG-RIF) [87].

5. Conclusions

Biodegradation is the naturally occurring degradation of complex substances into simple eco-friendly products by the action of microorganisms and plays an imperative role in sustainable development. One of the significant challenges of biodegradation includes the incomplete breakdown of materials due to the complexity of the materials arising from structure, molecular weight, crosslinking, shape, texture, and surface properties. Other setbacks include the screening and identifying of suitable microbes, nutrients, and environmental conditions.

Nanotechnology integrated biodegradation process has recently become an eco-friendly and cost-effective method of diminishing environmental pollutants due to the synergetic effects. The factors including the type of nanomaterials, type of the microorganism, and culture medium directly affect the involvement of nanomaterials in biodegradation. Common types of nanomaterials utilized in biodegradation processes include zero valent metals, oxides, sulfides, nanocomposites, nanoclay, carbon materials, biopolymers, and nanofibers. Biodegradable nanomaterials have been widely applied in food packaging, energy, environmental remediation and nanomedicine.

Acknowledgements

The authors are grateful for the financial and scientific support rendered by the Faculty of Technology, University of Sri Jayewardenepura.

Conflict of interest

The authors declare no conflict of interest.

Author details

Danushika C. Manatunga[1*], Rohan S. Dassanayake[1] and Renuka N. Liyanage[2]

1 Department of Biosystems Technology, Faculty of Technology, University of Sri Jayewardenepura, Nugegoda, Sri Lanka

2 Department of Materials and Mechanical Technology, Faculty of Technology, University of Sri Jayewardenepura, Nugegoda, Sri Lanka

*Address all correspondence to: danushi@sjp.ac.lk; danushi@15gmail.com

References

[1] Mensah J. Sustainable development: Meaning, history, principles, pillars, and implications for human action: Literature review. Cogent Soc Sci [Internet]. 2019; 5(1). Available from: https://doi.org/10.1080/23311886.2019.1653531

[2] Tahri N, Bahafid W, Sayel H, El Ghachtouli N. Biodegradation: Involved Microorganisms and Genetically Engineered Microorganisms. In: Biodegradation - Life of Science [Internet]. 2013. p. 289-320. Available from: https://www.intechopen.com/books/biodegradation-life-of-science/biodegradation-involved-microorganisms-and-genetically-engineered-microorganisms

[3] Verma R, Vinoda KS, Papireddy M, Gowda ANS. Toxic Pollutants from Plastic Waste- A Review. Procedia Environ Sci [Internet]. 2016;35:701-8. Available from: http://dx.doi.org/10.1016/j.proenv.2016.07.069

[4] Huang Y, Xiao L, Li F, Xiao M, Lin D, Long X, et al. Microbial degradation of pesticide residues and an emphasis on the degradation of cypermethrin and 3-phenoxy benzoic acid: A review. Molecules. 2018;23(9).

[5] Antipova T V., Zhelifonova VP, Zaitsev K V., Nedorezova PM, Aladyshev AM, Klyamkina AN, et al. Biodegradation of Poly-ε-caprolactones and Poly-l-lactides by Fungi. J Polym Environ. 2018;26(12):4350–9.

[6] Jawed K, Yazdani SS, Koffas MA. Advances in the development and application of microbial consortia for metabolic engineering. Metab Eng Commun [Internet]. 2019;9(November 2018):e00095. Available from: https://doi.org/10.1016/j.mec.2019.e00095

[7] Sahoo NK, Pakshirajan K, Ghosh PK, Ghosh A. Biodegradation of 4-chlorophenol by Arthrobacter chlorophenolicus A6: Effect of culture conditions and degradation kinetics. Biodegradation. 2011;22(2):275-86.

[8] Brzeszcz J, Kaszycki P. Aerobic bacteria degrading both n-alkanes and aromatic hydrocarbons: an undervalued strategy for metabolic diversity and flexibility. Biodegradation [Internet]. 2018;29(4):359-407. Available from: https://doi.org/10.1007/s10532-018-9837-x

[9] Singh B, Singh K. Microbial degradation of herbicides. Crit Rev Microbiol. 2016;42(2):245-61.

[10] Gogada R, Singh SS, Lunavat SK, Pamarthi MM, Rodrigue A, Vadivelu B, et al. Engineered Deinococcus radiodurans R1 with NiCoT genes for bioremoval of trace cobalt from spent decontamination solutions of nuclear power reactors. Appl Microbiol Biotechnol. 2015;99(21): 9203-13.

[11] N.Karak. Biodegradable polymers. Vegetable Oil-Based Polymers. 2012. 31-53 p.

[12] Vázquez-Núñez E, Molina-Guerrero CE, Peña-Castro JM, Fernández-Luqueño F, de la Rosa-Álvarez MG. Use of nanotechnology for the bioremediation of contaminants: A review. Processes. 2020;8(7):1-17.

[13] Mandeep, Shukla P. Microbial Nanotechnology for Bioremediation of Industrial Wastewater. Front Microbiol. 2020;11(November):1-8.

[14] Prasad R, Bhattacharyya A, Nguyen QD. Nanotechnology in sustainable agriculture: Recent developments, challenges, and perspectives. Front Microbiol. 2017;8(JUN):1-13.

[15] Fakruddin M, Hossain Z, Afroz H. Prospects and applications of nanobiotechnology: A medical perspective. J Nanobiotechnology. 2012;10:1-8.

[16] Mathew J, Joy J, George SC. Potential applications of nanotechnology in transportation: A review. J King Saud Univ - Sci [Internet]. 2019;31(4):586-94. Available from: https://doi.org/10.1016/j.jksus.2018.03.015

[17] Subramanian V, Lee T. Nanotechnology-based flexible electronics. Nanotechnology. 2012;23(34):12-4.

[18] Nile SH, Baskar V, Selvaraj D, Nile A, Xiao J, Kai G. Nanotechnologies in Food Science: Applications, Recent Trends, and Future Perspectives [Internet]. Vol. 12, Nano-Micro Letters. Springer Singapore; 2020. 1-34 p. Available from: https://doi.org/10.1007/s40820-020-0383-9

[19] S. Dassanayake R, Acharya S, Abidi N. Biopolymer-Based Materials from Polysaccharides: Properties, Processing, Characterization and Sorption Applications. In: Advanced Sorption Process Applications [Internet]. 2019. p. 1-24. Available from: https://www.intechopen.com/books/advanced-sorption-process-applications/biopolymer-based-materials-from-polysaccharides-properties-processing-characterization-and-sorption-

[20] Fernando MS, De Silva RM, De Silva KMN. Synthesis, characterization, and application of nano hydroxyapatite and nanocomposite of hydroxyapatite with granular activated carbon for the removal of Pb 2+ from aqueous solutions. Appl Surf Sci [Internet]. 2015;351(January 2020):95-103. Available from: http://dx.doi.org/10.1016/j.apsusc.2015.05.092

[21] A. K. D. Veromee Kalpana Wimalasiri, M. Shanika Fernando, Karolina Dziemidowicz, Gareth R. Williams, K. Rasika Koswattage, D. P. Dissanayake, K. M. Nalin de Silva and RM de S. Structure-Activity Relationship of Lanthanide-Incorpora.pdf. ACS Omega. 2021;6(21):13527-43.

[22] Wimalasiri AKDVK, Fernando MS, Williams GR, Dissanayake DP, de Silva KMN, de Silva RM. Microwave assisted accelerated fluoride adsorption by porous nanohydroxyapatite. Mater Chem Phys [Internet]. 2021;257(July 2019):123712. Available from: https://doi.org/10.1016/j.matchemphys.2020.123712

[23] Manatunga DC, de Silva RM, Nalin De Silva KM, de Silva N, Premalal EVA. Metal and polymer-mediated synthesis of porous crystalline hydroxyapatite nanocomposites for environmental remediation. R Soc Open Sci. 2018;5(1).

[24] Fernando MS, Wimalasiri AKDVK, Ratnayake SP, Jayasinghe JMARB, William GR, Dissanayake DP, et al. Improved nanocomposite of montmorillonite and hydroxyapatite for defluoridation of water. RSC Adv. 2019;9(61):35588-98.

[25] Fernando MS, Wimalasiri AKDVK, Dziemidowicz K, Williams GR, Koswattage KR, Dissanayake DP, et al. Biopolymer-Based Nanohydroxyapatite Composites for the Removal of Fluoride, Lead, Cadmium, and Arsenic from Water. ACS Omega. 2021;6(12):8517-30.

[26] Ionescu AM. Nanotechnology and Global Security. Connect Q J. 2016;15(2):31-47.

[27] Su S, Kang PM. Systemic review of biodegradable nanomaterials in nanomedicine. Nanomaterials. 2020;10(4).

[28] Dhillon GS, Kaur S, Verma M, Brar SK. Biopolymer-based

nanomaterials: Potential applications in bioremediation of contaminated wastewaters and soils [Internet]. 1st ed. Vol. 59, Comprehensive Analytical Chemistry. Elsevier B.V.; 2012. 91-129 p. Available from: http://dx.doi.org/10.1016/B978-0-444-56328-6.00003-7

[29] Pathak VM, Navneet Kumar. Implications of SiO 2 nanoparticles for in vitro biodegradation of low-density polyethylene with potential isolates of Bacillus, Pseudomonas, and their synergistic effect on Vigna mungo growth. Energy, Ecol Environ. 2017;2(6):418-27.

[30] Bhatia M, Girdhar A, Chandrakar B, Tiwari A. Implicating nanoparticles as potential biodegradation enhancers: A review. J Nanomedicine Nanotechnol. 2013;4(4).

[31] Rgpv B, Rgpv B, Rgpv B. Ldpe-Biodegradation Using Microbial Consortium By the Incorporation of Cobalt Ferrite Nanoparticle As the Enhancer for Biodegradation. Int J Adv Eng Res Dev. 2017;4(06):794-800.

[32] Misson M, Zhang H, Jin B. Nanobiocatalyst advancements and bioprocessing applications. J R Soc Interface. 2015;12(102):1-20.

[33] Koedrith P, Thasiphu T, Weon J Il, Boonprasert R, Tuitemwong K, Tuitemwong P. Recent trends in rapid environmental monitoring of pathogens and toxicants: Potential of nanoparticle-based biosensor and applications. Sci World J. 2015;2015:1-12.

[34] Li X, Xu H, Chen ZS, Chen G. Biosynthesis of nanoparticles by microorganisms and their applications. J Nanomater. 2011;2011:1-16.

[35] Das RK, Pachapur VL, Lonappan L, Naghdi M, Pulicharla R, Maiti S, et al. Biological synthesis of metallic nanoparticles: plants, animals and microbial aspects. Nanotechnol Environ Eng. 2017;2(1):1-21.

[36] Albanese A, Tang PS, Chan WCW. The effect of nanoparticle size, shape, and surface chemistry on biological systems. Annu Rev Biomed Eng. 2012;14:1-16.

[37] Khan I, Saeed K, Khan I. Nanoparticles: Properties, applications and toxicities. Arab J Chem [Internet]. 2019;12(7):908-31. Available from: https://doi.org/10.1016/j.arabjc.2017.05.011

[38] Singh M, Mitra CK, Morve RK. Rizwan, 2014. J Nanoparticles. 2014;2014(2):740-57.

[39] Shrivas K, Shankar R, Dewangan K. Gold nanoparticles as a localized surface plasmon resonance based chemical sensor for on-site colorimetric detection of arsenic in water samples. Sensors Actuators, B Chem [Internet]. 2015;220:1376-83. Available from: http://dx.doi.org/10.1016/j.snb.2015.07.058

[40] Alzahrani E. Colorimetric Detection Based on Localized Surface Plasmon Resonance Optical Characteristics for Sensing of Mercury Using Green-Synthesized Silver Nanoparticles. J Anal Methods Chem. 2020;2020:1-14.

[41] M. T. Amin, A. A. Alazba and UMA. A Review of Removal of Pollutants from Water/Wastewater Using Different Types of Nanomaterials. Adv Mater Sci Eng. 2014;2014:1-24.

[42] Dong H, Li L, Lu Y, Cheng Y, Wang Y, Ning Q, et al. Integration of nanoscale zero-valent iron and functional anaerobic bacteria for groundwater remediation: A review. Environ Int [Internet]. 2019;124(December 2018):265-77. Available from: https://doi.org/10.1016/j.envint.2019.01.030

[43] Debnath B, Majumdar M, Bhowmik M, Bhowmik KL, Debnath A,

Roy DN. The effective adsorption of tetracycline onto zirconia nanoparticles synthesized by novel microbial green technology. J Environ Manage [Internet]. 2020;261(September 2019):1-13. Available from: https://doi.org/10.1016/j.jenvman.2020.110235

[44] Mohanraj R, Gnanamangai BM, Poornima S, Oviyaa V, Ramesh K, Vijayalakshmi G, et al. Decolourisation efficiency of immobilized silica nanoparticles synthesized by actinomycetes. Mater Today Proc [Internet]. 2020;(In press). Available from: https://doi.org/10.1016/j.matpr.2020.04.139

[45] Kapri A, Zaidi MGH, Satlewal A, Goel R. SPION-accelerated biodegradation of low-density polyethylene by indigenous microbial consortium. Int Biodeterior Biodegrad [Internet]. 2010;64(3):238-44. Available from: http://dx.doi.org/10.1016/j.ibiod.2010.02.002

[46] Wang X, Pu J, Liu Y, Ba F, Cui M, Li K, et al. Immobilization of functional nano-objects in living engineered bacterial biofilms for catalytic applications. Natl Sci Rev. 2019;6(5):929-43.

[47] Biswas B, Sarkar B, Rusmin R, Naidu R. Bioremediation of PAHs and VOCs: Advances in clay mineral-microbial interaction. Environ Int [Internet]. 2015;85:168-81. Available from: http://dx.doi.org/10.1016/j.envint.2015.09.017

[48] Liu J, Morales-Narváez E, Vicent T, Merkoçi A, Zhong GH. Microorganism-decorated nanocellulose for efficient diuron removal. Chem Eng J [Internet]. 2018;354:1083-91. Available from: https://doi.org/10.1016/j.cej.2018.08.035

[49] He S, Zhong L, Duan J, Feng Y, Yang B, Yang L. Bioremediation of wastewater by iron Oxide-Biochar nanocomposites loaded with photosynthetic bacteria. Front Microbiol. 2017;8(MAY):1-10.

[50] Sah A, Kapri A, Zaidi MGH, Negi H, Goel R. Implications of fullerene-60 upon in-vitro LDPE biodegradation. J Microbiol Biotechnol. 2010;20(5):908-16.

[51] Fosso-Kankeu E, Mulaba-Bafubiandi AF, Mishra AK. Prospects for Immobilization of Microbial Sorbents on Carbon Nanotubes for Biosorption: Bioremediation of Heavy Metals Polluted Water. Appl Nanotechnol Water Res. 2014;9781118496(January 2018):37-61.

[52] Zommere Ž, Nikolajeva V. Immobilization of bacterial association in alginate beads for bioremediation of oil-contaminated lands. Environ Exp Biol. 2017;15(January):105-11.

[53] Garcia ACFS, Araújo BR, Birolli WG, Marques CG, Diniz LEC, Barbosa AM, et al. Fluoranthene Biodegradation by Serratia sp. AC-11 Immobilized into Chitosan Beads. Appl Biochem Biotechnol. 2019;188(4):1168-84.

[54] Sarioglu OF, Keskin NOS, Celebioglu A, Tekinay T, Uyar T. Bacteria encapsulated electrospun nanofibrous webs for remediation of methylene blue dye in water. Colloids Surfaces B Biointerfaces [Internet]. 2017;152:245-51. Available from: http://dx.doi.org/10.1016/j.colsurfb.2017.01.034

[55] San Keskin NO, Celebioglu A, Sarioglu OF, Uyar T, Tekinay T. Encapsulation of living bacteria in electrospun cyclodextrin ultrathin fibers for bioremediation of heavy metals and reactive dye from wastewater. Colloids Surfaces B Biointerfaces [Internet]. 2018;161:169-76. Available from: http://dx.doi.org/10.1016/j.colsurfb.2017.10.047

[56] Cai Y, Xu Z, Shuai Q, Zhu F, Xu J, Gao X, et al. Tumor-targeting peptide functionalized PEG-PLA micelles for efficient drug delivery. Biomater Sci. 2020;8(8):2274-82.

[57] Wang M, Zhan J, Xu L, Wang Y, Lu D, Li Z, et al. Synthesis and characterization of PLGA-PEG-PLGA based thermosensitive polyurethane micelles for potential drug delivery. J Biomater Sci Polym Ed. 2020;32(5):613-34.

[58] Łukasiewicz S, Mikołajczyk A, Błasiak E, Fic E, Dziedzicka-Wasylewska M. Polycaprolactone Nanoparticles as Promising Candidates for Nanocarriers in Novel Nanomedicines. Pharmaceutics. 2021;13(2):191.

[59] Unnati Garg, Swati Chauhan , Upendra Nagaich NJ. Current Advances in Chitosan Nanoparticles Based Drug Delivery and Targeting. Adv Pharm Bull [Internet]. 2019;9(2):195-204. Available from: http://dx.doi.org/10.15171/jcvtr.2015.24

[60] Mandal AK. Dendrimers in targeted drug delivery applications: a review of diseases and cancer. Int J Polym Mater Polym Biomater [Internet]. 2021;70(4):287-97. Available from: https://doi.org/10.1080/00914037.2020.1713780

[61] Beltrán-Gracia E, López-Camacho A, Higuera-Ciapara I, Velázquez-Fernández JB, Vallejo-Cardona AA. Nanomedicine review: Clinical developments in liposomal applications [Internet]. Vol. 10, Cancer Nanotechnology. Springer Vienna; 2019. 1-40 p. Available from: https://doi.org/10.1186/s12645-019-0055-y

[62] Patel A, Enman J, Gulkova A, Guntoro PI, Dutkiewicz A, Ghorbani Y, et al. Integrating biometallurgical recovery of metals with biogenic synthesis of nanoparticles. Chemosphere. 2021;263:1-23.

[63] Mahapatro A, Singh DK. Biodegradable nanoparticles are excellent vehicle for site directed in-vivo delivery of drugs and vaccines. J Nanobiotechnology [Internet]. 2011;9(1):55. Available from: http://www.jnanobiotechnology.com/content/9/1/55

[64] Jacob J, Lawal U, Thomas S, Valapa RB. Biobased polymer composite from poly(lactic acid): processing, fabrication, and characterization for food packaging [Internet]. Processing and Development of Polysaccharide-Based Biopolymers for Packaging Applications. Elsevier Inc.; 2020. 97-115 p. Available from: http://dx.doi.org/10.1016/B978-0-12-818795-1.00004-6

[65] Pandey VK, Upadhyay SN, Niranjan K, Mishra PK. Antimicrobial biodegradable chitosan-based composite Nano-layers for food packaging. Int J Biol Macromol [Internet]. 2020;157:212-9. Available from: https://doi.org/10.1016/j.ijbiomac.2020.04.149

[66] Sreelekha Ediyilyam, Bini George, Sarojini Sharath Shankar, Thomas Thuruthiyil Dennis Stanisław Wacławek, Miroslav ˇCerník and VVTP. Chitosan / Gelatin / Silver Nanoparticles Composites Films for Biodegradable Food Packaging Applications. Polymers (Basel). 2021;13(1680):1-18.

[67] Kumar S, Mudai A, Roy B, Basumatary IB, Mukherjee A, Dutta J. Biodegradable hybrid nanocomposite of chitosan/gelatin and green synthesized zinc oxide nanoparticles for food packaging. Foods. 2020;9(1143):1-13.

[68] K. SS, Indumathi MP, Rajarajeswari GR. Mahua oil-based polyurethane/chitosan/nano ZnO composite films for biodegradable food packaging applications. Int J Biol

Macromol [Internet]. 2019;124:163-74. Available from: https://doi.org/10.1016/j.ijbiomac.2018.11.195

[69] Peighambardoust SJ, Peighambardoust SH, Mohammadzadeh Pournasir N, Pakdel P. Properties of active starch-based films incorporating a combination of Ag, ZnO and CuO nanoparticles for potential use in food packaging applications. Food Packag Shelf Life [Internet]. 2019;22(October):100420. Available from: https://doi.org/10.1016/j.fpsl.2019.100420

[70] Mahmoodi A, Ghodrati S, Khorasani M. High-Strength, Low-Permeable, and Light-Protective Nanocomposite Films Based on a Hybrid Nanopigment and Biodegradable PLA for Food Packaging Applications. ACS Omega. 2019;4(12):14947-54.

[71] Ettefaghi E, Ghobadian B, Rashidi A, Najafi G, Khoshtaghaza MH, Rashtchi M, et al. A novel bio-nano emulsion fuel based on biodegradable nanoparticles to improve diesel engines performance and reduce exhaust emissions. Renew Energy [Internet]. 2018;125:64-72. Available from: https://doi.org/10.1016/j.renene.2018.01.086

[72] Abdalkarim SYH, Ouyang Z, Yu HY, Li Y, Wang C, Asad RAM, et al. Magnetic cellulose nanocrystals hybrids reinforced phase change fiber composites with highly thermal energy storage efficiencies. Carbohydr Polym [Internet]. 2021;254(December 2020):117481. Available from: https://doi.org/10.1016/j.carbpol.2020.117481

[73] Shaheen I, Ahmad KS, Jaffri SB, Ali D. Biomimetic [MoO3@ZnO] semiconducting nanocomposites: Chemo-proportional fabrication, characterization and energy storage potential exploration. Renew Energy [Internet]. 2021;167:568-79. Available from: https://doi.org/10.1016/j.renene.2020.11.115

[74] Aziz SB, Brza MA, Mishra K, Hamsan MH, Karim WO, Abdullah RM, et al. Fabrication of high performance energy storage EDLC device from proton conducting methylcellulose: Dextran polymer blend electrolytes. J Mater Res Technol [Internet]. 2020;9(2):1137-50. Available from: https://doi.org/10.1016/j.jmrt.2019.11.042

[75] Youssef AM, Hasanin MS, El-Aziz MEA, Turky GM. Conducting chitosan/hydroxylethyl cellulose/polyaniline bionanocomposites hydrogel based on graphene oxide doped with Ag-NPs. Int J Biol Macromol [Internet]. 2021;167:1435-44. Available from: https://doi.org/10.1016/j.ijbiomac.2020.11.097

[76] Kalita E, Baruah J. Environmental remediation [Internet]. Colloidal Metal Oxide Nanoparticles. Elsevier Inc.; 2020. 525-576 p. Available from: http://dx.doi.org/10.1016/B978-0-12-813357-6.00014-0

[77] Rajeswari A, Jackcina Stobel Christy E, Ida Celine Mary G, Jayaraj K, Pius A. Cellulose acetate based biopolymeric mixed matrix membranes with various nanoparticles for environmental remediation-A comparative study. J Environ Chem Eng [Internet]. 2019;7(4):103278. Available from: https://doi.org/10.1016/j.jece.2019.103278

[78] Pandey K, Saha S. Microencapsulated Zero Valent Iron NanoParticles in Polylactic acid matrix for in situ remediation of contaminated water. J Environ Chem Eng [Internet]. 2020;8(4):103909. Available from: https://doi.org/10.1016/j.jece.2020.103909

[79] Oliveira AG, Andrade J de L, Montanha MC, Ogawa CYL, de Souza Freitas TKF, Moraes JCG, et al. Wastewater treatment using Mg-doped ZnO nano-semiconductors: A study of

their potential use in environmental remediation. J Photochem Photobiol A Chem. 2021 Feb 15;407:113078.

[80] Lv D, Wang R, Tang G, Mou Z, Lei J, Han J, et al. Ecofriendly Electrospun Membranes Loaded with Visible-Light-Responding Nanoparticles for Multifunctional Usages: Highly Efficient Air Filtration, Dye Scavenging, and Bactericidal Activity. ACS Appl Mater Interfaces. 2019;11(13):12880-9.

[81] Barbosa RFS, Souza AG, Maltez HF, Rosa DS. Chromium removal from contaminated wastewaters using biodegradable membranes containing cellulose nanostructures. Chem Eng J [Internet]. 2020;395(January):125055. Available from: https://doi.org/10.1016/j.cej.2020.125055

[82] Far J, Abdel-Haq M, Gruber M, Abu Ammar A. Developing Biodegradable Nanoparticles Loaded with Mometasone Furoate for Potential Nasal Drug Delivery. ACS Omega. 2020;5(13):7432-9.

[83] Gai W, Hao X, Zhao J, Wang L, Liu J, Jiang H, et al. Delivery of benzoylaconitine using biodegradable nanoparticles to suppress inflammation via regulating NF-κB signaling. Colloids Surfaces B Biointerfaces [Internet]. 2020;191(March):110980. Available from: https://doi.org/10.1016/j.colsurfb.2020.110980

[84] Qin YT, Feng YS, Ma YJ, He XW, Li WY, Zhang YK, et al. Tumor-Sensitive Biodegradable Nanoparticles of Molecularly Imprinted Polymer-Stabilized Fluorescent Zeolitic Imidazolate Framework-8 for Targeted Imaging and Drug Delivery. ACS Appl Mater Interfaces. 2020;12(22):24585-98.

[85] Jacob EM, Borah A, Pillai SC, Kumar DS. Garcinol encapsulated pH-sensitive biodegradable nanoparticles: A novel therapeutic strategy for the treatment of inflammatory bowel disease. Polymers (Basel). 2021;13(6).

[86] Han X, Taratula O, Taratula O, Xu K, St Lorenz A, Moses A, et al. Biodegradable Hypericin-Containing Nanoparticles for Necrosis Targeting and Fluorescence Imaging. Mol Pharm. 2020;17(5):1538-45.

[87] Fernández-Gutiérrez M, Pérez-Köhler B, Benito-Martínez S, García-Moreno F, Pascual G, García-Fernández L, et al. Development of biocomposite polymeric systems loaded with antibacterial nanoparticles for the coating of polypropylene biomaterials. Polymers (Basel). 2020;12(8).

Chapter 22

Nanoparticles: Novel Approach to Mitigate Environmental Pollutants

Sushil Kumar Singh, Sakshi Singh, Ashutosh Singh Gautam, Virendra Kumar, Ravish Singh Rajput and Manish Singh Rajput

Abstract

Pollution is one of the biggest challenges of current times. For control of environmental pollutants, degradation of these contaminants is need of times. Degradation of pollutants can be achieved by various physical and chemical or by physicochemical approaches. Since these methods are in efficient, hence development of biological methods began. Bioremediation is the approach of using bacteria, fungi, plants, algae, etc. to degrade wide range of environmental pollutants. Nano-bioremediation is one of such method which has received lot of attention in past few years. Nano-sized particles have large surface area relative to their volumes and thus have enhanced chemical and biological reactivity. Nano-bioremediation aims at reducing the contaminant concentrations to low risk-based levels and alleviating environmental impacts simultaneously. It brings the benefits to both nanotechnology and bioremediation together to achieve remediation which is more efficient, less time taking and eco-friendly.

Keywords: Environmental Pollution, Bioremediation, Biodegradation, Nanoparticles, Nano-bioremediation

1. Introduction

Problem of Environmental pollution is not limited only to developed countries but the challenges of waste management are also being faced by developing countries which is gradually growing day by day. The situation of effective management of environmental contaminants is grim in developing countries. Automation of most of the industries, establishment of new industries, ineffective disposal mechanisms adopted by these and similar establishments have led the world to face serious problems of municipal solid waste management, air, water and land pollution, various types of dangerous organic and inorganic pollutants that have entered into water, soil, and air by these industrial and urban waste. The developments and technical progress have led to increased harmful effects by accumulated pollutants into the surroundings. The major kinds of pollution, usually classified on the basis of environment, are air pollution, water pollution, and land pollution (**Figure 1**). Modern society is also concerned about specific types of pollutants, such as noise pollution, light pollution, and plastic pollution. Pollution of all kinds has negative effects on the environment, wildlife and impacts human health and well-being.

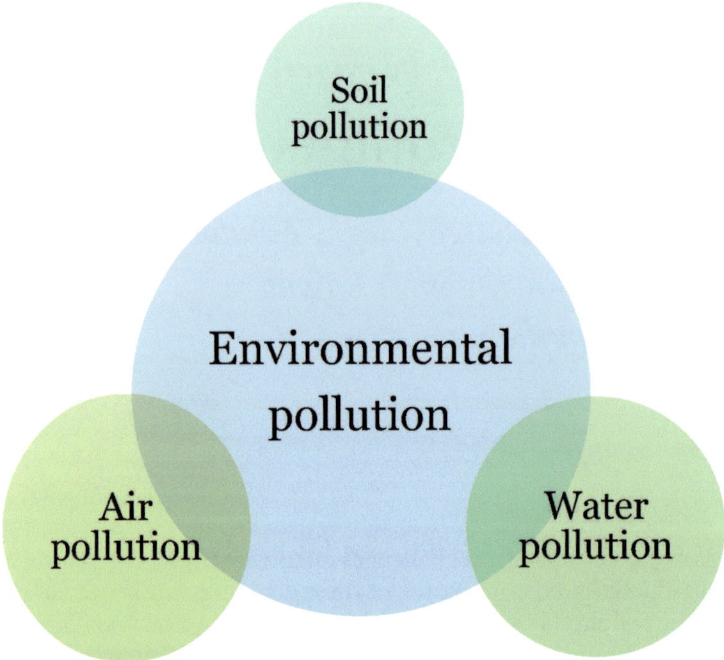

Figure 1.
Major environmental pollution.

Pollution is causing a lot of harm to human and animal health, plants and trees as well as the wider environment and has been harshly affecting the entire ecosystem and disrupting the marine life in lakes and streams causing depletion of the natural flora and fauna [1, 2].

Air pollution is increasing day by day due to release of smoke and toxic gases from various industries, transportation mediums, and various other human activities such as burning of wood, coal for fuels, etc., natural events such as forest fires. Air pollution is very common in large cities. Air pollutants cause significant damage to human health, vegetation, and cultural heritage [3].

Soil pollution is mainly caused by the disposal of municipal and industrial wastes as well as contaminated slurries from treatment plants at the waste disposal sites. Disposal and dumping of excessive use of single-use plastics, accumulation of chemicals such as pesticides, DDT, fertilizers, industrial use chemicals, raw materials for various manufacturing and production units, human activities such as deforestation, etc. have led to increased land pollution [4].

The problem of water pollution has grown with the increased amount of industries being set up and release of untreated industrial effluent in the water bodies such as ponds, lakes, riverine systems situated in the nearby vicinity of these industries. These industrial discharges include many toxic substances such as chemicals, heavy metals, metals and non-metals, floods runoffs, pesticides, fertilizers and acid rains, etc. [5]. In addition, greenhouse gas emissions such as methane and carbon dioxide continue to drive global warming and pose a great threat to biodiversity and public health [6].

The deposition of increased amount of toxicants and contaminants in our environment, have led to development and evolution of various technologies to reduce these pollutants and make our environment a better place to live. The developed technologies include use of various chemical, physical, physiochemical technologies to degrade environmental pollutants. In the last half decade bioremediation approaches has gained a lot of attention as the other technologies are either not much effective, are cost intensive and also leaves other accumulants into the environment [7].

Remediation of pollutants using nanotechnology based approaches is the recent advancements for the development of more effective, cost effective technologies for mitigating different type's pollutants form all kinds of environments [1]. Nanotechnology products, processes, and applications are expected to contribute significantly to the environment and eliminate protection by saving raw materials, energy and water as well as by reducing greenhouse gases and hazardous wastes (**Figure 2**). By the use of nanomaterials, therefore, promises certain environmental benefits and sustainability effects [1].

The word "nano" meaning "dwarf" in the Greek language refers to dimensions of the orders of magnitude 10^{-9}. Nanotechnology is the field of applied science, focuses on the design, synthesis, characterizations, and application of materials and devices on the nanoscale. It is also known as the study of phenomena and the manipulation of materials in the nanoscale. Nanotechnology can detect the presence of pathogens or toxic agents in air, water and soil are of great importance for human health and protection of the environment by offering equipment that has extremely sensitive and fast measurement with sensor equipment. Nanomaterials are less bulky, easy to operate, and having low cost [5, 8].

In addition to this, nanomaterials help to improve the quality and performance of many consumer products. Therefore, nanotechnology is very useful in solving environmental pollution [8].

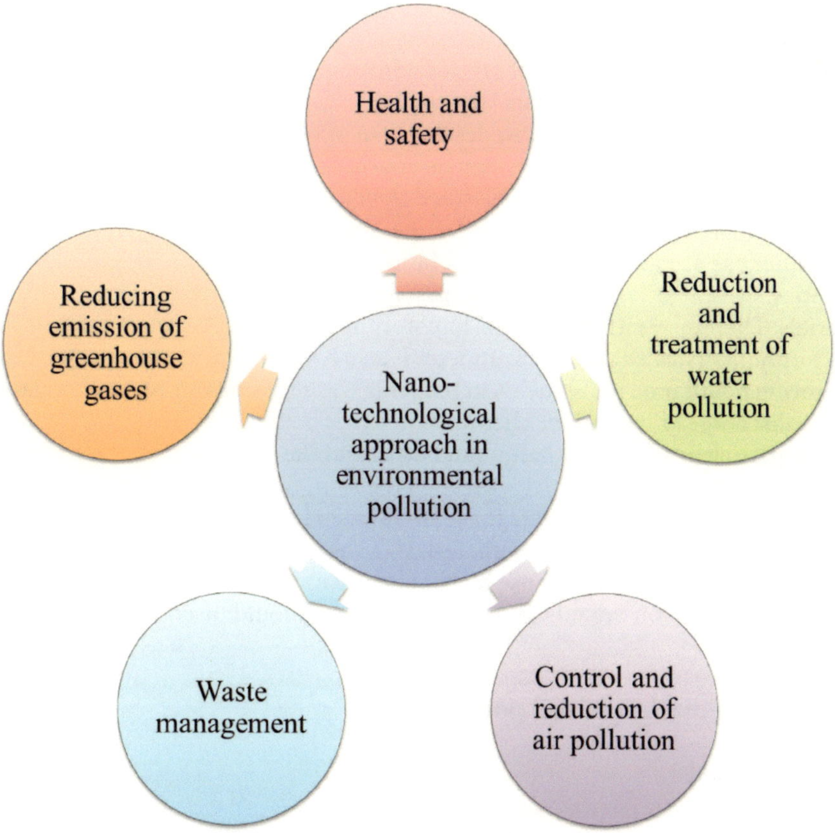

Figure 2.
Significance of nanoparticles in remediation of environmental pollution.

Accumulation of harmful air pollutants such as NOx, SOx, CO, etc., is increasing day by day. Nanotechnology is one of the current methods used in the world for controlling air pollutants using nanosensors, nanocatalysts, nanocomposites, nano filters, and nano-biomaterials [9].

Toxic gases in the air can be cleaned with help of nanotechnology but the most important thing is its continued control. Molecular level detection of toxic gases is done with the help of nanosensors such as nanocontact sensors and cantilever sensors. Nano contact sensors have been developed which can detect heavy metal ions and radioactive elements and cantilever sensors have been developed to sense VOCs, heavy metals, and pesticides. Nanocatalyst materials can be used as an environmental catalyst in the purification of automobile exhaust gases and air filtration. It is widely used in purification of water. Carbon nanotubes (CNT) have hollow ring structures composed of carbon atoms. CNT is divided into two parts, the first one is single-wall nanotubes and the second is multi-wall nanotubes which can be used to trace pollution and collect information on environmental pollutants [7].

As we know many toxic and non-biodegradable materials such as xenobiotics, heavy metals, leads, inorganic compounds cannot be eliminated by the biological and physicochemical method at environmental standards from wastes [1]. Nanofilters, nano-photocatalyst, and nanoporous catalysts are used for waste management. Nano filters can remove 60–70% of Chemical Oxygen Demand (COD) and 50% of ammonium in the leachate, anions, cations, arsenic, uranium, chromium, and pathogens from the wastewater. Nano-filter technology can be widely used in the separation and purification of gases and pollutant vapors in various industries and preventing their release into the environments. It is used in water treatments in homes, offices, and industries. Nano-photocatalyst is a nanotechnology that can absorb heavy metals from wastewater. Nanoporous catalysts can be used to convert the wastes into ethanol [10]. Another nonporous material, manganese oxide has great adsorption of toxic due to its large surface area and molybdenum disulphide (MoS_2) is used for energy-efficient desalination of water which filters five time more than conventional ones. To clean oil spills in the water bodies, a nanofabric paper towel has been developed which are woven from tiny wires of potassium manganese oxide that can absorb oil 20 times its weight [2, 8, 11].

Another use of nanotechnology is in human health and safety. It is used in many cosmetics, dyes, textiles, foods, medicines, medical equipments, and many other materials. Therefore with the help of nanotechnology, we can protect the sustainability of human health and environment. It also helps in waste management and monitoring pollutants. It also helps in reducing the emission of greenhouse gases and the discharge of hazardous chemicals in water bodies [2].

Nanoparticles can also harm human health and the environment. Due to the quick absorbance of nanoparticles through the skin, mucous and respiratory tract, causing new and unknown toxins that can threaten human health. Due to the lack of knowledge of the effect of nanomaterials in the environment and human society, we do not have a sufficient defense mechanism against them. Therefore, possible risks and its impact on organisms and environment should be considered [11].

2. Nano-remediation of soil pollution

2.1 Soil pollution

After air and water, soil is the third main components of an environment. Soil plays very important role in many activities like in plant growth (the prime activity of soil), gives anchorage to plants, supports many forms of life, provides proper environment to microbes for decaying the dead materials into simpler component which further goes in biological cycle perform by the living organism [4].

Soil pollution can be defined as the reduction in the productivity of soil due to presence of soil pollutants. Soil pollutants reduce the productivity of soil because it shows an adverse effect on the physical, chemical and biological properties of soil. Soil pollution occur due to pesticides, carcasses, tins-cans, fertilizers, organic manual, chemicals, bottles, plastics, paper, leather goods, discarded food, clothes and radioactive wastes (**Figure 3**) [12].

Sometimes soil pollution can be the reason of water pollution as well as air pollution, these chemicals leaches into groundwater and reaches to the water stream like lake, pond, sea and also the some volatile compound release harmful gases in the atmosphere respectively [13].

2.2 Causes of soil pollution

There are many causes of soil pollution and some are as follows:

2.2.1 Types of contaminants

Organic and inorganic are the two major groups of chemicals which cause soil pollution. Soil is polluted by many chemicals like simple inorganics ions to complex organic molecules.

2.2.2 Inorganic pollutants

In the environment due to industry, mining, transportation and urban activities inorganic pollutants are released. Inorganic pollutants can interact both extracellular and intracellular levels respectively and this make it high risk component for environment [14].

Figure 3.
Municipal solid waste pollution-municipal solid waste (MSW) on a beach. Such land pollution can contaminate the soil and water and is a health hazard to local communities.

Some elements known as Micronutrients are essentials for in small amount and become toxic when increase the concentration for example B, Cl, Cu, Fe, Mn, Mo and Zn. Metals like Cd, Cr, Cu, Hg, Ni, Pb, V and Zn, metalloids like As, Bo, and Sb, non- metals like Se, actinoids like U and halogens like I and F are the inorganic contaminants which causes soil pollution when exceed the certain threshold [15].

Some elements are toxic at all concentration for examples Hg, As and Tl and some form organomettalic compound which are highly toxic and lipophilic for example methylmercury and tributyl tin oxides [16].

2.2.3 Organic pollutants

Organic compounds in which carbon is the main elements in the structure with or without functional groups. Organic pollutants include several groups like pesticides, hydrocarbons, polycyclic aromatic hydrocarbon (PAHs), polychlorinated biphenyl (PCBs), polychlorinated dibenzo-p-dioxins (PCDDs) polychlorinated dibenzofurans (PCDF), polybrominated biphenyls (PBBs), polybrominated diphenyl ethers (PBDEs), surfactants, or pharmaceuticals. This group separately or together pollutes the soil for examples the groups of PCBs include 209 congeners [17, 18].

The compounds are used due to its wide range of physical properties like polarity, solubility, volatility etc. These compounds shows different behaviors in environment and toxicity in organism due to its physical properties even come in same group.

Some of the organic compounds can be degraded or bio transformed but many of them shows resistant to both chemical as well as biochemical transformation and have long half-lives for example polyhalogenated compounds. Due to its molecular stability it remains in soil for very long time and effect environment and can travel to long distance [19, 20].

2.3 Origin and sources of soil pollution

Soil contaminants can be naturally or anthropogenic. Natural activities like weathering of rocks and volcanic activities or forest fires can produces organic, inorganic or both types of pollutants respectively. Human can also cause soil pollution accidentally or deliberately. Mining, smelting, disposal of wastes, fossil fuel combustion, gaseous works, industries, sports shooting, and application of agrochemicals or sewage are the human activities which cause soil pollution. Nuclear accidents, flooding by rivers or seas, leaks from landfills, or accidental spills are the accidental pollution (**Figure 4**) [22]. Above all examples mining is one of the important sources of toxic elements.

2.4 Effects of soil pollution

Due to direct or indirect contact with contaminated soil, the health risks increases and causes many diseases. Ecological balance also disturb due to increase in pollution of soil as well as health of many living organisms is also under risk. Soil pollution affects the quality of soil which causes deaths of many organisms which are essential for the growth of plant (e.g., earthworm). Indirectly the soil pollution also affects the life of predators like birds which move to other places in search of food.

People who were living near polluted land are prone to health risk as polluted soil can cause poisoning directly and indirectly i.e. children playing near waste land come in direct contact and food which grow on polluted land cause disease indirectly like migraines, nausea, fatigue, skin disorders and even miscarriages are common symptoms seen in a people which are live near polluted land and drink polluted water [13].

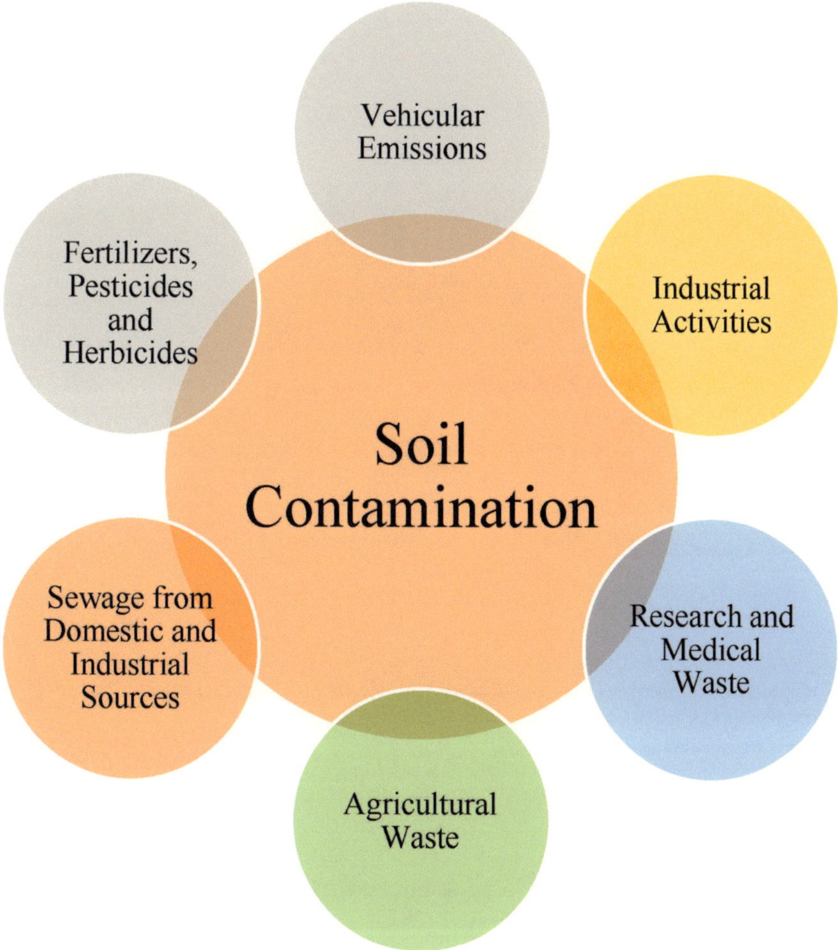

Figure 4.
Anthropogenic sources of soil contamination with toxic metalloids [21].

2.5 Nano remediation of soil pollution

There is several ways to remediate soil pollution but here we are discussing about remediation of soil pollution by nanoparticles.

In advancement of research, engineered nanoparticles plays a very important role in removal of environmental pollution as they are cheaper and more reactive.

For the treatment of environmental pollution engineered nanoparticles also enhance in situ method. Examples of some engineered nanoparticles used in soil remediation are given below:

- Nanoscale calcium peroxide – used for degrading organic compounds like gasoline
- Nanoscale zero valent iron – utilized for destroying organic compounds that are halogenated
- Nanoscale metal oxides – used for metal adsorption

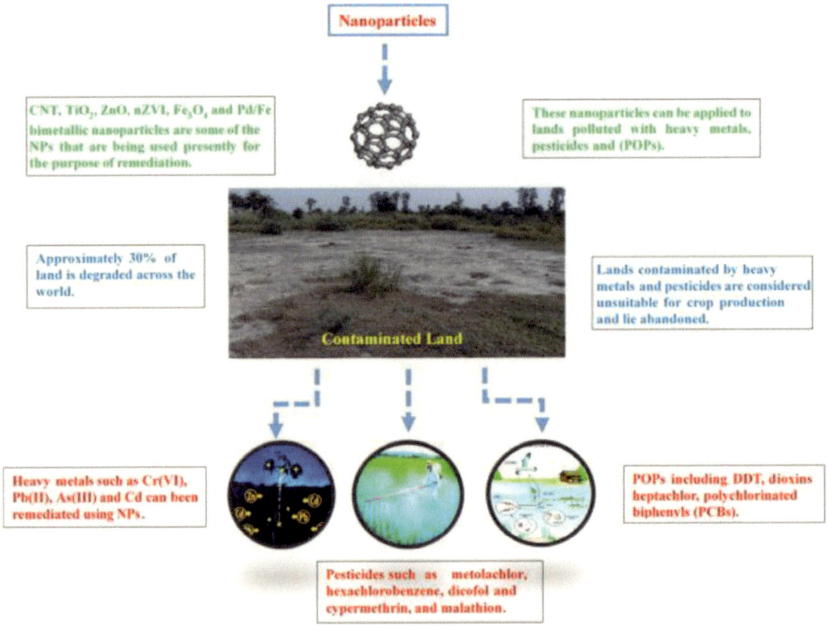

Figure 5.
The use of nanoparticles for remediation of soil contaminated with heavy metals, pesticides, and persistent organic pollutants (POPs) [23].

Other nanoparticles, such as carbon nanotubes, bio nanoparticles, polymeric nanoparticles, etc. used for the removal of aromatic and heavy metal contaminant (**Figure 5**) [24].

The above methods are considered as novel work but the effect of these engineered nanoparticles is yet to be studied [25].

From the above examples calcium peroxide is the best option because it releases peroxide very slowly which increases the attenuation time of the remediating reagent but due to its slow effective speed reaction, it shows some drawbacks. The speed of reaction of calcium peroxide nanoparticles can be increase by increase in surface to volume ratio [26].

2.6 Advantages of using nanoparticles

i. Large surface area – which helps in faster interaction.

ii. Presence of active sites – helps in increasing decontamination efficiency.

iii. Small size – helps in easy delivery at the contamination sites.

iv. Non-toxic at certain concentration.

v. Easily modified according to requirement.

vi. Pollutants can easily detected by use of nanomaterials.

vii. Low cost – the cost of remediation of soil pollutant is low.

viii. Effective and environmental friendly [27]

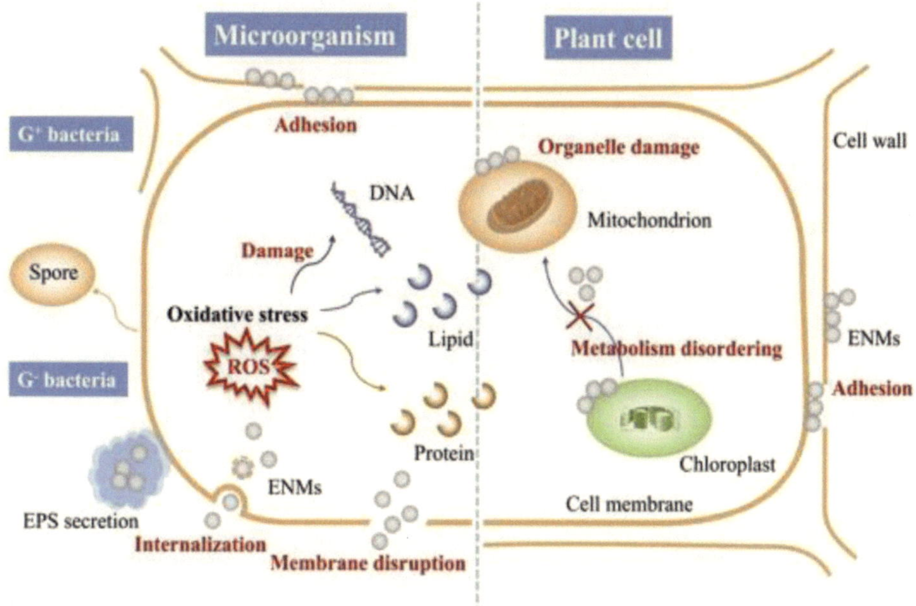

Figure 6.
Possible toxicity mechanisms of engineered nanomaterials (ENMs) at the cellular level (top left: Gram-positive (G + ve) bacteria; bottom left: Gram negative (G − ve) bacteria; right: Plant cells [29].

2.7 Disadvantages of using nanoparticles

i. Difference in function of natural and synthesized nanoparticles.

ii. Specific surface properties and chemistry for specific soil contaminants.

iii. Shows toxicity at certain size i.e. if size is less than 100 nm it shows pulmonary toxicity.

iv. Inhalation of nanoparticles can also cause damage inside body for examples FeO nanoparticles if inhaled, it release Fe(III) ion, also cause oxidative damage due to release of Fe (IV) ion.

v. Nanoparticles used in soil pollution degradation like Fe and magnetite responsible for oxidative stress response.

vi. Many studies shows adverse health effects in mammalian cells if expose to Fe oxide nanoparticles [28] (**Figure 6**).

3. Nanotechnology: an approach to improve water quality through nanoparticles

We need fresh water; it is a big task to have abundant fresh water for agriculture and industrial applications [30, 31]. More than one billion people in the world are facing challenges to have fresh water and in near future situation can be worst. It is estimated that the average supply of water to per person will drop by a factor of 1/3rd, which will result in premature death of millions of people. Non-contaminated water is also not in reach to meet proper agricultural practices [32]. The need of great amount of fresh water is leading towards the contamination of ground water through the use of pesticides, fertilizers and other chemicals.

Novel, sustainable and cost-effective approach is needed to tackle these problems. Nanotechnological researches have enabled us to find the solution for remediation and purification of waste water. Water resources within reach are usually contaminated with pathogens, salts, metals(Cu, Pb, As, etc.), pesticides, herbicides, pharmaceuticals and personal care products (PPCPs) and radioactive elements due to natural occurrences or anthropological activities [33–35].

Although the conventional methods are effective in many ways, still this method are quite outdated and not much cost effective and requires more man power and specificity and also, conventional methods are not applied to lower concentrations of pollutants (**Table 1**).

Methods	Process applied	Remarks
Physical Treatment Methods		
Precipitation	Suitable ion is applied to precipitate the metal salts.	Efficiency is affected by low pH and sludge disposal process makes it expensive.
Ion-exchange	Solid-phase matrix material is used which has ability to exchange cations or anions.	This method cannot handle high metal concentration and is relatively expensive.
Electrowinning	Electric current is passed from an inert anode through a leach solution for the recovery and removal of heavy metals.	The migration of ions to the cathode surface creates a 'zone of depletion', which hinders the process.
Electrocoagulation	Electro-chemical based approach when electric current is used to remove metals from solution.	It is an intensive process and places a lot of strains on electrodes, resulting in wear and tear.
Cementation	A type of precipitation involving an electrochemical approach in which metals with greater oxidation capacity flows into the solution.	Efficiency is influenced by various factors, such as pH, surface area of absorbent and its surface energy, etc.
Membrane filtration	Separation of metals from water takes place by semi-permeable membrane with pressure gradient.	Water pretreatment and monitoring makes the process expensive.
Electrodialysis	Analogous to RO except for Driving force in which electric field is applied across semi-permeable membrane.	Efficiency is affected by porosity, pH, flow rate, conductivity, etc.
Chemical Treatment Methods		
Reduction	Reductants (H_2S, diothionites) are injected deeply to polluted regions.	Toxic intermediate formation takes place.
Chemical washing	Heavy metals are directly removed by means of strong extrants like acids.	Deterioration of soil quality, which is hazardous to surrounding.
Chelate flushing	Chelating agents are used for extracting huge quantity of heavy metals.	Chelates are expensive and carcinogenic.

Table 1.
Physical and chemical methods used for remediation.

The conventional methods which are used in removing pollutants and particulate matters are broadly categorized into three categories:

1. Physical treatment methods

2. Chemical treatment methods

3. Biological treatment methods

The most promising methods can be said to be the biological treatment methods which includes bioremediation through different organisms, biofiltration, biosorption, etc. These are basically useful for in-situ treatment but these are slow processes and take much time but cheap and cost-effective. Biological treatment methods are quite effective but in current scenario these cannot fulfill the human requirements as pollution is many folds so a quick and highly efficient method is needed which can be fulfilled by use of nanoparticles which are quite fascinating in present days.

3.1 Nanoremediation

Nanoremediation has played an efficient and effective solution for challenges of site remediation and for the process of environmental cleanup in a cost effective manner. As their sizes are too small and have unique surface coating, these can pass through the smallest spaces and can remain suspended in ground water and then can go farther at larger distribution.

3.2 Nano bioremediation of heavy metals

The major global concern is contamination of heavy metals because of their toxicity and threat to flora and fauna [36]. Anthropogenic activities are the main cause for pollution of heavy metals. There are 65 groups of heavy metals which can be distinguished on a number of criteria, like cationic- hydroxide formation, specific gravity >5 g/ml, complex formation, hard- soft acids and bases, eutrophication and environmental toxicity. Higher doses can cause birth defects, cancer, skin lesions, retardation, major organ damage such as liver and kidneys [37]. Industries of electroplating, cement, paint, etc. discharges heavy metals such as Cd, Cu, Pb, Hg, Ni, Zn and As which are highly toxic to living organisms. The waste water is directly discharged to water bodies without any prior treatment. These can cause serious mutations and Physiological damages to organisms of habitat [38].

If possible, we will be pleased to have uses of microbes in bioremediation of toxic heavy metals with interest of nanofabrication of environmentally useful submicron scale particles. In that condition, microbes can be utilized in an eco-friendly and effective remediation of heavy metals [39].

There are mainly three attributes of nanotechnology which are being used in nanobioremediation-

a. Use of clean and green nanoparticles

b. To remove hazardous materials from composite sites

c. Act as an environmental sensor [40, 41]

Nanobioremediation prefers in-situ treatment of contaminants because it's economical, highly efficient and can be applied on larger scale [42].

3.3 Nanoiron and its derivatives

The nano zero-valent iron (nZVI) is most widely used nanomaterial. These are helpful in removing aqueous contaminants by reductive dechlorination or by reducing to an insoluble form, in case of chlorinated solvents and in case of aqueous metal ions respectively. Iron will also go for 'redox' reaction with dissolved oxygen and water. Heavy metals such as As and Cr, pesticides, chlorinated solvents and nitrates can be removed from ground water and water bodies using nZVI. The zero valent iron can be used to remove highly toxic, mobile and predominant arsenic species [As(III) and As(V)] in anoxic ground water (**Figure** 7) [44].

3.4 Single-enzyme nanoparticles

Enzymes are biocatalysts, can enhance the process of bioremediation. Their usefulness in economical way is limited as compared to synthetic catalysts, as the enzymes lack stability and longer catalytic life. To increase the stability in an effective way, magnetic iron nanoparticles can be made and on applying magnetic field these can be separated out from reactants or products [45].

3.5 Dendrimers

The word "*dendrimers*" is a Greek word, where "*dendri*" means like a branch of tree and "*meros*" means part of tree. These are highly branched and

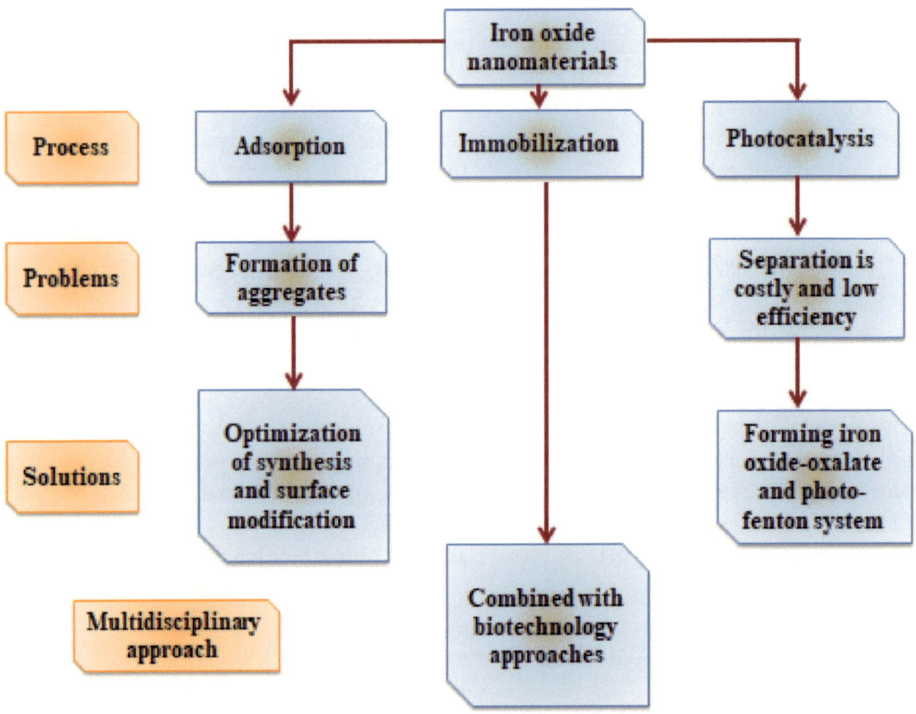

Figure 7.
Iron-oxide Nano-materials (NMs) for removal of heavy metals [43].

mono-disperse macromolecule of polymers [46–48]. These show a great potential in environmental applications. The composite, Dendrimers-NPs can be used to enhance catalytic activity due to less toxicity and more reactivity and more surface areas [49, 50].

3.6 Titanium dioxide (TiO_2) based nanoparticles (NPs)

Semiconducting, photolytic, energy-converting, electrical and gas sensing capabilities are found in TiO_2. Rutile, anatase and brookite are three polymorphs of TiO_2 crystals found in nature. These NPs are commonly available, affordable and non-toxic that's why these can be used to remove organic contaminants. The light energy is greater than band gap of semiconductor TiO_2 is necessary for the removal of different organic pollutants [51].

3.7 Bimetallic NPs

Halogenated Organic Compounds (HOCs) can be easily transformed to low or no toxic material by metals such as Zinc and Tin, which are more effective as compared to Iron. The complete dechlorination of many chlorinated aliphatic compounds to hydrocarbons can be achieved by superior catalytic ability of Palladized iron [52–54].

3.8 Photocatalytic NMs

Photocatalysis is used in the deterioration of organic contaminants in an effective manner, it is an advanced oxidation process (AOP). Even, trace amounts of pathogens and pollutants can be removed by oxidation process through photocatalysis. Catalyzation and other methods to modify NMs can improve the efficiency and remediation speed [55, 56].

3.9 Air pollution

Air pollution can be defined in many ways, one of the most used definitions of air pollution is the occurrence of chemical compounds in the atmospheric air that are toxic and present at concentration that may be injurious to animals, vegetation and humans. Air pollution decrease the quality of air due to presence of chemicals which are not present originally but resulting of human activity and it also change the lives on earth because of global warming and depletion of ozone layer [57].

3.10 Types of air pollutants

Air pollutants can be different types depending on the source, form and condition under which pollutants are generated.

3.10.1 Primary pollutants

Primary pollutants are the pollutants which are discharged directly in air.

For examples:

Particulate matter (PM10 and PM2.5); Carbon oxides (e.g. carbon monoxide); Oxides of sulfur; Ammonia; Light hydrocarbons; Volatile organic compounds; Metals (lead, mercury, cadmium) [58].

3.10.2 Secondary pollutants

Secondary pollutants are the pollutants which are formed when the chemical reaction occur between two gases present in the atmosphere.

For examples:

Sulfur dioxide, oxides of nitrogen, ammonia and non-methane volatile organic compound [58].

3.11 Causes of air pollution

Air pollution can be cause by naturally as well as man-made activities.

The examples of natural sources are emissions from plants, from the biomass of the ocean, volcanic gas and the re-suspension of dust in arid areas such as deserts.

The examples of man- made source are combustion engines (both diesel and petrol), household and industry solidfuel combustion for energy production (coal, lignite, heavy oil and biomass), other industrial activities (building, mining, manufacture of cement, smelting), agriculture, with the use of entrants, and the erosion of roads by vehicles and abrasion of brakes and tyres [59].

3.12 Other causes are

3.12.1 Urbanization and industrialization

Urbanization and Industrialization are one of the major cause of air pollution due to urbanization there is increase in use of automobiles and due to industrialization there is an increase in cutting of trees which increase the air pollution rapidly [59].

3.12.2 Burning of fossil fuels

Burning of fossil fuels is also the main cause of air pollution. Burning of fossil fuel can release a variety of primary and secondary pollutants as well as airborne oxide and also SO2, CO2, CO, hydrocarbons, organic compounds, chemicals, and nitrogen oxides (NOx). Due to burning of fossil fuel climates change and global warming is also increase due to release of major greenhouse gases (carbon dioxide, methane (CH4), nitrous oxide, and fluorinated gases) [59].

3.13 Effect of air pollution

The air pollution shows adverse effect on human health and causes many diseases like acute respiratory infection, pulmonary disease, cardiovascular disease etc.

3.13.1 Acute respiratory infection (ARI)

Acute respiratory infection is one of the common disease causes by air pollution. In this the pollutants can effects the lungs and show some symptoms like cough, sinusitis and ditis media [60].

3.13.2 Pulmonary disease

Pulmonary disease can cause due to inhalation of air pollutants which cause allergic disease and pulmonary disease such as asthma, atopic dermatitis and allergic rhinitis and lung cancer [60].

3.13.3 Cardiovascular disease (CVDs)

The pollutants like PM, PAHs, CO, heavy metal, and other organic pollutants are the main cause of cardiovascular disease [61].

Cardiovascular disease is severe when particulate matter increases and shows symptoms like ischemic stroke, myocardial infarction, cardiac arrhythmia, heart failure, and atrial fibrillation [62].

Particulate matter is one of the major factors which cause CVDs and induce the oxidative stress, systemic inflammation and increases the blood coagulability, and autonomic and vascular imbalance [60].

3.14 Nano-remediation of air pollution

Air pollution can be remediated in many ways and one of the best ways is use of nanoparticles like nanocatalyst, nano-membranes and photocatalyst semiconductor.

Nanocatalyst helps in gaseous exchange by increasing the surface area for the reaction of gases which increases the chemical reaction and help in conversion of harmful gases which are release from cars and industries into harmless gases. Nanofibre catalysts are one of the examples which are used in industries for the removal of organic compounds from the smoke stacks.

Nanostructured membrane is another way to separate methane or carbon dioxide from smoke due to having small pores. The best example is the Carbon nanotubes (CNTs) which are more efficient than other membrane for trapping greenhouse gases which is released by the mining and power generation. CNTs have major advantages as it can trap gases hundred times faster than other membranes [61].

The other approach for the remediation of air pollution is using of nanosize semiconductor photocatalyst. Photocatalyst are the catalyst that oxidized organic pollutants into nontoxic materials for examples titanium dioxide (TiO2), zinc oxide (ZnO), iron (III) oxide (Fe2O3) and tungsten oxide (WO3) and used in many ways [11].

Photocatalyst and semiconductor can use together to remediate air pollution by using the principle of semiconductor i.e. oxidizing of organic molecules by light. The process includes, at a sufficient light exposure the charge is moved from valence band to conduction band and causes the oxidation of surrounding substance. The semiconductor photocatalyst can be modified according to reactivity and selectivity [63].

4. Challenges of using NPs

There are several problems related in in-situ transport processes of nano-particles which depict the reactivity loss of zero valent ion-particles with time [64, 65]. Apart from this, several researchers have found that the in-situ application of nano-particles leads to clustering effect enhanced the deposition of nano-particles which blocks the pore of soil due to which difficulty faced to reach the targeted contaminated areas [66, 67]. Due to clogging effect of soil, filtration process becomes major issue in use of nZVI remediation. Besides this, NPs are denser than water molecules which lead to settle down of NPs in fluid medium, causing clogging effect [68].

Many studies have shown the effectiveness of nZVI application, but its effects to local microbes are still in early stages [69–72]. Lonf term effects and ethical issues related to use of NPs are hindrance to environmental agencies to implement NPs on field scale. Hence, application of NPs primarily depends on regulatory affairs and policies of each country, that's why use of NPs is limited to European countries and nZVI extensively used in US [73–75].

There is a great concern related to use of NPs due to their negative effect on microbes. Number of studies has been done to address the toxicity of NPs with microbes. However, these studies have been performed on controlled laboratory conditions [76–81]. However, contrary results have also been reported with some studies showing inhibitory effects on microbes where as some of the studies have showing stimulatory effects of NPs [82].

Thus, till date the existing results are controversial and lacks proper studies on field level as these studies are based on type of NPs and tested microbes, pollutants and also, several studies have used specific culture media in controlled lab environment [83].

5. Future prospects

The field of nanotechnology has gained attention among the researchers due to its number of beneficial effects like large surface area, multiple uses, tolerance to harsh conditions, easy and efficient manipulations, greater efficiency and many more.

We are marching towards a greener approach in management of industrial effluents due to the integration of NPs with microbes and enzymes [14, 84]. The residues left are either biocompatible or they can be easily removed by using simple filtration or precipitation techniques.

Continuous studies and its merits provide a vision for new discoveries into products. It is possible to produce synthetic 'living-like' or 'nano-bots' type things which will remediate hazardous pollutants and heavy metals from environments. So, the application of these organisms/products on a large scale will be a stepping stone for industries and environment.

Author details

Sushil Kumar Singh[1], Sakshi Singh[1], Ashutosh Singh Gautam[1], Virendra Kumar[1], Ravish Singh Rajput[2] and Manish Singh Rajput[1]*

1 Department of Biotechnology, Dr. Ambedkar Institute of Technology for Handicapped, Kanpur, UP, India

2 Rajkiya Engineering College, Kannauj, UP, India

*Address all correspondence to: msrbiotech@gmail.com

References

[1] Ms. Urmil, Environmental Application of Nanotechnology, International Journal of Engineering Research and Technology RACEE, (2015) Vol 4, Issue 03.

[2] Madkour L. H., Environmental Impact of Nanotechnology and Novel Applications of Nano materials and Nano devices. Nanoelectronic Materials. (2019) 605-699. Springer.

[3] McCormick R. A. and Holzworth, G.C., Air Pollution Climatology. Air Pollution (2013) 1976, 643-700.

[4] Bünemann E. K., Bongiorno G., Bai Z., Creamer R. E., De Deyn G., de Goede R., Fleskens L., Geissen V., Kuyper T. W., Mäder P., Pulleman M., Sukkel W., van Groenigen J. W., Brussaard L., Soil quality – A critical review, Soil Biology and Biochemistry(2018), 120, 105-125.

[5] Mojtaba T., Mohsen S., Nasser K. and Ali A., Benefits and Application of Nanotechnology in Environmental Science: an Overview. Biointerface Research in Applied Chemistry (2021) 11, 7860 – 7870.

[6] Nathanson, J. A. , *Pollution*. Encyclopedia Britannica (2021),

[7] Babatunde D. E., Iheanacho H. D., Babatunde O. M., Gbadamosi S. L., Babalola I. P. and Oluranti A., Environmental and Societal Impact of Nanotechnology, IEEE Access (2020) Vol 8, 4640-4667.

[8] Rani K. and Sridevi V., An Overview on Role of Nanotechnology in Green and Clean Technology. Austin Environ Sci. (2017) 2(3): 1026.

[9] Mansoori G. A., Bastami T. R., Ahmadpour A. and Eshaghi, Z., Environmental Application of Nanotechnology. Annual Review of Nano Research, (2008), 439-493.

[10] Chokshi N. P. and Bora L., An Overview of Nanotechnology in Waste Water Treatment. Resource recovery, water reuse and recycle for sustainable development, (2014), 1-5.

[11] Yunus I. S., Harwin, Kurniawan A., Adityawarman D. and Indarto A., Nanotechnologies in Water and Air Pollution Treatment. Environmental Technology Reviews (2012), 1:1, 136-148.

[12] Koul, B., and Taak, P., Soil Pollution: Causes and Consequences. Biotechnological Strategies for Effective Remediation of Polluted Soils. Springer, Singapore, (2018), 1-37.

[13] Mishra R. K., Mohammad N., and Roy C. N., Soil Pollution: Causes, Effects and Control. Tropical Forest Research Institute (2015). 3(1), 20-30.

[14] Saha J. K., Selladurai R., Coumar M. V., Dotaniya M. L., Kundu S. and Patra A. K., Major Inorganic Pollutants Affecting Soil and Crop Quality. Soil Pollution-An Emerging Threat to Agriculture Springer, Singapore (2017), 75-104.

[15] Yuvaraj, M., & Mahendran, P. P., Soil Pollution Causes and Mitigation Measures. Biotica Research Today, (2020), 2(7), 550-552.

[16] Megharaj M., Ramakrishnan B., Venkateswarlu K., Sethunathan N. & Naidu R., Bioremediation Approaches for Organic Pollutants: A Critical Perspective. Environment International, (2011), 37(8), 1362-1375.

[17] Chen D., Cheng Y., Zhou N., Chen P., Wang Y., Li K., Huo S., Cheng P., Peng P., Zhang R., Wang L., Liu H., Liu Y., Ruan R., Photocatalytic Degradation of Organic Pollutants using TiO2-based Photocatalysts: A Review. Journal of Cleaner Production, (2020), 268, *121725*.

[18] Rajput M. S., Dwivedi V. and Awasthi S. K., Biodegradation of Pyridine Raffinate by microbial laccase isolated from *Pseudomonas monteilii* & *Gamma proteobacterium* present in Woody Soil. Biocatalysis and Agricultural Biotechnology, (2020), 26, 101650.

[19] Cachada A., Rocha-Santos T. and Duarte A. C., Soil and Pollution: An Introduction to the Main Issues, Soil Pollution, Academic Press (2017).

[20] Rajput M. S., Dwivedi V. and Awasthi S. K., Enzymatic Degradation of Pyridine Raffinate using Response Surface and Artificial Neural Network Simulation. Indian Journal of Experimental Biology, (2020), 58, 584-592.

[21] Souza L. R. R., Pomarolli L. C. and da Veiga M. A. M. S., From Classic Methodologies to Application of Nanomaterials for Soil Remediation: An Integrated View of Methods for Decontamination of Toxic Metal(oid)s, Environmental Science and Pollution Research, (2020), 27(10), 10205-10227,

[22] Swartjes F. A., Introduction to Contaminated Site Management. Dealing with Contaminated Sites- From Theory towards Practical Application, Springer, Dordrecht (2018), 3-89.

[23] Bakshi M. and Abhilash P. C., Nanotechnology for Soil Remediation: Revitalizing the Tarnished Resource, Nano-Materials as Photocatalysts for Degradation of Environmental Pollutants, Elsevier (2020), 345-370.

[24] Sarkar A., Sengupta S. and Sen S., Nanoparticles for Soil Remediation, Nanoscience and Biotechnology for Environmental Applications, Springer Cham., (2019) 22, 249-262.

[25] Mueller N. C., and Nowack B., Nanoparticles for Remediation: Solving Big Problems with Little Particles, Elements, (2010), 6(6), 395-400,

[26] Khodaveisi J., Banejad H., Afkhami A., Olyaie E., Lashgari S. and Dashti R., Synthesis of Calcium Peroxide Nanoparticles as an Innovative Reagent for Insitu Chemical Oxidation. Journal of Hazardous Materials, (2011), 192(3), 1437-1440.

[27] Rabbani M. M., Ahmed I. and Park S. J., Application of Nanotechnology to Remediate Contaminated Soils. Environmental Remediation Technologies for Metal-Contaminated Soils, Springer, Tokyo (2016) 219-229.

[28] Srivastav A., Yadav K. K., Yadav S., Gupta N., Singh J. K., Katiyar R., and Kumar V., Nano-phytoremediation of Pollutants from Contaminated Soil Environment: Current Scenario and Future Prospects. In Phytoremediation, (2018), 383-401. Springer, Cham.

[29] Qian Y., Qin C., Chen M. and Lin S., Nanotechnology in Soil Remediation– Applications vs. Implications, Ecotoxicology and Environmental Safety, (2020), 201, 110815.

[30] Ditta A., How helpful is Nanotechnology in Agriculture? Adv. Nat. Sci. Nanosci. Nanotechnol. (2012), 3:33002.

[31] Cross KM, Lu Y, Zheng T, Zhan J, McPherson G, John V. Water Decontamination using Iron and Iron oxide Nanoparticles. Nanotechnology Applications for Clean Water. Norwich, NY, William Andrew Inc., 2014, 347-364.

[32] Dave S. and Sharma R., Use of Nanoparticles in Water Treatment: A Review. Int. Res. J. Environment Sci. (2015), 4(10), 103-106.

[33] Dasgupta N., Ranjan S. and Ramalingam C., Applications of

Nanotechnology in Agriculture and Water Quality Management, Environ Chem Lett (2017), 15, 591-605.

[34] Xue X. Y., Cheng R., Shi L., Ma Z., and Zheng X., Nanomaterials for Water Pollution Monitoring and Remediation. Environmental Chemistry Letters, (2017), 15(1), 23-27.

[35] Rajput M.S., Mishra B. N., Biodegradation of Pyridine Raffinate using Bacterial Laccase Isolated from Garden Soil. Biocatalysis and Agricultural Biotechnology. (2019), 17, 32-35.

[36] Chibuike G. U. and Obiora S. C. (2014) Heavy Metal Polluted Soils: Effect on Plants and Bioremediation Methods. Applied and Environmental Soil Science, 2014, 752708, 1- 12.

[37] Kaur S., and Roy A. Bioremediation of Heavy Metals from Wastewater using Nanomaterials. Environment, Development and Sustainability, (2020), 1-24.

[38] Dixit R., Malaviya D., Pandiyan K., Singh U. B., Sahu A., Shukla R., Singh B. P., Rai J. P., Sharma P. K., Lade H and Paul, D. Bioremediation of heavy metals from soil and aquatic environment: an overview of principles and criteria of fundamental processes. Sustainability, (2015), 7(2), 2189-2212.

[39] Rajendran P. and Gunasekaran P., Nanotechnology for Bioremediation of Heavy Metals. Environmental Bioremediation Technologies, Springer, Berlin, (2007a) 211-221.

[40] Rizwan M., Singh M., Mitra C. K., and Morve R. K., Ecofriendly Application of Nanomaterials: Nanobioremediation. Journal of Nanoparticles, 2014, 431787.

[41] Jagdale S., Hable A., and Chabukswar A., Nanobiotechnology for Bioremediation: Recent Trends. Biostimulation Remediation Technologies for Groundwater Contaminants, (2018), 259-284.

[42] Kumar S. R. and Gopinath P., Nano-Bioremediation Applications of Nanotechnology for Bioremediation. Remediation of Heavy Metals in the Environment, (2016) 27-48.

[43] Xu P., Zeng G. M., Huang D. L., Feng C. L., Hu S., Zhao M. H., Lai C., Wei Z., Huang C., Xie G. X. and Liu Z. F. Use of Iron oxide Nanomaterials in Wastewater Treatment: A Review. Sci Total Environ. (2012), 1(424), 1-10.

[44] Stevenson L. M., Adeleye A. S., Su Y., Zhang Y., Keller A. A., and Nisbet, R. M., Remediation of Cadmium Toxicity by Sulfidized Nano-iron: The Importance of Organic Material. ACS Nano, (2017), 11(10), 10558-10567.

[45] Şahutoğlu A. S. and Akgül C. One-phase Synthesis of Single Enzyme Nanoparticles (SENs) of *Trametes versicolor* Laccase by In situ Acrylamide Polymerisation. Biocatalysis and Biotransformation, (2020), 38:1, 64-74.

[46] Svenson S., and Tomalia, D. A., Dendrimers in Biomedical Applications—Reflections on the Field. Advanced Drug Delivery Reviews, (2012), 64, 102-115.

[47] Khin M. M., Nair A. S., Babu V. J., Murugan R., and Ramakrishna S., A Review on Nanomaterials for Environmental Remediation. Energy & Environmental Science, (2012), 5(8), 8075-8109.

[48] Undre S. B., Singh M. and Kale R. K., Interaction Behaviour of Trimesoyl Chloride Derived 1st Tier Dendrimers Determined with Structural and Physicochemical Properties Required for Drug Designing. J Mol Liq. (2013a), 182, 106-120.

[49] Undre S. B., Singh M., Kale R. K. and Rizwan M., Silibinin Binding and

Release Activities Moderated by Interstices of Trimesoyl, Tridimethyl, and Tridiethyl Malonate first-tier Dendrimers. J Appl Polymer Sci (2013b), (130), 3537-3554.

[50] Das S., Chakraborty J., Chatterjee S., and Kumar, H., Prospects of Biosynthesized Nanomaterials for the Remediation of Organic and Inorganic Environmental Contaminants. Environmental Science: Nano, (2018), 5(12), 2784-2808.

[51] Irshad M. A., Nawaz R., Zia ur Rehman M., Adrees M., Rizwan M., Ali S., Ahmad S., Tasleem S., Synthesis, Characterization and Advanced Sustainable Applications of Titanium Dioxide Nanoparticles: A Review. Ecotoxicology and Environmental Safety, (2021), (212), 111978.

[52] Liu W. J., Qian T. T., and Jiang, H., Bimetallic Fe Nanoparticles: Recent Advances in Synthesis and Application in Catalytic Elimination of Environmental Pollutants. Chemical Engineering Journal, 236, 448-463.

[53] Bhandari G., Environmental Nanotechnology: Applications of Nanoparticles for Bioremediation. Approaches in Bioremediation, Nanotechnology in the Life Sciences, Springer, Cham. (2018), 301-315.

[54] Ukaogo P. O., Ewuzie U. and Onwuka C. V., Environmental Pollution: Causes, Effects and the Remedies. Microorganisms for Sustainable Environment and Health, Elsevier (2020).

[55] Wahab R., Khan F., Kaushik N., Musarrat J. and Al-Khedhairy A. A., Photocatalytic TMO-NMs Adsorbent: Temperature-Time dependent Safranine Degradation, Sorption Study Validated under Optimized Effective Equilibrium Models Parameter with Standardized Statistical Analysis. Sci Rep (2017) 7, 42509.

[56] Balbi T., Caratto V., Fabbri R., Camisassi G., Villa S., Ferretti M. and Canesia L., Photocatalytic Fe-doped n-TiO2: From synthesis to Utilization of in vitro Cell Models for Screening Human and Environmental Nanosafety. Resource-Efficient Technologies (2017), 3(2), 158-165.

[57] Li M. and Mallat L., Health Impacts of Air Pollution, SCOR (2018), 1-42.

[58] Schraufnagel D. E., Balmes J. R., Cowl C. T., De Matteis S., Jung S. H., Mortimer K. and Wuebbles D. J., Air Pollution and Non-communicable Diseases: A Review by the Forum of International Respiratory Societies' Environmental Committee, Part 2: Air Pollution and Organ Systems. Chest, (2019), 155(2) 417-426.

[59] Tran V. V., Park D. and Lee Y. C., Indoor Air Pollution, related Human Diseases, and Recent Trends in the Control and Improvement of Indoor Air Quality, International Journal of Environmental Research and Public Health, (2020), 17(8), 2927.

[60] Raaschou-Nielsen O., Beelen R., Wang M., Hoek G., Andersen Z. J., Hoffmann B. and Vineis P., Particulate Matter Air Pollution Components and Risk for Lung Cancer. Environment International, (2016), 87, 66-73.

[61] Bhawana P. and Fulekar M., Nanotechnology: Remediation Technologies to Clean up the Environmental Pollutants. Res J Chem Sci ISSN, 2231, (2012), 606X,.

[62] Uzoigwe J. C., Prum T., Bresnahan E. and Garelnabi M., The Emerging Role of Outdoor and Indoor Air Pollution in Cardiovascular Disease. North American Journal of Medical Sciences, (2013), 5(8), 445.

[63] Folli A. (2011). Photocatalytic Cementitious Materials Containing Highly Active Nanosized TiO2:

Mechanisms of Air Pollution Remediation and the Effect of the Alkaline Environment. NORDIC Concr Res, (2011), 44, 11-24.

[64] Tosco T., Papini M. P., Viggi C. C., and Sethi R., Nanoscale Zerovalent Iron Particles for Groundwater Remediation: A Review. J. Clean. Prod. (2014), 1, 10-21.

[65] Yan W., Lien H. L., Koel B. E. and Zhang W. X., Iron Nanoparticles for Environmental Clean-up: Recent Developments and Future Outlook. Environ. Sci. Process & Impacts (2013) 63-77.

[66] Reddy K. R., Nanotechnology for Site Remediation: Dehalogenation of Organic Pollutants in Soils and Groundwater by Nanoscale Iron Particles. Proceedings of the 6th International Congress on Environmental Geotecnics, New Delhi, India (2010), 1,163e180.

[67] Kharangate-Lad A., and D'Souza N. C. Remediation of Toxic Environmental Pollutants Using Nanoparticles and Integrated Nano-Bio Systems. Rhizobiont in Bioremediation of Hazardous Waste. Springer, Singapore, (2021), 443-482.

[68] Berge N. D. and Ramsburg C. A., Iron-mediated Trichloroethene Reduction within Non-aqueous Phase Liquid. J. Contam, Hydrol. (2010), 118 105e116.

[69] Liu Z. G. and Zhang F. S., Nano-zerovalent Iron Contained Porous Carbons Developed from Waste Biomass for the Adsorption and Dechlorination of PCBs. Bioresour. Technol. (2010), 101, 2562-2564.

[70] Reddy K. R., Khodadoust A. P. and Darko-Kagya K., Transport and Reactivity of Lactate-modified Nanoscale Iron Particles in PCP-contaminated Soils. J. Hazard, Toxic, Radioact. Waste, (2012), 16, 68-74.

[71] Klaine S. J., Koelmans A. A., Horne N., Carley S., Handy R. D., Kapustka L., Nowack B., von der Kammer F., Paradigms to Assess the Environmental Impact of Manufactured Nanomaterials. Environ Toxicol Chem. (2012), 31(1):3-14.

[72] Lee C., Kim J. Y., Lee W. I., Nelson K. L., Yoon J. and Sedlak D. L., Bactericidal Effect of Zero-valent Iron Nanoparticles on *Escherichia coli*. Environ. Sci. Technol., (2008) 42, 4927-4933.

[73] Mueller N.C. and Nowack B., Application of Nanoscale Zero Valent Iron (NZVI) for Groundwater Remediation in Europe. Environ. Sci. Pollut. Res. (2012), 19, 550-558.

[74] Wilkin R. T., Lee T. R., Sexton M. R., Acree S. D., Puls R. W., Blowes D. W., Kalinowski C., Tilton J. M. and Woods L. L. Geochemical and Isotope Study of Trichloroethene Degradation in a Zero-Valent Iron Permeable Reactive Barrier: A Twenty-Two-Year Performance Evaluation. Environmental Science & Technology, (2019,)53(1), 296-306.

[75] Mirzajani F., Ghassempour A., Aliahmadi A., and Esmaeili M. A., Antibacterial Effect of Silver Nanoparticles on *Staphylococcus aureus*. Research in Microbiology, (2011), 162(5), 542-549.

[76] Lara H. H., Garza-Treviño E. N., Ixtepan-Turrent L., and Singh D. K., Silver Nanoparticles are Broad-spectrum Bactericidal and Virucidal compounds. Journal of Nanobiotechnology, (2011) 9(1), 1-8.

[77] Swain P., Nayak S. K., Sasmal A., Behera T., Barik S. K., Swain S. K., Mishra S. S., Sen A. K., Das J. K. and Jayasankar P. Antimicrobial Activity of Metal Based Nanoparticles against Microbes Associated with Diseases in Aquaculture. World Journal of

Microbiology and Biotechnology, (2014), 30(9), 2491-2502.

[78] Schacht V. J., Neumann L. V., Sandhi S. K., Chen L., Henning T., Klar P. J., Theophel K., Schnell S., and Bunge M., Effects of Silver Nanoparticles on Microbial Growth Dynamics. Journal of Applied Microbiology, (2013), 114(1), 25-35.

[79] Rudramurthy G. R., Swamy M. K., Sinniah U. R., and Ghasemzadeh A., Nanoparticles: Alternatives against Drug-resistant Pathogenic Microbes. Molecules, (2016), 21(7), 836.

[80] Griegar K. D., Fjordboge A., Hartmann N. B., Eriksson E., Bjerg P. L., Baun A., Environmental Benefits and Risks of Zerovalent Iron Nanoparticles, (nZVI) for in situ Remediation: Risk Mitigation or Trade-off? J. Contam. Hydrol (2010) 118, 165-183.

[81] Greulich C., Braun D., Peetsch A., Diendorf J., Siebers B., Epple M., and Köller M., The Toxic Effect of Silver Ions and Silver Nanoparticles towards Bacteria and Human Cells occurs in the Same Concentration Range. RSC advances, (2012), 2(17), 6981-6987.

[82] Lampron K. J., Cha D. K., and Chiu P., Microbial Reductive Dehalogenation of Chlorinated Ethenes Coupled with the Corrosion of Fe°. Hazardous and Industrial Waste Proceedings, 30th Mid-Atlantic Conference. CRC Press, 2014, 448.

[83] Zhang K., Yang W., Liu Y., Zhang K., Chen Y. and Yin X., Laccase Immobilized on Chitosan-coated Fe_3O_4 Nanoparticles as Reusable Biocatalyst for Degradation of Chlorophenol. J. Mol. Struct. (2020), 128769.

[84] Dixit M., Liu H., Luo J. and Shukla P., Effluents Detoxification from Pulp and Paper Industry using Microbial Engineering and Advanced Oxidation Techniques. J. Hazard. Mater (2020), 122998.

Section 4

Pollutant Analysis

Chapter 23

Mobile Phase Selection by Optimization for the Determination of Multiple Pesticides Using Liquid Chromatography-Tandem Mass Spectrometry

Abubakar Lawal and Lukman Bola Abdul'rauf

Abstract

The selection of the best mobile phase setup is one of the most important factors to be considered prior to quantitative instrumentation of multiple pesticides. Usually, mobile phases comprises of water (A) and an organic solvent (B) are the setup used in liquid chromatography instruments for the analysis of pesticide residues in various samples. Unfortunately, most of the analyses are being carried out without optimization and selection of the best mobile phase setup to improve the sensitivity of the instrument. For that reason, the comparative analysis of the reportedly used mobile phases and some few suggested ones was carried out on the multi-pesticide mixture of 0.1 mg/kg (100 μg/kg) standard solutions and quantified with liquid chromatography–tandem mass spectrometry (LC–MS/MS) instrument. Consequently, the best mobile phases setup that resulted in the sum of average total chromatographic peak areas (ATCPAs) and average total chromatographic peak heights (ATCPH) for the total ion chromatography (TIC) scans as an index that correspond to the concentration levels was selected [0.1% formic acid in H_2O (A) and 0.1% formic acid in acetonitrile (ACN) (B)]. And further optimization was successfully carried out on the selected mobile phase-A and the resulted setup [1% ACN and 0.1% formic acid in Milli-Q-water (mobile phase A) coupled with 0.1% formic acid in ACN (mobile phase-B)] improved the instrumental sensitivity on the targeted analytes. Thus, this justify the potential benefits of optimizing setup of the mobile phases prior to LC–MS/MS instrumentation of multi-pesticide analytes.

Keywords: Mobile phase, Analysis of multi-pesticide residues, Liquid chromatography–tandem mass spectrometry, Total ion chromatography, Total chromatographic peak areas

1. Introduction

Foods are contaminated through various activities performed by man such as the accidental or intentional discharge of chemicals or waste substances from domestic, industrial and agricultural sites into the environment [1, 2]. However, most of these contaminants are non-biodegradable, which can be easily transferred from

IntechOpen

the ground surface to the underground water because of their ability in dissolving sparingly in water [3, 4]. At long run, the contaminants pollute the foods through their respective circulatory movements in the environment [5]. The contaminants include inorganic matters such as heavy metals [6–8], as well as organic chemicals such as heat generated compounds [polycyclic aromatic hydrocarbons (PAHs) and acrylamide)] [9], organic polymers (bromodiphenyl ethers, chlorobiphenyls, chlorodibenzodioxins, chlorodibenzofurans etc), mycotoxins (aflatoxins), perfluoroalkyl acids [10–12]. Other contaminants with emerge-concerns include phthalates, bisphenol A, alkylphenols [13], phytosterols, estrogens, phytoestrogens [14], pharmaceuticals/veterinary drugs, synthetic dyes and pesticides [15–18].

Advantageously, pesticides have been used in domestic and agricultural practices for decades increasing the gross domestic products (GDP) of many countries around the globe. But their dangers in handling and excessive usage have been the issues of concern due to their residual accumulations in food chain resulting in many health problems that include cancers etc. However, there are challenging issues (problems) in the determination of multiple pesticide residues in food samples at lower concentration levels. These problems include extensive ranges of their chemical properties such as neutral, acidic and basic [19], vapor pressure/Henry's law constant [20], solubility [21], partition coefficient in octanol/water (log$P$) [22] and acid dissociation constant (pK_a) [23]. Besides, the analytical samples also play challenging roles for pesticides extraction during sample preparation because of their features that include non-polar, polar, fatty and waxy samples [24, 25].

Even though, the conventional methods such as liquid–liquid extraction (LLE), liquid-phase microextraction (LPME) as well as solid phase extraction (SPE) techniques were previously used as the sample preparation methods for the multiple pesticides analysis [16] but possesses poor efficiency and selectivity of the targeted, which were their major drawn backs [26]. Also, many detectors and quantification instruments were used previously for the analyses of multiple pesticide residues [26]. These instruments include the gas chromatography-atomic emission detector (GC-AED) [27] and the high performance liquid chromatography (HPLC) [28]. Others instruments include gas chromatography-tandem mass spectrometry (GC–MS/MS) [29] and liquid chromatography–tandem mass spectrometry (LC–MS/MS) [30]. Unfortunately, the poor sensitivity of these instruments is their major setbacks. Fortunately, the shortcomings of the conventional sample preparation techniques and that of the detecting and quantifying instruments could be corrected through optimization such as the use of response surface methodology (RSM) [26, 31].

Accordingly, these compel food safety analysts to improve better ways of analyzing multi-pesticide residues in food samples through effective sample preparations and instrumentation techniques. For instance, RSM optimization of the intrumental prameters for LC–MS/MS (advanced) instrument such as the setup of the mobile phases could overcome the afformentioned problems encountered in samples to obtain better results of multiple pesticides residues at the lower concentration levels.

Usually, mobile phases comprise of Milli-Q-water (A) and an organic solvent (B) setup are used in the liquid chromatography instruments for the analyses of pesticide residues in various samples of food materials [26, 32, 33]. In fact, the organic solvents such as acetonitrile (ACN) and methanol are significantly used in the reverse-phase of liquid chromatography (LC) due to their excellent compatibility [34].

Thus, the aim of this research is to comparatively study the most recently used (reported) setup of mobile phases and some few suggested ones (**Table 1**). The best mobile phases setup that provided highest average total chromatographic peak area (ATCPA) as an index that correspond to the concentration of analytes in the multi-pesticide mixture of standard solutions was selected after the LC–MS/MS instrumentation.

Pesticide	MF	MIM	TOP	COC	IM (ESI)	PI	MRM_1/MRM_2	CE_1/CE_2	ART
Dursban (Chlorpyrifos)	$C_9H_{11}Cl_3NO_3PS$	349	Insecticide & Nematicide	Organophosphorus	$[M + H]^+$	350	96.8/197.9	34/22	11.36
Diazinon	$C_{12}H_{21}N_2O_3PS$	304	Insecticide	Organophosphorus	$[M + H]^+$	305	96.9/169.1	42/22	10.22
Thiamethoxam	$C_8H_{10}ClN_5O_3S$	292	Insecticide	Neonicotinoid	$[M + H]^+$	292	132/211	26/10	2.68
Metalaxyl	$C_{15}H_{21}NO_4$	279	Fungicide	Xylylalanine	$[M + H]^+$	280	160.1/220.1	26/10	7.33
Thiobencarb	$C_{12}H_{16}ClNOS$	257	Herbicide	Thiocarbamate	$[M + H]^+$	258	89.1/125	54/26	10.34
Baycarb (Fenobucarb)	$C_{12}H_{17}NO_2$	207	Insecticide	Carbamate	$[M + H]^+$	208	77/95	42/10	8.34
Carbaryl	$C_{12}H_{11}NO_2$	201	Insecticide & Nematicide	N-Methyl Carbamate	$[M + H]^+$	202	127.1/145	30/6	7.16
Propamocarb	$C_9H_{20}N_2O_2$	188	Fungicide	Other Carbamate	$[M + H]^+$	189	74/102.1	26/14	1.36

PIN, pesticide identity number; MF, molecular formula; MIM, mono-isotopic mass; TOP, type of pesticide; COC, class of chemical; IM, ionization mode; ESI, electrospray ionization; PI, precursor ion (m/z); MRM, multiple reactions monitoring; CE, collision energy (eV); ART, average retention time (min).

Table 1.
Auto-tuning and mass-Hunter optimization results of the instrument using the multi-pesticides mixture of standard solutions.

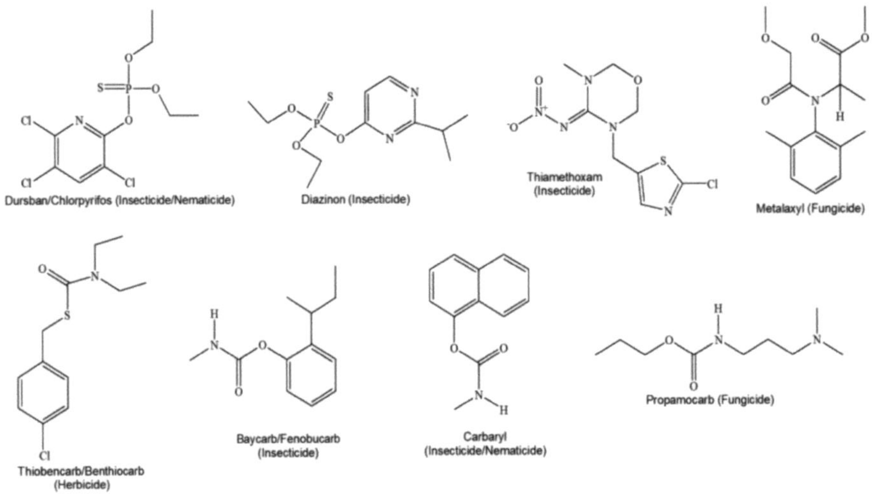

Figure 1.
Structural formula of the analyzed pesticide residues.

However, the multi-pesticides mixture of standard solutions of Dursban, Diazinon, Thiamethoxam, Metalaxyl, Thiobencarb, Baycarb, Carbaryl and Propamocarb (**Figure 1**) were analyzed for the purpose of the mobile phase optimization.

Thus, it is hoped that the result of this study would serve as a reference guide for the future studies, and the optimized mobile phase setup would be routinely used in LC–MS/MS for the determination of multiple pesticide residues in various food samples.

2. Material and methods

The chemicals and reagents such as the stock standard solution (100 mg/kg) for pesticide Baycarb, Carbaryl, Diazinon, Dursban, Metalaxyl, Propamocarb, Thiamethoxam and Thiobencarbwere purchased from AccuStandard® (New Haven, USA). The LC–MS grade organic solvents that include ACN and methanol were purchased from Merck (Germany). The formic acid was purchased from Fisher Scientific. The Millipore-filtered (deionized) water was obtained using Merck Millipore water purification system (Billerica, USA). While, the apparatus and equipments that include the 100 and 500 μL microsyringe were purchased from Agilent (Australia). The pH meter PB was purchased from Sartorius group (Germany). The HPLC autosampler vials were purchased from Agilent Technologies (USA). The Supelco HPLC column [Ascentis® Express C_{18} (5 cm x 2.1 mm, 2.7 μm)] was purchased from Sigma-Aldrich (USA). And theliquid chromatography–tandem mass spectrometry (LC–MS/MS) [triple quadrupole (G6490A) built in Electrosprays ESI (±) MS/MS Sensitivity and Jet stream Technology] instrument was purchased from Agilent (Singapore).

2.1 Conditioning of the LC–MS/MS

The following contributory parameters of the LC–MS/MS instrument were setup initially that include; analyte injection volume (5 μL), flow rate (0.1 mL/min), column temperature (30°C), gas temperature (200°C), nebulizer gas (45 psi), gas flow (14 L/min), sheath gas temperature (400°C), capillary voltage

	References	Water (A)	Organic Mobile Phase (B)
1.	1st suggested mobile phase	A	ACN
2.	Rajski *et al.* [35], Pérez-Ortega *et al.* [36]	A + 0.1% FA	ACN
3.	Economou *et al.* [37] and Lucas [38]	A + 0.1% FA	ACN + 0.1% FA
4.	Vázquez *et al.* [39]	A + 0.1% FA	ACN + 0.1% FA + 5% A
5.	2nd suggested mobile phase	A	MeOH
6.	Golge and Kabak [40]	A + 5 mM AF	MeOH +5 mM AF
7.	Zanella *et al.* [41]	A + 2% MeOH +0.1% FA + 5 mM AF	MeOH +0.1% FA + 5 mM AF
8.	3rd suggested mobile phase	A	MeOH/ACN (1:1)
9.	4th suggested mobile phase	A + 5 mM AF + 0.1%FA	MeOH/ACN (1:1) + 0.1% FA + 5 mM AF

Table 2.
The list of suggested and reported mobile phases used for the optimization.

(3000 V), sheath gas flow (11 L/min), and delta$^{(+)}$ EMV (200 V). However, these factors contributed in determining optimum fragmentor voltage and the four-fragmentor product ions with their respective retention time (RT) and collision energy (CE) (**Table 2**). Moreover, the instrumental default settings were further used for the development of the best gradient program runs for the mobile phase-B elution time by adopting and modifying the methods used by Rajski *et al.* [35] and Vázquez *et al.* [39] for analysis of similar multi-pesticide compounds. This results in the best shortest elution time, which provided the best total ion chromatography (TIC) peaks resolution for the LC–MS/MS instrumentation (**Figure 2**). However, TIC resolution providedan optimum condition for the attainment of higher total chromatographic peak area (TCPA) [42] and mathematically expressed in Eq. (1) [43].

Therefore,

$$TCPA = \sum CPA \tag{1}$$

Where $TCPA$: The total chromatographic peak area; CPA: The chromatographic peak areas.

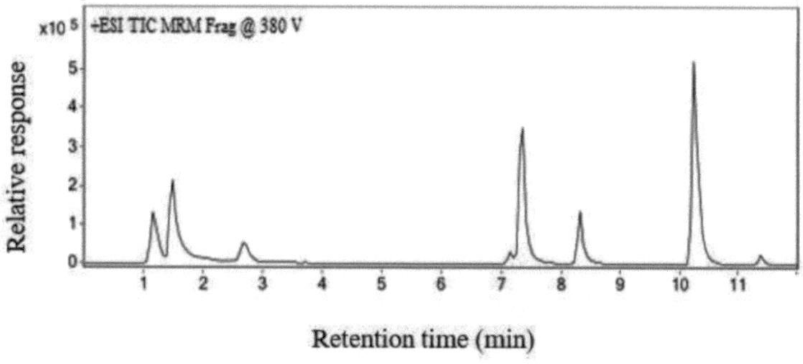

Figure 2.
The total ion chromatography (TIC) of the analyzed pesticide standards.

Notably, the best setup of mobile phases were also selected using the initial settings of the instrument. Therefore, the TCPA obtained from LC–MS/MS analysis serves as an index used for estimating the number of target analytes that are present in the analyzed samples [31]. It is because of the close similarities range of the resulted peak areas due to the log*P* of targeted analytes. Moreover, the peak areas maybe correlated and categorically suitable for multiple pesticides analysis using the LC–MS/MS instrument [44].

2.2 Sample treatment and methodology

The stock standard solution of 100 μg/mL that is equivalent to 100 mg/kg (i.e. 100,000 μg/kg) or parts per million (ppm) [45] for each pesticide was diluted to 10, 1 and 0.1 mg/kg (100 μg/kg) with appropriate volumes of methanol. The appropriate volumes were calculated using the dilution formula as expressed in Eq. (2) [46], separately. Afterward, the prepared working standard solutions were preserved in a refrigerator at 4°C before carrying out the LC–MS/MS analysis.

$$C_1C_2 = V_1V_2 \tag{2}$$

Ref codes	References	Water (A)	Organic M/Phase (B)	% M/Phase B	ATCPH ± STDEV	ATCPA ± STDEV
A	1st suggested mobile phase	A	ACN	25	$(361 \pm 2) \times 10^5$	$(47 \pm 3) \times 10^7$
B	Rajski *et al.* [35], Pérez-Ortega *et al.* [36]	A + 0.1% FA	ACN	30	$(349 \pm 3) \times 10^5$	$(46 \pm 1) \times 10^7$
C	Economou *et al.* [37] and Lucas [38]	A + 0.1% FA	ACN + 0.1% FA	15	$(50 \pm 1) \times 10^6$	$(72 \pm 9) \times 10^7$
D	Vázquez *et al.* [39]	A + 0.1% FA	ACN + 0.1% FA + 5% A	30	$(31 \pm 2) \times 10^6$	$(38 \pm 1) \times 10^7$
E	2nd suggested mobile phase	A	MEOH	30	$(17 \pm 1) \times 10^6$	$(23 \pm 2) \times 10^7$
F	Golge and Kabak [40]	A + 5 mM AF	MEOH +5 mM AF	30	$(26 \pm 2) \times 10^6$	$(30 \pm 1) \times 10^7$
G	Zanella *et al.* [41]	A + 2% MEOH +0.1% FA + 5 mM AF	MEOH +0.1% FA + 5 mM AF	10	$(58 \pm 3) \times 10^6$	$(60 \pm 7) \times 10^7$
H	3rd suggested mobile phase	A	MEOH/ACN (1:1)	30	$(27 \pm 1) \times 10^6$	$(30 \pm 4) \times 10^7$
I	4th suggested mobile phase	A + 5 mM AF + 0.1%FA	MEOH/ACN (1:1) + 0.1% FA + 5 mM AF	25	$(36 \pm 5) \times 10^6$	$(32 \pm 3) \times 10^7$

ATCPH, average total chromatographic peak height; ATCPA, average total chromatographic peak area; RT, retention time; AF, ammonium formate; FA, formic acid; STDEV, standard deviation; Ref, reference.

Table 3.
The ATCPH and ATCPA instrumental responses for the selection of mobile phase.

Where C_1 : The concentration of the stock standard solution, C_2 : The concentration of the working standard solution; V_1 : The volume of the stock standard solution; V_2 : The volume of the working standard solution.

Meanwhile, the selection of the LC–MS/MS mobile phase was carried out by optimization technique using one factor or variable at a time (OFAT or OVAT) based on the documentation of Sherma [47]. However, the multivariate optimization technique was not favorable for the selection because responses for each of the mobile phase is required individually without interaction to estimate the actual effect of the mobile phase setup. Moreover, the two setups of mobile (organic and aqueous) phases are involved with interactive percentage flow of organic/aqueous changes to create an optimum condition of analytes detection.

Thus, comparative analysis was carried out on some assumed and selected mobile phases reportedly used for analysis of pesticides in various samples. Experimentally, the comparative analysis was carried out on the multi-pesticide mixture of 0.1 mg/kg multi-pesticide mixture of standard solutions. Consequently, the TIC of the instrumental runs for each of the mobile phases resulted in chromatographic peak heights (ATCPH), and areas (ATCPAs) as presented in **Table 3**. Then again, the addition of organic solvent into aqueous mobile phase could provide the optimum condition of log$P$, which contributes to the attainment of good condition for the multi-pesticide residues analysis in food samples using LC–MS/MS instrument as revealed [41]. For this reason, optimization was carried out by serial addition of ACN into the aqueous mobile phase (0.1% FA milli-Q-water). Thus, the mobile phase setup that provided the best separation of analytes and the highest TCPA was selected for further optimization by adding 0, 0.5, 1.0, 1.5, 2.0, 2.5, 3.0, 3.5, 4.0, 4.5, 5.0, 7.5 and 10% ACN in mobile phase A. Moreover, the best pH solution was selected based on the results of the average TCPA responses of the LC–MS/MS instrument.

3. Results and discussion

The responses of the screened mobile phases were compared and recorded. The mobile phase setup [0.1% formic acid in Milli-Q-water (A) and 0.1% formic acid in ACN (B)] was the best based on the highest results obtained [ATCPAs ± standard deviation (STDEV) as well as ATCPH± STDEV)] in triplicates as tabulated and illustrated in **Table 3** and **Figure 3**, respectively. This result was also supported by other findings using the mobile phase for pesticides analysis [48, 49]. Meanwhile, further optimization result of mobile phase-A after addition of ACN (0–10%) revealed that the addition of 1% ACN into 0.1% FA Milli-Q-water at an average pH of 3.50 ± 0.07 STDEV (mobile phase A) coupled with 0.1% FA in ACN at pH 6.56 ± 0.04 STDEV (mobile phase-B) provided the highest ATCPA (**Table 4**). The results were supported by their respective pH readings as shown in **Table 4** and **Figure 4**, respectively. Moreover, the retention time (min) of the pesticide analytes were less than the results reported by some literatures such as thiamethoxam, 2.68 < 2.87 [50]; propamocarb, 1.36 < 1.47 [51]; carbaryl, 7.16 < 16.0 [52]; metalaxyl, 7.33 < 17.90 [53]; thiobencarb 10.34 < 10.76 [54], and dursban, 11.36 < 12.30 [55]. But the retention time (min) of baycarb (8.34) and diazinon (10.22) were more than 6.73 [56] and 7.09 [57] respectively. Fortunately, the optimized mobile phase contributes towards shortening the total run time (min) and improved the instrumental sensitivity of the LC–MS/MS towards better analysis of multiple pesticides.

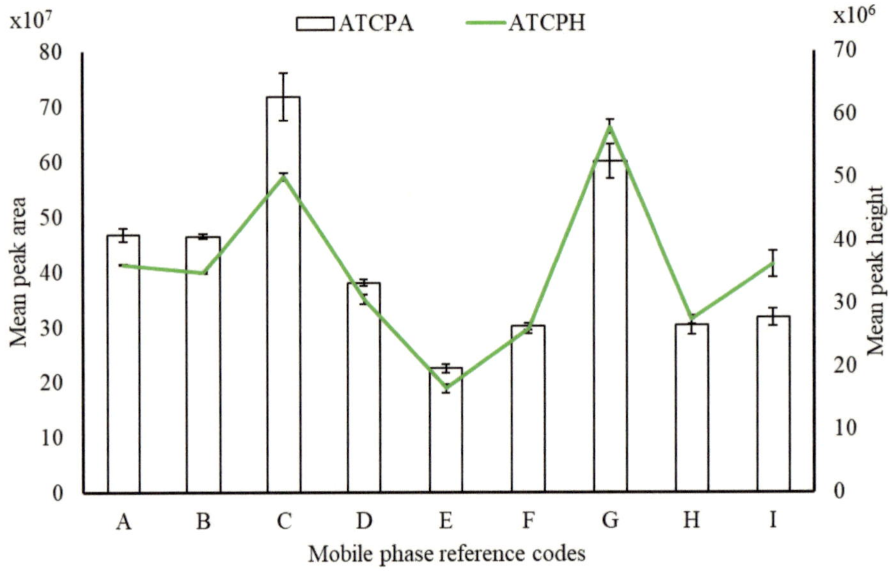

Figure 3.
The comparative studies of ATCPA and ATCPH results for the analyzed mobile phases.

Solution	% ACN in Aqueous Mobile Phase	ApH reading ± STDEV	Organic Mobile Phase	ATCPA ± STDEV
1	H_2O + 0.1% FA + 0% ACN	3.36 ± 0.00	ACN + 0.1% FA	(27 ± 2) x 10^6
2	H_2O + 0.1% FA + 0.5% ACN	3.37 ± 0.08	ACN + 0.1% FA	(27 ± 1) x 10^6
3	H_2O + 0.1% FA + 1.0% ACN	3.50 ± 0.07	ACN + 0.1% FA	(28 ± 2) x 10^6
4	H_2O + 0.1% FA + 1.5% ACN	3.48 ± 0.04	ACN + 0.1% FA	(27 ± 2) x 10^6
5	H_2O + 0.1% FA + 2.0% ACN	3.45 ± 0.01	ACN + 0.1% FA	(261 ± 3) x 10^5
6	H_2O + 0.1% FA + 2.5% ACN	3.47 ± 0.00	ACN + 0.1% FA	(265 ± 6) x 10^5
7	H_2O + 0.1% FA + 3.0% ACN	3.46 ± 0.01	ACN + 0.1% FA	(2652 ± 4) x 10^4
8	H_2O + 0.1% FA + 3.5% ACN	3.48 ± 0.00	ACN + 0.1% FA	(26 ± 1) x 10^6
9	H_2O + 0.1% FA + 4.0% ACN	3.45 ± 0.04	ACN + 0.1% FA	(26 ± 1) x 10^6
10	H_2O + 0.1% FA + 4.5% ACN	3.41 ± 0.00	ACN + 0.1% FA	(262 ± 5) x 10^5
11	H_2O + 0.1% FA + 5.0% ACN	3.38 ± 0.07	ACN + 0.1% FA	26 x 10^6 ± 0
12	H_2O + 0.1% FA + 7.5% ACN	3.37 ± 0.03	ACN + 0.1% FA	(259 ± 4) x 10^5
13	H_2O + 0.1% FA + 10.0% ACN	3.37 ± 0.03	ACN + 0.1% FA	(256 ± 4) x 10^5

FA, formic acid; ApH, average pH reading; ATCPA, average total chromatographic peak area; STDEV, standard deviation.

Table 4.
The instrumental responses for the optimization of the selected mobile phase.

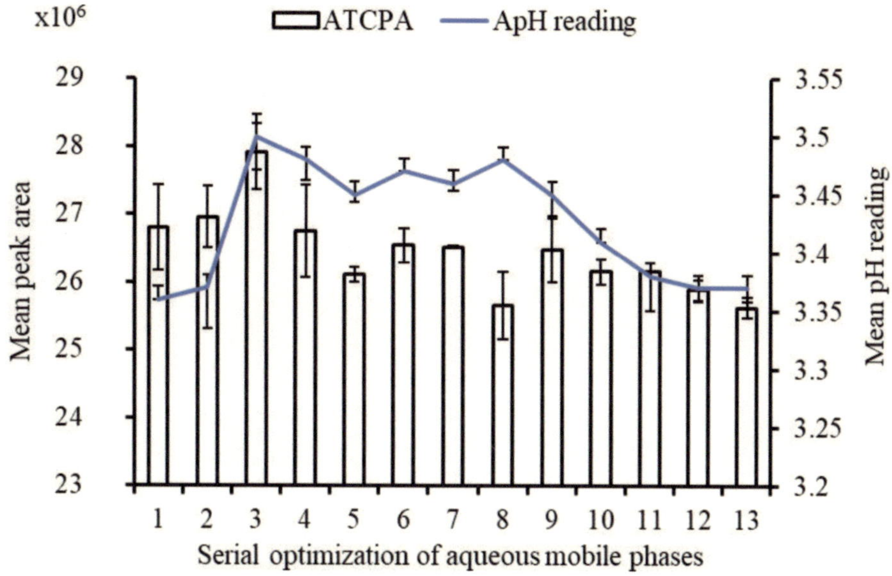

Figure 4.
Comparative illustration for the optimization of the selected aqueous mobile phase by ATCPA and ApH readings.

4. Conclusion

The selection and optimization of the best mobile phase setup was successfully carried out. Eventually, the optimized mobile phase setup [1% ACN and 0.1% FA in Milli-Q-water (mobile phase A) coupled with 0.1% FA in ACN (mobile phase-B)] improved the instrumental sensitivity on the targeted analytes. Thus, this justify the potential benefits of optimizing setup of the mobile phases prior to LC–MS/MS instrumentation of multi-pesticide analytes. Also, the selected and optimized mobile phase setup could be used for the analysis of other contaminants with similar properties to the analyzed pesticide compounds.

Acknowledgements

The authors acknowledge the Postgraduate Research Project (IPPP) for supporting this research under the grant no. PG 174-2014B, University of Malaya Kuala Lumpur, Malaysia.

Conflict of interest

The authors of this research agreed with no conflicts of interest.

Author details

Abubakar Lawal[1]* and Lukman Bola Abdul'rauf[2]

1 Department of Pure and Industrial Chemistry, Umaru Musa Yar'adua University, Katsina, Nigeria

2 Department of Chemistry, College of Pure and Applied Sciences, Kwara State University Malete, Ilorin, Nigeria

*Address all correspondence to: abubakarlawal360@yahoo.com

References

[1] Chapman, P. M. (2007). Determining when contamination is pollution—Weight of evidence determinations for sediments and effluents. Environment International, *33*(4), 492-501.

[2] Prasad, K. D., & Ramteke, P. (2013). Microbial contamination of important food borne pathogens in the production of an aseptically processed mango drink. Asian Journal of Dairy and Food Research, *32*(1), 56-64.

[3] Gong, Y., Tang, J., & Zhao, D. (2016). Application of iron sulfide particles for groundwater and soil remediation: A review. Water research, *89*, 309-320.

[4] McCarthy, J. F., & Zachara, J. M. (1989). Subsurface transport of contaminants. Environmental science & technology, *23*(5), 496-502.

[5] Lake, I. R., Hooper, L., Abdelhamid, A., Bentham, G., Boxall, A. B., Draper, A., Fairweather-Tait, S., Hulme, M., Hunter, P. R., & Nichols, G. (2012). Climate change and food security: Health impacts in developed countries. Environmental Health Perspectives, *120*(11), 1520.

[6] Baba, A., Garba, S. T., & Bello, H. S. (2020). Bioremediation potential of immobilized Corynebacterium kutsceri in the treatment of tannery industry effluent from Challawa industrial estate, Kano state, Nigeria. JOTCSA, *7*(2), 335-350.

[7] Koki, I. B., Lawal, A., & Taqui, S. N. (2018). Source identification and evaluation of surface water quality using factor and discriminant analysis. Bajopas, *11*(2), 169-175.

[8] Lawal, A. (2011). *Comparative analysis on selected bulb species.* (M.Sc. Applied Chemistry), Usmanu Danfodiyo University, Sokoto, Nigeria.

[9] Mohammed, A. A., Iniaghe, P. O., Abu, T. O., Bello, M. O., & Abdulkadir, M. D. (2020). Source analysis of heavy metals and polycyclic aromatic hydrocarbons from a popular dumpsite, Lagos state, Nigeria. JOTCSA, *7*(2), 489-504.

[10] Alsharif, A. M. A., Tan, G. H., Choo, Y.-M., & Lawal, A. (2016). Efficiency of hollow fiber liquid-phase microextraction chromatography methods in the separation of organic compounds: A review. Journal of chromatographic science, 1-14.

[11] Alsharif, A. M. A., Tan, G. H., Choo, Y. M., & Lawal, A. (2015). Liquid phase microextraction for analysis of mycotoxins in food samples: REVIEW. Research Journal of Chemical and Environmental Sciences, *3*(6), 5-21.

[12] Chan-Hon-Tong, A., Charles, M.-A., Forhan, A., Heude, B., & Sirot, V. (2013). Exposure to food contaminants during pregnancy. Science of the Total Environment, *458*, 27-35.

[13] Meador, J. P., Yeh, A., Young, G., & Gallagher, E. P. (2016). Contaminants of emerging concern in a large temperate estuary. Environmental Pollution, *213*, 254-267.

[14] Ribeiro, A. R., Maia, A., Santos, M., Tiritan, M. E., & Ribeiro, C. M. R. (2016). Occurrence of natural contaminants of emerging concern in the Douro River estuary, Portugal. Archives of environmental contamination and toxicology, *70*(2), 361-371.

[15] Lawal, A., Gwaram, N. S., & Suraj Abdulkarim, S. (2020). Spectrophotometric determination of Tartrazine in some selected beverages: A case study of Katsina town, Nigeria. FUDMA Journal of Sciences, *4*(3), 685-689.

[16] Lawal, A., Tan, G. H., & Alsharif, A. M. A. (2016). Recent advances in analysis of pesticides in food and drink samples

using LPME techniques coupled to GC-MS and LC-MS: A review. Journal of AOAC International, *99*(6), 1383-1394.

[17] Lawal, A., Wong, R. C. S., Tan, G. H., Abdulra'uf, L. B., & Alsharif, A. M. A. (2018c). Recent modifications and validation of QuEChERS-dSPE Coupled to LC–MS and GC–MS instruments for determination of pesticide/agrochemical residues in fruits and vegetables. *Journal of Chromatographic Science*, 1-14.

[18] McGrath, T., Elliott, C., & Fodey, T. (2012). Biosensors for the analysis of microbiological and chemical contaminants in food. Analytical and bioanalytical chemistry, *403*(1), 75-92.

[19] Yang, X. J., Du, Z., Lin, A., Yuan, Q., Wan, P., & Wong, C. (2013). Simultaneous determination of neutral, basic and acidic pesticides in aquatic environmental matrices by solid-phase extraction and liquid chromatography electrospray ionization mass spectrometry. Anal. Methods, *5*, 2083-2092.

[20] Jantunen, L. M., & Bidleman, T. F. (2000). Temperature dependent Henry's law constant for technical toxaphene. Chemosphere-Global Change Science, *2*(2), 225-231.

[21] Hijosa-Valsero, M., Bécares, E., Fernández-Aláez, C., Fernández-Aláez, M., Mayo, R., & Jiménez, J. J. (2016). Chemical pollution in inland shallow lakes in the Mediterranean region (NW Spain): PAHs, insecticides and herbicides in water and sediments. Science of the Total Environment, *544*, 797-810.

[22] Mamy, L., Patureau, D., Barriuso, E., Bedos, C., Bessac, F., Louchart, X., Martin-laurent, F., Miege, C., & Benoit, P. (2015). Prediction of the fate of organic compounds in the environment from their molecular properties: A review. Critical Reviews in Environmental Science and Technology, *45*, 1277-1377.

[23] Kortum, G., Vogel, W., & Andrussow, K. (2000). Dissociation Constants of Organic Acids and Bases. *CRC hand book of chemistry and physics*.

[24] Majors, R. E. (2007). QUEChERS-a New Technique for Multiresidue Analysis of Pesticides in Foods and Agricultural Samples: Advanstar Communications 131 W First ST, Duluth, MN 55802 USA.

[25] Orso, D., Martins, M. L., Donato, F. F., Rizzetti, T. M., Kemmerich, M., Adaime, M. B., & Zanella, R. (2014). Multiresidue determination of pesticide residues in honey by modified QuEChERS method and gas chromatography with electron capture detection. Journal of the Brazilian Chemical Society, *25*(8), 1355-1364.

[26] Lawal, A., Wong, R. C. S., Tan, G. H., & Abdulra'uf, L. B. (2018a). Determination of pesticide residues in fruit and vegetables by high-performance liquid chromatography–tandem mass spectrometry with multivariate response surface methodology. Analytical Letters, 1-18.

[27] Cook, J., Engel, M., Wylie, P., & Quimby, B. (1998). Multiresidue screening of pesticides in foods using retention time locking, GC-AED, database search, and GC/MS identification. Journal of AOAC international, *82*(2), 313-326.

[28] Aulakh, J., Malik, A., Kaur, V., & Schmitt-Kopplin, P. (2005). A review on solid phase micro extraction—High performance liquid chromatography (SPME-HPLC) analysis of pesticides. Critical Reviews in Analytical Chemistry, *35*(1), 71-85.

[29] Chang, C., Luo, J., Chen, M., Wu, K., Dong, T., He, X., Zhou, K., Wang, L., Chen, D., & Zhou, Z. (2016). Determination of twenty organophosphorus pesticides in blood serum by gas chromatography-tandem mass spectrometry. Analytical Methods, *8*(22), 4487-4496.

[30] Qin, Y., Chen, L., Yang, X., Li, S., Wang, Y., Tang, Y., & Liu, C. (2015). Multi-residue method for determination of selected neonicotinoid insecticides in traditional Chinese medicine using modified dispersive solid-phase extraction combined with ultra-performance liquid chromatography tandem mass spectrometry. Analytical Sciences, *31*(8), 823-830.

[31] Lawal, A., Wong, R. C. S., Tan, G. H., Abdulra'uf, L. B., & Alsharif, A. M. A. (2018b). Multi-pesticide residues determination in samples of fruits and vegetables using Chemometrics approach to QuEChERS-dSPE coupled with ionic liquid-based DLLME and LC–MS/MS. Chromatographia, *81*(5), 759-768.

[32] Lawal, A., & Abdulra'uf, L. B. (2020). Chemometrics approach to QuEChERS-dSPE for multi-standard determination of pesticides in blank samples of Milli-Q-water using high-performance liquid chromatography-tandem mass spectrometry (LC-MS/MS). ChemSearch Journal, *11*(1), 66-73.

[33] Lawal, A., & Koki, I. B. (2019). Determination of multi - pesticide residues in coconut water by QuEChERS - dSPE ionic liquid - based DLLME couple with high performance liquid chromatography - tandem mass spectrometry (LCMS/MS). *ChemSearch* Journal, *10*(1), 87 – 93.

[34] Anastassiades, M., Lehotay, S. J., Štajnbaher, D., & Schenck, F. J. (2003). Fast and easy multiresidue method employing acetonitrile extraction/partitioning and "dispersive solid-phase extraction" for the determination of pesticide residues in produce. Journal of AOAC International, *86*(2), 412-431.

[35] Rajski, Ł., Lozano, A., Uclés, A., Ferrer, C., & Fernández-Alba, A. R. (2013). Determination of pesticide residues in high oil vegetal commodities by using various multi-residue methods and clean-ups followed by liquid chromatography tandem mass spectrometry. Journal of Chromatography A, *1304*, 109-120.

[36] Pérez-Ortega, P., Gilbert-López, B., García-Reyes, J. F., Ramos-Martos, N., & Molina-Díaz, A. (2012). Generic sample treatment method for simultaneous determination of multiclass pesticides and mycotoxins in wines by liquid chromatography–mass spectrometry. Journal of Chromatography A, *1249*, 32-40.

[37] Economou, A., Botitsi, H., Antoniou, S., & Tsipi, D. (2009). Determination of multi-class pesticides in wines by solid-phase extraction and liquid chromatography-tandem mass spectrometry. Journal of Chromatography A, *1216*(31), 5856-5867. doi:http://dx.doi.org/10.1016/j.chroma.2009.06.031

[38] Lucas, D. (2013). Optimizing sample preparation for LC/MS/MS of pesticide residues in herbal teas. Agilent Technologies, Inc.,, 1-14.

[39] Vázquez, P. P., Lozano, A., Uclés, S., Ramos, M. G., & Fernández-Alba, A. (2015). A sensitive and efficient method for routine pesticide multiresidue analysis in bee pollen samples using gas and liquid chromatography coupled to tandem mass spectrometry. Journal of Chromatography A, *1426*, 161-173.

[40] Golge, O., & Kabak, B. (2015). Determination of 115 pesticide residues in oranges by high-performance liquid chromatography–triple-quadrupole mass spectrometry in combination with QuEChERS method. Journal of Food Composition and Analysis, *41*, 86-97. doi:http://dx.doi.org/10.1016/j.jfca.2015.02.007

[41] Zanella, R., Munaretto, J. S., & Martins, M. L. (2013). Determination of Pesticide Multiresidues in Apple, Pear and Grape using modified QuEChERS

and analysis by LC-QTOF-MS. *Agilent Technologies, Inc.*, 1-12.

[42] Scientific, C. (2014). The Theory of HPLC. Chromatographic parameters. *e-Learning for the Analytical Chemistry Community, LC/GC, Chromacademy*, 1-21.

[43] Bramston-Cook, R. (2009). Peak detection with varian star work station for varian 3800 and 450 gas chromatographs (pp. 1-24). Lotus Consulting 5781 Campo Walk Long Beach, USA: Lotus Flower, Inc. .

[44] Lazartigues, A., Fratta, C., Baudot, R., Wiest, L., Feidt, C., Thomas, M., & Cren-Olivé, C. (2011). Multiresidue method for the determination of 13 pesticides in three environmental matrices: Water, sediments and fish muscle. Talanta, *85*(3), 1500-1507.

[45] Data-handling. (2018). Units and concentrations. Retrieved 24-09-2018, from Environmental Protection (Water) Policy 2009 https://www.ehp.qld.gov.au/water/monitoring/sampling-manual/pdf/data-handling-units-and-concentrations.pdf

[46] Koenig, K. (2010). How To Find Dilution In Chemistry - Calculate Dilution. Retrieved 24-09-2018 http://www.tutapoint.com/knowledge-center/view/calculating-dilution/

[47] Sherma, J. (2001). Recent advances in thin-layer chromatography of pesticides. Journal of AOAC International, *84*(4), 993-999.

[48] Chen, M., Yi, Q., Hong, J., Zhang, L., Lin, K., & Yuan, D. (2015). Simultaneous determination of 32 antibiotics and 12 pesticides in sediment using ultrasonic-assisted extraction and high performance liquid chromatography-tandem mass spectrometry. Analytical Methods, *7*(5), 1896-1905.

[49] Pastor-Belda, M., Garrido, I., Campillo, N., Viñas, P., Hellín, P., Flores, P., & Fenoll, J. (2016). Determination of spirocyclic tetronic/tetramic acid derivatives and neonicotinoid insecticides in fruits and vegetables by liquid chromatography and mass spectrometry after dispersive liquid–liquid microextraction. Food Chemistry, *202*, 389-395.

[50] Friedrich, M. T., Martins, M. L., Prestes, O. D., & Zanella, R. (2016). Use of factorial design in the development of multiresidue method for determination of pesticide residues in wheat by liquid chromatography-tandem mass spectrometry. Food Analytical Methods, *9*(9), 2541-2551.

[51] Martínez-Domínguez, G., Nieto-García, A. J., Romero-González, R., & Frenich, A. G. (2015). Application of QuEChERS based method for the determination of pesticides in nutra-ceutical products (Camellia sinensis) by liquid chromatography coupled to triple quadrupole tandem mass spectrometry. Food Chemistry, *177*, 182-190.

[52] Morais, D. C., Collins, E. H., Jardim, C. H., & Fontes, I. C. S. (2018). Pesticide determination in sweet peppers using QuEChERS and LC–MS/MS. Food Chemistry, *249*, 77-83.

[53] Miliadis, G., Tsiantas, P., & Siragakis, G. (2017). Problems encountered in LC-MS/MS analysis for the determination of pesticide residues in food. Journal of the Hellenic Veterinary Medical Society, *68*(4), 635-640.

[54] Rebelo, A. M., Dolzan, M. D., Heller, M., Deschamps, F. C., Abate, G., Micke, G. A., & Grassi, M. T. (2016). Simultaneous determination of herbicides in rice by QuECHERS and LC-MS/MS using matrix-matched calibration. Journal of the Brazilian Chemical Society, *27*(1), 186-193.

[55] Bordin, A. B., Minetto, L., Filho, I. D. N., Beal, L. L., & Moura, S. (2016). Determination of pesticide residues in

whole wheat flour using modified QuEChERS and LC–MS/MS. Food Analytical Methods, 1-9.

[56] Zheng, W., Park, J.-A., Zhang, D., El-Aty, A. A., Kim, S.-K., Cho, S.-H., Choi, J.-M., Shim, J.-H., Chang, B.-J., & Kim, J.-S. (2017). Determination of fenobucarb residues in animal and aquatic food products using liquid chromatography-tandem mass spectrometry coupled with a QuEChERS extraction method. Journal of Chromatography B, *1058*, 1-7.

[57] Lopez, S. H., Lozano, A., Sosa, A., Hernando, M. D., & Fernandez-Alba, A. R. (2016). Screening of pesticide residues in honeybee wax comb by LC-ESI-MS/MS. a pilot study. Chemosphere, *163* 44-53.

Chapter 24

Mass Spectrometry Coupled with Chromatography toward Separation and Identification of Organic Mixtures

Asmae Bouziani and Mohamed Yahya

Abstract

Mass spectrometers can provide information about molecular composition and chemical structure. However, with complex mixtures, superpositions and even suppression of signals may occur. On the other hand, Chromatography is an ideal technique for separating complexes but is often insufficient for compound identification. Hence, coupling both techniques in order to eliminate the limitations of each technique makes perfect sense. In this contribution, a brief description of mass spectrometry coupled with chromatography in the gas and liquid phase will be discussed to explain the advantages of coupling the two methods. The ionization techniques are also reported and followed by application areas of these techniques. Finally, the recording and treatment of the results are reviewed.

Keywords: Gas Chromatography, Liquid chromatography, Ionization, EI, CI, ESI, APCI, MALDI

1. Introduction

One of the shortcomings of mass spectroscopy (MS) is the identification of a complex mixture. However, to overcome this limitation, MS could be coupled with a separation technique such as liquid chromatography (LC) or gas chromatography (GC). The sample injected into the MS ought to be separated first. The injected samples could be in the liquid phase for LC/MS or the gas phase for GC/MS. The injection of the sample into MS could be done in two ways: either the sample is collected and then analyzed off-line, or the MS is linked to the chromatograph, and the mass spectrum is obtained as the mixture is eluted [1–4]. Though the primary benefit of the separation technique coupling with MS is the obtention of a spectrum that allows identifying the separated product, it is not the only advantage that may be attained. The detector must display the following properties:

- The products separated before the detector need to stay separated, meaning that the detector does not interfere with the chromatographic resolution.

- Highly sensitive.

- Can detect all product eluted.

- Provide enough information about the structure to be able to identify the compounds eluted.
- Selectivity: allows the identification of a specific product in the mixture.
- The output signal must be proportional to the concentration.
- The response factor must be constant or at least foreseeable.
- The performance/cost ratio must be as small as possible.
- Do not damage the product.
- The deconvolution of chromatographic peaks needs to be possible.

The last parameter is important because of the possibility that one chromatographic peak may correspond to two products.

In this contribution, the MS coupling with GC and LC will be discussed, focusing on the ionization techniques used for the coupling. The most important application of the GC–MS and LC–MS are also given in brief. Finally, the recording and treatment of the outcome are reviewed.

2. Mass spectrometry coupled with gas chromatography

A complex mixture can be separated via GC, and MS can identify these compounds. Hence combining these two techniques can be advantageous. Moreover, GC and MS can both run in the gas phase making the linking straightforward, the performance stable, and good reproducibility.

The GC separates and introduces molecules into the MS via direct injection or after heating. The separation depends on the difference of the thermodynamic properties (boiling points and selective absorption in the stationary phase) and the difference in the distribution in the stationary phase and the mobile phase (carrier gas). In this case, MS acts as a detector, which includes an ionization source, mass analyzer, and electron multiplier tubes. First, the analyzed molecules are injected into MS via GC, and the ionization source ionizes them into gaseous ions, then they enter into the mass analyzer. The separation of ions occurs based on the variance of the mass-to-charge ratios, and then the separated ions reach the electron multiplier, which produces an electrical signal and giving a 3D output of the analyzed molecules. A schematic figure of the main parts of GC–MS is given in **Figure 1**.

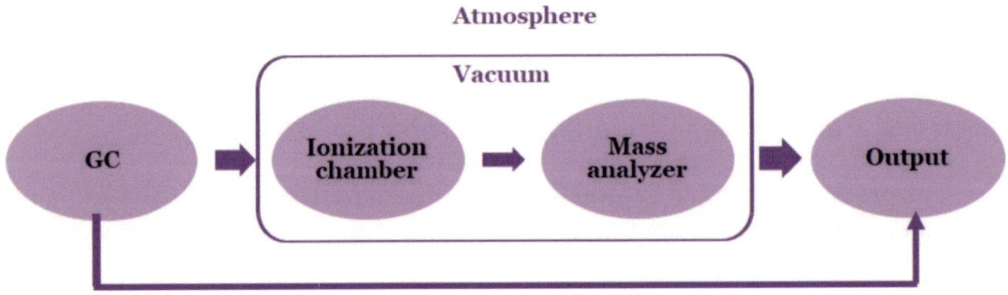

Figure 1.
The schematic of the components of GC–MS.

2.1 Coupling GC with MS

2.1.1 Open coupling

In the open-coupling system, the chromatographic column is connected to the MS with a T-shaped tube encompassing a smaller diameter tube (**Figure 2**). A deactivated fused silica or platiniumcapillary also leads to this tube and goes into the source of the MS. In order to evade condensation, the capillary needs to be kept under a vacuum and heated. The pressure inside the T-shaped tube must be equal to the atmospheric pressure, so the tube is closed at the edges but not sealed. The oxygen can oxidize the eluted molecules; in order to avoid that, helium is used. The diameter of the tube that enters the MS is essential. It needs to supply an adequate flow with regards to the gaseous conductance and pumping capacity. Hence the diameter of the capillary is 0.15 mm, and the length is 50 cm heated to 250°C will carry 2.5 ml/min of the eluted gas into the source. In practice, this is enough to pump everything coming out of a capillary column. The eluted molecules are not enhanced in an open-coupling system. The experiment is carried under the typical chromatographic environments, with one end of the column is under atmospheric pressure. The advantages of this system are the easiness of the column changing and the simplicity of the settings (no unique settings are needed). This system is generally used when no enhancement is required.

2.1.2 Direct coupling

In this system, the capillary column enters the spectrometer source directly through a set of vacuum-sealed connections. No pumping is needed since the capillary is essentially very lengthy. The column inside diameter of 0.25 mm with a length of 15 m minimum is needed (**Figure 3**). The major downside of this system is not permitting the solvent's removal, and the column change is complex. When the column is sufficiently long, the chromatography is conducted between an atmospheric and vacuum at the opening and the other end of the column, respectively.

2.2 The ionization techniques

The ionization techniques such as electron impact (EI), Chemical ionization (CI), and field ionization (FI) have been accessible for several years, which makes the GC–MS the oldest coupling technique [5]. The ionization occurs inside the

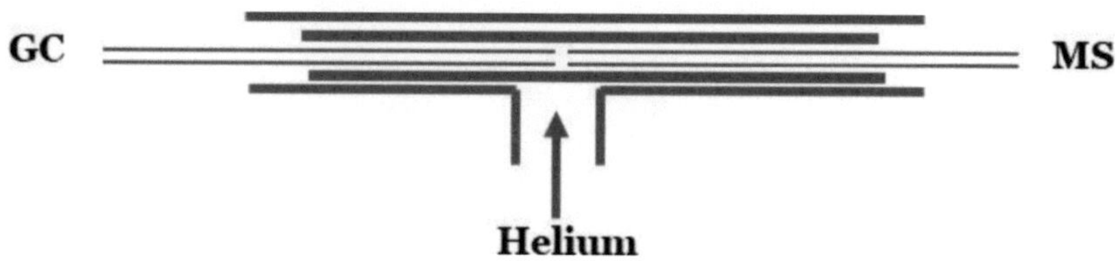

Figure 2.
Schematic representation of an open-coupling system.

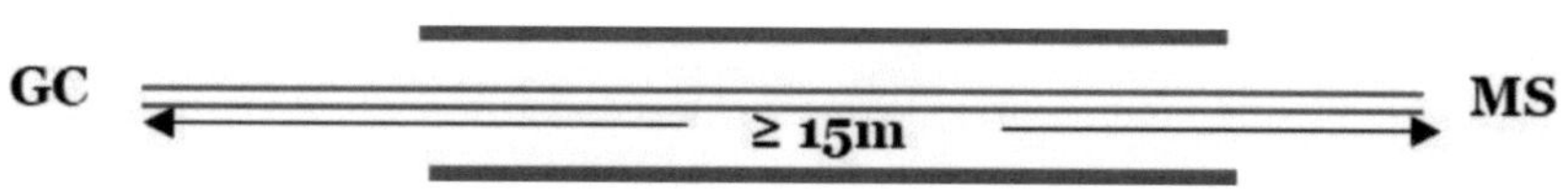

Figure 3.
Schematic representation of a direct coupling system.

instruments for the three mentioned techniques, namely in the high vacuum of the MS. For the hydrophobic and small hydrophobized molecules to be analyzed with GC–MS, the MS must vaporize undecomposed analytes [6]. The gas is led into the ionization chamber of MS via the outlet of the GC separating capillary. The MS must be kept under a vacuum because of the capillary columns, which operate at low flow rates; for this reason, the carrier gas that emerges from the GC column into the chamber of ionization needs to be pumped out.

2.2.1 Electron impact ionization

A hot-cathode discharges electrons through the EI, resulting in an electron beam-forming at the ionization chamber between the glowing cathode and the capture anode (**Figure 4**). Once the molecules go through the electron beam, an electron is bumped out of the molecule's surface, which gives a radical cation. The obtained ions at an electron energy of 70 eV are unstable and deteriorate rapidly, generating characteristic fragments that are automatically identified through the spectrum libraries. The GC–MS allows easy and reliable identification as well as the quantification of the molecules existing in the user database. Currently, NSIT 20 Mass Spectral Library has 350.643 carefully evaluated spectra.

2.2.2 Chemical ionization

The CI is similar to the EI, except that the reactant gas molecules are ionized and not the analyte molecules (**Figure 5**). Ammonia, methane, or isobutane may be used as a reactant gas. The charge transfer due to the deprotonation (negative ion mode) or protonation (positive ion mode) occurs between the analyte molecules and the ionized reactant gas. The negative CI is particularly very sensitive. The detection of a quantity of octafluoronaphthalene corresponding to 200,000 molecules was successfully reached in 1992 when McLafferty and Michnowicz used negative CI [7]. The CI generates fewer fragment ions contrary to EI.

2.2.3 Field ionization

The FI almost does not generate any fragments. A high voltage is applied to a carbon-activated metal fiber in the source chamber. The excavating of separate

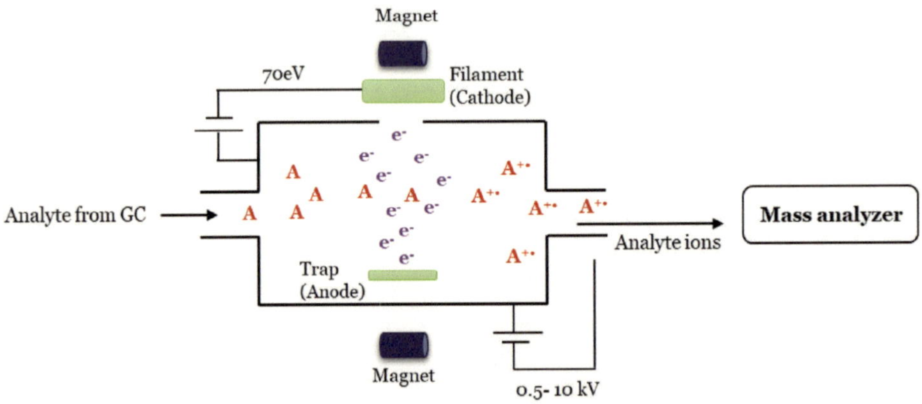

Figure 4.
Schematic representation of EI source.

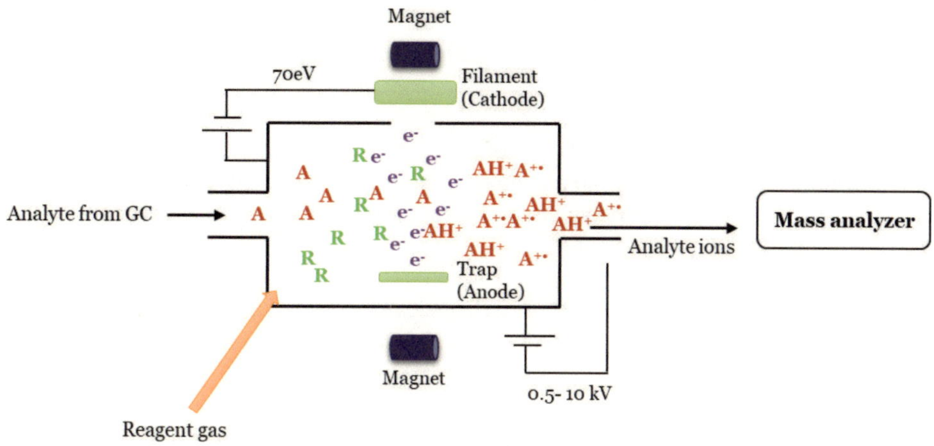

Figure 5.
Schematic representation of CI source.

electrons from the analyte molecules occurs due to the high field strengths that form at the tips of the branches of the carbon dendrites [8, 9]. The FI is a less sensitive ionization technique compared to the EI and CI (**Figure 6**).

2.3 Domaine of GC: MS applications

Coupling GC with MS opened the door to several applications [10]. In this contribution, we will limit to the most important ones.

- Environmental monitoring: The major application of GC–MS is monitoring environmental pollutants. Dibenzofurans, herbicides, dioxins, phenols, sulfur, and chlorophenol are all detected via GC–MS in air, soil, and water.

- Medicine: the detection of numerous congenital metabolic diseases is possible due to GC–MS usage for the screening tests. If the subject has a genetic metabolic disorder, a specific compound is detected in the urine.

- Food: GC–MS can analyze aromatic compounds present in food or beverages, including ester, alcohols, and fatty acids. It is mainly used to detect contamination or spoilage. Oils, perfumes, and essential oils also can be analyzed.

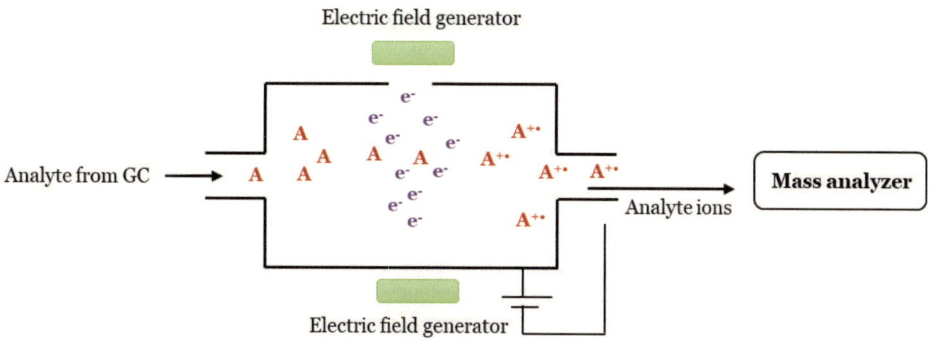

Figure 6.
Schematic representation of FI source.

- Forensic: especially forensic toxicology, GC–MS finds a wide application identifying poisons and steroids (anabolic steroids) in biological samples and anti-doping labs.

- Pharmaceutical: The GC–MS is used primarily for the identification of impurities in the active pharmaceutical ingredients. Furthermore, in the field of medicinal chemistry, it can be used to characterize the synthesized compounds.

- Biological: Narcotics, alcohols, and drugs can be detected in the body fluid via GC–MS. Moreover, it allows the detection of pollutants and metabolites in serum.

- Geochemical research: GC–MS finds a vital application in geochemical research because of the structured mass spectral peaks and low volatile sample analyzability. The atmosphere of Venus was analyzed using GC–MS.

- Chemical war: The detection of chemical warfare agents in public places is performed using GC–MS.

- Industrial: Aromatic solvent and inorganic gases can be analyzed via GC–MS to detect impurities in cosmetics.

3. Mass spectrometry coupled with liquid chromatography

High-performance liquid chromatography (HPLC) is an innovative type of LC used in various fields, including food analysis and pharmaceuticals. It is primarily beneficial for low or non-volatile organic compounds that are not suitable for GC. The main difference between HPLC and LC is the solvent's mobility. In the case of LC, the solvent moves by force of gravity, while in HPLC, it moves under high pressure obtained through pumps. The use of the pumps ensures the overcome of the pressure drop in the column and reducing the separation time. The combined technique between MS and HPLC is generally identified as LC–MS (**Figure** 7).

LC coupling with MS is more complicated than with GC because of the need to generate gas-phase ions for the MS. Furthermore, the necessity to eliminate the elution solvent is another downside of LC–MS. In the case of water, if the column used has a small diameter permitting a maximum flow rate of 0.1 ml min^{-1}, which is equal to 0.1 g min^{-1} of water, generating a flow rate of 135 cm^3 min^{-1} of gas at atmospheric pressure. This flow is too high to be injected under a vacuum into a source. In order to overcome this downside, numerous methods are used [11–13].

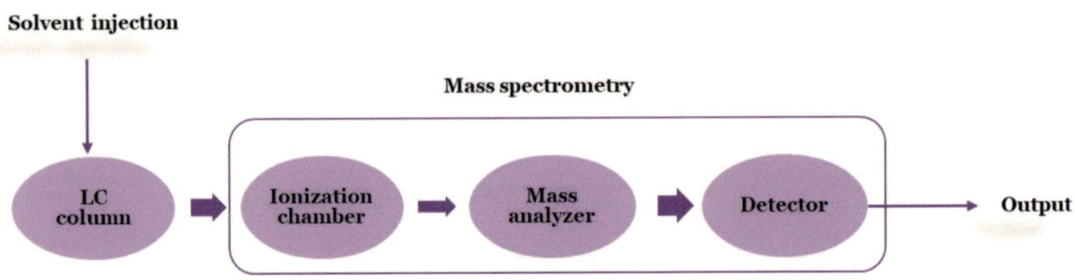

Figure 7.
Schematic representation of LC–MS.

3.1 Ionization and ions source

The coupling of HPLC and MS became possible with the installation of electrospray ionization (ESI) and atmospheric pressure chemical ionization (APCI) in the commercial apparatus [14, 15].

3.1.1 Electrospray ionization

Electrospray ionization (ESI) consists of pressing the analytes present in the solution through a capillary. The charged droplets form when a high voltage is applied (between 1.5 and 5 kV) [16]. The charge density improves with the elimination of solvents from the droplets via continuous evaporation. In addition, the surface area increases due to splitting droplets into smaller droplets at a specific charge density (Coulomb explosion). At the end of this process, the remaining microdroplets emit single ions, or the droplets only contain single solvated ions that will be entirely desolvated upon further drying [17–19]. The transfer of the ions into the high vacuum of the MS is carried out via a capillary or small hole in the front plate through electric fields (**Figure 8**).

ESI is applicable for various compounds such as proteins and peptides, oligosaccharides, bio-organic molecules, polymers, and non-covalent complexes.

3.1.2 Atmospheric pressure chemical ionization

APCI has attracted considerable attention due to its ability to produce ions from solution and analyzing rather nonpolar compounds. Like electrospray, the liquid analyte is directly injected into the ionization chamber via an APCI probe (**Figure 9**). The analyte solution is submitted to a nebulization to produce fine droplets of aerosol spray, which will undergo rapid heating in the nitrogen stream and then emerge at the end of the probe as a stream of a vaporized analyte. In the area of the corona discharge needle, the reagent ions are formed. The analyte molecules react with these ions and form protonated or deprotonated analyte ions that are singly charged [20, 21].

Generally, the transfer of proton happens in the positive mode to generate $[A^{+}H]^{+}$ ions. However, the negative mode may also occur, and the M^{-} and $[A^{-}H]^{-}$ are formed from electron transfer or proton loss, respectively. During ionization, the solvent clusters and high gas pressure influence the reagent ions resulting in reduced fragmentation and intact quasi-molecular ions. The process is considered more energetic than ESI, which results in the absence of multiple charging [22].

3.1.3 Matrix-assisted laser desorption ionization

The matrix-assisted laser desorption ionization (MALDI) is another ionization technique, which permits high molecular weight molecules injection into the

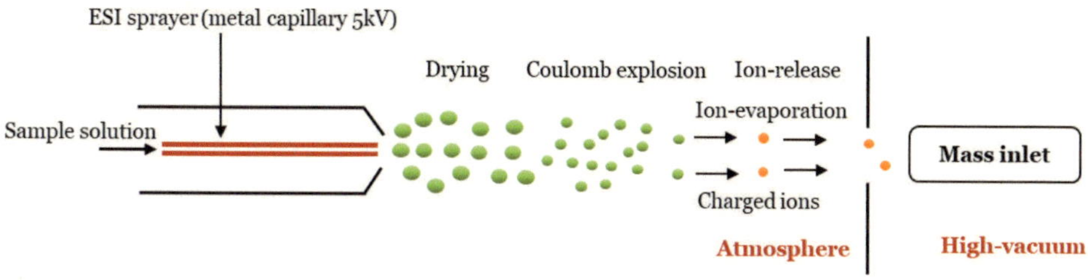

Figure 8.
Schematic representation of ESI.

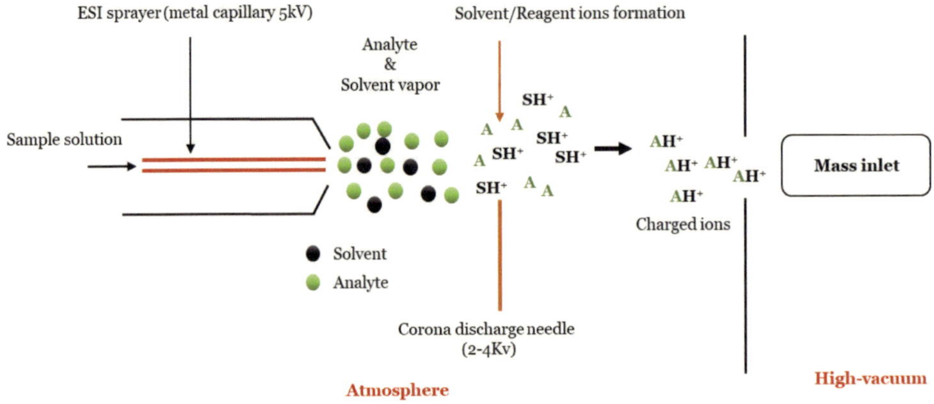

Figure 9.
Schematic representation of APCI.

gas phase as intact ions. MALDI technique gives desorbed analyte with a relative mass of 300KDa. In MALDI, the analytes are crystallized using an excess matrix compound (DBA, Sinapic acid, etc). Then the crystalşized analyte is carried into the high vacuum of the MS and irradiated via laser. Finally, the analyte molecules are carried into the gas phase after the matrix evaporated the absorbed laser energy (**Figure 10**). The transfer of protons between the matrix and analyte molecules is responsible for ionization [23–25]. The downside of this technique is the connection to the chromatography, which needs to be indirect either manually or through robotics. Currently, MALDI is limited to scanning applications where a matrix sprayed sample is scanned in two-dimension via a laser beam to get a mass distribution to produce false-color images [26, 27].

No fragmentations due to ionization are obtained when ESI, APACI, and MALDI are used, hence the "soft" reference. Furthermore, because of their covered polarity and molecular weight array, ESI and MALDI are perfect for bio-molecules analysis.

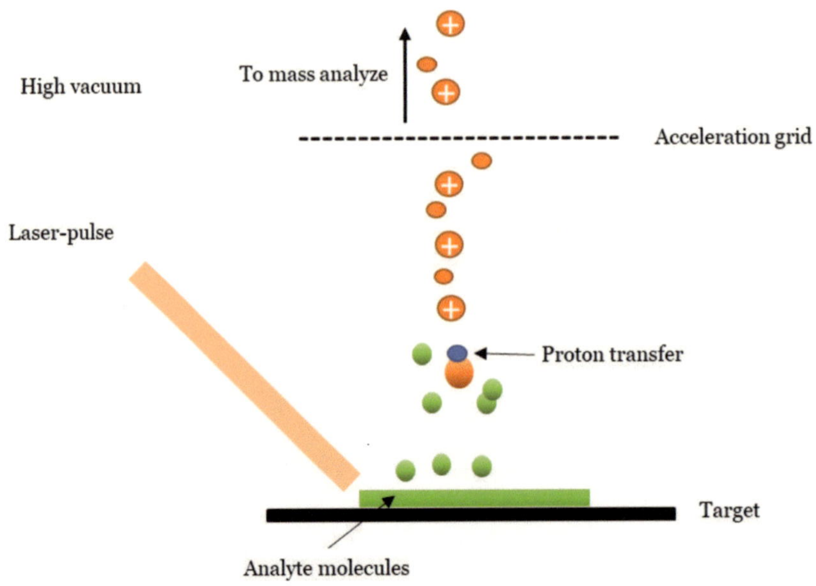

Figure 10.
Schematic representation of MALDI.

ESI and MALDI, in particular, are ideal for bioanalytics (proteins, peptides, etc.) due to their covered polarity and molecular weight range.

3.2 LC: MS domaine of applications

The LC–MS found application in numerous fields. In this section, the most crucial area will be discussed.

- Forensic: LC–MS could be used to determine toxicity in food and beverages, also in drug analysis. The LC–MS can detect trace amounts of toxins in numerous materials [28].

- Doping: LC/EDI-MS in negative mode can detect doping agents such as 4-Methyl-2-hexaneamine in the urine [29, 30].

- Environmental: Phenyl urea-based herbicides are detected via LC–MS as well as trace amounts of carbaryl in food [31].

- Pharmacology: LC–MS is used to quantify and elucidate the structure of drugs in biological samples (urine, saliva, plasma, etc). It can also be used for the study of the metabolism of drugs [30].

4. Outcomes recording and treatment

Regardless of GC–MS or LC–MS, an online data system is present, containing an acquisition processor, a magnetic recorder, and a computer.

4.1 Outcomes recording

As a function of time, the spectrometer offers two series of outputs: the number of ions detected and, at the same time, the mass of these ions is given. The mass of each ion emerges with a particular distribution over some time, as displayed in **Figure 11**. Thus, the number of ions detected can be computed from the area under the curve, whereas the centroid of the peak displays the ion's mass. The mass determination is effectuated via the acquisition processor, where the signal related to the number of ions accumulates quickly.

For instance, in 1 s, a spectrometer covers 500 mass, which means in 2 ms 1 mass. For this period, eight measurements of the number of ions ought to be conducted, meaning 0.25 ms assigned for each sample. In other words, 4000 samples ought to be measured per second, and the frequency of the sampling is 4 kHz. The ions detector's current goes through a resistance 4000 times a second, and at the end of the resistance, the acquisition processor is responsible for reading the potential difference relative to the number of ions detected and then digitalize it. The obtained output value corresponds to the y axis of the mass spectrum. The x-axis value corresponds to the reading of the mass indicator. The bar graph is the result of an algorithm that permits the processor to define the limits of the peak and centroid. The number of ions corresponds to the sum of the values read within these limits, whereas the ion's mass corresponds to the interruption of the indicator value at the centroid. A representative obtained bar graph is given in **Figure 12**.

In the case of a broader mass range scanning or high-resolution usage, increasing the sampling speed is needed, increasing the data points per unit time.

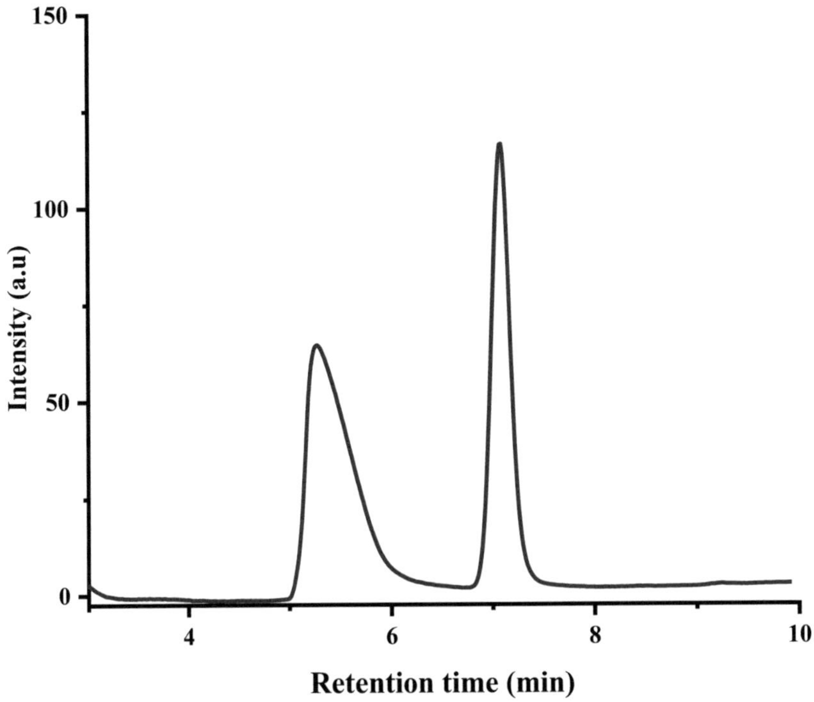

Figure 11.
Schema of a chromatogram.

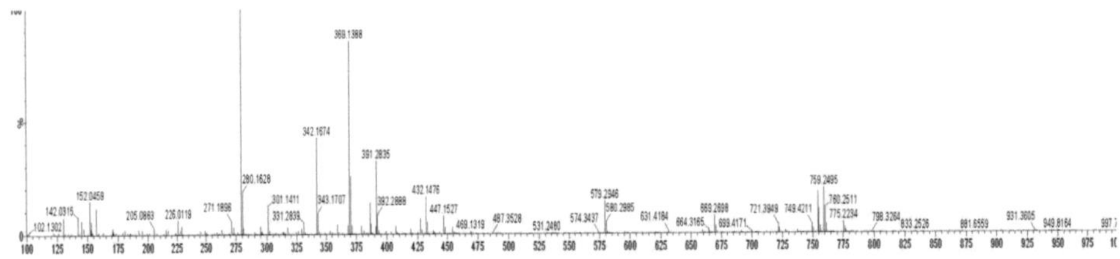

Figure 12.
MS spectra.

One of the essential characteristics of the data acquisition process is the dynamic range, connected in part to the signal digitization possibilities. For instance, an ion detector may identify one to a million ions that reach the detector at once.

Its dynamic range, the largest to the lowest measurable signal ratio, is equivalent to 10^6. In an ADC with 16 bits, the numerical obtained values are between 1 and 2^{16}. Therefore the dynamic range is considerably lower than that of the detector. However, this problem can be overcome by reading different value ranges consecutively.

4.2 Instrument monitor and treatment of outcomes

The subsequent operations are possible because of the newly available programs.

4.2.1 Outcomes acquisition management

The operators can select numerous parameters such as the scan mode or the selected-ion monitoring mode, the array of the scanned masses, low or high resolution, primarily due to the acquisition program. The acquisition processor settings are arranged to correspond to the data supplied by the operator regarding the analysis to be performed. The parameter offered by the operator wholly controls the recent apparatus. Automatic injectors allow the performance of numerous successive chromatographic analyses with no interference from the operator. The most current systems allow the programing of tuning modifications or Changement in the type of measurement. For instance, the operator can program the system to measure the spectrum in a negative mode if an ion with a given m/z value is spotted at a set retention time. The MS was revolutionized with these options.

4.2.2 Interpretation of the outcomes

The operator can intervene and modify the parameters at any time due to the interactive program. Additionally, the following operations may be done:

- Reconstruction of the ion chromatogram based on the sum of the intensities of the ions detected.

- The chromatogram can be enlarged or vertically amplified to highlight the low-intensity peaks.

- Multiple spectra can be displayed on the screen. Therefore, the comparison of spectra, one at the beginning and the other at the end of elution, is possible.

- Detecting compounds that may not be noticeable on the chromatogram.

Coupling an MS to chromatography leads to an enhanced dynamic range of the chromatography as well as an improved resolution.

4.2.3 Other programs

Other programs can be used; they are given below:

- Individual program: Can be used to draw spectra with several formats, comparing spectra, and 2D or 3D spectra drawing.

- A subtraction program is used to eliminate the background noise from a spectrum or highlight the variations between 2 spectra.

- Library search programs: for the identification of the obtained spectrum.

- Labeling a mass to an elemental composition can be effectuated by limiting the search to acceptable chemical formulas. For instance, a mass of 40 Da can be ascribed to C_2H_5O and CHO_2. Again, low or high resolution can be used for calculation. The downside of the low resolution is that the number of possibilities is too high.

- The calculated isotopic abundances can be compared with experimental values.

Utility programs can be used to extract spectra from analysis and then delete the others.

5. Conclusions

In summary, it can be said that the most popular separation techniques (GC, HPLC) can be coupled with MS applying suitable ionization techniques. Coupling essentially removes current constraints of the single methods; thus, chromatography coupled with MS has become crucial in many analytics fields. Mainly in the area of bioanalytics, "proteomics" has launched an entirely different area of work over the past 20 years.

Conflict of interest

The authors declare no conflict of interest.

Author details

Asmae Bouziani[1]* and Mohamed Yahya[2]

1 Chemical Engineering Department, Middle East Technical University, Ankara, Turkey

2 Faculty of Science, Department of Chemistry, Gazi University, Yenimahalle, Ankara, Turkey

*Address all correspondence to: asmae@metu.edu.tr

References

[1] Kostiainen R, Kotiaho T, Kuuranne T, Auriola S. Liquid chromatography/atmospheric pressure ionization - Mass spectrometry in drug metabolism studies [Internet]. Vol. 38, Journal of Mass Spectrometry. John Wiley & Sons, Ltd; 2003 [cited 2021 Jun 10]. p. 357-72. Available from: www.interscience.wiley.com

[2] Gelpí E. Interfaces for coupled liquid-phase separation/mass spectrometry techniques. An update on recent developments. J Mass Spectrom [Internet]. 2002 Mar 1 [cited 2021 Jun 10];37(3):241-53. Available from: www.interscience.wiley.com

[3] Mastovska K, Lehotay SJ. P ractical approaches to fast gas chromatography-mass q spectrometry. J Chromatogr A [Internet]. 2003 [cited 2021 Jun 10]; 1000:153-80. Available from: www.elsevier.com/locate/chroma

[4] Klampfl CW. Review coupling of capillary electrochromatography to mass spectrometry. Vol. 1044, Journal of Chromatography A. Elsevier; 2004. p. 131-144.

[5] Gohlke RS, McLafferty FW. Early gas chromatography/mass spectrometry. J Am Soc Mass Spectrom [Internet]. 1993 [cited 2021 Jun 14];4(5):367-71. Available from: https://link.springer.com/article/10.1016/1044-0305(93)85001-E

[6] Hubsehmann H-J. Subject Index. In: Handbook of GC/MS [Internet]. Wiley-VCH Verlag GmbH; 2007 [cited 2021 Jun 14]. p. 575-83. Available from: https://onlinelibrary.wiley.com/doi/full/10.1002/9783527612857.indsub

[7] Fred W McLafferty JAM. State-of-the-art GC/MS. Chemtech. 1992;22(3): 182-189.

[8] Gross JH. Field Ionization and Field Desorption. In: Mass Spectrometry [Internet]. Springer Berlin Heidelberg; 2004 [cited 2021 Jun 14]. p. 355-80. Available from: https://link.springer.com/chapter/10.1007/3-540-36756-X_8

[9] Bursey MM. H. D. Beckey. Principles of field ionization and field desorption mass spectrometry. Pergamon Press, Oxford, 1977. Biol Mass Spectrom [Internet]. 1978 Jul 1 [cited 2021 Jun 14];5(7):iii–iii. Available from: https://onlinelibrary.wiley.com/doi/full/10.1002/bms.1200050710

[10] Medeiros PM. Gas Chromatography - Mass Spectrometry (GC--MS). In: White WM, editor. Encyclopedia of Geochemistry: A Comprehensive Reference Source on the Chemistry of the Earth [Internet]. Cham: Springer International Publishing; 2018. p. 530-5. Available from: https://doi.org/10.1007/978-3-319-39312-4_159

[11] Niessen WMA. State-of-the-art in liquid chromatography-mass spectrometry. Vol. 856, Journal of Chromatography A. Elsevier; 1999. p. 179-197.

[12] Ramarao NT, Vidyadhara S, Sasidhar RLC, Deepti B, Yadav RS. Development and Validation of LC-MS/MS Method for the Quantification of Chiral Separated R-Bicalutamide in Human Plasma. Am J Anal Chem. 2013;04(02):63-76.

[13] Ardrey RE. Liquid Chromatography - Mass Spectrometry: An Introduction [Internet]. John Wiley & Sons; 2003. (Analytical Techniques in the Sciences (AnTs) *). Available from: https://books.google.com.tr/books?id=f1QiHP3wsAcC

[14] Holčapek M, Jirásko R, Lísa M. Recent developments in liquid chromatography-mass spectrometry and related techniques. Vol. 1259, Journal of Chromatography A. Elsevier; 2012. p. 3-15.

[15] Thomson BA. Atmospheric pressure ionization and liquid chromatography/mass spectrometry - Together at last. J Am Soc Mass Spectrom [Internet]. 1998 [cited 2021 Jun 15];9(3):187-93. Available from: https://link.springer.com/article/10.1016/S1044-0305(97)00285-7

[16] Yamashita M, Fenn JB. Electrospray ion source. Another variation on the free-jet theme. J Phys Chem [Internet]. 1984 Sep 1;88(20):4451-9. Available from: https://doi.org/10.1021/j150664a002

[17] Nguyen S, Fenn JB. Gas-phase ions of solute species from charged droplets of solutions. Proc Natl Acad Sci [Internet]. 2007 Jan 23;104(4):1111 LP – 1117. Available from: http://www.pnas.org/content/104/4/1111.abstract

[18] Chiarinelli J, Bolognesi P, Avaldi L. Ion optics simulation of an ion beam setup coupled to an electrospray ionization source, strengths, and limitations. Rev Sci Instrum. 2020 Jul;91(7):73203.

[19] Majuta SN, DeBastiani A, Li P, Valentine SJ. Combining Field-Enabled Capillary Vibrating Sharp-Edge Spray Ionization with Microflow Liquid Chromatography and Mass Spectrometry to Enhance 'Omics Analyses. J Am Soc Mass Spectrom [Internet]. 2021 Feb 3;32(2):473-85. Available from: https://doi.org/10.1021/jasms.0c00376

[20] Thurman EM, Ferrer I, Barceló D. Choosing between atmospheric pressure chemical ionization and electrospray ionization interfaces for the HPLC/MS analysis of pesticides. Anal Chem. 2001 Nov;73(22):5441-5449.

[21] Byrdwell WC. Atmospheric pressure chemical ionization mass spectrometry for analysis of lipids. Vol. 36, Lipids. American Oil Chemists Society; 2001. p. 327-46.

[22] Cai SS, Syage JA. Comparison of atmospheric pressure photoionization, atmospheric pressure chemical ionization, and electrospray ionization mass spectrometry for analysis of lipids. Anal Chem. 2006 Feb;78(4):1191-1199.

[23] Karas M, Gï M, Urgen Schäfer J¨. Ionization in matrix-assisted laser desorption/ionization: singly charged molecular ions are the lucky survivors. Vol. 35, JOURNAL OF MASS SPECTROMETRY J. Mass Spectrom. 2000.

[24] Wilkendorf LS, Bowles E, Buil JB, Van der Lee HAL, Posteraro B, Sanguinetti M, et al. Update on matrix-assisted laser desorption ionization-time of flight mass spectrometry identification of filamentous fungi. J Clin Microbiol. 2020 Dec;58(12):1263-1283.

[25] De Cesare V, Moran J, Traynor R, Knebel A, Ritorto MS, Trost M, et al. High-throughput matrix-assisted laser desorption/ionization time-of-flight (MALDI-TOF) mass spectrometry–based deubiquitylating enzyme assay for drug discovery. Nat Protoc. 2020 Dec;15(12):4034-4057.

[26] Zaima N, Hayasaka T, Goto-Inoue N, Setou M. Matrix-assisted laser desorption/ionization imaging mass spectrometry. Vol. 11, International Journal of Molecular Sciences. Molecular Diversity Preservation International; 2010. p. 5040-5055.

[27] Angel PM, Caprioli RM. Matrix-assisted laser desorption ionization imaging mass spectrometry: In situ molecular mapping. Biochemistry. 2013 Jun;52(22):3818-3828.

[28] Di Stefano V, Avellone G, Bongiorno D, Cunsolo V, Muccilli V, Sforza S, et al. Applications of liquid chromatography-mass spectrometry for food analysis. Vol. 1259, Journal of Chromatography A. Elsevier B.V.; 2012. p. 74-85.

[29] Pitt JJ. Principles and applications of liquid chromatography-mass

spectrometry in clinical biochemistry. Clin Biochem Rev. 2009 Feb;30(1):19-34.

[30] Zhang T, Watson DG. A short review of applications of liquid chromatography mass spectrometry based metabolomics techniques to the analysis of human urine. Vol. 140, Analyst. Royal Society of Chemistry; 2015. p. 2907-2915.

[31] Pérez-Magariño S, Revilla I, González-Sanjosé ML, Beltrán S. Various applications of liquid chromatography-mass spectrometry to the analysis of phenolic compounds. J Chromatogr A. 1999 Jun;847(1-2):75-81.

Chapter 25

Gas Chromatographic: Mass Spectrometric Mining the Volatilomes Associated to Rhizobiota Exposed to Commonly Used Pharmaceuticals

Emoke Dalma Kovacs and Melinda-Haydee Kovacs

Abstract

Rhizobiota are involved in plant protection through plant development facilitation and plant defense against stress factors. Pressures of global change either as abiotic or biotic stress factor could modify rhizobiota abundance, community structure, or functioning. Such change could result in anomalies of plant development. Human and veterinary medicines are widely used pharmaceuticals. Their active ingredients are not fully adsorbed and metabolized by living organisms and are therefore excreted unmodified. As current technologies of wastewater treatment plants are not designed to remove these contaminants, pharmaceuticals may be discharged into the environment and reach the soil in multiple ways. At present, there are no standard procedures or methodologies that could be easily applied and cover pharmaceuticals impact on soil microbiota. Besides that, available molecular and genetic approach through which soil microdiversity abundance, structure, and functions are evaluated involves high and expensive technology, which is not easily available to laboratories widespread. In this chapter, we propose an effortless way to address this issue by using gas chromatography–mass spectrometry (GC–MS) approaches to assess soil microbiota responses to commonly used pharmaceuticals. The chapter will refer to gas chromatographic techniques applied in assessment of soil microbiota diversity structure, abundance, and health status.

Keywords: microorganisms, volatile organic compounds, anthropogenic stress

1. Introduction

Microorganisms support essential roles in soil environment with important effects on ecosystem functioning and stability. They assure soil fertility, sustainability, and plant development [1]. However, global change pressures with increasing contaminant inputs and changing climate and environment have influenced native microbiota in different extent. Changes in soil microbiota community abundance, structure, and functioning could limit soil-provided ecosystem services that they mediate [2, 3].

Rhizosphere refers to the plant roots and soils adhering to them. It is considered the most dynamic and biologically active region of soil. Through the large variety and quantity of metabolites that are released by plant root fibrous system or root hair [4], rhizosphere sustains the large diversity of rhizobiota. This includes microorganisms such as algae, protozoa, slime molds, fungi, bacteria, archaea, viruses, etc. [5]. Most of these microorganisms are responsible for plant protection against pathogenic organisms, plant growth and development facilitation, and plant defense against abiotic stress factors [6].

Rhizobiota abundance and community structure modify once with plant root development and changes of soil environment property. Such changes could convert their functions either positively or negatively. Further, that potentially could impact the plant development [7]. There have been developed several approaches that allow assessment and monitoring of soil microbiota diversity and activity to better understand the soil ecology. These could be culture-based or culture-independent approaches. Culture-based techniques have been shown unable to isolate and grow a large domain of soil microorganisms [1]. Phospholipid-derived fatty acids (PLFA) profile analysis for monitoring soil microbiota phenotypic structure is a common culture-independent approach [8]. Additionally, in situ analysis of nucleic acids, direct analysis of DNA/RNA and polymerase chain reaction (PCR)-amplified segment of DNA molecules are frequently applied culture-independent methods [9, 10]. These culture-independent analytical tools are applied for microbial biomass, diversity, and activity assessment based on taxon richness and evenness. Commonly cited disadvantages are those related to sample storage and sample handling, which could limit results' accuracy. Soil samples' physicochemical properties also could restrict DNA/RNA extraction efficiency because of potential presence of inhibiting organic compounds or due to binding properties of nucleic acid molecules to soil particles. Further, DNase and RNase contamination could be easily acquired, which reduces results' accuracy too [1, 9]. Therefore, in the context of global changes that resulted in imminent abiotic and biotic pressures, as well of the importance of microorganism's key role in assuring soil-provided ecosystem services, it has become a requirement to find optimal evaluation tools for soil microbiota abundance, structure, and functioning assessment. Such biomonitoring tools help to improve ecosystem management strategies and consequently to protect biodiversity and conserve its functionality before loss of delivered ecoservices.

Biomonitoring tools are important to provide quick answer to changes in soil system offering insight on microbiota and its activity without disturbing soil. Gas chromatographic approaches could provide such information. Soil microbiota profiling based on phospholipids-derived fatty acids profile allows quantitative evaluation of living microbiota abundance and phenotypic structure. Bacterial released volatile organic compounds assessment permits evaluation of microbial metabolism status. Therefore, using such analytical approaches is possible to obtain a wider view on microbiota evolution and function under abiotic and biotic pressures raised by global change.

In the following chapter two chromatographic approaches will be presented that allow soil rhizosphere microorganisms' assessment and monitoring under a common abiotic pressure, the increased presence of pharmaceuticals in our surrounding environment.

2. Rhizobiota functions and related ecosystem services

Rhizosphere could comprise either free living or symbiotic microorganisms. Rhizobiota-related ecosystem services are the benefic outputs for human well-being resulted from microbial activities of the rhizosphere. These benefits are the end results of biotic and abiotic interactions and processes. Microbial communities are involved in organic matter decomposition, nutrient cycling, and pollutants degradation. They could stimulate or inhibit plants through the metabolites that they release. Those, they are directly involved in soil regulating, supporting, and provisioning services.

Rhizobiota involvement in soil regulating services: Main regulating services managed by soil microorganism are diseases and pest regulation, organic waste matter degradation, and pollutants degradation.

Diseases and pest regulation: Biological factors induced plant diseases cause decline of crop yields and quality. In the context of increased food demand once with global population increases, this resulted in frequent use of chemicals that prevent and control plant diseases and that warrant required fertilizers. Although extensive use of these chemicals has achieved the proposed objectives, their use has also resulted in unwanted side effects such as environment pollution, food products contamination and ecosystem alteration. The potential amplitude in time of these side effects ended in the necessity to find alternative eco-friendly solutions that minimize these associated health risks. Studies revealed that use of biological control agents could become a suitable alternative [11]. This started from the evidences that there are numerous mechanisms throughout microorganisms, including rhizobiota also, that influence directly or indirectly pathogen diseases. Soil bacteria and fungi life support functions could be summarized as antagonism, resistance, competition, and stimulation of plant defenses. At the moment is an increased interest in identifying and further the applying of beneficial plant-associated microorganisms instead of classical pesticides for pest and disease management. There are studies that identified different bacterial or fungal species that could control or act against several pests or diseases. *Pseudomonas* strains, which could be located also in rhizosphere, act against *Phytophthora infestans* oomycetes, which are a frequent damaging pathogen of potato [12, 13]. *Trichoderma harzianum* biocontrol *Rhizoctonia solani*, a soil-born pathogen, and *Fusarium oxysporum* f.sp., a fungus, act against crops of the Solanaceae family and other plants [12, 14]. The biocontrol ability of these microorganisms against pests and diseases is directly correlated with production of numerous organic and inorganic molecules.

Organic waste degradation: Organic waste degradation is considered a two-stage biotic process. Through this process, organic waste is first fragmented into smaller pieces by deprives, followed by deconstruction of these fragments into organic and inorganic molecules by microorganisms [15]. Rhizosphere microbiota are involved in decomposition of organisms and plants waste. Obtained components after decomposition processes are easily taken up by living organisms or removed from soil environment through leaching and runoff processes. Their ability to enhance decomposition processes of organic wastes resulted in raising the use in different fields of organic waste management areas (municipal solid waste, agricultural waste, etc.). They started to be used successfully in different biotechnology fields as renewable biogas production or composting for soil fertility enrichment [16].

Pollutant degradation: Presence of pollutants increased over years in all environmental compartments. Although abiotic treatment processes (physical and chemical) are widely applied and considered most of the time efficient in decontamination processes, these processes are often associated with potential few constraints. One of such constraints is the generation of secondary pollutants through the decontamination process. Also, the high cost required in such treatment processes also limits their applicability in low-income regions. Use of biological treatment was found as an advantageous approach [17]. It is widely used all over the world for emerging pollutants removal from wastewater. Also, different microbial strains started to prove their efficiency in degradation of different pollutants. Mycobacterium sp. and Bacillus megaterium, for example, are efficient in polycyclic aromatic hydrocarbons degradation in the presence of specific enzymes (e.g., ring hydroxylating and ring cleavage dioxygenase) [18]. *Acidisphaera*, *Burkholderia*, *Geobacillus*, *Pseudomonas*, and *Rhodococcus* bacteria are considered alkane-degrading bacteria [19]. *Sphingomonas* sp. and *Burkholderia* sp. degrade fenitrothion pesticide [20].

Rhizobiota involvement in soil supporting services: Supporting services are that services that hold up ecosystem functions that are indirectly used by humans. Soil microorganisms are critically involved in soil nutrients cycling and primary production.

***Nutrients cycling*:** Plant development and production require adequate nutrient resources. Rhizosphere bacterial and fungal community are involved in organic matter breakdown and recycling. Through their related catabolic reaction, they breakdown, transform, and mineralize macro- and micro-nutrients [21]. *Proteobacteria* and *Rhizobia* bacteria are involved in nitrogen fixation. *Cianobacteria* and *Eubacterium* enhance Zn and Fe translocation into plant.

***Primary production*:** Plant production is influenced by numerous microbial processes. *Proteobacteria* and *Firmicutes* enhance plant root system proliferation through produced organic compounds. *Rhizobium* sp., and *Frankia* sp. are involved in nitrogen fixation. *Streptomyces* and *Pseudomonas* produce iron-chelating compounds that facilitate Fe availability for plants. *Agrobacterium sp*. and *Bacillus* sp. are involved in phosphate solubilization. Numerous microorganisms increase plant stress tolerance through their produced organic and inorganic compounds [22].

Rhizobiota involvement in soil provisioning services: Provisioning services are those products that are obtained from ecosystem. Generally, these refer at food, fiber, genetic resources, chemicals, pharmaceuticals, etc. Microorganisms are considered important resources for numerous chemicals and pharmaceuticals with a broad range of applications.

***Bioresource*:** Bacterial and fungal population of rhizosphere influences plant communities, pathogens abundance, nutrient acquisition, and stress tolerance [15]. In most cases these are controlled by the produced bacterial and fungal origin molecules. Rhizosphere microorganisms were acknowledged as important bioresources for bioactive substances. They produce antibiotics, bacteriocins, lipopeptides, toxins, siderophores, enzymes, biosurfactants, osmoprotective substances, and other secondary metabolites [23].

3. Microbial volatilome: challenges under pressures of pharmaceuticals' presence in environment

Rhizosphere microorganisms either free-living or biofilm-forming or root-colonizing emit numerous volatile organic compounds through metabolic processes and biochemical processes that synthesize. Microbial volatile organic compounds are small molecules with low boiling points. These compounds diffuse quickly through soil particles. As many microbial volatile organic compounds are acknowledged for their pathogen's suppression activity, the physicochemical properties of these molecules increase their efficiency [24].

Garcia-Delgado et al. [25] evidenced that repeated application of herbicides reduced significantly fungal community abundance while that of *Actinobacteria* has been increased. Presence of metal pollutants also could exert pressures on soil microbial community phenotypic structure [26, 27]. Studies revealed that presence of pollutants changes soil microbiology through microbiota community structure and activity, lowering both their abundance and metabolic activity.

Human and veterinary pharmaceutical's presence in environment increased over the years. Nonsteroidal anti-inflammatory drugs (NSAIDs) are popular over-the-counter drugs used for mild-to-moderate pain such as headache, muscle, or other inflammation issues [28]. These medicines reach environment through wastewater treatment plants' end products, treated water, and sludge. This is because current wastewater treatment plants are not adequate enough to remove pharmaceuticals from wastewater body [29]. NSAIDs could pollute soil environment through sludge dispersal as fertilizer on agricultural soils or through wastewater reuse for irrigation purposes. Among NSAIDs, diclofenac and ibuprofen are the most reported drugs. Both ibuprofen and diclofenac were reported in environmental samples within ranges of ppb–ppm [28]. Concern of NSAIDs' presence in environmental compartments is heightened by their potential chronic adverse effect on nontargeted organisms because of long-term exposure. Studies present that diclofenac may induce changes in the physiology of *Hediste diversicolor* and *Solea senegalensis* and other marine species [30]. Kidney and liver damage was reported by Hussain et al. [31] in case of *Gallus gallus*, *Columba livia*, *Coturnix japonica,* and *Acridotheres tristis* exposed to diclofenac.

Earlier studies have shown that management practices influence microbial community structure and abundance. This in turn could change microbial volatile organic compounds in composition as well as in quantity. However, at the moment there are no data reported on how the presence of these pharmaceuticals could affect rhizosphere microbiota and its functioning. Based on the importance of rhizobiota-emitted volatile organic compounds in plant protection and development enhancement and potential toxic effects of such pharmaceuticals on rhizosphere microbiota, it has become important to understand microbiota community change and behavior under such challenges.

4. Key study: rhizobiota volatilome changes under challenges of pharmaceuticals

4.1 Experimental setup

Based on the frequency of their presence in the environment, the following nonsteroidal anti-inflammatory drugs were choosing for experiment: ibuprofen, ketoprofen, and diclofenac. Commonly consumed aromatic plants such as sage (*Salvia officinalis* L.), dill (*Anethum graveolens*), and rosemary (*Rosmarinus officinalis* L.) were used in the experiment. Argic phaeozem soil, free of studied pharmaceuticals, was used in this experiment. Soil material was contaminated individually with each of these pharmaceuticals with the following theoretical concentrations: 0.7 $mg{\cdot}kg^{-1}$ diclofenac, 0.5 $mg{\cdot}kg^{-1}$ ibuprofen, and 0.2 $mg{\cdot}kg^{-1}$ ketoprofen. Ten-day seeds of the selected plant materials were planted in pots containing approximately 1.2 kg of contaminated soils and allowed for development until maturity. Plant growth was performed in laboratory in a climate chamber with the following conditions: day – 24°C, 12 h of light; night – 18°C, 12 h; soil water-holding capacity was adjusted to 58% during the experiment. Control samples without soil contamination were grown in similar conditions. Each experiment was performed with three pots in parallel. Rhizosphere soil samples were collected after each plant has reached maturity. The schematic presentation of experiment setup is shown in **Figure 1**. From each rhizosphere 1 g of soil was collected for soil microbiota community assessment and 1 g of soil for microbial volatile organic compounds analysis.

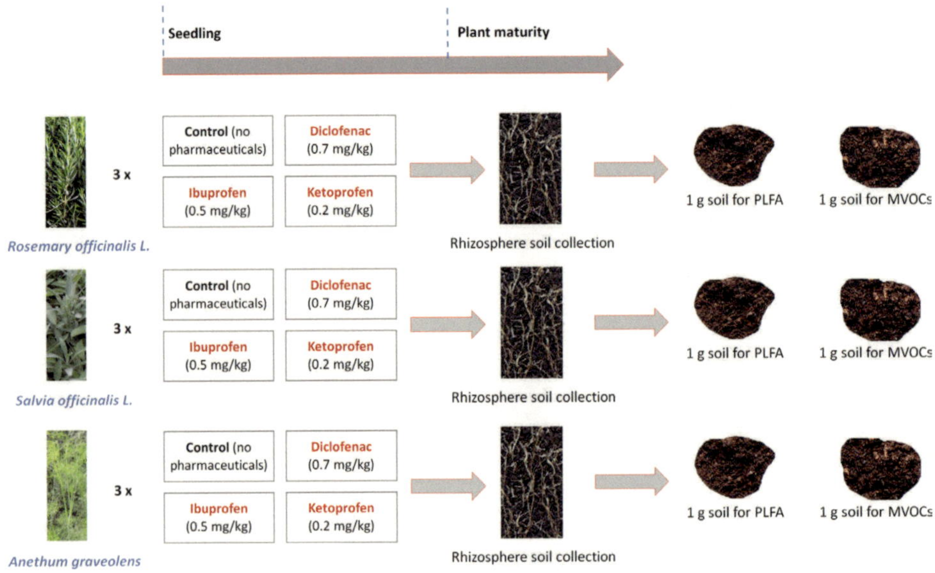

Figure 1.
Experiment setup.

4.2 Gas chromatographic assessment of rhizosphere microbiota community

One gram of lyophilized rhizosphere soil was used for microbiota phenotypic structure and abundance analysis. Assessment was performed applying phospholipids-derived fatty acids (PLFA) gas chromatographic approach. Phospholipids-derived fatty acids were extracted according to the method presented by Blight and Dyer [32], and Frostegard et al. [33] and derivatized for gas chromatographic analysis (7890A GC-FID, Agilent Technologies, Santa Clara, CA, USA). Their detection was done with flame ionization detector. Separation of all fatty acid methyl esters from each extract was done with a 5% phenyl-methyl polysiloxane column (25 mm × 0.2 mm id., 0.33 μm film thickness, HP-Ultra 2, J&W Scientific, Folsom, CA, USA). Helium was used as carrier with 1.2 mL·min^{-1} flow. Detector and injector temperature was set at 300 and 280°C, respectively. Oven temperature program starts at 170°C followed by an increase with 28°C·min^{-1} until 288°C, continued with an increase with 60°C·min^{-1} until 310°C. This final temperature was maintained isotherm for 1.25 min. Phenotypic profile of rhizobiota based on PLFA profile was determined using the MIDI SherlockTM Microbial Identification System software (Microbial ID, Inc., Newark, DE, USA). Also, the following PLFA biomarkers were used to identify saprotrophic fungi, ectomycorrhizal fungi, nitrogen-reducing bacteria and sulfur-reducing bacteria: 18: 2ω6c – saprotrophic fungi; 18:2ω9c – ectomycorrhizal fungi; 18:2ω6c and 18:3ω3 – nitrogen-reducing bacteria; and 17:1ω7c, 10Me16:0, 17:1ω6, 15:1, i17:1ω7c, cy18:0ω7.8, i15:1ω7c and i19:1ω7c for sulfur-reducing bacteria [32, 33].

4.3 Gas chromatographic–mass spectrometric in profiling rhizobiota volatilomes

Volatile organic compounds emitted by rhizobiota were assessed through headspace-solid-phase microextraction sampling using 85 μm polyacrylate fiber (Supelco Inc., Bellefonte, PA, USA). For this analysis, 1 g of rhizosphere soil was diluted with 2 mL of PBS solution in 20 mL headspace glass vials

(Agilent Technologies). The tightly capped headspace vials were incubated for 72 h in dark at 25°C. After this period, the vials were equilibrated for 30 min at 60°C using a TriPlus RSH autosampler (Thermo Scientific, Austin, TX, USA). Thermally activated SPME fiber was inserted in the vial headspace surface and kept for 15 min to allow the adsorption of volatile organic compounds on the fiber extractive phase. Rhizobiota volatilome analysis was conducted on gas chromatography–mass spectrometry (GC–MS/MS, Trace 1310, TSQ 9000, Thermo Scientific, Austin, TX, USA). Ionization was carried out in electron impact mode at 70 eV ionization energy. Volatile organic compounds were separated on a HP-5MS capillary column 30 m × 0.25 mm, 0.25 μm). The carrier gas was He with 1.2 $mL \cdot min^{-1}$ flow. The SPME fiber with adsorbed volatile organic compounds was inserted into the GC injection port at 250°C for 5 min to allow the desorption of analytes. The volatile organic compounds were identified by comparison of their mass spectra with compounds corresponding to mass spectra library (NIST/EPA/NIH, Chromeleon 7.2 CDS Software, Thermo Scientific, Austin, TX, USA). All identified volatile organic compounds were expressed in percentages as a normalized amount of each volatile organic compound resulted after the division of peak areas of identified volatile organic compounds by total peak area of all identified volatile organic compounds.

4.4 Rhizobiota differentiation between studied plant species rhizosphere

Total abundance of control samples of rhizosphere soil of *Rosmarinus officinalis* L., *Anethum graveolens,* and *Salvia officinalis* L. microbiota varies within the range of 216.6–191.8 $nmol \cdot g^{-1}$. Higher abundance was identified in *R. officinalis* L. rhizosphere, followed by *A. graveolens* and *S. officinalis* L. Bacterial dominance was observed in all cases, bacterial PLFA:total PLFA being higher than 0.8. Representative bacterial groups were Gram-negative bacteria, followed by Gram-positive and aerobe bacteria group. Gram-negative bacteria abundance represented 82.3% in *R. officinalis* L., while in rhizosphere of *A. graveolens* and *S. officinalis* L. represented 79.1 and 68.5% (see **Figure 2**).

Fungal community represented approximately 14% of the total microbial abundance, with higher abundance in *S. officinalis* L. – 16.2%.

4.5 Rhizobiota emitted volatile organic compounds variation studied rhizosphere

In control rhizosphere of the three aromatic plants, the main emitted volatile organic compounds measured through GC–MS were terpenes, alcohols, aromatic compounds, ketones, and organic acids. Identified volatile organic compounds percentage amount is listed in **Table 1**.

Terpenes were determined in higher amount in all rhizosphere soil with an average amount of 26%. Higher content of terpene compounds was measured in *R. officinalis* L. (30%) followed by *A. graveolens* (28%) and *S. officinalis L.* (20%). In case of *S. officinalis* L. rhizosphere, the second representative group of volatile organic compounds was organic acids (13.4%) followed by alkane compounds (12.6%). In the rhizosphere soil of *A. graveolens* and *R. officinalis* L., the second prevalent group was alkane for both cases. In all rhizosphere soils, ester compounds were found in lower amount (< 5%).

4.6 Pharmaceuticals' influence on rhizobiota community and volatilome profile

Rhizosphere microbiota total PLFA ranged between 216.6 and 167.6 $nmol \cdot g^{-1}$ dry weight soil. Between studied plant species during all experiment cases,

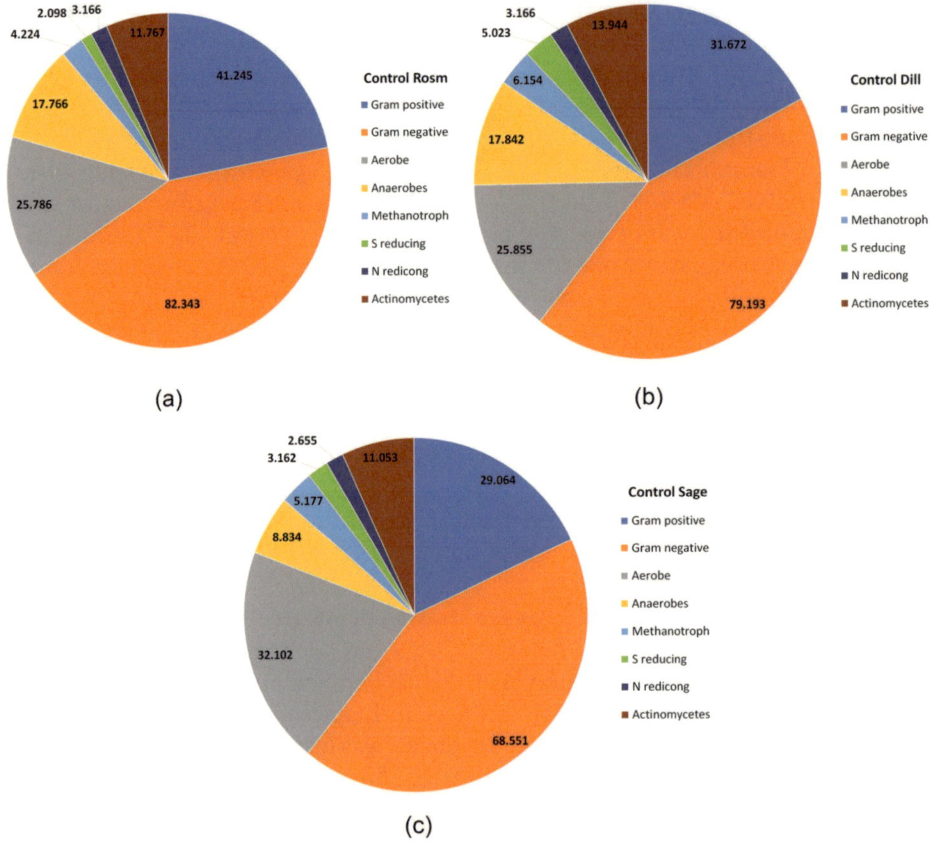

Figure 2.
Bacterial communities' abundance variation in studied rhizosphere soils. a.) Rosmarinus officinalis L.; b.) Anethum graveolens; c.) Salvia officinalis L.

it was observed that the rhizosphere microbiota abundance was higher in case of *R. officinalis* L. rhizosphere soil followed by *A. graveolens* and *S. officinalis* L. (see **Figure 3**). Rhizobiota community phenotypic profile revealed bacterial dominance, bacterial PLFA:total PLFA >0.837. Gram-negative bacteria were the most representative bacterial group in studied rhizosphere soils. Their abundance was within 56.8–82.3 nmol·g^{-1}. They were followed by Gram-positive bacteria and aerobe bacteria. PLFA ratio among aerobe bacteria and anaerobe bacteria was higher than 3.6 for *S. officinalis* L. rhizosphere and 1.4 for *A. graveolens* and *R. officinalis* L., these data clearly evidence aerobic bacteria dominance. Fungal community was represented within 9.5–16.3% of the total rhizobiota abundance in studied experiments.

Influence of studied NSAIDs on rhizosphere microbial community was observed from experimental data. Compared with control samples, in all cases was seen a decrease in total abundance. *Rosmarinus officinalis* L. rhizosphere total abundance decreased with 8.5% under exposure to ketoprofen, followed by a decrease with 7.4% at diclofenac exposure and 5.1% at ibuprofen exposure. Compared with control samples, *Anethum graveolens* rhizosphere microbiota decreased with 13% in case of exposure to diclofenac and ketoprofen and with 11% in case of exposure to ibuprofen. *Salvia officinalis* L. rhizosphere microbial community did not decrease when it was exposed at diclofenac but presented a lower abundance with 12.6% and 5.6% when it was exposed to ibuprofen and ketoprofen, respectively.

Group	Compound name	Volatile organic compounds (%)		
		***Salvia officinalis* L.**	***Anethum graveolens***	***Rosmarinus officinalis* L.**
Alcohol	butan-1-ol	0.78	0.25	0.44
	heptan-1-ol	1.84	3.14	2.67
	hexan-1-ol	4.18	1.45	2.01
	1-butoxypropan-2-ol	0.88	0.25	0.35
	Ethanol	0.95	2.17	0.44
	butane-2.3-diol	2.15	0.55	0.84
	1-octen-3-ol	1.35	1.14	2.17
Aromatic compounds	2-phenylethanol	3.15	2.65	0.78
	Phenol	1.42	3.75	2.86
	Phenylacetaldehyde	0.77	1.55	0.54
	Acetophenone	0.89	2.65	1.64
	2-amino-1-phenylethanol	1.23	—	—
Ketone	Acetone	0.98	1.74	0.89
	butan-2-one	1.64	2.88	1.65
	nonan-2-one	1.58	2.87	0.59
	heptan-2-one	3.04	1.62	1.02
	octan-3-one	0.67	2.15	0.46
	undecan-2-one	0.28	0	0.35
	decan-2-one	1.11	0.64	0.77
	pentadecan-2-one	2.56	0.24	0.15
	tridecan-2-one	0.28	—	—
Terpene	Geranylacetone	1.66	2.78	3.15
	germacrene D	1.06	4.66	0.95
	Geosmin	0.89	0.68	1.18
	Terpineol	4.08	2.54	5.12
	germacradien-1 l-ol	1.25	0.88	1.62
	2-methylisoborneol	1.51	3.54	5.98
	p-cymene	3.14	5.56	5.45
	methyl eugenol	4.48	6.75	4.11
	Pentalenene	1.78	0.44	2.15
Organic acids	acetic acid	0.84	3.51	0.32
	dodecanoic acid	1.55	0.65	1.71
	propanoic acid	3.68	0.54	2.21
	2-methylbutanic acid	1.75	1.62	0.55
	pentanoic acid	1.59	0.84	0.94
	butanoic acid	2.15	1.54	3.01
	2-methylpentanoic acid	1.88	0.56	1.84

Group	Compound name	Volatile organic compounds (%)		
		Salvia officinalis L.	*Anethum graveolens*	*Rosmarinus officinalis* L.
Aldehyde	Decanal	2.45	2.07	3.15
	3-methylbutan-1-al	3.05	0.65	0.87
	2-methylbutan-1-al	2.14	1.31	2.07
	Nonanal	2.78	4.15	1.89
Ester	ethyl acetate	2.78	2.33	4.15
	Methylbutanoate	0.68	0.34	0.78
Alkane	Tetradecane	1.18	2.08	1.45
	Undecane	2.05	4.67	1.74
	Dodecane	1.85	1.88	2.05
	Hexadecane	3.84	1.68	4.55
	Nonane	3.66	1.86	3.15
Nitrogen compounds	Indole	2.15	2.08	3.54
	methyl pyrazine	0.44	0.24	0.44
	2-methylquinoxaline	0.75	0.91	0.95
	2.5-dimethyl pyrazine	0.58	0.64	2.15
Alkene	Ethylene	1.17	0.95	1.24
	Isoprene	3.15	2.84	4.14
	undec-1-ene	0.28	0.54	0.78

Table 1.
Percentage value of emitted volatiles in control samples rhizosphere.

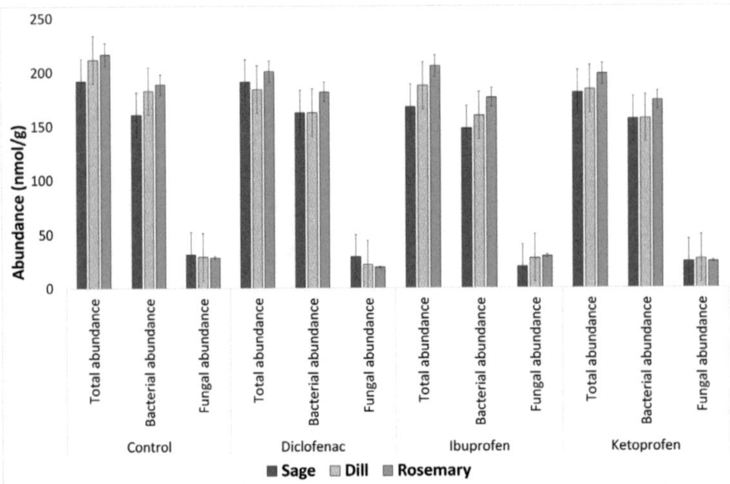

Figure 3.
Total microbial, bacterial, and fungal abundance variation in studied aromatic plants rhizosphere exposed at different NSAIDs.

Principal component analysis revealed that rhizosphere microbial community structure could change under exposure to specific NSAIDs (**Figure 4**). Gram-positive bacteria, fungi, and methanotroph bacteria in *S. officinalis* L. rhizosphere

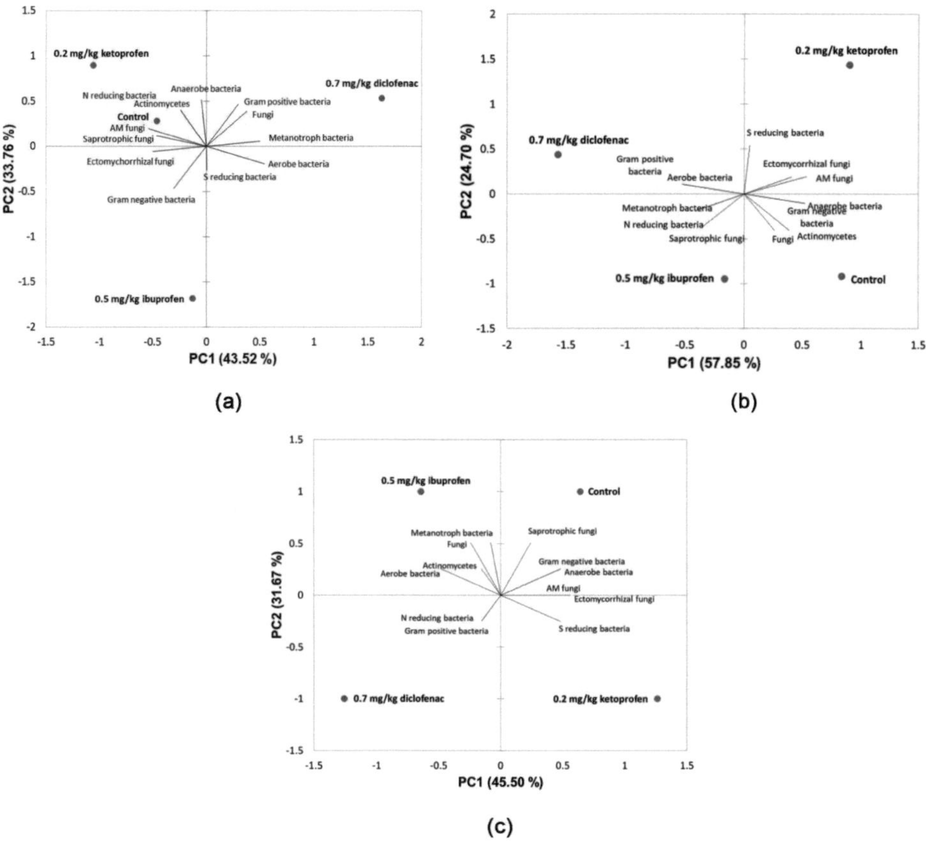

Figure 4.
Principal component analysis of NSAIDs' impact on rhizosphere soils microbial communities. a.) Salvia officinalis L.; b.) Rosmarinus officinalis L.; c.) Anethum graveolens.

presented a positive correlation in the presence of diclofenac while Gram-negative bacteria and ectomycorrhizal fungi decreased in the presence of ibuprofen (**Figure 4a**). This was explained by principal component analysis in 77.3%. Ibuprofen presence has a negative impact on methanotroph bacteria, N-reducing bacteria, and saprotrophic fungi in *R. officinalis* L. rhizosphere. Diclofenac's presence did not show significant impact on Gram-positive and aerobe bacteria (PC1 58%, PC2 25%; **Figure 4b**). Diclofenac's presence was also negatively correlated with N-reducing bacteria and Gram-positive bacteria abundance in *A. graveolens* rhizosphere (**Figure 4c**).

Emitted volatile organic compounds in studied rhizosphere changed over exposure to NSAIDs. In case of *S. officinalis* L. rhizosphere, the higher decrease was observed for alcohol compounds and organic acids in the presence of ibuprofen (< 15%), especially in case of hexan-1-ol, butane-2.3-diol and propanoic acid. Under contamination with diclofenac, aromatic compounds increased slightly (> 2%). Rhizosphere emitted volatile organic compounds of *A. graveolens* presented an increase with approximately 5–10% in case of phenol, germacrene, methyl eugenol, butanoic acid, and nonanal in the presence of ketoprofen while indole decreased with approximately 30% in case of exposure to ibuprofen and diclofenac. *R. officinalis* L. rhizosphere emitted volatile organic compounds changed significantly in the presence of ibuprofen and diclofenac. The following decrease was observed compared with control samples: terpene content <12%, organic acids <7% and alkane <4%.

5. Conclusion

Changes in both rhizosphere microbiota abundance and structure as well of emitted volatile organic compounds in the presence of commonly reported NSAIDs contaminants make us to suppose that microbiota rhizobiota functioning could be also changed. However, more studies in this area should be conducted to better understand the impact of pharmaceutical residues' presence on rhizosphere microbiota mediated soil ecosystem services. These are important to conserve soil ecosystem provided ecosystem services.

Acknowledgements

This research was funded by ANCS, Program NUCLEU, grant number PN 19-18.01.01-18 N/08.02.2019 and project financed by the Ministry of Research, Innovation and Digitization through Program 1- Development of the national research and development system, Subprogram 1.2 - Institutional performance - Projects that finance the RDI excellence, Contract no. 18PFE/30.12.2021.

Conflict of interest

The authors declare no conflict of interest.

Author details

Emoke Dalma Kovacs[1,2*] and Melinda-Haydee Kovacs[1]

1 Research Institute for Analytical Instrumentation, INCDO-INOE 2000, Cluj-Napoca, Romania

2 Faculty of Chemistry and Chemical Engineering, Babes-Bolyai University, Cluj-Napoca, Romania

*Address all correspondence to: dalma_kovacs2005@yahoo.com

References

[1] Lahlali R, Ibrahim DSS, Belabess Z, Roni KZ, Radouane N, Vicente CSL, et al. High-throughput molecular technologies for unraveling the mistery of soil microbial community: Challenges and future prospects. Heliyon. 2021;**7**:308142. DOI: 10.1016/j.heliyon.2021.e08142

[2] Yang T, Lupwayi N, Marc SA, Siddique KHM, Bainard LD. Anthropogenic drivers of soil microbial communityies and impacts on soil biological functions in agroecosystems. Global Ecology. 2021;**27**:e01521. DOI: 10.1016/j.gecco.2021.e01521

[3] Ngososng C, Buse T, Ewald M, Richter A, Glaser K, Schoning I, et al. Influence of management intensity and environmental conditions on microbiota in biological soil crusts and crust-free soil habitats of temperate forests. Soil Biology and Biochemistry. 2020;**144**:107761. DOI: 10.1016/j.soilbio.2020.107761

[4] Nihorimbere V, Ongena M, Maite S, Thonart P. Beneficial effect of the rhizosphere microbial community for plant growth and health. Biotechnologie, Agronomie, Société et Environnement. 2011;**15**:327-337

[5] Pitt TL, Barer MR. Classification, identification and typing of micro-organisms. Journal of Medical Microbiology. 2012:24-38. DOI: 10.1016/B978-0-7020-4089-4.00018-4. Available from: https://www.sciencedirect.com/science/article/pii/B9780702040894000184?via%3Dihub

[6] Lu T, Ke M, Lavoie M, Jin Y, Fan X, Zhang Z, et al. Rhizosphere microorganisms can influence the timing of plant flowering. Microbiome. 2018;**6**:231. DOI: 10.1186/s40168-018-0615-0

[7] Ai C, Liang G, Sun J, Wang X, Zhou W. Responses of extracellular enzyme activities and microbial community in both the rhizosphere and bulk soil to long-term fertilization practices in a fluvo-aquic soil. Geoderma. 2012;**173-174**:330-338. DOI: 10.1016/j.geoderma.2011.07.020

[8] Lee J, Kim SH, Jo HJ, Kwon MJ. Revisiting soil bacterial counting methods: Optimal soil storage and pretreatment methods and comparison of culture-dependent and -independent methods. PLoS One. 2021;**16**:e0246142. DOI: 10.1371/journal.pone.0246142

[9] Ricon-Florez VA, Carvalhais LC, Schenk PM. Culture-independent molecular tools for soil and rhizosphere microbiology. Diversity. 2013;**5**:581-612. DOI: 10.3390/d5030581

[10] Jo J, Oh J, Park C. Microbial community analysis using high-throughput sequencing technology: A beginner's guide for microbiologists. Journal of Microbiology. 2020;**58**:176-192. DOI: 10.1007/s12275-020-9525-5

[11] Zehra A, Raytekar NA, Meena M, Swapnil P. Efficiency of microbial bio-agents as elicitors in plant defense mechanism under biotic stress: A review. CRMICR. 2021;**2**:100054. DOI: 10.1016/j.crmicr.2021.100054

[12] Cianco A, Pieterse CMJ, Mercado-Blanco J. Editorial: Harnessing useful rhizosphere microorganisms for pathogen and pest biocontrol – Second edition. Frontiers in Microbiology. 2019;**2**:1935. DOI: 10.3389/fmicb.2019.01935

[13] Vrieze MD, Germanier F, Vuille N, Weisskopf L. Combining different potato-associated Pseudomonas strains for improved biocontrol of *Phytophthora infestans*. Frontiers in Microbiology. 2018;**2**:2573. DOI: 10.3389/fmicb.2018.02573

[14] Manganiello G, Sacco A, Ercolano MR, Vinale F, Lanzuise S, Pascale A, et al. Modulation of tomato response to *Rhizoctonia solani* by *Trichoderma harzinum* and its secondary metabolite harzianic acid. Frontiers in Microbiology. 2018;**2**:1966. DOI: 10.3389/fmicb.2018.01966

[15] Coats VC, Rumpho ME. The rhizosphere microbiota of plant invaders: An overview of recent advances in the microbiomics of invasive plants. Frontiers in Microbiology. 2014;**5**:364. DOI: 10.3389/fmicb.2014.00368

[16] Rastogi M, Nandal M, Khosla B. Microbes as vital additives for solid waste composting. Heliyon. 2020;**6**:e03343. DOI: 10.1016/j.heliyon.2020.e03343

[17] Ahmad HA, Ahmad S, Cui Q, Wang Z, Wei H, Chen X, et al. The environmental distribution and removal of emerging pollutants, highlighting the importance of using microbes as a potential degrader: A review. Science of The Total Environment. 2022;**809**:151926. DOI: 10.1016/j.scitotenv.2021.151926

[18] Xie X, Yuan K, Yao Y, Sun J, Lin L, Huang Y, et al. Identification of suspended particulate matters as the hotspot of polycyclic aromatic hydrocarbon degradation-related bacteria and genes in the Pearl River estuary using metagenomic approaches. Chemosphere. 2022;**286**:131668. DOI: 10.1016/j.chemosphere.2021.131668

[19] Li H, Wang XL, Mu BZ, Gu JD, Liu YD, Lin KF, et al. Molecular detection, quantification and distribution of alkane-degrading bacteria in production water from low temperature oilfields. International Biodeterioration & Biodegradation. 2013;**76**:49-57. DOI: 10.1016/j.ibiod.2012.06.007

[20] Katsuyama C, Nakaoka S, Takeuchi Y, Tago K, Hayatsu M, Kato K. Complementary cooperation between two syntrophic bacteria in pesticide degradation. Journal of Theoretical Biology. 2009;**256**:644-654. DOI: 10.1016/j.jtbi.2008.10.024

[21] Yadav AN, Kour D, Kaur T, Devi R, Yadav A, Dikilitas M, et al. Biodiversity, and biotechnological contribution of beneficial soil microbiomes for nutrient cycling, plant growth improvement and nutrient uptake. Biocatalysis and Agricultural Biotechnology. 2021;**33**:102009. DOI: 10.1016/j.bcab.2021.102009

[22] Chandra P, Wunnava A, Verma P, Chandra A, Sharma RK. Strategies to mitigate the adverse effect of drought stress on crop plants – Influences of soil bacteria: A review. Pedosphere. 2021;**31**:496-509. DOI: 10.1016/S1002-0160(20)60092-3

[23] Sehrawat A, Sindhu SS, Glick BR. Hydrogen cyanide production by soil bacteria: Biological control of pests and promotion of plant growth in sustainable agriculture. Pedosphere. 2022;**32**:15-38. DOI: 10.1016/S1002-0160921060058-9

[24] Agtmaal M, Straathof AL, Termoshuizen A, Lievens B, Hoffland E, Boer W. Volatile-mediated suppression of plant pathogens is related to soil properties and microbial community composition. Soil Biology and Biochemistry. 2018;**117**:164-174. DOI: 10.1016/j.soilbio.2017.11.015

[25] Garcia-Delgado C, Barba-Vicente V, Marin-Benito MJ, Iqual M, Sanchez-Martin MJ, Rodriquez-Cruz S. Influence of different agricultural management practices on soil microbial community over dissipation time of two herbicides. Science of The Total Environment. 2019;**646**:1478-1488. DOI: 10.1016/j.sscitotenv.2018.07.395

[26] Cellini A, Spinelli F, Donati I, Ryu CM, Klopper JW. Bacterial volatile compound-based tools for crop management and quality. Trends in Plant Science. 2021;**26**:968-983. DOI: 10.1016/j.tplants.2021.05.006

[27] Tripathi S, Purchase D, Al-Rashed S, Chandra R. Microbial community dynamics and their relationship with organic and metal pollutants of sugarcane molasses-based distillery wastewater sludge. Environmental Pollution. 2022;**292**:118267. DOI: 10.1016/j.envpol.2021.118267

[28] Marchlewicz A, Guzik U, Wojcieszynska D. Over-to-counter monocyclic non-steroidal anti-inflammatory drugs in environment – Source, risks, biodegradation. Water, Air, and Soil Pollution. 2015;**226**:355. DOI: 10.1007/s11270-015-2622-0

[29] Jiang C, Geng J, Hu H, Ma H, Gao X, Ren H. Impact of selected non-steroidal anti-inflammatory pharmaceuticals on microbial community assembly and activity in sequencing batch reactors. Plos One. 2017;**12**:e0179236. DOI: 10.1371/journal.pone.0179236

[30] Nunes B, Daniel D, Canelas GC, Barros J, Correia AT. Toxic effects of environmentally realistic concentrations of diclofenac in organisms from two distinct trophic levels, Hediste diversicolor and Soleaa senegalensis. Comparative Biochemistry and Physiology Part C: Toxicology & Pharmacology. 2020;**231**:108722. DOI: 10.1016/j.cbpc.2020.108722

[31] Hussain I, Khan MZ, Khan A, Javed I, Saleemi KM. Toxicological effects of diclofenac in four avian species. Avian Pathology. 2008;**37**:315-321. DOI: 10.1080/03079450802056439

[32] Blight EG, Dyer WJ. A rapid method for total lipid extraction and purification. Canadian Journal of Biochemistry and Physiology. 1959;**37**: 911-917. DOI: 10.1139/o59-099

[33] Frostegard A, Tunlid A, Baath E. Use and misuse of PLFA measurements n soil. Soil Biology and Biochemistry. 2011;**43**:1621-1625. DOI: 10.1016/j.soilbio.2010.11.021